Praise for
Blu-ray Disc Demystified

"***BD Demystified*** is an essential reference for designers and developers building within Blu-ray's unique framework and provides them with the knowledge to deliver a compelling user experience with seamlessly integrated multimedia."
— *Lee Evans, Ambient Digital Media, Inc., Marina del Rey, CA*

"Jim's Demystified books are the definitive resource for anyone wishing to learn about optical media technologies."
— *Bram Wessel, CTO and Co-Founder, Metabeam Corporation*

"As he did with such clarity for DVD, Jim Taylor (along with his team of experts) again lights the way for both professionals and consumers, pointing out the sights, warning us of the obstacles and giving us the lay of the land on our journey to a new high-definition disc format."
— *Van Ling, Blu-ray/DVD Producer, Los Angeles, CA*

"***Blu-ray Disc Demystified*** is an excellent reference for those at all levels of BD production. Everyone from novices to veterans will find useful information contained within. The authors have done a great job making difficult subjects like AACS encryption, BD-Java, and authoring for Blu-ray easy to understand."
— *Jess Bowers, Director, Technical Services, 1K Studios, Burbank, CA*

"Like its red-laser predecessor, ***Blu-ray Disc Demystified*** will immediately take its rightful place as the definitive reference book on producing BD. No authoring house should undertake a Blu-ray project without this book on the author's desk. If you are new to Blu-ray, this book will save you time, money, and heartache as it guides the DVD author through the new spec and production details of producing for Blu-ray."
— *Denny Breitenfeld, CTO, NetBlender, Inc., Alexandria, VA*

"An all in one encyclopedia of all things BD."
— *Robert Gekchyan, Lead Programmer/BD Technical Manager*
Technicolor Creative Services, Burbank, CA

About the Authors

Jim Taylor is chief technologist and general manager of the Advanced Technology Group at Sonic Solutions, the leading developer of BD, DVD, and CD creation software. Jim is the author of the first two editions of *DVD Demystified*, *Everything You Ever Wanted to Know About DVD* and the lead author the *DVD Demystified Third Edition*.

Jim is active in the DVD Forum working groups, serves as Chairman of the DVD Association and was named one of the 21 most influential DVD executives by DVD Report. He received the 2000 DVD Pro Discus Award for Outstanding Contribution to the Industry and was an inaugural inductee into the Digital Media Hall of Fame. And, he was named one of the pioneers of DVD by One to One magazine.

Formerly a Microsoft DVD Evangelist and Chief Technology Officer with Daikin US, Jim is recognized worldwide as an expert on DVD and associated technologies. Jim lives on an island near Seattle, Washington.

Charles G. Crawford is the Co-Founder of Television Production Services, Inc and Heritage Series, LLC, production companies specializing in traditional and interactive production and title development. He has been involved with DVD technology since 1998 when he wrote the Operations and Reference Manuals for Matsushita/Panasonic's award winning DVD Authoring System.

Chuck has been involved in video production for over 35 years at the local, network, and international levels, and has received three national EMMY awards for his technical and directorial expertise, plus numerous other industry awards for his production skills. He is a co-author of *DVD Demystified Third Edition* and a co-author of *High Definition DVD Handbook: Producing for HD DVD and Blu-ray Disc*. Chuck lives in Washington, DC.

Christen M. Armbrust, is a leader in the development and implementation of new digital media. He joined Technicolor Creative Services, a Thomson Company in Burbank, CA to define, develop and implement their new encoding, authoring and mastering processes for the next generation of DVDs (HD DVD and Blu-ray).

Chris is the Founder of Marin Digital, a company which focuses on the "new" technology of DVD and has produced over 1,100 DVD titles. He has delivered white papers and has spoken at SMPTE, NAB/BEA, Replitech Europe, DVDA and other DVD related conferences. Chris is a co-author of *High Definition DVD Handbook: Producing for HD DVD and Blu-ray Disc*. Chris lives in Sausalito, CA.

Michael Zink is director of advanced technology at Technicolor, Burbank, CA. Mike helped set up the Blu-ray Disc production units within Technicolor as well as accessing and developing tools to enhance high definition DVD production. Mike represents Technicolor in standards associations such as the DVD Forum, Blu-ray Disc Association and AACS.

Mike attended the University of Applied Sciences, Mittweida/Saxony, Germany and wrote his dissertation on Web DVD applications. Prior to joining Technicolor in 2003, he worked and consulted with several firms in Germany on television production and DVD title development. Mike is a co-author of *Programming for HD DVD and Blu-ray Disc: The HD Cookbook*. Mike resides in Burbank, CA.

Blu-ray Disc Demystified

Blu-ray Disc Demystified

Jim Taylor
Charles G. Crawford
Christen M. Armbrust
Michael Zink

New York   Chicago   San Francisco   Lisbon   London   Madrid
Mexico City   Milan   New Delhi   San Juan   Seoul
Singapore   Sydney   Toronto

The McGraw Hill Companies

Cataloging-in-Production Data is on file with the Library of Congress

McGraw-Hill books are available at special quantity discounts to use as premiums and sales promotions, or for use in corporate training programs. To contact a special sales representative, please visit the Contact Us page at www.mhprofessional.com.

Copyright © 2009 by The McGraw-Hill Companies, Inc. All rights reserved. Printed in the United States of America. Except as permitted under the United States Copyright Act of 1976, no part of this publication may be reproduced or distributed in any form or by any means, or stored in a data base or retrieval system, without the prior written permission of the publisher.

1 2 3 4 5 6 7 8 9 0   DOC/DOC   0 1 4 3 2 1 0 9 8

ISBN: P/N 978-0-07-159094-5 of set
         978-0-07-159092-1
MHID: P/N 0-07-159094-3 of set
         0-07-159092-7

This book is printed on acid-free paper.

Information contained in this work has been obtained by The McGraw-Hill Companies, Inc. ("McGraw-Hill") from sources believed to be reliable. However, neither McGraw-Hill nor its authors guarantee the accuracy or completeness of any information published herein, and neither McGraw-Hill nor its authors shall be responsible for any errors, omissions, or damages arising out of use of this information. This work is published with the understanding that McGraw-Hill and its authors are supplying information but are not attempting to render engineering or other professional services. If such services are required, the assistance of an appropriate professional should be sought.

Dedications

To all my brothers and sisters, who put up with me and my need to know everything – or, at least, to act as if I did.

— Jim Taylor

To my partner in work and in life, Samantha Cheng, for her perseverance, patience and fortitude in bringing this book to life.

— Chuck Crawford

To my wife, Grace, for her understanding, patience, support and insistence to get this book project completed.

— Chris Armbrust

To my beautiful fiancée, Sheila, the only one who really matters! Without your loving support and understanding, this would not have been possible.

— Michael Zink

Contents

List of Figures and Tables	xv
Foreword	xix
Preface	xxii
Acknowledgements	xxiii
Introduction	**I-1**
About Blu-ray Disc™	I-1
Movies, Television Shows, and More	I-1
About This Book	I-4
Units and Notation	I-6
Other Conventions	I-9
Chapter 1 – The Development of Blu-ray Disc	**1-1**
A Brief History of Storage Technology	1-1
TV's Digital Face-Lift	1-4
The CD Revolution	1-5
The Long Gestation of DVD	1-6
Format Wars, the Next Generation	1-14
Chapter 2 – Technology Primer	**2-1**
Understanding Digital and Analog	2-1
Pits and Marks and Error Corrections	2-2
Layers	2-4
The Two HDs – High Definition and High Density	2-6
The World's Television Systems	2-6
Frame Rates	2-8
High-Definition Image Resolutions	2-10
Chroma Subsampling and Bit Depth	2-12
High-Definition Data Streams	2-12
Bird Over the Phone: Understanding Video Compression	2-14
Advanced Video Codecs	2-23
Birds Revisited: Understanding Audio Compression	2-26
Speakers Everywhere	2-34
A Few Timely Words About Jitter	2-34
Pegs and Holes: Understanding Aspect Ratios	2-39
Widescreen Displays	2-45
Why 16:9	2-48
The Transfer Tango	2-51
More on Interlaced vs. Progressive Scanning	2-52
New Display Technologies	2-54
HD Discs Meet HD Video Meet HD Television	2-56

Contents

Chapter 3 – Features   3-1
    Building on DVD Features   3-1
    The Makeover from DVD to Blu-ray Disc   3-2
    Network Connection   3-9
    Persistent Storage and Local Storage   3-11
    Authorized Copying and Authorized Recording   3-11

Chapter 4 – Content Protection, Licensing, and Patents   4-1
    Implementing Content Protection   4-2
    Advanced Access Content System (AACS)   4-4
    AACS Managed Copy   4-8
    BD-ROM Mark   4-8
    BD+   4-9
    Content Protection on Recordable Discs   4-10
    Watermarking   4-12
    Digital Transmission Content Protection (DTCP)   4-13
    High-bandwidth Digital Content Protection (HDCP)   4-13
    Summary of Protection Schemes   4-15
    The Analog Sunset   4-15
    Regional Management   4-16
    Ramifications of Content Protection   4-18

Chapter 5 – Physical Disc Formats   5-1
    BD-ROM Mastering   5-5
    BD-ROM Composition and Production   5-5
    BD-RE Composition   5-6
    BD-R Composition   5-6
    BD Error Correction   5-7
    BD Data Modulation   5-8
    BD-R/RE Recording   5-8
    Phase-Change Recording   5-9
    Burst Cutting Area (BCA)   5-10
    Hybrid Discs   5-11
    Media Storage and Longevity   5-11
    Improvement over DVD   5-14

Chapter 6 – Application Details   6-1
    BDAV   6-3
    Organizational Structure   6-4
    BDMV   6-5
    Presentation Data   6-7
    Paths and Subpaths   6-8
    BD-ROM Application Types   6-9

Contents

Presentation Planes — 6-10
Organization Structure — 6-11
Navigation Data — 6-12
User Interaction — 6-16
Video Formats — 6-18
Advanced Video Applications — 6-21
Audio Formats — 6-26
Presentation Graphics — 6-29
Parental Management — 6-31
Metadata — 6-31
HDMV Details — 6-32
HDMV Graphics Limitations — 6-34
Interactive Audio — 6-34
Browsable Slideshow — 6-35
BD-Java (BD-J) Details — 6-35
System Overview — 6-36
BD-J Application Programming Interface (API) — 6-37
BD-J Menus — 6-40
Graphics Drawing — 6-41
BD-J Text — 6-43
Advanced BD-J Features — 6-43
Local Storage — 6-45
Virtual File System — 6-46
BD-Live Functionality — 6-46
Application Signing — 6-48
Credential Process — 6-50

Chapter 7 – Players — 7-1
High-Definition Multimedia Interface (HDMI) — 7-3
DisplayPort Interface Standard — 7-4
Upscaling DVD — 7-5
Player Types — 7-6
Blu-ray Disc Recorders — 7-7
Player Connections — 7-7
BD Player User Tools — 7-19
How to Get the Best Picture and Sound — 7-21
Viewing Distance — 7-21
Compatibility — 7-22

Chapter 8 – Myths — 8-1
Myth: "Blu-ray is Revolutionary" — 8-1
Myth: "Blu-ray Will Fail" — 8-1

Contents

Myth: "We'll Soon Get Everything from the Internet and Discs Will Go the Way of Dinosaurs" 8-2

Myth: "Some Discs Won't Play If the Player Doesn't Have an Internet Connection" 8-2

Myth: "BD-Live Discs Don't Work on All Players" 8-2

Myth: "Profile 2 (BD-Live) Players and Discs Make Previous Players Obsolete" 8-3

Myth: "Profile 3 and Profile 4 or Future Profiles Will Make Previous Players Obsolete" 8-3

Myth: "The BDA Will Soon Mandate that All BD Players Have an Internet Connection" 8-3

Myth: "If You Unplug Your Profile 2 (BD-Live) Player From the Internet It Will Stop Working" 8-3

Myth: "AACS, BD+, and BD-Live Allow Studios to Spy on Consumers" 8-4

Myth: "Blu-ray Doesn't Look Any Better Than Upconverted DVD" 8-4

Myth: "Blu-ray Players Downconvert Analog Video" 8-4

Myth: "DVD Players Can Be Upgraded to Play Blu-ray Discs" 8-5

Myth: "Older Blu-ray Players Can Be Upgraded to New Profiles" 8-5

Myth: "Existing Receivers with Dolby Digital and DTS Decoding Work Perfectly With Blu-ray Players" 8-5

Myth: "Analog Connections from DVD Players and Blu-ray Players Won't Work with US TVs After the February 2009 Analog Cutoff" 8-6

Myth: "Blu-ray Manufacturing Is Too Intricate and Too Expensive" 8-6

Myth: "The Blu-ray Disc Association Prohibits Adult Content" 8-7

Myth: "Blu-ray Is a Worldwide Standard" 8-7

Myth: "1080p Video is Twice the Resolution of 1080i" 8-7

Myth: "All Blu-ray Titles are (or Must Be) Encoded in 1080p" 8-8

Myth: "Blu-ray Players Only Output 1080i Video" 8-8

Myth: "Blu-ray Does Not Support Mandatory Managed Copy" 8-8

Myth: "Managed Copy Means Every Blu-ray Disc Can be Copied For Free" 8-9

Myth: "BD+ Interferes With Managed Copy" 8-9

Myth: "Region Codes Don't Apply to Computers" 8-9

Myth: "Blu-ray Players Can't Play CDs or DVDs" 8-9

Myth: "Blu-ray Is Better Because It Is Digital" 8-10

Myth: "Blu-ray Video is Poor Because It Is Compressed" 8-10

Myth: "Video Compression Does Not Work for Animation" 8-11

Myth: "Discs Are Too Fragile to Be Rented" 8-11

Myth: "Dolby Digital or DTS Means 5.1 Channels" 8-12

Myth: "The Audio Level from Blu-ray Players Is Too Low" 8-12

Myth: "Downmixed Audio Is Not Good Because the LFE Channel Is Omitted" 8-13

Myth: "Blu-ray Lets You Watch Movies as They Were Meant to Be Seen" 8-13

Myth: "Java and JavaScript Are the Same Thing" 8-13

Myth: "All Blu-ray Discs Must Use AACS" 8-13

Contents

Myth: "AACS Is Required for HDMV, BD-J, Network Access, or Local Storage Access 8-14

Chapter 9 – What's Wrong with Blu-ray Disc" 9-1
Copy Protection 9-2
Regional Management 9-3
Hollywood Baggage on Computers 9-4
NTSC versus PAL Is Also 60 Hertz versus 50 Hertz 9-4
Connection Incompatibilities 9-5
Playback Incompatibilities 9-5
Poor Performance 9-7
Feeble Support of Parental Choice Features 9-7
Not Better Enough 9-8
No Reverse Play 9-9
Only Two Aspect Ratios 9-9
No Barcode Standard 9-10
No External Control Standard 9-10
Poor Computer Compatibility 9-10
Limited Web Standard 9-11
Too Many Encoding Formats 9-11
Too Many Inputs 9-12
Too Many Channels 9-12
Not Enough Interactivity 9-13
Too Much Interactivity 9-13
Conclusion 9-13

Chapter 10 – Interactivity 10-1
Recreating DVD's Success 10-1
A New Kind of Interactivity 10-2
The Seamless User Experience 10-3
Target Applications 10-4
Additional Features 10-6
Specialized User Input 10-9

Chapter 11 – Use in Business and Education 11-1
The Appeal of Blu-ray 11-1
The Appeal of Blu-ray for Video 11-4
The Appeal of Blu-ray for Data 11-5

Chapter 12 – Production Essentials 12-1
Blu-ray Disc Project Examples 12-1
General BD Production 12-2
Tasks and Skills 12-5
The BD Production Process 12-7
Persistent Storage 12-11
Network Connected Features – BD-Live 12-12

Contents

Project Design	12-14
Menu Design	12-16
Navigation Design	12-20
Balancing the Bit Budget	12-23
Asset Preparation	12-26
The Zen of Subwoofers	12-31
Slipping Synchronization	12-32
Preparing Subtitles	12-33
Preparing Graphics	12-34
Video Artifacts	12-37
Putting It All Together (Authoring)	12-38
Formatting and Output	12-39
Testing and Quality Control	12-40
Replication, Duplication, and Distribution	12-43
Disc Labeling	12-43
Package Design	12-43
Production Maxims	12-44
BD-ROM Production	12-45
Hybrid Discs	12-47

Chapter 13 – Blu-ray and Beyond — 13-1

Looking Back at DVD	13-1
Peering Forward...Into the Digital Fog	13-4
Beyond Blu-ray	13-5
The Death of DVD?	13-6
Seeing Double	13-6
The Changing Face of Home Entertainment	13-7
The Far Horizon	13-8

Appendix A – About the Disc — A-1
Appendix B – Reference Data — B-1
Appendix C – Related Standards and Specifications — C-1
Appendix D – References and Information Sources — D-1
Glossary — G-1
Index — Index-1

List of Figures and Tables

Chapter 1 – Introduction
    Table I.1 Meaning of Prefixs   I-6
    Table I.2 Media Gigabyte Conversions   I-7
    Table I.3 Notations Used in This Book   I-8

Chapter 1 – The Development of Blu-ray Disc
    Figure 1.1 Punched Card and Optical Disc   1-1

Chapter 2 – Technology Primer
    Figure 2.1 Optical Disc Pits   2-2
    Figure 2.2 Number Squares   2-3
    Figure 2.3 Optical Disc Layers   2-5
    Figure 2.4 Run-length Compression Example   2-15
    Figure 2.5 Color and Luminance Sensitivity of the Eye   2-17
    Figure 2.6 Block Transforms and Quantization   2-19
    Figure 2.7 Typical MPEG Picture Sequence   2-21
    Figure 2.8 MPEG Video Compression Example   2-22
    Figure 2.9 Frequency Masking and Hearing Threshold   2-27
    Figure 2.10 Effects of Interface Jitter   2-35
    Figure 2.11 Effects of Sampling Jitter   2-36
    Figure 2.12 TV Shape vs. Movie Shape   2-40
    Figure 2.13 Peg and Hole   2-40
    Figure 2.14 Shrink the Peg   2-40
    Figure 2.15 Slice the Peg   2-41
    Figure 2.16 Squeeze the Peg   2-41
    Figure 2.17 Soft Matte Filming   2-42
    Figure 2.18 Pan and Scan Transfer   2-42
    Figure 2.19 The Anamorphic Process   2-43
    Figure 2.20 Aspect Ratios, Conversions, and Displays   2-44
    Figure 2.21 Wide (Full) Mode on a Widescreen TV   2-46
    Figure 2.22 Expand (Theater) Mode on Widescreen TV   2-46
    Figure 2.23 Opening the Frame from 1.85 to 1.78   2.48
    Figure 2.24 Common Aspect Ratios   2-49
    Figure 2.25 Display Sizes at Equal Heights   2-50
    Figure 2.26 Relative Display Sizes for Letterbox Display   2-51
    Figure 2.27 Interlaced Scan and Progressive Scan   2-52
    Figure 2.28 Converting Film to Video   2-54
    Figure 2.29 DTV and HD Ready Logos   2-57
    Table 2.1 Image Data Stream Examples   2-13
    Table 2.2 Compression Ratios for Disc Technologies   2-24

Chapter 3 – Features
    Table 3.1 Hollywood Studio Requirements for Next-Generation Formats   3-2

List of Figures and Tables

Chapter 4 - Content Protection, Licencing, and Patents   4-1
- Figure 4.1 Copy Protection   4-4
- Figure 4.2 Content Protection Systems   4-5
- Figure 4.3 Sequence Key Example   4-7
- Figure 4.4 AACS Authentication Process   4-10
- Figure 4.5 Overview of BD-ROM Content Protection Systems   4-17
- Figure 4.6 Blu-ray Disc Regions   4-16
- Table 4.1 Allowed Analog Output Devices for AACS   4-17
- Table 4.2 Blu-ray Disc Regions

Chapter 5 – Physical Disc Formats
- Figure 5.1 BD Disc Structure   5-4
- Figure 5.2 BD Disc Cartridges   5-5
- Figure 5.3 BD Dual-layer Construction   5-6
- Figure 5.4 Phase-Change Recording   5-9
- Figure 5.5 Burst Cutting Area   5-10
- Table 5.1 Physical Characteristics of BD   5-1
- Table 5.2 Blu-ray Disc Capacities   5-4
- Table 5.3 BD-RE Characteristics   5-7
- Table 5.4 BD-R Characteristics   5-7

Chapter 6 – Application Details
- Figure 6.1 Blu-ray Disc Association Organizational Structure   6-1
- Figure 6.2 Structural Organization of BDAV Content   6-4
- Figure 6.3 Directory Structure of BDMV Disc   6-6
- Figure 6.4 Multiplexing Process   6-8
- Figure 6.5 Blu-ray Disc Presentation Planes   6-10
- Figure 6.6 Structural Organization of BDMV Content   6-11
- Figure 6.7 Display Formats of 4:3 Aspect Ratio Content   6-19
- Figure 6.8 Display Formats of 16:9 Aspect Ratio Content   6-19
- Figure 6.9 Interlaced versus Progressive Display Modes   6-20
- Figure 6.10 Camera Angles Example   6-22
- Figure 6.11 Seamless Playback Example   6-23
- Figure 6.12 Speaker Positioning for 7.1 Channel Configuration   6.28
- Figure 6.13 HDMV Multipage Menu Example   6-33
- Figure 6.14 BD-J System Overview   6-36
- Figure 6.15 BD-J Specification Structure   6-38
- Figure 6.16 Graphics Drawing in Source Mode   6-41
- Figure 6.17 Graphics Drawing in Source-over Mode   6-42
- Figure 6.18 Audio Mixer Components   6-45
- Figure 6.19 Workflow for Application Signing   6-49
- Figure 6.20 Credential Creation Process   6-50
- Figure 6.21 Generalized Block Diagram of an MPEG-2 Encoder   6-51

List of Figures and Tables

Figure 6.22 Generalized Block Diagram of a VC-1 Encoder 6-52
Figure 6.23 Generalized Block Diagram of an AVC Encoder 6-53
Figure 6.24 Block Diagram of a Dolby TrueHD Decoder 6-54
Figure 6.25 Block Diagram of a DTS-HD Lossless Decoder 6-55
Figure 6.26 Example of Virtual Package 6-56
Table 6.1 Blu-ray Disc Association Technical Expert Groups 6-1
Table 6.2 Blu-ray disc Association Specification Books 6-2
Table 6.3 Guide to BD Alphabet Soup 6-3
Table 6.4 General Characteristics of BDMV Presentation Data 6-7
Table 6.5 Comparison of HDMV and BD-J Features 6-9
Table 6.6 Navigation Commands 6-13
Table 6.7 Player Status Registers (PSRs) 6-14
Table 6.8 User Operations (UO) 6-17
Table 6.9 Virtual Key (VK) Events 6-18
Table 6.10 Supported Resolutions and Frame Rates for Primary Video on Blu-ray Disc 6-20
Table 6.11 Recommended Data Rate Limitations for Camera Angles 6-23
Table 6.12 Allowed Combinations of Primary and Secondary Video Codecs 6-24
Table 6.13 Allowed Combinations of Primary and Secondary Video Formats 6-24
Table 6.14 Supported Primary Audio Formats for Blu-ray 6-26
Table 6.15 Supported Secondary Audio Formats for Blu-ray 6-29
Table 6.16 Correlation of PSD 13 Settings to MPAA Ratings 6-31
Table 6.17 BD-J Memory Overview 6-37
Table 6.18 BD-J API Overview 6-39

Chapter 7 – Players

Figure 7.1 RCA Phono Connector 7-9
Figure 7.2 BNC Connector 7-10
Figure 7.3 Phono/Miniphone Connector 7-10
Figure 7.4 DIN-4 (S-video) Connector 7-10
Figure 7.5 Toslink Connector 7-10
Figure 7.6 IEEE 1394 Connector 7-11
Figure 7.7 DB-25 Connector 7-11
Figure 7.8 SCART Connector 7-11
Figure 7.9 Type F Connector and Adapters 7-11
Figure 7.10 RJ-45 Connector 7-12
Figure 7.11 DVI Connector 7-12
Figure 7.12 HDMI Connector 7-12
Figure 7.13 DisplayPort Connector 7-12
Figure 7.14 Example BD Remote Control 7-20
Table 7.1 Blu-ray Disc Player Profiles and Features 7-1
Table 7.2 Examples of Compatibility Problems 7-22

List of Figures and Tables

Chapter 10 – Interactivity
    Figure 10-1 Studio Revenue in 2007 — 10-2

Chapter 11 – Use in Business and Education
    Figure 11.1 The Authoring Environment Spectrum: Utility vs. Ease of Use — 11-7

Chapter 12 – Production Essentials
    Figure 12.1 Sample Disc Flowchart — 12-15
    Figure 12.2 Navigation Flowchart Example — 12-21
    Figure 12.2 Blu-ray Player Remote Control Example — 12-23
    Figure 12.4 Data Rates vs Capacity — 12-24
    Figure 12.5 Video Safe Areas — 12-36
    Figure 12.6 Package Icons and Identifiers — 12-44
    Figure 12.7 Sample PNG Mosaic Graphic — 12-48
    Table 12.1 BD Project Examples — 12-2
    Table 12.2 The BD Production Process — 12-7
    Table 12.3 Disc Capacities for Bit Budgeting — 12-25
    Table 12.4 Sample Bit Budget — 12-25
    Table 12.5 Typical Project Assets — 12-27
    Table 12.6 Video Graphics Checklist — 12-38
    Table 12.7 Testing Checklist — 12-42

Chapter 13 – Blu-ray and Beyond
    Figure 13-1 US DVD and BD Penetration in the First Three Years — 13-2
    Figure 13.2 DVD and BD Title Releases — 13-3
    Figure 13.3 DVD and BD Disc Shipments — 13-3
    Figure 13.4 Media Convergence in the Digital Age — 13-10
    Table 13.1 Technology Penetration Rates-Years to Reach 50% of US Homes — 13-1

Appendix A – About the Disc
    Figure A.1 Disc Flowchart — A-4

Appendix B – Reference Data
    Figure B.1 Conversion Formulas for Playing Times, Data Rate, and Size — B-1
    Figure B.2 Relationships of BD Formats — B-1
    Table B.1 BD, DVD, and CD Capacities and Playing Times — B-2
    Table B.2 Data Rates at Various Playing Times — B-3
    Table B.3 BD Stream Data Rates — B-4
    Table B.4 Limits on BD Elements — B-5
    Table B.5 Download Time for Various Payloads and Bandwidths — B-6
    Table B.6 Bandwidth Requirements for Desired Download Times — B-7
    Table B-7 DVD, HD DVD, BD, and CD Characteristics Comparison — B-8
    Table B.8 Video Resolutions — B-10
    Table B.9 ISO 3166 Country Codes, BD and DVD Regions — B-12
    Table B.10 ISO 639 Language Codes — B-18

Foreword

"So what sort of business are you in?" I am asked by a bored traveler sitting next to me on an airplane trip, sometime in 2004.

"A new technology called Blu-ray Disc™," I answer.

"Blue.....uh......what?"

"Blu-ray Disc. It'll be a next generation disc format – similar to DVD – but this one is going to be high definition. You know, HDTV on a disc that looks and feels just like a DVD when you hold it in your hand. You won't believe how great the picture looks on a decent HDTV set." A flicker of hopeful recognition at "DVD" and "HDTV". Then, suddenly, a dark look.

"Wait a minute. Does that mean I have to throw away all of my DVDs? Already?? Why do we need another disc format, anyway?"

This exchange illustrates a pretty common set of questions I've encountered in recent years in my role as a spokesperson for the Blu-ray Disc format: Do we really need a new disc format? Isn't DVD good enough?

To make things even more challenging, a confusing format war between Blu-ray and another high-definition disc format broke out, causing many pundits to proclaim that unlike DVD, that overnight sensation from the mid-nineties, a newfangled "blue" format would likely have a tough time even making it to market, let alone succeeding.

There is a great deal of irony in this thinking. If you have followed DVD's progress since the format first appeared in late 1996 – perhaps by being a faithful reader of **DVD Demystified**, the predecessor of this book – you already know that it actually took years for that format to become the "overnight sensation" that many people seem to remember now. In fact, about two years after DVD hit the market, one analyst concluded that VHS had a "long and healthy life ahead", and another determined that DVD sales were "slow to take off, partly due to consumer confusion". The thinking was that since few people had heard of DVD and no one had asked for it, why did we need a new format?

Actually, just about everyone needed it, as it soon turned out. After about three uncertain years on the market, consumers began to fall in love with DVD, and sales of players and discs increased dramatically. DVD quickly became the fastest growing consumer format in history, and everyone lived happily ever after.

So if DVD appears to have a long and healthy life ahead of it and no one was asking for Blu-ray Disc, why do we need a new format? As you can see, it's a familiar question, and the fact that you are reading this book shows you already know the answer.

DVD managed to squeeze a remarkable degree of picture quality out of broadcast TV standards that were introduced more than fifty years ago. But, by early 2009, the United States' venerable old "480i" NTSC broadcast system will be shut off and replaced with high-definition signals containing up to six times more picture information. The long dream of upgrading from an old, noise-prone analog TV system to a new, crystal clear digital one is finally happening, and the electronics industry is embracing the change with all its might. Walk into a typical electronics retail store, and you'll notice that every single TV is displaying beautiful high definition images on newly affordable flat-panel TVs that you can hang on your wall. Recent advances in video projection technology are bringing the "theater" to home theaters like never before, since HDTV enables stunning picture quality on much larger screens than standard definition could ever support.

In short, high-definition technology is bringing the full visual and sonic impact of filmed entertainment into your home with a mesmerizing clarity that was only a distant promise ten years ago. In the home video world, this is a Very Big Deal – a tectonic change of a magnitude we have not experienced since the introduction of color TV decades ago.

Given this momentous high-definition revolution, should we all just keep watching programs on the same old standard- definition DVD format forever? Or, should we also have a new high-definition disc format that shows just what that marvelous 1080p display you recently bought can really do? Such a new format is already here, and it's called Blu-ray Disc.

Blu-ray's designers realized that the opportunity to create a new worldwide disc format doesn't come along every day. They knew that, historically, optical disc formats have always provided the very best picture and sound that the technology of the day could deliver, from analog LaserDisc in the '80's to digital DVD in the '90's. So the new Blu-ray Disc format first needed enough high-def horsepower to exceed what anyone had ever seen or heard before, including even the best 1080i broadcasts. This is why so much effort was made to provide a lavish amount of storage capacity and data throughput: the ultimate, archive quality version of a program deserves no less.

But Blu-ray Disc also needed to offer more than the absolute best picture and sound. It needed to provide new ways to go far deeper into a film-viewing experience than DVD could ever deliver. Don't get me wrong – in 1996, DVD was an exciting new technology that greatly improved the home video experience compared to tape. But just look at some of the features the Blu-ray format offers using today's technology: a full-fledged programming language for in-movie games, education and interactive exploration; two simultaneous video channels that allow film-school-quality director commentaries; far superior graphics with less disruptive "popup" menus; and, with BD-Live, even a connection to the Internet that can extend the original linear film into something much more immersive, interactive and social. If I've learned anything in nearly 30 years in this business, it's that giving powerful new tools to highly creative people almost always results in something ingenious and compelling. Give it some time – you'll see.

Meanwhile, the timing of this new book, ***Blu-ray Disc Demystified***, is superb. The industry has settled on a single high-definition disc format, thereby removing any need to hedge bets or sit on a fence. As I write this, Blu-ray is already beginning to see substantial growth in the marketplace, so it's an ideal time to learn all about how the format works and what it can do for you.

If you've had the pleasure of reading any of Jim Taylor's earlier ***DVD Demystified*** books, you already know that they have been a treasure for those seeking more than a just a thorough technical analysis. This new book continues the highly successful blend of carefully researched factual information and industry insight that the other books have become known for, and, as always, no punches are pulled along the way. You will find a unique perspective of the history, practical applications, and even some of the common myths surrounding the format, all written in an accessible style that makes for an enjoyable and engaging read. Jim and his co-authors have done it again, so whether your profession requires you to know about Blu-ray Disc or you just have an interest in knowing how this wonderful new technology works, you've come to the right place.

— Andy Parsons
SVP, Advanced Product Development
Pioneer Electronics (USA) Inc.

Preface

I've concluded that writing a book is a bit like having a baby (although clearly not in the same league). It takes a few years before you forget just how hard it was — enough that you rashly sign up for another one. And, as with a child, you hope that it comes out all right and that people like it when you let it loose on the world. Luckily, I had a team of similarly rash co-authors who made it all possible. I am indebted to Chuck, Chris, and Mike for their knowledge, their hard work, and their good natures. Special thanks again to Samantha Cheng for keeping everything together and everyone on track. I'd like to acknowledge my colleagues at Sonic Solutions, who considerately support my literary endeavors.

I wrote the original **DVD Demystified** in 1997 because I was captivated by DVD, an extraordinary new technology that I wanted to tell everyone about. I knew almost nothing when I started, so I began asking questions of anyone I could find who knew more than I did. Literally hundreds of people, far too many to list, contributed to the collection of know-how that I synthesized and disseminated over the years. I'm gratified that, three editions later, **DVD Demystified** is still useful as a guide and reference for those who are just crossing the threshold of DVD as well as those who are venturing deeper into the maze of twisty passages.

I hope this new book is also useful for anyone eager to learn more about Blu-ray Disc technology. Once again I am obliged to many people who have taken time to explain many things. The art of optical media, interactivity, and digital video and audio progressed immensely in the decade after DVD debuted. Descriptions of Blu-ray often come out sounding like a portrayal of a medieval ruler: rich, powerful, renowned, inscrutable, versatile, acclaimed, byzantine, stimulating, dominating. But Blu-ray genuinely is a world-changing innovation — it will reign for decades as the flag-bearer of highest quality home entertainment. If this book benefits anyone who wants to take advantage of this important medium then my time writing it will have been well spent.

As always, my deep thanks to Steve Chapman at McGraw-Hill for putting up with far more than anyone should ever have to. And to my congenial colleague, Andy Parsons, who always manages to start things off with an astute and thoughtful foreword. I express my gratitude and my apologies to my wife and children, who allowed me to hole up for days on end when deadlines loomed, and came, and passed, and were finally met. And thanks to Fleischman and Arthur, who, as always, have been solidly supportive.

— Jim Taylor
Fox Island, Washington

Acknowledgments

Team work, that's what it's all about. This book, without a doubt, could not have been written without the generosity of spirit, time and expertise from our friends and colleagues. On behalf of the authors Jim Taylor, Chuck Crawford, Chris Armbrust and Mike Zink, thank you.

To the team at McGraw-Hill Professional – Steve Chapman, David Fogarty, Pamela Pelton, Alexis Richard, Jeff Weeks, and David Zielonka, thank you.

Our friends at Technicolor/Thomson – Bob Michaels, Chuck Null, and Miles Del Hoyo in Burbank; Jobst Hörentrup and Ralf Ostermann in Germany; and Masaru Yamamoto in Japan, thank you.

Thank you to Jim Burger, Benn Carr, Alan Bell, and Brad Collar for your advice and assistance.

To the incredible team of technical reviewers, thank you for keeping us honest — Scott Bates, Jess Bowers, Denny Breitenfeld, Chris Brown, Augusto Cardoso, Annie Chang, Brad Collar, Yoshiharu Dewa, Roger Dressler, Lee Evans, Bill Foote, Roger Gekchyan, Greg Gewicky, Bill Hofmann, Koen Holtman, Jinha Kim, Van Ling, Phillip Maness, Kyle Prestenback, Joe Rice, Daniel Robertson, Craig Smithpeters, Phil Starner, Paul Wenker and Bram Wessel.

Thank you to Mark Johnson and Shu-Ping Lu at Javelin Ventures in Glendale.

A very special thank you to Phil Starner – if not for Phil's commitment to the format and this project there would be no demonstration disc.

Thank you to the Blu-ray Disc Association for their development support of the demonstration disc; the production team in Washington, DC – Keith Wood, Nathan John Best, Jamie Pickell; Chris Cardno, Visual Edge; Ari Zagnit, and Aaron Rehm, Henninger Media Services; and Thomas Bennett; the Graphics and Blu-ray teams at Technicolor – Kristine Bailey-Mayers, Eric Bashore, Paul Batrikian, Xian Chen, Jay Clinton, Marcus Clinton, Marc Delafin, Steve Ehrgott, Attila Fruttus, Robert Gekchyan, Andrew Giacumakis, Daniel Haffner, Tom Jaeger, Zheng Lu, Chengwu Luo, Guozhi Ma, Paul Resendi, Rinzi Ruiz, Vahe Stambultsyan, Ryan Whearty and John Town, HES Camarillo; and the Blu-ray team at Sonic Solutions – Paul Wenker, Kevin Baribault and the entire QC department.

It was my pleasure to work with all of you.

— E. Samantha Cheng, Managing Editor

A very special thank you to Catherine E. McDermott for her friendship, guidance, and support.

— Samantha Cheng and Chuck Crawford

Blu-ray Disc Demystified

Introduction

About Blu-ray Disc™

We have entered the third age of digital optical disc technology, or fourth age if you start with analog laserdisc. Laserdisc led to CD which led to DVD which led to Blu-ray Disc (BD). Milestones along the way include the transition from analog to digital (laserdisc to CD), stereo audio to video plus multichannel audio (CD to DVD), and standard definition to high definition (DVD to BD). Storage capacity leapt from 650 megabytes (a little more than half a gigabyte) on CD to 50 gigabytes on dual-layer BD (and more, if three, four, or more-layer BD experiments ever escape from the laboratory).

DVD is commonly put on a pedestal as the most successful consumer electronics product of all time. It has clearly changed the way many of us watch movies at home, adding menus, commentaries, widescreen video, multichannel audio, and more. DVD was a sea change from analog videotapes, a hugely compelling improvement that led to over one billion DVD playback devices finding their way into homes, automobiles, and offices in ten swift years.

Although the triumph of DVD set very high expectations for BD as a successor format, BD isn't a similar quantum leap, so it will never achieve the same world-altering success. However, the shift to high-definition video is inevitable, and BD will undoubtedly be along for the ride.

Many people will require a working knowledge of BD, including its capabilities, strengths, and limitations. This book provides a significant amount of this knowledge. DVD, and now BD, affects a remarkably diverse range of fields. Advances in the arts and technologies of film and video, combined with breakthroughs in recording, display, and interactivity, have vaulted shiny discs into a leading role in a broad spectrum of uses. A few are illustrated in the following pages.

Movies, Television Shows, and More

High-definition video moves the home entertainment experience closer to the theater experience. According to the Consumer Electronics Association (CEA), by the end of 2007 approximately 50 million HDTV sets had been sold to consumers in the U.S. alone since the format debuted in 1998.

Music and Audio

Because audio CD is well established and satisfies the needs of most music listeners, DVD and BD have less of an effect in this area. Unlike DVD, BD has the capacity to eas-

Introduction

ily hold uncompressed or losslessly compressed multichannel audio tracks. DVD-Audio and SACD (Super Audio Compact Disc) never achieved mainstream success, and the audio-only playback features of BD were wisely rolled back into the video format. DVD-Audio and SACD appealed to music labels because they included content protection features that CD never had. The same is provided by AACS for BD, but BD will never become a dominant music distribution medium, especially in the face of Internet music distribution medium.

BD is a boon for audio books and other spoken-word programs. Literally hundreds of hours of stereo audio can be stored on a single disc — a disc that is cheaper to produce and more convenient to use than cassette tapes or multi-CD sets.

Music Performance Video

Music performance video was one of the most successful categories for DVD, and should prove equally popular for BD. With high-definition, long-playing video and multichannel surround sound, BD music video appeals to a range of fans from opera to ballet to New Age to acid rock. Music albums on BD can improve their fan appeal by adding such tidbits as live performances, interviews, backstage footage, outtakes, video liner notes, musician biographies, and documentaries. And of course BD players can still play existing CD audio discs and DVD music discs.

Training and Special Interest

Until it is eventually replaced by video delivered over broadband Internet, DVD has become the medium of choice for video training. The low cost of hardware and discs, the widespread use of players, the availability of authoring systems, and a profusion of knowledgeable DVD developers and producers made DVD - in both DVD-Video and multimedia DVD-ROM form - ideal for industrial training, teacher training, sales presentations, home education, and any other application where full video and audio are needed for effective instruction.

Videos for teaching skills from accounting to TV repair to dental hygiene, from tai chi to guitar playing to flower arranging become vastly more effective when they take advantage of the on-screen menus, detailed images, multiple soundtracks, selectable subtitles, and other advanced interactive features. Consider an exercise video that randomly selects different routines each day or lets you choose the mood, the tempo, and the muscle groups on which to focus. Or a first-aid training course that slowly increases the difficulty level of the lessons and the complexity of the practice sessions. Or an auto-mechanic training video that allows you to view a procedure from different angles at the touch of a remote control (preferably one with a grease-proof cover). Or a cookbook that helps you select recipes via menus and indexes and then demonstrates with a skilled chef leading you through every step of the preparation. All this on a small disc that never wears out and never has to be rewound or fast-forwarded.

Discs are cheaper and easier to produce, store, and distribute than videotape. Earlier products such as laserdisc, CD-i, and Video CD did well in training applications, but they required expensive or specialized players. Given the widespread availability of DVD players, many training titles will continue to be created on DVD, but BD is the perfect solution for applications that require high-definition video or advanced interactivity.

Education

Filmstrips, 16-mm films, VHS tapes, laserdiscs, CD-i discs, Video CDs, CD-ROMs, and the Internet have all had roles in providing images and sound to supplement textbooks and teachers. But those technologies lacked the picture quality and clarity that is so important for classroom presentations. The unprecedented image detail, high storage capacity, and low cost of BD make it an excellent candidate for use in classrooms, especially since it integrates well with computers. Although BD players are not widely adopted in education, computers with BD-ROM drives will become more commonplace in the classroom. CD-ROM infiltrated all levels of schooling from home to kindergarten to college and then passed the baton to DVD as new computers with built-in DVD-ROM drives were purchased. The same will happen with BD as educational publishers discover how to make the most of the format, creating truly interactive applications with the sensory impact and realism needed to stimulate and inspire inquisitive minds.

Computer Software

The Internet is supplanting CD-ROM and DVD-ROM as the computer software distribution medium of choice. Still, multi-gigabyte applications that would take days or weeks to download can be distributed quickly and efficiently on DVD-ROM or BD-ROM. These include large application suites, clip art collections, software libraries containing dozens of programs that can be unlocked by paying a fee and receiving a special code, specialized databases with hundreds of millions of entries, and massive software products such as network operating systems and document collections.

Video Games

The capacity to add high-quality, real-life video and full surround sound to three-dimensional game graphics has proven to be highly attractive to video game manufacturers. A combination video game/CD/DVD/BD player is quite appealing. Sony pushed the envelope with BD drives in the PlayStation 3 console, putting over 9 million devices into the worldwide marketplace by the Spring of 2008.

Information Publishing

The Internet is a wonderfully effective and efficient medium for information publishing, but it currently lacks the bandwidth needed to do justice to large amounts of data rich with graphics, audio, and motion video. BD, with storage capacity far surpassing CD and DVD, high-definition video, and the power of Java programming, is perfect for publish-

Introduction

ing and distributing information in our ever more knowledge-intensive and information-hungry world.

Organizations can use BD to quickly and easily disseminate reports, training material, manuals, detailed reference handbooks, document archives, and much more. Portable document formats such as Adobe Acrobat and HTML are perfectly suited to publishing text and pictures on BD. Recordable BD is available for custom publishing of discs created on the desktop.

Marketing and Communications

BD is well suited for carrying information from businesses to their customers and from businesses to businesses. A BD can hold an exhaustive catalog that elaborates on each product with full-color illustrations, video clips, demonstrations, customer testimonials, and more, at a fraction of the cost of printed catalogs.

About This Book

BD Demystified is an introduction and reference for anyone who wants to understand Blu-ray Disc. It is not a production guide, nor is it a detailed technical handbook[1], but it does provide an extensive technical grounding for anyone interested in BD technology. This book borrows a few elements from previous publications — *DVD Demystified*, *DVD Demystified Second Edition*, and *DVD Demystified Third Edition*. This book can be considered as the "fourth edition" in the *DVD Demystified* family, but we gave it a new title to match the winner of the next-generation format war. This book is divided into sections, which are subdivided into chapters.

Section I - History and Background

Chapter 1, "The Development of Blu-ray Disc," provides historical context and background, and describes the growing pains of BD as it evolved from DVD. Any top analyst or business leader will tell you that extrapolating from prior technologies is the best way to predict technology trends. This chapter takes a historical stroll through the developments leading up to the introduction of BD.

Section II - Fundamentals and Features

Chapter 2, "Technology Primer," explains concepts such as aspect ratios, digital compression, progressive scan, and describes associated display and signal technologies. This chapter is a gentle technical introduction for nontechnical readers, and it will be useful for technical readers as well.

---

[1] See *High-Definition DVD Handbook, Producing for HD DVD and Blu-ray Disc* and *The HD Cookbook: Programming HD-DVD and Blu-ray Disc*, for more technical details of next generation optical disc production, compression, and authoring.

Introduction

Chapter 3, "**Features,**" covers the basic features of the BD video formats. In addition to feature details, topics such as interoperability, network capabilities, and programmability are discussed in this chapter.

Chapter 4, "**Content Protection, Licensing, and Patents,**" details the mechanisms that have been developed to protect the assets of the content owner. This chapter covers AACS, BD+, region playback control, watermarking, managed recording, and managed copying, as well as transmission protection technologies such as DTCP and HDCP. It also attempts to make sense of the dozens of format and patent licenses that are necessary for BD product makers.

Section III - Formats

Chapter 5, "**Physical Disc Formats,**" breaks down the physical characteristics of blue-laser optical disc technologies, from the mechanics to the data handling attributes.

Chapter 6, "**Application Details,**" reveals the particulars of the video and audio specifications. It lays out the data structures, stream composition, navigation information, and other elements of the application formats.

Section IV - Implementation

Chapter 7, "**Players,**" provides information on BD players and their connections.

Chapter 8, "**Myths,**" does a reality check on the myths and the misunderstood characteristics of BD. Be careful what you believe.

Chapter 9, "**What's Wrong with Blu-ray Disc,**" explores the shortcomings of the format and how to deal with them.

Section V - Uses

Chapter 10, "**Interactivity,**" investigates what happens when you add the power of Java programming and Internet-style multimedia with multiple video planes, mixable audio streams, downloadable content, and flexible user settings.

Chapter 11, "**Use in Business and Education,**" helps you decide what disc flavor is right for you. This chapter explains how BD can be integrated in the home, in the workplace, and in training, teaching, and learning situations.

Chapter 12, "**Production Essentials,**" dips into the most important mysteries of producing content for BD. This chapter provides a thorough grounding for anyone interested in creating discs or simply learning more about how the creation process works.

Section VI - The Future

Chapter 13, "**Blu-ray and Beyond,**" is a peek into the crystal ball to see what possibilities lie ahead for BD and the world of digital video. Keep your arms in the vehicle at all times.

Introduction

Appendixes and Glossary

Appendixes provide information about the disc accompanying this book and provide reference details on the data and standards for the optical disc formats.

A glossary of esoteric, arcane, and mundane words and terms is provided to aid those who know they saw a word somewhere but cannot remember what it means or where they saw it.

Units and Notation

BD is yet another casualty of an unfortunate collision between the conventions of computer storage measurement and the norms of data communications measurement. The SI[1] abbreviations of k (thousands), M (millions), and G (billions) usually take on slightly different meanings when applied to bytes, in which case they are based on powers of 2 instead of powers of 10 (see Table I.1).

Table I.1 Meanings of Prefixes

Symbol	Prefix SI	Prefix IEC	IEC Symbol	Common Use	Computer Use	Usage Difference
k or K	kilo	kibi	Ki	[k] 1,000 (10^3)	[K] 1024 (2^{10})	2.4%
M	mega	mebi	Mi	1,000,000 (10^6)	1,048,576 (2^{20})	4.9%
G	giga	gibi	Gi	1,000,000,000 (10^9)	1,073,741,824 (2^{30})	7.4%
T	tera	tebi	Ti	1,000,000,000,000 (10^{12})	1,099,511,627,776 (2^{40})	10%
P	peta	pebi	Pi	1,000,000,000,000,000 (10^{15})	1,125,899,906,842,624 (2^{50})	12.6%

The problem is that there is no universal standard for unambiguous use of these prefixes. One person's 50 GBs is another person's 46.57 Gbytes and one person's 6 MB/s is another's 5.72 Mbytes/s. Can you tell which is which?[2]

The laziness of many engineers who mix notations such as KB/s, kb/s, and kbps with no clear distinction and no definition compounds the problem. And since divisions of 1000 look bigger than divisions of 1024, marketing mavens are much happier telling you that a dual-layer BD holds 50 gigabytes rather than a mere 46.57 gigabytes. It may seem trivial, but at larger denominations the difference between the two usages — and the resulting

---

[1]Système International d'Unités — the international standard of measurement notations, such as, millimeters and kilograms.

[2]50 GB is the typical data capacity given for a dual layer BD – fifty billion bytes, measured in magnitudes of 1000. The 46.57 Gbyte value is the true data capacity of BD, measured using the computing method in magnitudes of 1024. The 6 MB/s value is the 48 Mbps BD reference data transfer rate in thousands of bytes per second. The 5.72 Mbytes/s value is the rate measured in computer units of 1024 bytes per second. If you don't know what any of this means, don't worry, by the end of Chapter 5 it should all make sense.

Introduction

potential error — becomes significant. There is almost a five percent difference at the mega level and more than a seven percent difference at the giga level. If you are planning to produce a Blu-ray Disc and you take pains to make sure your data takes up just under 25 gigabytes, you will be surprised and annoyed to discover that only 23.28 gigabytes fit on the disc. Things will get worse down the road with a ten percent difference at the tera level.

For the mathematically challenged, Table I.2 provides a ready reference reflecting the measurement values in decimal and in binary computation.

Table I.2 Media Gigabyte Conversions

Media Type	Decimal Size Claim	Actual Binary Value
DVD-5 (SL)	4.7 billion bytes	4.37 gigabytes
DVD-9 (DL)	8.5 billion bytes	7.9 gigabytes
BD-25 (SL)	25 billion bytes	23.28 gigabytes
BD-50 (DL)	50 billion bytes	46.57 gigabytes

Note: SL is the abbreviation for single layer, and DL is the abbreviation for dual layer.

Because computer memory and data storage usually are measured in megabytes and gigabytes (as opposed to millions of bytes and billions of bytes), this book uses 1024 as the basis for measurements of data size and data capacity, with abbreviations of KB, MB, and GB. However, since these abbreviations have become so ambiguous, the term is spelled out when practical. In cases where it is necessary to be consistent with published numbers based on the alternative usage, the words thousand, million, and billion are used, or the abbreviations k bytes, M bytes, and G bytes are used (note the small k and the spaces).

To distinguish kilobytes (1024 bytes) from other units such as kilometers (1000 meters), common practice is to use a large K for binary multiples. Unfortunately, other abbreviations such as M (mega) and m (micro) are already differentiated by case, so the convention cannot be applied uniformly to binary data storage. And in any case, too few people pay attention to these nuances.

In 1999, the International Electrotechnical Commission (IEC) produced new prefixes for binary multiples[3] (see Table I.1). Although the new prefixes have not caught on, they are a valiant effort to solve the problem. The main strike against them is that they sound a bit silly. For example, the prefix for the new term gigabinary is gibi, so a BD-25 can be said to hold 23.28 gibibytes, or GiB. The prefix for kilobinary is kibi, and the prefix for terabinary is tebi, yielding kibibytes and tebibytes. Jokes about "kibbles and bits" and "teletebis" are inevitable.

As if all this were not complicated enough, data transfer rates, when measured in bits per second, are almost always multiples of 1000, but when measured in bytes per second

---

[3]The new binary prefixes are detailed in IEC 60027-2-am2 (1999-01): Letter symbols to be used in electrical technology. Part 2: Telecommunications and electronics, Amendment 2.

Introduction

are sometimes multiples of 1000 and sometimes multiples of 1024. For example, a 1X BD drive transfers data at 48 million bits per second (Mbps), which might be listed as 6 million bytes per second or might be listed as 5.722 megabytes per second. A 1X DVD drive transfers data at 11.08 million bits per second (Mbps), which might be listed as 1.385 million bytes per second or might be listed as 1.321 megabytes per second. The 150 KB/s 1X data rate commonly listed for CD-ROM drives is "true" kilobytes per second, equivalent to 153.6 thousand bytes per second.

This book uses 1024 as the basis for measurements of byte rates (computer data being transferred from a storage device such as a hard drive or DVD-ROM drive into computer memory), with notations of KB/s and MB/s. For generic data transmission, generally measured in thousands and millions of bits per second, this book uses 1000 as the basis for bitrates, with notations of kbps and Mbps (note the small k) (See Table I.3).

Keep in mind that when translating from bits to bytes, there is a factor of 8, and when converting from bitrates to data capacities in bytes, there is an additional factor of 1000/1024.

Table I.3 Notations Used in This Book

Notation	Meaning	Magnitude	Variations	Example
b	bit	(1)		
kbps	thousand bits per second	10^3	Kbps, kb/s, Kb/s	56 kbps modem
Mbps	million bits per second	10^6	mbps, mb/s, Mb/s	11.08 Mbps DVD data rate
Gbps	billion bits per second	10^9	gbps, gb/s, Gb/s	
B	byte	(8 bits)		
KB	kilobytes	2^{10}	Kbytes, KiB	2 KB per DVD sector
KB/s	kilobytes per second	2^{10}	KiB/s	150 KB/s CD-ROM data rate
MB	megabytes	2^{20}	Mbytes, MiB	650 MB in CD-ROM
M bytes	million bytes	10^6	Mbytes, MB	682 M bytes in CD-ROM
MB/s	megabytes per second	2^{20}	MiB/s	1.32 MB/s DVD data rate
GB	gigabytes	2^{30}	Gbytes, GiB	4.37 GB in a DVD
G bytes	billion bytes	10^9	Gbytes, GB	4.7 G bytes in a DVD
TB	terabytes	2^{40}	Tbytes, TiB	
T bytes	trillion bytes	10^{12}	Tbytes, TB	
PB	petabytes	2^{50}	Pbytes, PiB	
P bytes	quadrillion bytes	10^{15}	Pbytes, PB	

Introduction

Other Conventions

Spelling

The word disc, in reference to optical discs, should be spelled with a c, not a k. The generally accepted rule is that optical discs are spelled with a c, whereas magnetic disks are spelled with a k. For magneto-optical discs, which are a combination of both formats, the word is spelled with c because the discs are read with a laser. The New York Times, after years of head-in-the-sand usage of k for all forms of data storage, revised its manual in 1999 to conform to industry practice. Standards bodies such as ECMA and the International Organization for Standardization (ISO) persist in spelling it wrong, but what can you expect from bureaucracies? Anyone writing about DVD who spells it "disk" instead of "disc" immediately puts the reader on notice that the author is clueless.

PC

The term PC means "personal computer." If a distinction is needed as to the specific type of hardware or operating system, brand names such as Windows, Mac OS, and Linux are added.

Player

The term "player" used in a general sense refers to standalone players, game consoles, and software playback applications.

Title

The word title fulfills two primary roles in optical disc parlance. In a general sense, title may be used to indicate the entire contents of a disc, as in, the disc accompanying this book is a BD title. In a specific sense, title is the highest-level unit of a BD disc, and a single disc may contain up to 1,001 titles.

Aspect Ratios

This book usually normalizes aspect ratios to a denominator of 1, with the 1 omitted most of the time. For example, the aspect ratio 16:9, which is equivalent to 1.78:1, is represented simply as 1.78. Normalized ratios such as 1.33, 1.78, and 1.85 are easier to compare than unreduced ratios such as 4:3 and 16:9. Note also that the ratio symbol (:) is used to indicate a relationship between width and height rather than the dimension symbol (×), which implies a size.

Widescreen

When the term widescreen is used in this book, it generally means an aspect ratio of 1.78 (16:9). The term widescreen, as traditionally applied to movies, has meant anything wider than the standard 1.33 (4:3) television aspect ratio, from 1.5 to 2.7. Since the 1.78 ratio has been chosen for DVD, BD, digital television, and widescreen TVs, it has become the commonly implied ratio of the term widescreen.

Blu-ray Disc Demystified

Introduction

Disc Format Names

The term BD is often applied both to the BD family as a whole and specifically to the HDMV and BD-J video application formats, which together are often called BDMV. This book follows the same convention for simplicity and readability but only when unambiguous.

When a clear distinction is needed, this book uses precise terms, such as, BD-ROM for replicated (pre-recorded) discs; HDMV, BD-J (BD-Java), or BDAV for application formats; and BD-R or BD-RE for write-once and rewritable formats, respectively.

HD

The initialism HD can refer to both high definition and high density. This book generally uses HD for high definition, which should be clear from the context.

Television Systems

There are basically two mutually incompatible standard-definition television recording systems in common use around the world. One system uses 525 lines scanned at 60 fields per second with NTSC color encoding and is used primarily in Japan and North America. The other system uses 625 lines scanned at 50 fields per second with PAL or SECAM color encoding and is used in most of the rest of the world. This book generally uses the technically correct terms of 525/60 (simplified from 525/59.94), and 625/50, but also uses the terms NTSC and PAL in the generic sense.

When the world moved to high-definition the number of scan lines were standardized at 720 and 1080, but there are still two common frame rates of 50 Hz and 60Hz that vary among regions. In addition, the "film" frame rate of 24 Hz has become common for both encoding and display. This book uses terms 720p and 1080p to refer to progressive-scan picture formats regardless of frame rate, and 1080i to refer to the interlaced variation. Terms such as 720/30p, 1080/24p, and 1080/30i are used when more precision is needed. The number after the slash is always the frame rate, not the field rate, to avoid confusion caused by sloppy notation such as 60i or 50i that all careful writers should eschew.

Colorspaces

BD uses three standards for video — for high-definition video, ITU-R BT.709 component digital video 4:2:0 colorspace, and for standard-definition video, ITU-R BT.601 component digital video 4:2:0 colorspace. BD allows for the JPEG background plane to be either $Y'C_bC_r$ 4:2:0 or $Y'C_bC_r$ 4:2:2 colorspace. The Java graphics plane uses the RGBA colorspace.

Some BD video playback systems include analog component video output in Y'PbPr format, which is also correctly called Y'/B'-Y'/R'-Y'. Digital component output is correctly called $Y'C_bC_r$, although it's often referred to as HDMI or DVI, the digital connection format. When a technical distinction is not critical or is clear from the context, this book uses the general term YUV to refer to the component video signals in nonlinear color-difference format. This book also uses the general term RGB to refer to nonlinear R'G'B' video.

Chapter 1
The Development of Blu-ray Disc™

A Brief History of Storage Technology

In 1801, Joseph-Marie Jacquard devised an ingenious method for weaving complex patterns using a loom controlled by punched metal cards. The same idea was borrowed over 30 years later by Charles Babbage as the program storage device for his mechanical computer, the Analytical Engine.[1] Ninety years after Jacquard, Herman Hollerith used a similar system of punched cards for tabulating the US Census, after getting the idea from watching a train conductor punch tickets. Fifty years later, the first electronic computers of the 1940s also employed punched cards and punched tape — using the difference between a surface and a hole to store information. The same concept is used in modern optical storage technology. Viewed through an electron microscope, the pits and lands of a DVD or Blu-ray Disc would be immediately recognizable to Jacquard or Hollerith (see Figure 1.1). However, the immensity and variety of information stored on these miniature pockmarked landscapes would truly amaze them. Hollerith's cards held only 80 characters of information and were read at a glacial few per second.[2]

Figure 1.1 Punched Card and Optical Disc

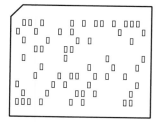
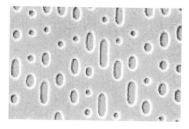

Another essential method of data storage — magnetic media — was developed for the UNIVAC I in 1951. Magnetic tape improved on the storage density of cards and paper tape by a factor of about 50 and could be read significantly faster. Magnetic disks and drums appeared a few years later and improved on magnetic tape, but cards and punched tape were still much cheaper and remained the primary form of data input until the late 1970s. IBM

---

[1]Unlike Jacquard, whose system enjoyed widespread success, Babbage seemed incapable of finishing anything he started. He never completed any of his mechanical calculating devices, although his designs were later proven correct when they were turned into functioning models by other builders.

[2]Modern card readers of the 1970s still used only 72 to 80 characters per card but could scan over a thousand cards per second.

The Development of Blu-ray Disc

introduced 8-inch flexible floppy disks in 1971. These gave way to smaller 5.25-inch disks in 1976 and, finally, in 1983, 3.5-inch diskettes small enough to fit in a shirt pocket. Each successive version held two or three times more data than its predecessor despite being smaller.

Magnetic tape was adapted for video in 1951 by Bing Crosby Enterprises. Videotape recording was first used only by professional television studios, but a new era for home video was ushered in when Philips and Sony produced black-and-white, reel-to-reel videotape recorders in 1965. However, at $3,000 (the equivalent of about $19,000 in 2007) they did not grace many living rooms. Sony's professional 3/4-inch U-matic videocassette tape appeared in 1972, the same year as the video game Pong. More affordable color video recording reached home consumers in 1975 when Sony introduced the Betamax videotape recorder. The following year JVC introduced VHS (Video Home System)[3], which was slightly inferior to Betamax but won the battle of the VCRs by virtue of longer recording time, lower costs, and extensive licensing agreements with equipment manufacturers and video distributors. The first Betamax VCR cost $2,300 ($8,800 in 2007 dollars), and a 1-hour blank tape cost $16 (about $60 at today's value). The first VHS deck was cheaper at $885 (over $3,200 today), and Sony quickly introduced a new Betamax model for $1300, but both systems were out of the financial reach of most consumers for the first few years.

MCA/Universal and Disney studios sued Sony in 1976 in an attempt to prevent home copying with VCRs. Eight years later, the courts upheld consumer recording rights by declaring Sony the winner. Established content owners, such as record labels and motion picture studios, have fought a similar battle with each new introduction of recording technology, going back to player pianos and extending to MP3, DVD, and Blu-ray, as they try to control the technology and influence the law.

In 1978, Philips and Pioneer introduced videodiscs, which actually had been first developed in rudimentary form in 1928 by John Logie Baird.[4] The technology had improved in 50 years, replacing wax discs with polymer discs and delivering an exceptional analog color picture by using a laser to read information from the disc. The technology got a big boost from General Motors, which bought 11,000 players in 1979 to use for demonstrating cars. Videodisc systems became available to the home market in 1979 when MCA joined the laserdisc camp with its DiscoVision brand. A second videodisc technology, called capacitance electronic disc (CED), introduced just over two years later by RCA, used a diamond stylus that came in direct contact with the disc — as with a vinyl record. CED went by the brand name SelectaVision and was more successful initially, but eventually failed because of its technical flaws. CED was abandoned in 1984 after fewer than 750,000 units were sold, leaving laserdisc to overcome the resulting stigma. JVC and Panasonic (Matsushita) developed a similar technology called *video high density* or *video home disc* (VHD), which used a groove-

---

[3]It is sometimes claimed that VHS originally stood for *vertical helical scan* and was later changed to a more consumer-friendly expansion. The backronyms *Victor Home System* and *Victor helical scan* are also occasionally seen.

[4]Baird was a brilliant inventor who made major contributions to the development of television. He used standard gramophone music recording equipment to record a 30-line video signal from a Nipkow disc. In the early 1980's, Donald McLean was able to digitize, process, and recover the video from Baird's surviving Phonovision discs, making it possible to see the very first television recordings with remarkable clarity.

The Development of Blu-ray Disc

less disc that was read by a floating stylus. VHD limped along for years in Japan and was marketed briefly by Thorn EMI in Great Britain, but it never achieved significant success.

For years, laserdisc was the Mark Twain of video technology, with many exaggerated reports of its demise as customers and the media confused it with the defunct CED. Adding to the confusion was the addition of digital audio to laserdisc, which in countries using the PAL television system required the re-launch of a new, incompatible version. Laserdisc persevered, but because of its lack of recording, the high price of discs and players, and the inability to show a movie without breaks (a laserdisc cannot hold more than one hour per side), it was never more than a niche success, catering to videophiles and penetrating less than 2 percent of the consumer market in most countries. In some Asian countries, laserdisc achieved as much as 50 percent penetration, almost entirely because of its karaoke features.

Despite the exceptional picture quality of both laserdisc and CED, they were quickly overwhelmed by the eruption of VHS VCRs in the late 1970s, which started out at twice the price of laserdisc players but soon dropped below them. The first video rental store in the United States opened in 1977, and the number grew rapidly to 27,000 in the late 1990s. Direct sales of prerecorded movies to customers, known as sell-through, began in 1980. In 2008, the home video market was a $24 billion business in the US and generated over $50 billion worldwide.

Optical media languished during the heyday of magnetic disks, never achieving commercial success other than for analog video storage (i.e., laser videodiscs). It was not until the development of the *compact disc* (CD) in the 1980s that optical media again proved its worth in the world of bits and bytes, setting the stage for DVD.

Another innovation of the 1980s was erasable optical media based on *magneto-optical* (MO) technology. Magneto-optical discs use a laser to heat a polyphase crystalline material that can then be aligned by a magnetic field. Features of MO technology were later adapted for recordable DVD and BD.

In the late 1980s, a new video recording format based on 8-mm tape with metal particles was introduced. The reduced size and improved quality were not sufficient to displace the well-established VHS format, but 8-mm and Hi8 tapes were quite successful in the camcorder market, where smaller size is more significant. 8mm tape was later adapted by Sony to store digital video in the DV format.

Around this time, minor improvements were made to television, with stereo sound added in the United States in 1985 and, later, closed captions.

In 1987, JVC introduced an improved "super VHS" system called *S-VHS*. Despite being compatible with VHS and almost doubling the picture quality, S-VHS was never much of a success because there were too many barriers to customer acceptance. Oddly, S-VHS foreshadowed very similar problems that DVD-Audio would encounter a decade later. Players and tapes were much more expensive, and VHS tapes worked in S-VHS players but not vice versa. A special s-video cable and an expensive TV with an s-video connector were required to take advantage of the improved picture, and S-VHS was not a step toward high-definition television (HDTV), which was receiving lavish attention in the late 1980s and was expected to appear shortly. The major contribution of S-VHS to the industry was the popularization of the s-video connector, which carried better-quality signals than the standard composite format. S-video connectors are still mistakenly referred to as S-VHS connectors.

The Development of Blu-ray Disc

TV's Digital Face-Lift

As television began to show its age, new treatments appeared in an attempt to remove the wrinkles. European broadcasters rolled out *PALplus* in 1994 as a stab at *enhanced-definition television* (EDTV) that maintained compatibility with existing receivers and transmitters. PALplus achieved a widescreen picture by using a letterboxed[5] image for display on conventional TVs and hiding helper lines in the black bars so that a widescreen PALplus TV could display the full picture with extended vertical resolution.

Other ways of giving television a face-lift included *improved-definition televisions* (IDTV) that doubled the picture display rate or used digital signal processing to remove noise and to improve picture clarity.

None of these measures were more than stopgaps while we waited for the old workhorse to be replaced by high-definition television. HDTV was first demonstrated in the United States in 1981, and the process of revamping the "boob tube" reached critical mass in 1987 when 58 broadcasting organizations and companies filed a petition with the FCC (Federal Communications Commission) to explore advanced television technologies. The *Advanced Television Systems Committee* (ATSC) was formed and began creeping toward consensus as 25 proposed systems were evaluated extensively and either combined or eliminated.

In Japan, a similar process was underway that was unfettered by red tape. An HDTV system called *HiVision*, based on MUSE compression, was developed quickly and put into use in 1991.[6] Ironically, the rapid deployment of the Japanese system was its downfall. The MUSE system was based on the affordable analog transmission technology of its day, but soon the cost of digital technology plummeted at the same time as its capability skyrocketed. Back on the other side of the Pacific, the lethargic US HDTV standards creation process was still in motion as video technology graduated from analog to digital. In 1993, the ATSC recommended that the new television system be digital. The Japanese government and the Japanese consumer electronics companies — which would be making most of the high-definition television sets for use in the United States — decided to wait for the American digital television standard and adopt it for use in Japan, as well. A happy by-product of HiVision is that the technology-loving early Japanese adopters served as guinea pigs, supporting the development of high-resolution widescreen technology and paving the way for HDTV elsewhere. A European HDTV system, *HD-MAC*, which combined widescreen analog video with digital audio, was even more short-lived; it was demonstrated at the 1992 Albertville Olympics but was abandoned shortly thereafter.

Discussions had begun in Europe around creating a digital television format, which lead to the launch of the DVB (Digital Video Broadcast) Project in 1993. It became clear that the satellite and cable industries would be able to bring digital television to market before the terrestrial broadcast industries, which led to the development and launch of DVB-S in 1995, followed quickly by DVB-C, and finally by DVB-T in 1998.

---

[5]Letterboxing is the technique of preserving a wide picture on a less wide display by covering the gap at the top and bottom with black bars. Aspect ratios and letterboxing are covered in **Chapter 2**.

[6]The HiVision system used 1125 video scan lines. In the charmingly quirky style of Japanese marketing, HiVision was introduced on November 25 so that the date (11/25) corresponded with the number of scan lines.

The Development of Blu-ray Disc

The ATSC made its proposal for a digital television standard to the US FCC in November 1995, and it was approved at the end of 1996. HDTV, originally promised for the early 1990s, finally concluded its long gestation period in December 1998, only to begin an even longer battle for ascendancy.

During the gestation of HDTV, numerous formats were developed to store video digitally and convert it to a standard analog television signal for display. The elephantine size of video is a problem. Uncompressed standard-definition digital television video requires at least 124 million bits per second (Mbps).[7] Obviously, some form of compression is needed. Many proprietary and incompatible systems were developed, including Intel's DVI in 1988 (which had been developed earlier by the Sarnoff Institute). That same year, the *Moving Picture Experts Group* (MPEG) committee was created by Leonardo Chairiglione and Hiroshi Yasuda with the intent to standardize video and audio for CDs. In 1992, the *International Organization for Standardization* (ISO) and the *International Electrotechnical Commission* (IEC) adopted the standard known as MPEG-1. Audio and video encoded by this method could be squeezed to fit the limited data rate of the single-speed CD format. The notorious MP3 format is a nickname for *MPEG-1 Layer III* audio.

The MPEG committee extended and improved its system to handle high-quality audio/video at higher data rates. *MPEG-2* was adopted as an international standard in 1994 and is used by many digital video systems, including DVB and the ATSC's DTV. An updated version of Video CD, called Super Video CD, used MPEG-2 for better quality. MPEG-2 is also the basis of DVD-Video, augmented with the Dolby Digital (AC-3) multichannel audio system, developed as part of the original work of the ATSC.

The CD Revolution

Sony and Philips had reinvigorated optical storage technology when they introduced *Compact Disc Digital Audio* in 1982. This was known as the "Red Book standard" because of the red cover on the book of technical specifications. The first CD players cost around $1,000 (over $2,000 in 2007 dollars).

Three years later, a variation for storing digital computer data — CD-ROM — was introduced in a book with yellow covers. As CD-ROM entered the mainstream, original limitations such as slow data rates and glacial access times were overcome with higher spin rates, bigger buffers, and improved hardware. CD-ROM became the preeminent instrument of multimedia and the standard for delivery of software and data.

After introducing CD-ROM, Sony and Philips continued to refine and expand the CD family. In 1986, they produced *Compact Disc Interactive* (CD-i, the "Green Book standard"), intended to become the new system for interactive home entertainment. CD-i incorporated specialized file formats and custom hardware with the OS-9 operating system running on the Motorola 68000 microprocessor. Unfortunately, CD-i was obsolete before it was finished,

---

[7]124 Mbps is the data rate of active video at ITU-R BT.601 4:2:0 sampling with 8 bits, which provides an average of 12 bits per pixel (720 × 480 × 30 × 12 or 720 × 576 × 25 × 12). The commonly seen 270 Mbps figure comes from studio-format video that uses 4:2:2 sampling at 10 bits and includes blanking intervals. At higher 4:4:4 10-bit sampling, the data rate is 405 Mbps.

The Development of Blu-ray Disc

and the few supporting companies dropped out early, leaving Philips to stubbornly champion it alone before finally giving up in the mid 1990s.

The "Orange Book standard" was developed in 1990 to support magneto-optical (MO) and *write-once* (WO) technology. MO had the advantage of being rewritable but was incompatible with standard CD drives and remained expensive. CD-WO was superseded by "Orange Book Part II," *recordable CD* (CD-R). Recordable technology revolutionized CD-ROM production by enabling developers to create fully functional CDs for testing and submission to disc replicators. As prices dropped, CD-R became widely popular for business and personal archiving.

MPEG-1 video from a CD was demonstrated by Philips and Sony on their CD-i system in 1991. The CD-i MPEG-1 Digital Video format was used as the basis for Karaoke CD, which became *Video CD*, a precursor to DVD. About 50 movies were released in CD-i Digital Video format before a standardized version appeared as Video CD (VCD) in 1993, based on proposals by JVC, Sony, and Philips and documented as the "White Book specification." Not to be confused with *CD-Video* (CDV), Video CD used MPEG-1 compression to store 74 minutes of near-VHS-quality audio and video on a CD-ROM XA bridge disc.[8] *Video CD 2.0* was finalized in 1994. Video CD did quite well in markets where VCRs were not already established, but Video CD never fared as well in Europe, and almost qualified as an endangered species in the United States.

Erasable CD, part III of the "Orange Book standard," did not appear until 1997. The standard was finalized in late 1996, and the official name of *Compact Disc Rewritable* (CD-RW) was chosen in hopes of avoiding the disturbing connotations of "erasable." Around 1999, the prices of CD-R/RW drives and blank CD-R discs dipped enough to provoke their widespread use for recording custom music CDs. This, plus the proliferation of MP3 music files across the Internet, drove a worldwide demand for 1.3 billion CD-Rs in 1999, with an estimated 30 to 40 percent being used for music.

The Long Gestation of DVD

Shortly after the introduction of CDs and CD-ROMs, prototypes of their eventual replacement began to be developed. Systems using blue lasers achieved four times the storage capacity, but were based on large, expensive gas lasers because cheap semiconductor lasers at that wavelength were not available. In 1993, 10 years after the worldwide introduction of CDs, the first next-generation prototypes neared realization. Some of these first attempts simply used CD technology with smaller pits to create discs that could hold twice as much data. Although this far exceeded original CD tolerances, the optics of most drives were good enough to read the discs, and it was possible to connect the digital output of a CD player to an MPEG video decoder. The reality that CD technology tended to falter when pushed too far, however, soon became apparent. Philips reportedly put its foot down and said it

---

[8]This is actually quite amazing when you consider the difference in data bandwidth between audio and video. VCD, using MPEG-1 compression, made it possible to store just as much video (and its accompanying audio) on a compact disc as CD audio can store for audio alone. The difference, of course, is that VCD used data compression while CD audio did not.

would not allow CD patent licensees to market the technology because it could not guarantee compatibility and a good user experience. When all things were considered, it seemed better to develop a new system.

The stage was set in September 1994 by seven international entertainment and content providers — Columbia Pictures (Sony), Disney, MCA/Universal (Panasonic), MGM/UA, Paramount, Viacom, and Warner Bros., who called for a single worldwide standard for the new generation of digital video on optical media. These studios formed the Hollywood Advisory Committee and requested the following —

- Room for a full-length feature film, about 135 minutes, on one side of a single disc
- Picture quality superior to high-end consumer video systems such as laserdisc
- Compatibility with matrixed surround and other high-quality audio systems
- Ability to accommodate three to five languages on one disc
- Content protection
- Multiple aspect ratios for wide-screen support
- Multiple versions of a program on one disc, with parental lockout

Preparations and proposals commenced, but two incompatible camps soon formed. Like antagonists in some strange mechanistic mating ritual, each side boasted of its prowess and attempted to enlist the most backers. At stake was the billion-dollar home video industry, as well as millions of dollars in patent licensing revenue.

1995 – The Pre-DVD Formats Duke it Out

On December 16, 1994, Sony and its partner Philips independently announced their new format — a red laser based, single-sided, 3.7-billion-byte *Multimedia CD* (MMCD, renamed from HDCD, which, it turned out, was already taken). The remaining cast of characters jointly proposed a different red laser standard one month later on January 24, 1995. Their *Super Disc* (SD) standard was based on a double-sided design holding five billion bytes per side.

The SD Alliance was led by seven companies — Hitachi, Panasonic (Matsushita), Mitsubishi, Victor (JVC), Pioneer, Thomson (RCA/GE), and Toshiba (business partner of Time Warner) — and attracted about ten other supporting companies, primarily home electronics manufacturers and movie studios. Philips and Sony assembled a rival gang of about 14 companies, largely peripheral hardware manufacturers. Neither group had support from major computer companies. At this stage, the emphasis was on video entertainment, with computer data storage as a subordinate goal.

Sony and Philips played up the advantages of MMCD's single-layer technology, such as lower manufacturing costs and CD compatibility without the need for a dual-focus laser, but the SD Alliance was winning the crucial support of Hollywood with its dual-layer system's longer playing time. On February 23, 1995, Sony played catch-up by announcing a two-layer, single-side design licensed from 3M that would hold 7.4 billion bytes.

The scuffling continued, and the increasing emphasis on data storage by consumer electronics companies began to worry the computer industry. At the end of April 1995, five computer companies — Apple, Compaq, HP, IBM, and Microsoft — formed a technical working

The Development of Blu-ray Disc

group that met with each faction and urged them to compromise. The computer companies flatly stated that they did "not plan to choose between these proposed new formats" and provided a list of nine objectives for a single standard —

- One format for both computers and video entertainment
- A common file system for computers and video entertainment
- Backward compatibility with existing CDs and CD-ROMs
- Forward compatibility with future writable and rewritable discs
- Costs similar to current CD media and CD-ROM drives
- No mandatory caddy or cartridge
- Data reliability equal to or better than CD-ROM
- High data capacity, extensible to future capacity enhancements
- High performance for video (sequential files) and computer data

Sony refused to budge, and one month later said there would be "no adjustment" in its MMCD format. On August 14, 1995, the computer industry group, now up to seven members with the addition of Fujitsu and Sun, concluded that the most recent versions of the two formats essentially met all their requirements except the first — a single, unified standard. In order to best support the requirements for a cross-platform file system and read/write support, the group recommended adoption of the *Universal Disk Format* (UDF) developed by the *Optical Storage Technology Association* (OSTA). Alan Bell, at IBM, reportedly told Sony and Philips that the group intended to settle on the SD format and gave the squabbling companies a few weeks to produce a compromise. Faced with the peril of no support from the computer industry or the worse prospect of a standards war reminiscent of Betamax versus VHS, the two camps announced at the Berlin IFA show that they would discuss the possibility of a combined standard. The companies officially entered into negotiations on August 24. The computer companies expressed their preference that the MMCD data storage method and dual-layer technology be combined with SD's bonded substrates and better error-correction method.

On September 15, 1995, the SD Alliance announced that "considering the computer companies' requests to enhance reliability," it was willing to switch to the Philips/Sony method of bit storage despite a capacity reduction from 5 billion to 4.7 billion bytes. Sony and Philips made a similar conciliatory announcement, and thus, almost a year after they began, the hostilities officially ended. The two DVD groups continued to hammer out a consensus that finally was announced on December 12, 1995. The combined format covered the basic DVD-ROM physical and video elements, taking into account the recommendations made by movie studios and the computer industry. A new alliance was formed — the DVD Consortium — consisting of Philips and Sony, the big seven from the SD camp, and Time Warner. When all was said and done, Panasonic held roughly 25 percent of the technology patents; Pioneer and Sony each had 20 percent; Philips, Hitachi, and Toshiba were left with 10 percent of the pie. Many people at Sony felt they had lost the war, since only a small part of their format was adopted and they were left with a smaller patent position than expected. A group of engineers decided to start work on a successor format with the goal of being more in the driver's seat. This was the birth of *Blu-ray*.

The Development of Blu-ray Disc

1996 – DVD is Born, Barely

In January 1996, companies began to announce their DVD plans for the coming year. The road ahead looked smooth and clear until the engineers in their rose-colored glasses, riding forecast-fueled marketing machines, crashed headlong into the protectionist paranoia of Hollywood.

As the prospect of DVD solidified, the movie studios began to obsess about what would happen when they released their family jewels in pristine digital format with the possibility that people could make high-quality videotape copies or even perfect digital dubs. Rumors began to surface that DVD would be delayed because of copyright worries. On March 29, the *Consumer Electronics Manufacturers Association* (CEMA) and the *Motion Picture Association of America* (MPAA) announced that they had agreed to seek legislation that would protect intellectual property and consumers' rights concerning digital video recorders. They hoped their proposal would be included in the *Digital Recording Act of 1996* that was about to be introduced in the US Congress.

Their recommendations were intended to —

- allow consumers to make home video recordings from broadcast or basic cable television,
- allow analog or digital copies of subscription programming, with the qualification that digital copies of the copy could be prevented, and
- allow copyright owners to prohibit copying from pay-per-view, video-on-demand, and prerecorded material.

The two groups hailed their agreement as a landmark compromise between industries that often had been at odds over copyright issues. They added that they welcomed input from the computer industry, and input they got! A week later, an agitated group of 30 computer and communications companies fired off a "list of critiques" of the technical specifications that had been proposed by Hollywood and the consumer electronics companies. They were less than thrilled that MPAA and CEMA had attempted to unilaterally dictate hardware and software systems that would keep movies from being copied onto personal computers. The computer industry said it preferred voluntary standards for content protection and objected to being told exactly how to implement things. Hollywood countered that the computer industry had been invited to participate early on but did not, either due to laziness or arrogance. The news media was filled with reports of DVD being "stalled," "embattled," and "derailed."

In anticipation of the imminent release of DVD, and with content protection details still unresolved, the MPAA and CEMA announced draft legislation called the *Video Home Recording Act* (VHRA), designed to legally uphold content protection, which included the insertion of a 2-bit identifier in the video stream to mark content as copy-prohibited. The two groups had been working on the legislation since 1994, apparently without directly consulting the computer industry, which filed a letter of protest against the proposed legislation, requesting that an encryption and watermarking scheme be used as a technical alternative to legislation. The content protection technology eventually became the much more complex CSS, and the legislation mutated into the *Digital Millennium Copyright Act* (DMCA).

On June 25, Toshiba announced that the 10 companies in the consortium had agreed to

The Development of Blu-ray Disc

integrate standardized content protection circuits in their players, including a regional management system to control the distribution and release of movies in different parts of the world. Many people understood this to mean that the content protection issue had been laid to rest, not realizing that this content protection agreement dealt only with analog copying by using the Macrovision signal-modification technology to prevent recording on a VCR. The manufacturers, still clutching their dreams of DVD players under Christmas trees and hoping to keep enthusiasm high, conveniently failed to mention that Hollywood was holding out for digital content protection as well.

At the end of July, more than a month after a content protection settlement was to have been made, the DVD Consortium — the hardware side of the triangle — agreed to support a content protection method proposed by Panasonic. The *Copy Protection Technical Working Group* (CPTWG), representing all three sides of the triangle, agreed to examine the proposal, which used content encryption to prevent movie data from being copied directly using a DVD-ROM drive.

The big news finally arrived on October 29, 1996 — the Copy Protection Technical Working Group announced that a tentative agreement had been reached. The modified copy protection technology developed by Panasonic and Toshiba, called *Content Scramble System* (CSS), would be licensed through a nonprofit entity, later formed and named the DVD CCA (Copy Control Association).

Panasonic and Toshiba debuted DVD players in Japan on November 1, a dismal rainy day with lackluster player sales and a handful of discs, mostly music videos. Warner Home Video began sales in Japan of four major movie titles — *The Assassin*, *Blade Runner*, *Eraser*, and *The Fugitive* — on December 20. Players from Panasonic and Pioneer finally began to appear in the United States in February as more movies and music performance videos were announced from Lumivision, Warner Bros. Records, and others. Eager customers bought the players, only to discover that no titles other than those from Lumivision were scheduled to appear before the end of March. Warner, the primary supplier, limited its release to seven test cities.

DVD had finally embarked, late and lacking some of its early luster, on the rocky road to acceptance.

1997 – Ups and Downs

The DVD roller coaster ride continued into the fall of 1997. Disney announced that it would finally enter the market with DVD discs, but the planned titles did not include any of the coveted animated features.

In October, the 10-member DVD Consortium changed its name to the *DVD Forum* and opened membership to all interested companies. By the time the first DVD Forum general meeting was held in December, the organization had grown to 120 members.

The first DVD-R drives appeared from Pioneer. In spite of the $17,000 price tag, DVD developers, desperate for an easy way to test their titles, snapped them up. After all, compared to the $150,000 price of the first CD-R recorders, DVD-R recorders were a bargain.

In November 1997, the first public breach of DVD content protection occurred. A program called *softDVDcrack* was posted to the Internet, allowing digital copies of movies to be made on computers. The press reported that CSS encryption had been cracked but, in actuality, the program hacked the Zoran software DVD player to get to the decrypted, decompressed video.

At the end of December 1997, 340,000 DVD players had been sold in the US. This was far below "expectations" of 1.2 million. Dozens of reports appeared that disparaged DVD's performance. Few bothered to mention that most of the expectations had been unreasonably optimistic.

1998 – DVD Matures

In 1998, the new Toshiba SD-7108 progressive scan DVD players, able to display almost twice the resolution as standard players from the same DVD, did not make it out of the warehouse. Concerns about lack of content protection on progressive scan output kept them there for another year.

Demolishing opinions that dual-layer, double-sided discs (DVD-18s) would never be achieved, Warner Advanced Media Operations (WAMO) announced a new process to make them. This was immediately followed by rumors that the 195-minute *Titanic* would be the first DVD-18 release, since director James Cameron was insisting that the DVD contain both fullscreen and widescreen versions. It was no surprise that the rumors were false.

1999 – The Year of DVD

It was, by most reckonings, the year of DVD.

The *Digital Display Working Group* (DDWG) announced on February 23 the completion of the *Digital Visual Interface* (DVI) specification for final draft review. Studios were looking forward to this replacement for computer VGA output, since DVI included an optional content protection mechanism.

On March 14, director John Frankenheimer hosted a live web event for owners of the *Ronin* DVD. As he answered questions over the Internet, he played back hidden content from the disc to illustrate behind-the-scenes events and to explain details of producing the movie. Since it was impossible to stream DVD video over the Internet, the trick was to use InterActual's *PCFriendly* software to remotely control the DVD-ROM drives of the tens of thousands of chat participants. This was one of the first attempts to extend the features of DVD by connecting it to the Web. InterActual pioneered other interactive adaptations of DVD, many of which were the genesis for some of the advanced interactive features in HD DVD and Blu-ray, including *BD-Live*.

On July 6, 1999, Pioneer Entertainment announced its completed transition out of the laserdisc business. After years of being the leading laserdisc supplier, Pioneer Entertainment had shifted to DVD and VHS only, which by then accounted for more than 90 percent of the company's title business. Although laserdisc players were still made by Pioneer New Media Technologies for education and industrial applications, the decision by Pioneer Corporation's entertainment group to abandon laserdiscs marked the end of an era.

The Development of Blu-ray Disc

Panasonic, hoping to create a new era of digital tape for consumers, announced again that it would begin to sell its HD-capable D-VHS VCR.

In September 1999, *The Matrix* became the first DVD to sell a million copies.[9] *The Matrix* gained notoriety as buyers reported problems playing it on dozens of different player models. While there were some UDF filesystem errors on the disc itself, it was discovered that some players had not been properly engineered to handle a disc that aggressively exercised DVD features and included extra content for use on PCs. The success of *The Matrix* amplified the magnitude of the problem. Under pressure from disgruntled customers, manufacturers released firmware upgrades to correct the flaws in their players. Years later, when these same manufacturers began working on vastly more complex Blu-ray players, the lessons they learned from DVD upgrades led them to add an Ethernet port to many Blu-ray player models so they could download firmware updates over the Internet.

Also in September, Toshiba shipped its progressive scan DVD player, which had been languishing in ware-houses for a year. Panasonic released its progressive scan player in October. The true potential of DVD video quality was finally being unlocked in places other than computers.

At the last minute, Sony, Philips, and HP cancelled their planned *DVD+RW* launch. The DVD+RW camp had decided to retrench, abandoning the 3.0G format — which would have been incompatible with every existing DVD-ROM drive and player — and focusing on the improved 4.7G version that promised backward compatibility with most DVD readers.

Pioneer announced that it would release a DVD-Audio player in Japan without content protection, since that was the only part that was unresolved. The player would only be able to play unprotected DVD-Audio discs until it was updated with final content protection support. This turned out to be a reasonable compromise because decisions on DVD-Audio encryption and watermarking were almost a year away.

In November and December, fallout began from a Windows-based software program called *DeCSS* that had spread across the Internet in late October. The program was designed to remove CSS encryption from discs and to copy the video files to a hard drive. DeCSS was written by 16-year-old Jon Johansen of Norway, based on code created by a German programmer who was a member of an anonymous group called *Masters of Reverse Engineering* (MoRE). The MoRE programmers reverse-engineered the CSS algorithm and discovered that the Xing software DVD player had not encrypted the key it used to unlock protected DVDs. Because of weak security in the CSS design, additional keys were quickly generated by computer programs that guessed at values and tried them until they worked. Johansen claimed his intent in turning the MoRE code into an application was to be able to play movies using the Linux operating system, which had no licensed CSS implementation.

Anyone familiar with CSS was surprised that it had taken so long for the system to be cracked. After all, the first edition of ***DVD Demystified*** (the predecessor to this book) had predicted three years earlier that CSS would be compromised. In spite of this, frenetic press reports portrayed a "shocked" movie industry, taken aback by the failure of the system that was supposed to protect its assets. DVD-Audio player manufacturers announced a six-month

---

[9] It took two years for DVD to reach the first million-copy point for a title. Audio CD took four years with George Michael's *Faith*, and it took 11 years before a million VHS copies of *Top Gun* were shipped.

The Development of Blu-ray Disc

delay, presumably to counter the threat of DeCSS by reworking the planned *CSS2* copy protection system. Those familiar with the lack of DVD-Audio titles and the incompleteness of CSS2 saw the announcement as a convenient excuse to delay products that would not have been ready in any case.

2000 – The Medium of the New Millennium

On February 14, Jack Valenti, then the head of the MPAA, speaking of DeCSS and other threats to Hollywood's intellectual property, told Salon that, "...some obscure person sitting in a basement can throw up on the Internet a brand new motion picture, and with the click of a button have it go with the speed of light to 6 billion people around the world, instantaneously." This led people to wonder where they could get these new ISL modems[10] that 6 billion people already had.

Region coding, part of the CSS license, came under attack in London when British supermarket group Tesco wrote to Warner Home Video demanding an end to the policy.

Sony's *PlayStation 2* shipped more than 1 million units in its first 12 days, and in less than a month doubled the number of DVD players in Japan. DVD industry analysts worldwide quickly updated their forecasts to account for the expected impact of this new kind of DVD player.

JVC announced a consumer version of D-VHS. The digital tape format, originally designed only to record data, had been reworked with a standard way of recording and playing back digital video, along with now requisite content protection.

In July 2000, more than three years after the introduction of DVD-Video and DVD-ROM in the US, DVD-Audio players belatedly shipped. It was not what anyone would call a grand coming-out. There was no marketing or PR hoopla and the players were in short supply and hard to find.

In November, Sony announced that it was working on *Ultra Density Optical* (UDO), a new blue-laser optical disc format intended to replace magneto-optical discs. In October, Sony showed UDO at the Ceatec show in Japan and also jointly with Pioneer unveiled a sister format called *DVR Blue*, which eventually became first-generation Blu-ray Disc rewritable format (BD-RE).

2001 – DVD Turns Five

On October 29, 2001, the DVD format officially turned 5 years old. During this one year, over 12.7 million DVD-Video players shipped to form an installed base of more than 26 million players in the US alone, with over 14,000 titles to chose from.[11] There were more than 45 million DVD-ROM drives in the US, with approximately 90 million installed worldwide. And, for the first time, DVD player sales exceeded VCR sales. Not bad for a format with only five candles on its cake!

The *D-Theater* format was released, which provided copy-protected high-definition movies

---

[10]ISL = instantaneous speed of light, of course.

[11]This figure did not include adult video titles, which accounted for an estimated additional 15%.

The Development of Blu-ray Disc

at excellent quality on D-VHS tape. However, after a limited test run by Universal and others, the format never took hold.

The next several years saw DVD flourish as sales of DVD players in the US continued to grow to form an installed base of over 73 million players by the fall of 2003. By that time, more than 27,000 DVD-Video titles were available, of which over 1.5 billion copies had been shipped. Within a year, 40,000+ DVD-Video titles were available as the industry topped $22 billion. (That's right, "billion" is the one with nine zeros after it.) By this point, home video releases on DVD began to overtake theatrical revenue for the same titles, marking the start of a new paradigm in the content community.

In December, the DVD Forum set up an ad-hoc group to study new, efficient HD codecs — including Microsoft's *Corona*, the code name for what became *Windows Media Video 9* and *VC-1* — with an eye toward fitting HD movies on dual-layer DVDs. Meanwhile, work had been steadily progressing on next-generation optical disc technology. Unfortunately, factions were forming, and it soon became clear that there would be a repeat of the early days of DVD, with more than one contender for the title of successor.

Format Wars, the Next Generation

On January 7, 2002, Toshiba announced its next-generation read/write optical disc technology using 405-nm blue-laser technology, able to store up to 30 Gbytes on a single side, capacity enough to record three hours of high-definition video. Toshiba demonstrated the technology at the *International Consumer Electronics Show* (CES) in Las Vegas, and said it would propose the technology to the DVD Forum.

On February 19, 2002, nine companies — Hitachi, LG, Panasonic, Pioneer, Philips, Samsung, Sharp, Sony, and Thomson Multimedia — announced that they had established preliminary specifications for the *Blu-ray Disc* format, a new, high-density, rewritable disc format with up to 50-Gbyte capacity based on solid-state blue-laser technology.

The war had begun.

Of course, blue-laser announcements did not make DVD instantly obsolete, and on February 26 the DVD Forum voted to move ahead with a proposal from Warner Bros. and other motion picture studios to put high-definition video on dual-layer DVD-9 discs. This was informally dubbed *HD DVD9*.

Seeing a storm on its horizon, the DVD Forum (which included Toshiba and the nine Blu-ray companies) announced in April that it would investigate next-generation standards for high-capac-ity, blue-laser discs. Many in the industry predicted that Blu-ray would soon be adopted by the DVD Forum and there would be a smooth road to the next-generation of optical disc formats. Insiders quickly realized, however, that the beginning of a politically charged conflict was being staged that would create new friends and new enemies in a fight for the multi-billion dollar legacy from DVD.

In May, the nine Blu-ray companies announced the formation of the *Blu-ray Disc Founders* group and stated that they planned to release version 1.0 of the recordable disc specifications in June. Incredibly, in spite of the clear lesson from DVD that entertainment content could make or break a format, the group was focused on recording technology and seemed to

The Development of Blu-ray Disc

ignore the behemoth industry in the hills of Los Angeles.

On June 20, 2002, Circuit City, the largest electronics retailer in the US, became the first major VHS retailer to announce that it was eliminating VHS movies from its stores to make room for faster-selling DVDs.

In August, Toshiba and NEC formally proposed their blue-laser optical disc format, now called *Advanced Optical Disc* (AOD), to the DVD Forum. The AOD format was twice voted down in meetings of the DVD Forum Steering Committee, with Blu-ray supporters mostly voting against or abstaining (which had the effect of a "no" vote).

In November 2002, amidst charges of anti-trust behavior against the Blu-ray companies that had blocked previous votes, the DVD Forum voting rules were changed (so that abstentions no longer counted) and AOD was finally approved by the DVD Forum Steering Committee. It was later renamed to *HD DVD*.

Also in November, the Chinese government announced that *enhanced versatile disc* (EVD), a project that started in 1999 and experienced several cancelled launches, would be launched by Christmas 2003. EVD had been created as an alternative to DVD in part to decrease the flow of licensing royalties to Japan and other DVD patent holders. The plan was for EVD to use a new optical disc format and compression technologies developed in China, but the anticipated new technologies were too slow in coming, so the format ended up using red-laser DVD, flirted with proprietary video compression technologies (VP5 and VP6, developed by On2 in the US), and finally settled on MPEG-2. The EVD format supported HD video resolutions, including 1280×720 at 60 fps progressive (720p) and 1920×1080 at 30 fps interlaced (1080i). In April 2004, another group led by Amlogic and several major Chinese television manufacturers, announced development of the HVD format, and in May the DVD Forum began investigating a proposal from the *Industrial Research Technology Institute* (ITRI) of Taiwan for the FVD (*forward versatile disc*) format, intended specifically for the low-cost Chinese market.

In October, prototypes of both dueling blue-laser formats were unveiled at the Ceatec exhibition in Tokyo. Sony, Panasonic, Sharp, Pioneer, and JVC showed prototype Blu-ray Disc recorders while Toshiba showed an AOD prototype.

2003 – Blu-ray: Take One

In early 2003, at least five candidates were vying to be anointed as the new high-definition DVD format — Blu-ray Disc, HD DVD-9 and AOD from the DVD Forum, Blue-HD-DVD-1 and Blue-HD-DVD-2 from the *Advanced Optical Storage Research Alliance* (AOSRA) in Taiwan, and WMV HD (WMV 9 files on standard DVD) from Microsoft. Of course most bets were on Blu-ray and AOD, where the real battle lines were drawn.

In February, the Blu-ray Disc Founders group began licensing the Blu-ray Disc format. On April 7, Sony confused a lot of BD watchers when it announced a new blue-laser Professional Disc format for data archiving applications, which was not part of the Blu-ray format, at all.

On April 10, 2003, Sony opened the book on an entire new generation of optical technology when it put on sale in Japan the world's first consumer Blu-ray Disc recorder, the BDZ-S77 at a rather steep price of ¥450,000 ($3,800 in 2003 dollars, $4,200 in 2007 dollars). The

The Development of Blu-ray Disc

recorder used a single-layer, 23-Gbyte BD-RE disc in a cartridge, to record about 2 hours of high-definition MPEG-2 video direct from the built-in BS digital broadcast tuner, or 3 to 12 hours of NTSC analog video encoded internally to MPEG-2.

On May 28th, Mitsubishi Electric joined the BDF, bringing total membership to ten.

In the middle of March 2003, statistics from the *Video Software Dealers Association* (VSDA) indicated that DVD had reached a major milestone in the US — rental revenue had surpassed VHS tape. DVD sales had passed VHS sales in 2001. DVD players were now in more than 40 million households in the US, and shipments of DVD titles had passed 700 million units. DVD was nearing its peak just as Blu-ray began to slowly creep up from behind.

On September 12, Microsoft announced that it had submitted Windows Media Series 9 to the *Society of Motion Picture Television Engineers* (SMPTE) for standardization. This was a strategic move to distance the codec technology from Microsoft in order to get the DVD Forum (and later the Blu-ray camp) to consider adopting it.

2004 – Blue-laser Specs and Battle Lines Are Drawn Up

Toshiba started out the year by unveiling its first prototype HD DVD player at CES in January. The following week, Hewlett-Packard and Dell put their support behind Blu-ray Disc.

In April, Sony pulled off a cute publicity stunt by announcing that it had worked with Toppan Printing to make a Blu-ray disc that was 51 percent paper.

On May 18, 2004, the ten members of the Blu-ray Disc Founders group, plus HP and Dell, announced the formation of the *Blu-ray Disc Association* (BDA) to continue development of technical specifications and promote the format. The group welcomed other companies to join, in a clear bid to set up a rival organization to the DVD Forum. Many companies jumped on the bandwagon, but now had to send engineers and managers to two sets of regular technical meetings. Sun Microsystems joined immediately to stump for Java.

On June 10, the DVD Forum approved version 1.0 of the HD DVD read-only (ROM) physical specifications.

Meanwhile, not wanting to repeat either the format delays or the content protection breaches that had plagued DVD, an unlikely group of bedfellows from the PC, CE, and motion-picture industries — IBM, Intel, Microsoft, Sony, Panasonic, Toshiba, Disney, and Warner Bros. — had been working on a new content protection system to apply to upcoming HD video recorders, and players. On July 14, 2004, they announced the formation of the *Advanced Access Content System* (AACS) group. The goal was to finish the specification by December. Old hands in the industry chuckled at the aggressive timeline, but no one had an inkling of how long it would really take for the sometimes fiercely contentious group to produce the final documents.

At the end of July, Panasonic improved on Sony's Blu-ray debut with the DMR-E700BD Blu-ray recorder, which used a dual-layer 50-Gbyte BD-RE cartridge that could record 4.5 to 6 hours of HD digital satellite (BC) or terrestrial (BS) broadcast, or up to 63 hours of analog NTSC video. The Panasonic model could play discs recorded on the Sony, but the Sony could not read the 50-GByte discs. The Panasonic recorder was relatively less inexpensive at

The Development of Blu-ray Disc

$2,800 ($3,200 in 2007 dollars). The 50-Gbyte discs cost around $69 and the 25-Gbyte discs cost around $32.

On August 11, 2004, the BDA approved version 1.0 of the physical format specifications for the read-only (ROM) version of Blu-ray Disc so that manufacturers could begin designing components. Physically, the BD-ROM format was similar to the existing BD-RE version, but the file system and application specs for playback were being extensively reworked, and video codec selection had yet to be determined. Players and discs were expected to be available in late 2005.

On September 21, 2004, Sony announced that the *PlayStation 3* (PS3) would use Blu-ray Disc. This was a huge boost for the Blu-ray format and, certainly, one of the biggest factors in its eventual success.

At the beginning of October, 20th Century Fox joined the BDA but stayed mum about which format it would support. At the end of November, HBO, New Line Cinema, Paramount, Universal, and Warner Bros., announced support for HD DVD. With only Sony Pictures and MGM (which was controlled by Sony) officially behind Blu-ray, the balance was clearly tipping toward HD DVD. The significant holdout was Disney. If Disney chose HD DVD — which many expected, since Disney had worked with Microsoft to develop the proposal for HD DVD advanced interactivity — then the war would be all but over. The camp with the most content was destined to win. But on December 8, 2004, Disney announced that it would support Blu-ray Disc. The balance was restored and the game was back in full swing.

2005 – Pits and the Pendulum

On December 4, 2004 in Japan, followed in the US on March 5, 2005, Sony launched the PlayStation© Portable (PSP™) and its new *universal media disc* (UMD™) format, a 60-mm, red-laser disc that held 1.8 Gbytes, plenty of room for games and music, and just large enough to hold a two-hour movie encoded in AVC format. In spite of predictions that no one really wanted to watch an entire movie holding a small game player in front of their face, PSP movie titles began to sell briskly. Hollywood studios scrambled to ramp up production and accelerate planned title releases. *DVD Release Report* statistics showed that within five months of launch, over 240 movie and TV titles were in the market or announced, significantly more than games. People held up UMD as the example of the right way to launch a format.

At the CES show, backers of both blue-laser formats promised players and movies in North America by the end of the year. They were only off by several months.

TDK announced that they had developed a hard-coat process for Blu-ray discs. Although it could be just as easily applied to HD DVD discs (or DVDs), it was touted as a major advantage of Blu-ray. It did mean that the BDA was finally able to drop requirements for cartridges, which no one wanted anyway.

On March 10, Apple Computer announced that it had joined the BDA and would support Blu-ray.

During March, as a result of strong industry pressure, talks started between Toshiba, Sony, and Panasonic in an attempt to unify the formats. Ryoji Chubachi, president of Sony, said,

The Development of Blu-ray Disc

"Listening to the voice of the consumers, having two rival formats is disappointing and we haven't totally given up on the possibility of integration or compromise." Within weeks, however, it was reported that the talks had completely broken down with neither side willing to compromise. Those hoping for a single next-generation format watched the situation go from bad to worse. A format war raged in China between EVD, HVD, and FVD, and convergence between HD DVD and BD seemed impossible. Meanwhile Microsoft's WMV-HD had sold a few titles in the US and was actually doing quite well in Europe. A few AVC-based formats such as *DivX HD* and *Nero Digital* attempted to make a land grab amidst the turmoil, but with little backing and no distinct disc specification to tie their formats to, they fell by the wayside.

On May 10, Toshiba announced that it had developed a triple-layer HD DVD-ROM disc with a data capacity of 45 Gbytes, 50 percent more than the 30-Gbyte dual-layer disc. This brought them within spitting distance of Blu-ray's 50-Gbyte disc. Proponents of each camp argued incessantly that the other camp's highest-capacity disc was a fantasy and would never see the light of day. And in a somewhat glib response, Blu-ray promoters announced a theoretical four-layer version that could hold up to 100 Gbytes.

The pendulum swung toward the HD DVD camp in June 2005 when Microsoft and Toshiba said they would consider working together on the development of HD DVD players using Microsoft Windows software. Microsoft chairman Bill Gates hinted that the Xbox 360 might add an HD DVD drive. "The initial shipments of Xbox 360 will be based on today's DVD format," he said. "We are looking at whether future versions of Xbox 360 will incorporate an additional capability of an HD DVD player or something else." It was not exactly a ringing endorsement of the format, but the Internet reality distortion machine was soon churning out reports that Gates had chosen HD DVD for the Xbox. Not surprisingly, "considering" and "looking" turned into doing, and an HD DVD drive was eventually released for the Xbox. Likewise, Toshiba's first HD DVD player was co-developed with Microsoft.

On July 29, 2005, the pendulum swung back to the Blu-ray side when 20th Century Fox officially announced its support. Fox had cleverly hung back until it was the last holdout, so it was able to wrangle promises of increased content protection mechanisms in Blu-ray in return for its support. The BDA soon announced that it was adding *ROM Mark* (a mechanism to deter professional piracy of replicated discs) and *BD+* (a system to implant special code on discs to check player security).

The pendulum inched a bit more toward Blu-ray in August when Lionsgate and Universal Music Group announced their support.

On August 22, the *Yomiuri Shimbun* newspaper in Japan reported that yet another round of unification negotiations between the Blu-ray Disc Association and the DVD Forum had failed.

Ironically, there was quite a bit of informal "merging" of specifications occurring behind the scenes. Some people attended spec-development meetings in both groups. Other companies kept separate delegates for each format, but people working on one format shared ideas and concerns with colleagues working on the other format. For example, in August an ad-hoc group of the DVD Forum published the Application Requirements document, a thorough investigation of scenarios desired by Hollywood studios and others in the industry. Features to support may of these scenarios soon appeared in the Blu-ray format.

The Development of Blu-ray Disc

Page one of the *Washington Post* on Sunday, August 28, 2005, carried an obituary for VHS, "the beloved videotape format that bravely won the war against Betamax and charmed millions of Americans by allowing them to enjoy mindless Hollywood entertainment without leaving their homes." Although 94.7 million American households still owned VCRs, and more than $3 billion had been spent on VHS rentals and purchases in 2004, revenue from DVD sales and rentals was more than eight times greater. The unprecedented success of DVD was pushing VHS off store shelves, out of customer storage cabinets, and into oblivion.

The pendulum swayed back on September 27, when Microsoft finally officially announced that it would back HD DVD along with Intel. In an interview with the *Daily Princetonian*, Bill Gates sniped at Blu-ray, saying, "it's that the protection scheme on Blu is very anti-consumer," referring to the added locks and lack of commitment to managed copy, unlike the HD DVD camp. Gates — and many other Blu-ray detractors — glossed over the crucial point that managed copy was a feature of AACS so it applied equally to both formats. He followed that with a prognostication that "this is the last physical format there will ever be. Everything's going to be streamed directly or on a hard disk. So, in this way, it's even unclear how much this one counts." Once again, having Bill Gates as a cheerleader resulted in faint praise indeed.

In October, Paramount decided it would firmly straddle the fence and offer movies on both HD DVD and Blu-ray Disc. Warner Bros. also climbed out of the HD DVD camp onto the fence, announcing it would also release titles in both formats.

At the same time, HP suddenly and publicly called for changes to the Blu-ray format, strongly suggesting that it would defect to the HD DVD format if its demands were not met. HP specifically requested two features that were already part of HD DVD — *mandatory managed copy* (to allow disc owners to copy content from their discs onto PCs and portable devices) and *iHD* (the XML and ECMAScript-based advanced interactivity format that Microsoft and others had developed in the DVD Forum). HP's cover story was that it did not want customers to have to choose between dissimilar competing formats. Conspiracy theories blossomed — Microsoft was paying HP off or had threatened to shortchange HP's access to the Windows OS.

The BDA's response came from spokesman Andy Parsons on November 16, 2005, "Mandatory managed copy will be part of the Blu-ray format, but while HP's request [for interactivity] is being considered, at this point in time, the Blu-ray group is still proceeding down the path of Java," he told Reuters. HP decided this was not good enough, and on December 16 announced that it would support both HD DVD and Blu-ray.

This was good news for Toshiba, which two days earlier had said that design work on its first HD-DVD player was complete, with factories ready to produce the players, but that it was held up waiting for the final version of AACS. Remember AACS? The content protection system that was going to be done the year before? It was not for lack of effort, since the group had been gathering regularly for over a year, with meetings often dragging late into the night, but they had been unable to reach consensus on items such as managed copy and requests to add BD+ into the mix.

The Development of Blu-ray Disc

2006 – On Your Marks, Get Set, Go

The big news at CES in January was from Bill Gates, who announced that Microsoft would offer an add-on HD DVD drive for the Xbox 360 console. HD DVD supporters who expected a new version of the Xbox with an HD DVD drive built in were not exactly dancing in the streets, but at least it was some ammunition against the specter of millions of Blu-ray-equipped PS3s.

The AACS group met on February 10 to crank out the final specifications, but once again came up empty-handed. Reports stated that "an important member of the Blu-ray Disc Association is still voicing concerns about the interaction of AACS and the additional BD+ protection for Blu-ray movies." But there was a massive hitch — Toshiba had planned for a March launch date, with a 40-city tour starting on February 22 to kick it off. Likewise, Samsung and Pioneer had picked May launch dates for their Blu-ray players. If AACS was not ready to go, everyone would be trapped at the starting gate. Finally, on February 15, after working feverishly for days to come up with a compromise, the AACS group agreed to release an interim license that would permit hardware makers and disc manufacturers to sign up for the encryption keys they needed to make the discs and players work together with protected content. The interim license would leave out elements such as managed copy and watermark detection. On February 21, the ACCS LA (license administrator) announced availability of the interim agreement documents, saying, "The final license agreement for AACS is expected to be announced in the coming months." Hope sprang eternal.

In related news, on March 14 the *Nihon Keizai Shimbun* newspaper in Japan reported that Sony had pushed back the *PlayStation 3* release date from early spring to November. This was allegedly because of the delay in finalizing AACS, but some opined that it was just a handy excuse to cover Sony's inability to pull everything together for the complicated game system.

Meanwhile, the meteoric rise of UMD for the PSP (that's movies on tiny discs for Sony's tiny game player, for readers whose acronym buffer has overflowed) had turned into a disastrous tailspin. Initial sales demand from a year earlier had evaporated, and in February, Sony Pictures, Paramount, Warner, and other smaller studios began delaying or cancelling UMD movie releases. Sony had shipped more than 10 million PSP units worldwide since its release a year earlier, but average UMD sales of only 50,000 units mean that less than 1 percent of PSP owners were interested. Various theories were put forth — PSP owners bought a few movies when they first got their system to try it out, but it turned out that people really did not want to watch a movie on a tiny screen with tinny speakers while their arm muscles cramped from holding it; the first rush of titles purchased were from mothers and grandmothers who were overjoyed that they could buy something that was not a violent videogame for their teenager; prices were simply too high compared to DVDs; or people realized they should not have to buy what they already owned on DVD. The latter theory was rooted in

---

[12] These may seem like high prices, but for an introductory high-tech product they are stupendously low. Consider other introductions in the US (each adjusted for inflation to 2006): 1939 TVs, $2900 to $8700; 1975 Betamax VCR, $8600; 1976 VHS VCR, $3100; 1978 laserdisc player, $2300; 1982 CD player, $2000. *iSuppli*, a market analysis company specializing in "teardowns" of products to inspect the components, estimated that the bill of materials for the HD-A1 came to $674, which revealed that Toshiba was heavily subsidizing the price in hopes of winning sales.

The Development of Blu-ray Disc

the growing realization within Hollywood that solutions such as managed copy, to allow consumers to legally move purchased content between media and devices, were needed.

By the end of March, Universal, New Line, and others had cut their UMD releases completely or to just a trickle. One unnamed studio executive told *Hollywood Reporter*, "It's awful. Sales are near zilch. It's another Sony bomb — like Blu-ray." Just over 600 titles had been released by the time the format neared a total eclipse two years later. In a classic case of making lemonade out of lemons, many UMD titles wasting in inventory were added as free "extra bonus discs" in DVD packages.

Toshiba won the race to store shelves, shipping the world's first HD DVD player, the HD-XA1, on March 31, 2006 in Japan for ¥110,000 ($940). The HD-XA1 arrived in the US on April 18, priced at $800, along with its simpler sibling, the $500 HD-A1.[12] Each model was, in fact, a high-powered PC in a box, running Linux on a 2.5-GHz Intel Pentium CPU augmented with a Broadcom decoder chip. The HD-XA1 added four SHARC digital signal processors (DSPs) to handle the audio processing load. Disc load time was interminable and the interactive features were sluggish, but the high-definition picture was widely praised in spite of complaints that it was only 1080i. Toshiba had the market to itself for three months.

On June 20, 2006, Sony and MGM released the first Blu-ray titles — *50 First Dates*, *The Fifth Element*, *Hitch*, *House of Flying Daggers*, *The Terminator*, *Underworld: Evolution*, and *XXX*. All contained MPEG-2 video, although VC-1 encoders were being used in HD DVD production. Five days later, the first Blu-ray player appeared in the US — the Samsung BD-P1000, priced at $1,000. Sony and Pioneer had planned to release Blu-ray players in June, but pushed their release dates out by a few months. A few days earlier, a Samsung representative had confessed that the company was considering releasing a "universal" player that would take both Blu-ray and HD DVD discs.[13]

At Toshiba's annual shareholder meeting on June 27, 2006, President Atsutoshi Nishida said, "We have not given up on a unified format. We would like to seek ways for unifying the standards if opportunities arise."

In July, the first mass-market Blu-ray rewritable drive for the PC, the $700 BWU-100A, was released by Sony. It recorded both single- and dual-layer BD-R and BD-RE discs.

On September 26, 2006, the first titles using the more efficient VC-1 codec hit the streets in the US — *Corpse Bride*, *Swordfish*, *Space Cowboys*, *The Fugitive*, *Lethal Weapon 2*, and *House of Wax*. The next month, Sony Pictures debuted *Click*, the first title on a dual-layer disc. Detractors who had up to this point maintained that dual-layer discs were too problematic and too expensive to be used for mass production quietly melted back into the crowd.

On November 11, 2006, Sony's PlayStation 3 went on sale in Japan. The US launch followed on the 17th, but the PS3 didn't make it to Europe and elsewhere until March 23, 2007. The two US models were priced at $500 and $600 (each with an estimated $250 price subsidy), making them the cheapest available Blu-ray players. One million units sold in the US in less than 2 months, with sales in Japan passing one million units shortly after. By April 2007, over 5 million units had sold worldwide. This was a massive boost to the installed base

---

[13]Samsung eventually did release the $1000 BD-UP5000 at the end of 2007, but not before LG beat them to the punch in February with the $1200 Super Multi Blu BH100, which was rushed to market with only a partially implementation of HD DVD. Many reviewers remarked that you could buy the Blu-ray-capable PS3 and an HD DVD player together for less than the price of the combo player and get a game console "for free."

The Development of Blu-ray Disc

of Blu-ray players, and although game titles obviously competed with Blu-ray video titles for mindshare and play time, later data from Sony indicated that 87 percent of PS3 owners watched Blu-ray discs on their system.

Just in time for the holidays, a hacker under the alias of *muslix64* made a December 27 posting on the *Doom9* Internet discussion boards claiming success in breaking part of the AACS content protection scheme with a program called *BackupHDDVD*. He (or she) did point out that, "This software don't [sic] provide any cryptographic keys, so you have to add your own keys." It turned out that the crack was not all that helpful, as it required someone very knowledgeable to figure out where in memory a Blu-ray or HD DVD software player kept the AACS decryption key for a particular volume, extract the keys, and then plug them into *BackupHDDVD* to decrypt the disc. Since every title has a different key, the sleuthing had to be repeated for each title, although, of course, lists of keys soon appeared on the Internet and, as expected, the AACS LA then used the legal muscle of the Digital Millennium Copyright Act (DMCA) to send cease-and-desist letters to Web sites posting "stolen" AACS keys. The game of cat and mouse was afoot.

2007 – HD Formats Take their Second Lap

At CES, *LG Electronics* unveiled their dual-format player. Many people in the industry predicted that if the format war did not end within a few years, dual-format players would become the norm, and the battle would end in détente rather than producing a single surviving format. In a similar vein, Warner Bros. showed off a prototype disc, dubbed *THD* for *Total HD*, carrying both an HD DVD and a Blu-ray layer to be compatible with either kind of player. (It turned out that THD never made it past the demo stage.)

On January 24, AACS LA released a statement in response to the December news, saying, "AACS LA has confirmed that AACS Title Keys have appeared on public websites without authorization. Such unauthorized disclosures indicate an attack on one or more players sold by AACS licensees. This development is limited to the compromise of specific implementations, and does not represent an attack on the AACS system itself." In a restrained jab at the software developers whose products had exposed the keys, the statement went on to say, "it illustrates the need for all AACS licensees to follow the Compliance and Robustness Rules set forth in the AACS license agreements to help ensure that product implementations are not compromised." The software developers were privately told they had a few months to fix their players. The story picks up again in April....

In February, VHS turned 30 years old. As if echoing the theme from *Logan's Run*, Wal-Mart — the world's biggest video retailer, with over 30 percent of DVD and VHS sales — finally abandoned VHS. Back in June 2005, long after Circuit City and others had pushed VHS tapes off the shelves, there had been a spurious report that Wal-Mart was exiting the VHS business, but this time it was for real. DVD reigned supreme.

In April 2007, step-by-step instructions on how to reset region code change counts in the four major software players (*CyberLink PowerDVD*, *Corel WinDVD*, *Arcsoft Total Media Theater*, and *Nero Showtime*) began to appear on websites. The process was usually as simple as deleting a file or stopping a service to be able to change the region code an unlimited number of times.

The Development of Blu-ray Disc

On April 16, AACS LA announced that it had taken action to "expire the encryption keys" (is expire now a transitive verb?) of specific implementations of AACS-enabled software. The revocation feature of AACS was being activated, which allowed the encryption key structure to be changed on new discs so that they would not play on compromised players. The AACS LA posted links to websites of CyberLink and Corel, where users could download security updates that kept their players working with new discs and did a better job of obfuscating AACS keys. Of course, the new versions of the software applications were soon cracked as well, and cracking websites with lists of keys popped up in new places like moles in a carnival game, but at least AACS was able to stay ahead of the game, unlike CSS, which lost most of its efficacy once all the player keys were compromised.

In May, the new head of the *Motion Picture Association of America* (MPAA), Dan Glickman, said that the AACS support behind managed copy would be ready before the end of the year. Once again, the pessimists in the audience were doubtful and, once again, even their extensions of Glickman's assertion turned out to be madly optimistic.

On June 30, 2007, *BD+ Technologies*, the company licensing the additional content protection layer for Blu-ray, announced that it had completed the specifications for BD+. Luckily the industry had not waited for them.

In August, the HD DVD backers worked to swing the pendulum back in their direction. Microsoft cut the price of its HD DVD add-on drive for the Xbox 360 from $199 to $179 and threw in five free movies. Toshiba allegedly offered Paramount and Dreamworks each $75 million to drop Blu-ray in favor of HD DVD, which they both did with announcements on August 20.

On September 7, in response to another round of attacks because of inadequate security in the *CyberLink PowerDVD* software player, AACS LA announced it was again revoking keys. AACS also announced that it was activating another feature of AACS, proactive renewals, which required software products to periodically release updates that refreshed the AACS encryption keys. Software that was not updated would no longer be able to play newly released discs. The software companies — who knew it was coming — grumbled but complied, since by this time it was common to provide automatic downloadable software updates.

In October 2007, Panasonic released the DMP-BD30K, the first Blu-ray *Profile 1.1* (*BonusView*) player on the market, beating its competitors by a few months, even Sony's PS3, which did not receive a software update to Profile 1.1 until December 18. However, discs using Profile 1.1 features were nowhere to be found.

In November 2007, apparently feeling pressure from the gradually increasing sentiment that Blu-ray was winning the format war, Toshiba held closeout sales of old-model HD DVD players at several major retailers in the US. Limited quantities of players were sold for less than $100, causing a few stampedes and minor injuries. The HD DVD promotion group announced that 750,000 HD DVD players had been sold, which included standalone players and the Xbox 360 add-on.

Also in November, *SlySoft*, the makers of the popular but, in many cases, illegal *AnyDVD* ripping software, released a new version that could crack and rip the few Blu-ray discs protected by BD+. Of course BD+ was designed to be changed when the algorithms were cracked, but SlySoft claimed they'd figured out the basics and could handle any variation that BD+ threw at them.

The Development of Blu-ray Disc

On December 10, 2007, the world's first Blu-ray disc to use the Profile 1.1 picture-in-picture feature, *Neues vom Wixxer*, was released in Germany from *Imagion AG*. Unfortunately, no one could try out the feature since there were no Profile 1.1 players in Germany. On the other side of the pond, the US had Profile 1.1 players but no discs. Finally, on December 20, New Line released *Rush Hour 3* in the US, the first disc there to use picture in picture.

The year closed out with reports of multi-region Blu-ray players appearing for sale on websites outside of the US. Firmware modifications had apparently been made to the Samsung BD-P1400, Sony BDP-S300, and Sony BDP-S500 to make them switchable between Blu-ray regions A and B as well as DVD regions 1 and 2. The modified players were not cheap, ranging from $900 to $1,300.

2008 – The War Is Over, Long Live Blu-ray

On the first day of 2008, Lionsgate released *War*, the first BD-Live disc. It was a fitting title given the state of the industry and the chest-thumping leading up to CES. Rumors were circulating that Warner Bros. would announce its allegiance to one format. Everyone knew what this meant if it were true. Given the stalemate between the formats, and with Warner representing roughly 40 percent of available studio content but releasing most of its titles in both formats, any move in either direction would drastically tip the balance. But some rumors pegged HD DVD while others named Blu-ray. Then, on January 4, 2008, Warner dropped its bombshell — it would phase out HD DVD and move exclusively to Blu-ray. A shell-shocked HD DVD Promotion Group canceled its CES news conference. Toshiba executives immediately flew back to Japan. Apparently they had been expecting Warner to announce it would be exclusive to HD DVD. Unsubstantiated reports declared that Sony and/or the BDA had paid Warner anywhere from $250 to $500 million dollars, beating Toshiba's highest offer. Some reports claimed that Fox switching to HD DVD was to have been part of the deal, which was improbable in the extreme, given how hard Fox had worked to get BD+ into Blu-ray. Warner Home Entertainment President Kevin Tsujihara denied that there had been a payoff, maintaining that it was about choosing the best format for consumers. Warner had been in the expensive position of releasing two versions of every title, and Blu-ray discs had been outselling HD DVD discs roughly 2:1 for the previous year.

Toshiba responded with a statement that began, "Toshiba is quite surprised by Warner Bros.' decision to abandon HD DVD in favor of Blu-ray, despite the fact that there are various contracts in place between our companies concerning the support of HD DVD," and ended with, "We will assess the potential impact of this announcement with the other HD DVD partner companies and evaluate potential next steps. We remain firm in our belief that HD DVD is the format best suited to the wants and needs of the consumer." Translation: "Warner broke our agreements. If we weren't a polite Japanese company we would sue them. We don't know what to do now, but we can repeat our mantra."

Flurries of articles, blogs, and opinion pieces announced the imminent end of the format war, laying short odds on how long Toshiba would last. In a sense it became a self-fulfilling prophecy. By this time the industry, the press, and the public were so tired of the delays, the uncertainty, and the waste of time and money dealing with two irreconcilable formats that they welcomed any excuse to get it all over with and move on. The fat lady was warming up

The Development of Blu-ray Disc

in the dressing room.

Toshiba tried measures such as cutting the price of the HD-A3 player to $150 and running a $2.7 million Super Bowl ad. But the bad news kept coming —

- New Line, HBO, and National Geographic Presents all followed in the footsteps of their parent Warner and announced they would phase out HD DVD.
- The BDA Promotions committee announced that roughly 3.5 million Blu-ray players had been sold to date in the US (three million PS3 consoles and 500,000 standalone players). Over 10.5 million PS3s had been sold worldwide. This compared to roughly 850 thousand HD DVD players in the US and just over one million worldwide (300,000 PCs with HD DVD drives, 300,000 Xbox 360 drives, and 430,000 standalone players).
- An Xbox group marketing manager said Microsoft would consider a Blu-ray add-on drive for the Xbox 360. Oops, hang on — the next day Microsoft reaffirmed its belief that HD DVD was the best optical disc for consumers and that it had no plans to support Blu-ray Disc on the Xbox 360.
- Nielsen VideoScan data showed that for the week ending January 13, all ten of the top ten high definition discs in the US home video sales and rental charts were Blu-ray discs.
- Analyst firm NPD said that for the week ending January 12, Blu-ray players outsold HD DVD players 9 to 1. HD DVD player sales sank from 14,558 two weeks before to a meager 1,758. Blu-ray sales jumped from 15,257 to 21,770 for a 93 percent share. NPD analysts cautioned that one week of data did not a trend make.[14]
- Sonic Solutions, the main provider of professional authoring tools for both HD DVD and Blu-ray formats announced that its Professional Products Group would focus on Blu-ray.
- Woolworths in the UK announced that beginning in March it would sell only Blu-ray discs.
- NetFlix announced it would phase out HD DVD discs by the end of the year. (Blockbuster had decided in June the year before to stock only Blu-ray titles.)
- BestBuy stated it would focus on Blu-ray.
- Rumors that Circuit City would follow suit were dispelled when a representative stated that the company would remain platform agnostic. A statement from the President of Philips Consumer Electronics that Target would go exclusively Blu-ray was unsubstantiated.
- Wal-Mart, the world's largest retailer, said it would phase out HD DVD products by June.

---

[14]In fact, the next week Blu-ray player sales were back at a 65 percent share. Then at the end of April, NPD stated that Blu-ray player unit sales in the US had dropped 40 percent from January to February while HD DVD player unit sales dropped only 13 percent. In the next period, from February to March, Blu-ray players saw a 2 percent increase while HD DVD entered its death spiral, dropping 65 percent. Blu-ray proponents pointed out that the January to March period followed a very successful holiday season and sales were down partly because of low inventory.

Blu-ray Disc Demystified

The Development of Blu-ray Disc

There were a few glimmers of good news. Universal stated that it had no plans to abandon HD DVD. An online "Save HD DVD" petition garnered over 31,600 signatures by February 8, while the competing "Let HD DVD Die" petition had reached only 11,300.

The fat lady was clearing her throat. Rumors surfaced from *The Hollywood Reporter* and elsewhere that Toshiba was preparing to throw in the towel. Japanese public broadcaster *NHK* reported on February 18 that Toshiba was closing its HD DVD production factories. The *Nikkei Business Daily* reported that Toshiba would announce tomorrow the discontinuation of HD DVD. Sure enough, on February 19, 2008, Toshiba formally announced it would quickly phase out the production of HD DVD players and recorders by the end of March. "We concluded that a swift decision would be best," said Toshiba President Atsutoshi Nishida.

The fat lady had sung. The format war was over, exactly six years after it stepped onto the public stage.[15]

The dominos fell quickly after that, with Universal Studios announcing the same day that it would be releasing movies on the Blu-ray format, and Paramount making the same announcement two days later.

In March, Toshiba projected that it would lose $665 million (¥65 billion) at the upcoming end of its fiscal year, "reflecting the discontinuation of the HD DVD business and the decline in sales prices of NAND flash memories." Goldman Sachs had earlier said that pulling out of the HD DVD business would improve Toshiba's profitability by $370 to $460 million a year.

In retrospect, the competition between Blu-ray and HD DVD was beneficial, to a point. It drove both groups to push harder for advanced features, sped the pace of development, and kept prices low. However, it would have been much better had there been a reconciliation around the middle of 2005, when the prospect was still workable.

On March 25, 2008, Sony's PlayStation 3 became the first *Profile 2* (*BD-Live*) player by virtue of a software upgrade.

On May 27 the first Blu-ray audio-only release, *Divertimenti*, performed by the Trondheim Solistene ensemble, was released in Norway by 2L. Although there were as yet no *Profile 3* (audio-only) players, the disc played on standard Blu-ray players but showed just how many of them were incapable of reproducing 192-kHz audio.

May 27 also marked the release of the last US HD DVDs from a major studio, *P.S. I Love You* and *Twister* from Warner Bros.

In June, the first standalone BD-Live player arrived, the Panasonic DMP-BD50. At a launch party on June 9 for *Sleeping Beauty* (which would not appear until four months later), Disney Studios president Bob Chapek said, "Every subsequent Disney title will have BD-Live." Contrary to Disney's approach with DVD, which had been to hang back until the format was established, then cautiously ensure that every title played on as many players as possible, Disney was pushing the envelope with dozens of groundbreaking features. It was a clever strategy that forced player makers to fix deficiencies in their players and prodded the

---

[15]In one of those cosmic coincidences, Toshiba originally announced its next-generation technology in January 2002, followed by the first Blu-ray announcement on February 19.

The Development of Blu-ray Disc

others studios to keep up. A few weeks later, Rich Marty of Sony Pictures stated that all the studio's titles from the end of June on would have some level of BD-Live interactivity. Marty said that so far one in five BD-Live disc buyers had activated and used the online features. Given the difficulty of connecting players to the Internet, 20 percent was a promising number.

On July 8, the BDA announced that China's *DigiRise DRA* digital audio coding technology had passed technical evaluation, paving the way for it to be formally adopted for the China version of Blu-ray. China remained the last battleground for HD DVD, which had been adapted to form the CBHD (formerly CH-DVD) specification for use in China.

At the International Symposium on Optical Memory in July, a team of Pioneer engineers presented results showing that a 20-layer Blu-ray discs was viable, potentially providing a whopping 500 Gbytes on a single disc.

Meanwhile, the final version of AACS was getting closer to takeoff but was still being delayed on the tarmac. In September, reports from a BDA news conference indicated that the final license was delayed yet again. Published articles referred to industry officials who said that managed copy complications were causing the delay because of concerns that a company could own the rights to release a movie on disc but not as a digital sale, which might or might not encompass a managed copy. In private conversations fingers were again pointing to Fox for slowing things down with its almost fanatically cautious scrutiny of details. To be fair, there had been plenty of complications and external demands hampering progress throughout the process, such as adding BD+, accommodating the China HD format, and getting the key generation facilities up and running. In spite of that, draft documents were circulated with a hoped-for final date in mid October, which would clear the way for managed copy to be available in July 2009. This would also require Blu-ray players released after May 2009 to check for audio watermarks.

On September 18, 2008, Memorex signaled the inevitable price drops heading into the holiday sales season with the first Blu-ray player priced below $270.

Three weeks later, players began to appear at Amazon.com and RadioShack.com on sale for $199, blasting through the $250 barrier that few people expected to be broken before 2009. In spite of the near-meltdown of economies around the world, hopes were high for healthy holidays for Blu-ray, heading into the crucial shopping season for once without a conflicting format to confuse and deter potential buyers. Hit titles were coming close to selling a million copies each — a breakthrough often considered to herald the coming of age of a new format. At long last, Blu-ray was beginning to fulfill its considerable potential.

Chapter 2
Technology Primer

This chapter[1] explains key technologies developed for or used by Blu-ray Disc™. Many of the fundamental technologies in use with DVD are being used by the Blu-ray Disc (BD) format. We feel it is beneficial to use the advances instituted with DVD as aids in understanding the Blu-ray format.

Understanding Digital and Analog

We live in an analog world. Our perceptions are stimulated by information that is received in smooth, unbroken form, such as sound waves that apply varying pressure on our eardrums, a mercury thermometer showing infinitely measurable detail, or a speedometer dial that moves continuously across its range. Digital information, on the other hand, is a series of snapshots depicting analog values coded as numbers, such as a digital thermometer that reads 71.5 degrees or a digital speedometer that reads 69 mph.[2]

The first recording techniques used analog methods — changes in physical material such as wavy grooves in plastic discs, silver halide crystals on film, or magnetic oxides on tape. After transistors and computers appeared, it was discovered that information signals could be isolated from their carriers if stored in digital form. One of the major advantages of digital information is that it is infinitely malleable. It can be processed, transformed, and copied without losing a single bit of information.

Analog recordings always contain noise (such as, tape hiss) and random perturbations, so each generation of recording or transmission decreases in quality. Digital information can pass through multiple generations, such as, from a digital video master, through a network, over the Internet, into a computer, out to a recordable DVD or BD, back into a computer, through the computer graphics chips, out over an HDMI (High Definition Multimedia Interface) connection, and into a digital monitor, all with no loss of quality. Digital signals representing audio and video can also be numerically processed. Digital signal processing is what allows AV receivers to simulate concert halls, surround sound headphones to simulate multiple speakers, and studio equipment to enhance video or correct colors.

---

[1]Please note, several of the sections in this chapter are lifted from our previous publications — **DVD Demystified Third Edition**, **High Definition DVD Handbook**, and **The HD Cookbook**. We hope this neither confuses nor disturbs the reader.

[2]There are endless debates about whether the "true" nature of our world is analog or digital. Consider again the thermometer. At a minute level of detail, the readings can't be more accurate than a molecule of mercury. Physicists explain that the sound waves and photons that excite receptors in our ears and eyes can be treated as waves or as particles. Waves are analog, but particles are digital. Research shows that we perceive sound and video in discrete steps, so our internal perception is actually a digital representation of the analog world around us. There are a finite number of cones in the retina, similar to the limited number of photoreceptors in the CCD of a digital camera. At the quantum level, all of reality is determined by discrete energy states that can be thought of as digital values. However, for the purposes of this discussion, referring to gross human perception, it's sufficiently accurate to say that sound and light, and our sensation of them, are analog.

Technology Primer

When storing analog information in digital form, the trick is to produce a representation that is very close to the original. If the numbers are exact enough (such as a thermometer reading of 71.4329 degrees) and repeated often enough, they closely represent the original analog information.[3]

Digital video is a sheet of dots, called *pixels*, each holding a color value. This is similar to drawing a picture by coloring in a grid, where each square of the grid can only be filled in with a single color. If the squares are small enough and there is sufficient range of colors, the drawing becomes a reasonable facsimile of reality. For Blu-ray, each grid of 1920 squares across by 1080 squares down represents a still image, called a *frame*. Twenty-four to sixty frames are shown each second to convey motion.

Digital audio is a series of numbers representing the intensity, or amplitude, of a sound wave at a given point. For BD, these numbers are "sampled" over 48,000 times a second (sampling may be performed as frequently as 192,000 times a second for super-high–fidelity audio), providing a much more accurate recording than is possible with the rough analogues (pun intended) of vinyl records or magnetic tape. When a digital audio recording is played back, the stream of numerical values is converted into a series of voltage levels, creating an undulating electrical signal that drives a speaker.

Pits and Marks and Error Correction

Data is stored on optical discs in the form of microscopic *pits* (see Figure 2.1). The space between two pits is called a *land*. On writable discs, pits and lands are often referred to as *marks* and *spaces*. Read-only discs are stamped in a molding machine from a liquid plastic such as polycarbonate or acrylic and then coated with a reflective metallic layer. Writable discs are made of material designed to be physically changed by the heat of a laser, creating marks. As the disc spins, the pits (or marks) pass under a reading laser beam and are detected according to the change they cause in the intensity of the beam. These changes happen very fast (over 980,000 times per second on Blu-ray Discs) and create a stream of transitions spaced at varying intervals: an encoded digital signal.

Figure 2.1 Optical Disc Pits

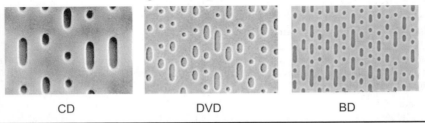

CD   DVD   BD

---

[3]Ironically, digital data is stored on analog media. The pits and lands on a DVD or a BD are not of a uniform depth and length, and they do not directly represent ones and zeros. They produce a waveform of reflected laser light that represents coded runs of zeros and transition points. Digital tape recordings use the same magnetic recording medium as analog tapes. Digital connections between AV components (digital audio cables, HDMI cables, etc.) encode data as square waves at analog voltage levels. However, in all cases, the digital signal threshold is kept far above the noise level of the analog medium so that variations do not cause errors when the data is retrieved.

Technology Primer

> **NOTE** To help put the burning process in perspective, consider that at 16x DVD recording speeds a disc makes 10,800 rotations per minute, corresponding to a linear velocity of 56 meters per second (over 200 km/h), while marks are burned with a precision of less than 0.05 micrometer!

Many people assume that the digital ones and zeros that comprise the data stored on the disc are encoded directly as pits and lands, but the reality is actually much more complicated. Pits and lands both represent strings of zeros of varying lengths, and each transition between them represents a one, but neither is a direct representation of the contents of the disc. Almost half of the information has been used to pad and rearrange (*modulate*) the data in sequences and patterns designed to be accurately readable as a string of pulses. Modulation makes sure strings of zeros aren't too long (no more than 10) or too short (no fewer than two) and that there is only a single one between them (since ones are represented by "edges" and it's not physically possible to have two edges together.)

About 13 percent of the digital signal before modulation is extra information for correcting errors. Errors can occur for many reasons, such as imperfections on the disc, dust, scratches, a dirty lens, and so on. A human hair is about as wide as 150 pits, so even a speck of dust or a minute air bubble can cover a large number of pits. However, the laser beam focuses past the surface of the disc so the spot size at the surface is much larger and is hardly affected by anything smaller than a few millimeters. This is similar to the way dust on a camera lens is not visible in the photographs because the dust is out of focus. As the data is read from the disc, the error correction information is separated and checked against the remaining information. If it doesn't match, the error correction codes are used to try to correct the error.

The error correction process is like a number square, where you add up columns and rows of numbers (see Figure 2.2). You could play a game with these squares where a friend randomly changes a number and challenges you to find and correct it. If the friend gives you the sums along with the numbers, you can add up the rows and columns and compare your totals against the originals. If they don't match, then you know that something is wrong—either a number has been changed or the sum has been changed.[4] If a number has been changed,

Figure 2.2 Number Squares

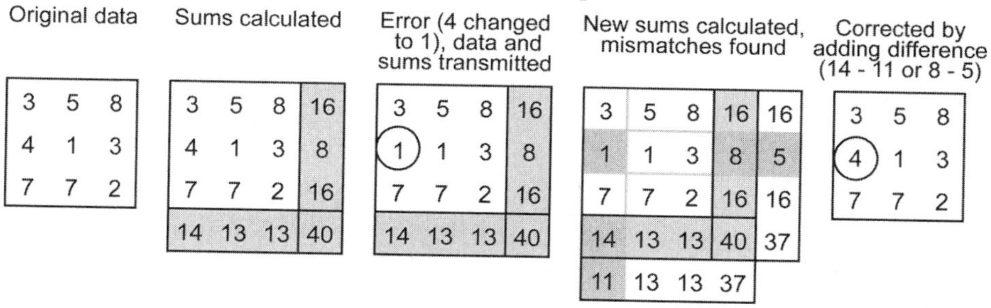

Original data	Sums calculated	Error (4 changed to 1), data and sums transmitted	New sums calculated, mismatches found	Corrected by adding difference (14 - 11 or 8 - 5)
3 5 8 4 1 3 7 7 2	3 5 8 16 4 1 3 8 7 7 2 16 14 13 13 40	3 5 8 16 (1) 1 3 8 7 7 2 16 14 13 13 40	3 5 8 16 16 1 1 3 8 5 7 7 2 16 16 14 13 13 40 37 11 13 13 37	3 5 8 (4) 1 3 7 7 2

---

[4]It is possible for more than one number to be changed in such a way that the sum still comes out correct, but the BD encoding format makes this an extremely rare occurrence.

Technology Primer

then a corresponding sum in the other direction also will be wrong. The intersection of the incorrect row and incorrect column pinpoints the guilty number and, in fact, by knowing what the sums are supposed to be, the original number can be restored. The error correction scheme used by DVD and BD is a bit more complicated than this but operates on the same general principle.

It is always possible that so much of the data is corrupted that error correction fails. In this case, the player must try reading the section of the disc again. In the very worst cases, such as, an extremely damaged disc, the player will be unable to read the data correctly after multiple attempts. At this point, a movie player will continue on to the next section of the disc, causing a brief glitch in playback. A DVD-ROM or a BD-ROM drive, on the other hand, cannot do this. Computers will not tolerate missing or incorrect data, so the -ROM drive must signal the computer that an error has occurred so that the computer can request that the drive either try again or give up.

Layers

One of the innovations of DVD and Blu-ray Disc is to use layers on the disc to increase storage capacity. Commercially, there are only single-layer and dual-layer discs, but research is ongoing and discs containing four, eight, and more layers are under development.

The laser that reads the disc can focus at different levels so that it can look through a layer to read the layer beneath. With a dual-layer disc, the outside layer is coated with a semireflective material that enables the laser to read through it when focused on the inner layer. When the player reads a disc, it starts at the inside edge of the track area and moves toward the outer edge, following a spiral path. When the laser reaches the end of the first layer, it quickly refocuses onto the second layer and starts reading in the opposite direction, from the outer edge toward the inner. This is known as *opposite track path* (OTP) or *reverse spiral dual layer* (RSDL). Unlike DVD, BD does not allow *parallel track path* (PTP), where the second layer can optionally spiral from the hub to the outer diameter. Refocusing happens very quickly, but on most DVD players the video and audio pause for a fraction of a second as the player searches for the resumption point on the second layer. For BD, if the player has a large enough buffer and the disc is carefully designed to lower the data rate at the layer switch point (so that the buffer will take longer to empty), the laser pickup may have time to refocus and retrack without causing a visible break.

The DVD specification requires backwards compatibility with the earlier CD format and the BD specification does not require players to read DVDs, but manufacturers recognize have a snowball's chance in Hollywood of surviving. The difficult part is that the pits on CDs, DVDs, and BDs are all at different levels (see Figure 2.3). In essence, the newer format player must be able to focus a laser at different distances. This problem has various solutions, including using lenses that switch in and out, and holographic lenses that are actually focused at more than one distance simultaneously. Further, each format uses a different wavelength —780nm infrared for CD, 650nm red for DVD, and 405nm blue for BD — so multiple lasers or multi-wavelength diodes are needed.

Technology Primer

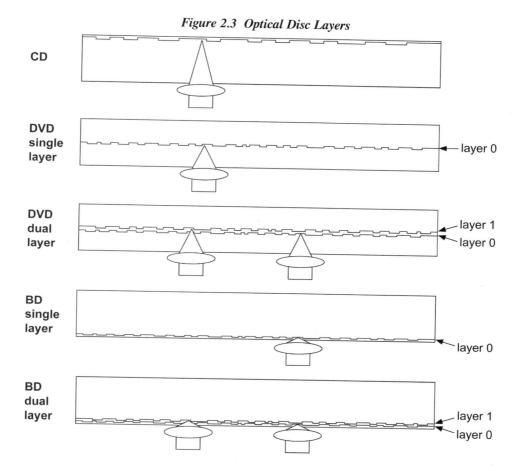

Figure 2.3 Optical Disc Layers

The remaining task to ensure CD and DVD compatibility requires an extra bit of circuitry and firmware for reading format-specific data. However, the CD family is quite large and includes some odd characters, not all of which fit well with DVD and BD. The prominent members of the CD family are audio CD, Enhanced CD (or CD Plus), CD-ROM, CD-R, CD-RW, CD-i, Photo CD, CDV, and Video CD. It would be technically possible to support all these yet most of them require specialized hardware. Therefore, some manufacturers choose to support only the most common or easiest-to-support versions. Disc types, such as Enhanced CD and Video CD, are easy to support with existing hardware. Others, such as CD-i and Photo CD, require additional hardware and interfaces, so they are not commonly supported. Given that the native data on a CD or a DVD can be read by any compatible BD system, conceivably any CD and DVD format could be supported. Computers support more CD formats than DVD and BD players partly because some are designed for computer applications and also because specialized CD systems can be simulated with computer software.

Technology Primer

The Two HDs — High Definition and High Density

With Blu-ray Disc, the acronym "HD" has more than one meaning. Although this book regards "HD" as shorthand for "high definition" because the disc can contain high definition video, the acronym can also mean *high density* when refering to the physical disc. The BD format uses the same basic technology as DVD and CD, but one of the key differences is that the pits and lands are recorded at a much higher density. As a result, a shorter wavelength blue laser must be used to read the disc, just as DVD required a shorter wavelength (635-650 nm) laser relative to CD (780 nm).

Backward compatibility with DVD and CD is a *de facto* requirement for BD, just as DVD players needed to support CD formats to succeed in the marketplace. This results in significant challenges for the player manufacturers, who must design pick-up heads that contain three lasers, one for each wavelength, all aiming down the same emission path. For Blu-ray Disc, this is particularly challenging because BD uses a different type of lens than its fellow travellers. Instead of the 0.65 NA (*numeric aperture*) lens that CD and DVD both use, Blu-ray hosts a 0.85 NA lens, which corresponds to a much shorter focal length. As a result, the pickup head and the drive mechanism have to be much more sophisticated in order to adjust for the difference in focal length between BD and the other disc formats.

Another effect of the Blu-ray 0.85 NA lens is that it results in a much smaller spot on the surface of the disc. As a result, dust, hair, fingerprints and scratches all have a much greater negative impact on the player's ability to read the disc. The size of a particle or a scratch is considerably larger for the laser spot diameter of a BD player than that of a DVD or CD. As a result, there is far less of the defocusing effect described earlier and the resulting signal is much more degraded by a surface aberration.

The BD format has taken two steps to address this issue. First, disc caddies (for BD-RE) and special hard surface coats have been adopted to help protect the surface of the discs from fingerprints, scratches, or anything that may threaten the readability of the disc. Second, BD uses much stronger error correction, allowing it to more easily detect and correct errors that may have occured while reading the disc. One interesting technique used is to almost double the size of the error control block such that a read error from a hair or finerprint has much less significance relative to the size of the error control block. In essence, the larger error control block in BD offers essentially the same increase in resiliency as the defocusing effect for DVD and CD.

The World's Television Systems

It is important to understand how the world's television systems were developed. Not to get the reader bogged down in minutiae, but without a basic knowledge of how and why US television operates differently from other countries, it is very difficult to explain, let alone understand, the variety of frame rates and video resolutions that are now available with high definition television.

There are three standard definition television (SDTV) systems in the world. The NTSC

Technology Primer

standard is used in Canada, Mexico, the U.S, and Japan. NTSC is shorthand for *National Television Systems Committee*. The PAL standard and its variations PAL-M and PAL-N are used in most of Europe, as well as China, India, South America, Africa, and Australia. PAL is shorthand for *Phase Alternating Line*. A third standard, known as SECAM, is in use in Russia and France. SECAM is the acronym for *Système Électronique pour Couleur avec Mémoire*. The English translation of SECAM is Sequential Color with Memory. Production for SECAM is accomplished using PAL format equipment. These television standards are adapted from the electrical standards of the respective countries.

Long before the invention of television, the US standard for electricity was developed so that manufacturers would be able to make products that would work when plugged in to a common electrical source. The US standard was set at 110/120 volts, with the system changing polarity from positive to negative and back again, 60 times a second. This sequence is referred to as *alternating current*, with each positive/negative polarity change termed a *cycle*. In Europe, the standard was set at 220/240 volts and 50 cycles per second.

When first introduced, television broadcasts were in black and white, and the NTSC standard mandated a rate of 30 frames per second with two fields comprising a frame, for 60 fields per second. NTSC uses 525 scan lines to compose a frame, with the odd lines displayed as field one and the even lines displayed as field two. The fields interlace on presentation to compose a frame. The PAL standard adopted a rate of 25 frames per second, also comprised of two fields, for 50 fields per second, and uses 625 scan lines to compose a frame.

The scan line counts are the product of a string of small integer factors that, at the time, could only be reliably supported by vacuum tube divider circuits — 525 is $7 \times 5 \times 5 \times 3$, and 625 is $5 \times 5 \times 5 \times 5$. These divider circuits derive the field rate from the power line rate. These factors combined to give monochrome NTSC 525/60 television a line rate of $30 \times (7 \times 5 \times 5 \times 3)$, or exactly 15.750 kHz. Monochrome PAL 625/50 television has a line rate of $25 \times (5 \times 5 \times 5 \times 5)$, or exactly 15.625 kHz. With the introduction of color for television, the NTSC number takes on immense significance, whereas the PAL signal structure was mostly unaffected.

In 1953, when adding color to the monochrome signal, a second NTSC determined that it would be appropriate to choose a color subcarrier in the region of 3.6 MHz. This subcarrier would imbed the color information within the television signal. But there was a technical concern with adding the color information without disturbing the sound subcarrier element of the signal. Too close a relationship in the subcarrier frequencies would result in distortion that would become visible in the luminance component of the signal.

Remember that the responsibility for setting broadcast standards resides with the Federal Communications Commission (FCC). If the FCC had altered the sound subcarrier frequency to be increased by the fraction 1001/1000 (as recommended by the second NTSC) — that is, increased by 4.5 kHz to about 4.5045 MHz — then the color subcarrier in NTSC could have been exactly 3.583125 MHz, the line and the field rates would have been unchanged, and we would have retained exactly 30 frames per second!

But true to form, alas, the FCC refused to alter the sound subcarrier for fear of possible minor audio interference on existing black and white television receivers. Instead, the FCC

Blu-ray Disc Demystified

Technology Primer

chose to reduce both the line rate and the field rate by the fraction 1001/1000, effectively dropping those rates to about 15.734 kHz and 59.94 Hz, respectively. Thus the 59.94 rate was born![5]

Unfortunately the field rate of 60 divided by 1.001, rounded to 59.94 for convenience, means that 60 fields consumes slightly more than one second, so 30 frames no longer agrees with one second of clock time. As a side note, this precipitated the invention of dropframe timecode that, ostensibly, alleviated this disparity.

Regrettably, the shortsighted, although pragmatic, FCC decision helped to bring about the frame and video rate complexity that exists today. This is further amplified by the continuing convergence of video and computer technologies and terminologies, let alone the emergence of high definition standards and formats. As noted in the Introduction, a billion might mean something other than a thousand million.

Frame Rates

With the exceptions of cinema, computer displays, and HDTVs, we are in a universe of NTSC and PAL television systems. Even with cinema, though, that may be the initial display, only to be rapidly followed by conversion to video for alternate market exploitation. Thus, no matter the acquisition, television boundaries will be encountered by virtually every production.

The explosion in frame rate flexibility of video acquisition cameras has been generated by the desire to match film acquisition settings, so that a video camera can achieve the *cine-look* of film. That is a highly subjective goal. One person's "cine-" is another person's "(fill in your own word here)". At the upper reaches of the technology, that is to say the more expensive end of the spectrum, the "cine-look" attained by a video camera is indistinguishable from a film source. And there are tools that can be used that will enhance footage from a lower-end camera that may render the final product compatible with a satisfactory "film-like" look.

But, when using video tools in place of film, extreme care must be exercised in all aspects of image preparation — lighting, set design, cast choices, costumes, makeup, time of day and phase of the moon. It must be understood that video tools have limitations in extreme conditions that may preclude their use. If the image acquired by the source is less than desired, no amount of manipulation after the 'get' can save the picture.

But what exactly is "frame rate"? Is it a component of acquisition or of display, or both? You're probably not gonna love us for this one…it is both, and more.

Frame rate in acquisition establishes the speed with which action is transcribed to media, capturing moments in time either as two fields or as a frame. Frame rate in display establishes the frequency with which frames are presented for viewing. Frame rate is expressed in the number of frames per second, or, by the number of frames in Hertz. What's "Hertz"? Hertz is a unit of frequency equivalent to cycles per second. Alternating current of 60 cycles per second can be called 60 Hertz. Hertz is abbreviated "Hz".

---

[5]Actually, the number 59.94 is rounded off, as the fraction results in an incomplete number that repeats the integers in the sequence 59.94005994005994005994005994...forever.

Technology Primer

Additionally, there are two methods of frame construction — interlace and progressive. Interlace is the method of interweaving the two fields that comprise a frame. Progressive is the method of contiguously presenting all of the lines that comprise a frame, as in a snapshot. Interlace is represented as "i" and progressive is represented as "p".

There are five primary frame rate standards for television and movies —

- **30i:** 30 frames interlaced, comprised of 60 fields. For NTSC, this is really 59.94 fields, interlaced, composing 29.97 frames per second.
- **30p:** 30 frames, progressive, for video cameras, producing an image without interlaced field artifacts. For NTSC, this is really 29.97 frames per second.
- **25i:** 25 frames interlaced, comprised of 50 fields.
- **25p:** 25 frames progressive.
- **24p:** 24 frames, progressive, for film and capable video cameras.

Depending on the camera, there are any number of frame rates that may be used for the acquisition rate. With a film camera, the standard acquisition rate is either 24 or 25 frames per second (fps), depending on the region of display. Film cameras can be run off-speed, either sped up or slowed down, to accomplish various special effects and/or to capture scenes under unique conditions.

Since their invention, NTSC color video cameras have captured images at the rate of 59.94 fields per second. Nowadays, high definition capable video cameras can be set to record either 59.94 fields per second or 59.94 frames per second, as well as rates from 2 frames per second to 60 frames per second. Unfortunately, we identify all of these rates with the abbreviation "fps".

Very often, the goal is to give video images a "cine-look", so the opportunity to shoot footage at a rate equivalent to the film rate is highly prized. However, 59.94 fields per second is equal to shooting 29.97 frames per second. We have to get from 29.97 to 24, and you cannot get there from there. But you can get to 23.976[6], which is sometimes rounded to 23.98. Employing the frame rate of 23.98 also allows for an easy expansion to the 29.97 or 59.94 rates by the use of what are called "pulldown" techniques, wherein fields or frames are duplicated to meet the higher rate. This allows for the integration of "cine-look" video with other video assets, as well as the presentation of the footage via television. (Although that might defeat the quest for "cine-", wouldn't it? But we digress...)

Further, with sophisticated circuitry, sensors and integrated chips, some video cameras may be set to record 24 frames per second, also known as "true 24". However, 24 fps is not natively compatible with an NTSC display, but video shot at 24 frames per second can readily integrate with film originals that are shot at 24 frames per second. It is this fluidity of medium that accommodates the extensive use of computer generated imagery and augmentation that is evident in motion pictures today.

As noted earlier, if you are producing for theatrical presentation exclusively you can establish 24 fps as the production criteria, whether shooting film or video. The follow-on production steps would all respect the 24 fps rate, and all will remain as intended.

---

[6]23.976 is shorthand for the result of dividing 24 by 1,001. The actual quotient is the incomplete number 23.976002397600239760023976...forever, akin to the result that gives us 59.94.

Blu-ray Disc Demystified

Technology Primer

But, "true 24" and "23.98" are not the same. In fact, 23.98 can become something called *progressive segmented frame* (PsF), wherein the odd lines from the frame are segregated from the even lines for the frame. This is not the same as interlace fields because the progressive frame segments are not presented sequentially. Rather, the two segments are recombined prior to presentation as a single frame. The PsF technique allows for easier integration with interlace technologies.

Beyond the imaging source is the recording device for saving and storing the images acquired. The recording rate for devices may be set independently from the frame rate, depending on the technology employed. Once again, it is the ultimate display criteria that should govern the settings for recording. The 24 fps rate may be used when the footage is destined for cinema, whereas a rate of 29.97 Hz or 59.94 Hz should be used when footage will appear on television. Put another way, when the intent is to shoot cine-like video, set the camera to image 23.98 fps and the recording device to match NTSC display criteria.

And what about frame rate for display, as in the frequency that images are presented for viewing? The methods for presentation vary between platforms — projectors, television receivers, and displays. A projector runs at a speed of 24 frames per second, but shows each frame twice, thereby displaying the film at the rate of 48 Hz. This action is done to minimize the perception of flicker between frame images.

Television receivers present the image standard for their format, PAL or NTSC, as either 25 or 29.97 frames per second, respectively, but given the interlaced field construction of the frames, the display rate is either 50 or 59.94 Hz.

Technology has provided new ways to display images that are not bound to television standards and do not use a cathode-ray tube (CRT) for presentation. These displays are capable of presenting standard definition television, high definition television, and images from a variety of sources — disc players, hard drives, computers, et cetera. The circuitry employed by these technologies provides vastly improved control over the data being displayed. And, we are now able to view images with display rates that are multiples of either television or film frame rates — 48, 60, 72 Hz and higher (120 Hz is highly regarded as it is a multiple of both 24 and 30). These increased frame display rates can significantly improve the perceived clarity, depth, and color characteristics of a presentation, depending on the breadth of features included in the display.

High Definition Image Resolutions

Image resolution is determined by several factors — image size, frame style, color or chroma sampling, and data bit rate or depth. When multiplied by frame rate, we get the data rate required for streaming images to a display and/or to storage. Picture sizes for high definition and beyond generate exceptionally large numbers, both in terms of data rate and cumulative data for storage.

The high definition digital television formats defined by the ATSC (Advanced Television Systems Committee)[7] are —

---

[7]The ATSC standards are intended to replace the NTSC system and are being adopted by many other countries besides the US. The ATSC has defined systems for both standard and high definition television, but only the high definition formats are included here.

Technology Primer

- 1280 pixels per line × 720 lines, in 16:9 widescreen aspect, with progressive frames, at frame rates of 23.976, 24, 29.97, 30, 59.94 and 60 per second
- 1920 pixels per line × 1080 lines, in 16:9 widescreen aspect, with either —
    - interlace frames, at frame rates of 29.97 (59.94 fields) and 30 (60 fields) frames per second, or
    - progressive frames, at frame rates of 23.976, 24, 29.97 and 30 frames per second

Shorthand is applied to these standards, resulting in the terms 720p, 1080i, and 1080p. Adding frame rates to those terms, creates terms like 720/24p or 1080/30i. Although, in the case of 720, the "p" is unnecessary as all of the frame rates for 720 lines are progressive.

Also, please note that the much ballyhooed image rates of 1080/60p and 1080/59.94p are not included in the ATSC high definition standards. Another component of the ATSC standards is the use of MPEG-2 compression in the transmission of data for the ATSC formats. All of the ATSC image formats are compressed somewhat to accommodate the bandwidth allocation for television channel transmission. With the conversion to digital television, the channel bandwidth allocation is only 19.39 megabits per second for over-the-air transmission. The high data rates generated by images at 1080/60p (59.94) would result in poor image quality when compressed, transmitted, uncompressed, and displayed.

The ATSC standards notwithstanding, Blu-ray Disc specifications define high definition video as —

- picture sizes, in 16:9 widescreen aspect —
    - 1920 pixels per line × 1080 lines
    - 1440 pixels per line × 1080 lines
    - 1280 pixels per line × 720 lines
- display frame rates —
    - for 1080 lines — frames rates of 23.976p, 24p, 25i, 29.97i per second
    - for 720 lines — frame rates 23.976p, 24p, 50p, 59.94p per second

The Blu-ray Disc frame rate of 23.976 for both 1080 and 720 includes a 2-3 pulldown instruction in order to display at 29.97 frames per second. Depending on the connection type and the display type, this frame type may be either interlace or progressive.

Typically, Blu-ray Discs will have content that is encoded at either 1080/24p or 1080/23.976p for film-based content, and 1080/29.97i or 1080/25i for video-based content. And, please note that marketing jargon will refer to these numbers as 1080/24p and 1080/30i (or 1080/60i, which does not mean 60 frames per second).

Whereas ATSC incorporates MPEG-2 compression in meeting the delivery and display requirements for television broadcast, Blu-ray can use either MPEG-2 or MPEG-4 AVC or VC-1 for compression, and is not restricted by the digital television channel bandwidth limitation. These expanded compression choices allow for improved data rates, at higher compression levels. The Blu-ray Disc maximum data rate for the video component of the transport stream is 40.0 Mbps. Thus, the seeds for confusion are sown when comparing data rates for broadcast high definition television with that of high definition discs.

Technology Primer

Chroma Subsampling and Bit Depth

Another component of image data is the sampling that is performed on an image. This is referred to as *chroma subsampling*, and is generally expressed as a three number ratio, such as, 4:4:4 or 4:2:2 or 4:2:0, among others. These numbers reflect the sampling rate and method that is applied to the *luminance* and *chrominance* components of an image. The theory behind this sampling process is that the human eye is more sensitive to changes in brightness (luminance) than changes in color (chrominance), and there is no perceived loss when sampling color details at a lower rate. The signal is divided into a luminance element (Y′) and two color difference elements that are derived as red minus luminance (C_r) and blue minus luminance (C_b)[8]. But, the three numbers are not directly related to the three signal elements and instead reflect the methodology of the sample (otherwise, the 4:2:0 variant would have no data for one of the color channels...doh!).

The terms 4:4:4 and 4:2:2 were developed for standard definition digital video. Nowadays, the sampling rates for high definition video are 22:22:22 and 22:11:11, respectively, but casual use of the SD terms has been adopted for the HD rates.

Technically, sampling is performed using rates that are multiples of 3.375 MHz, and 4 times 3.375 is 13.5 MHz and 22 times 3.375 is 74.25 MHz. Thus, 4:4:4 means that the signal elements are each sampled at 13.5 MHz (74.25 MHz for HD). Sampling at 4:4:4 provides close to real-life color representation and is considered to be lossless. Taking advantage of the different sensitivity to color perception, 4:2:2 samples the luminance at 13.5 MHz (74.25 MHz), but the color difference signals are each sampled at 6.75 MHz (37.125 MHz) or one half the frequency of 4:4:4. The ratio 4:2:0 means that the luminance is still sampled at 13.5 MHz, but the color difference channels are subsampled by a factor of 2 both horizontally and vertically, in a two-line grouping structure. The 4:2:0 sampling scheme is the method in use for BD, as well as standard DVD.

Bit depth reflects the number of bits used to store information about each sample. The higher the bit depth means that more bits per pixel are used for an image, and the more pixels used for an image then the larger the image file. Eight bits provide up to 256 color gradations from black to white, while 10 bits provide up to 1024 color gradations from black to white. And, there are 12-bit schemes that, in conjunction with 4:4:4, provides for the best image quality currently available.

High Definition Data Streams

The goal is capturing data digitally at the highest possible quality given the current capabilities of technology. Translating these image format sizes and frame rates to continuous data streams requires advanced degrees in mathematics or an association with some extremely knowledgeable and quick partners. In our case, we have a couple of authors who can rattle off data rate computations on request (see Table 2.1).

---

[8] The terms Y′, C_b, and C_r denote digital encoded component video signals. The terms Y, R-Y, and B-Y are used to denote analog component signals. Technically, these terms are not interchangeable, although tech talk frequently crosses the term boundaries.

Technology Primer

Table 2.1 Image Data Stream Examples

Format	Frames per sec	Frame Size	Number of Pixels	Bits per Pixel	Data Rates Mb/s	Data Rates MB/s
Sampling Rate: 4:4:4 RGB 12 bit						
1080/60i	29.97	1920 × 1080	2,073,600	36	2,237.2	266.7
1080/24p	23.976	1920 × 1080	2,073,600	36	1,789.8	213.4
720/60p	59.94	1280 × 720	921,600	36	1,988.7	237.1
720/24p	23.976	1280 × 720	921,600	36	795.5	94.8
Sampling Rate: 4:2:2 YUV 10 bit						
1080/60i	29.97	1920 × 1080	2,073,600	20	1,242.9	148.2
1080/24p	23.976	1920 × 1080	2,073,600	20	994.3	118.5
720/60p	59.94	1280 × 720	921,600	20	1,104.8	131.7
720/24p	23.976	1280 × 720	921,600	20	441.9	52.7
Sampling Rate: 4:2:0 YUV 8 bit						
1080/60i	29.97	1920 × 1080	2,073,600	12	745.7	88.9
1080/24p	23.976	1920 × 1080	2,073,600	12	596.6	71.1
720/60p	59.94	1280 × 720	921,600	12	662.9	79.0
720/24p	23.976	1280 × 720	921,600	12	265.2	31.6

Table 2.1 provides a sampling of the data rates that are generated when working with high definition data streams. What is important to note is that even before compression is executed on an image data stream, the amount of data can be reduced dramatically by choosing a different sampling scheme, and/or by resampling the image data to meet the requirements of BD playout, which is 4:2:0 YUV 8-bit.

There is a concerted effort underway to address the issue of producing HD images in the 1080 line format with 60 progressive frames per second. Termed 1080/60p, the hope is that cameras will be capable of acquiring images in that image format and productions will be able to take advantage of the perceived superior resolution. But this quest overlooks the hard fact of exactly how much data can be transmitted via an HD Serial Digital Interface (HD SDI) link, the standard for connecting cameras to recording devices, monitors, et cetera. The maximum data rate for a single HD SDI link is 1.485 Gbps, including audio. As you can see in Table 2.1, the data rate for 1080/60 interlaced video, sampled at 4:4:4 12 bit, is 2.24 Gbps, and the data rate at 4:2:2 YUV 10 bit is 1.24 Gbps. This means that HD SDI cannot accommodate 4:4:4 12-bit data, but can accommodate 4:2:2 10-bit data. When these data rates are doubled to accommodate 60p, it becomes obvious that some other interconnection is required.

Standards have been established for a dual-link HD SDI structure which ties two HD SDI links together in parallel, which is intended to accommodate 4:4:4 sampling or 1080/60p. Yet another standard was adopted in June 2006 for a 2.97 Gbps interface that uses a single cable, which may be used to replace the dual-link approach.

Technology Primer

That, then, is the conundrum. Even if cameras and sensors are created that are capable of 1080/60p acquisition, the production infrastructure will have to be changed. New tools need to be and are being developed, as well as taking on the massive tasks of updating editing and workstation interfaces to accommodate the realtime needs when presenting images at these gargantuan data rates. All of these are considerable undertakings.

And yes, these 1080/60p data rates could be dramatically reduced with compression, but wouldn't that be self-defeating? We'll leave it to others to decide.

Birds Over the Phone: Understanding Video Compression

After compact discs (CDs) appeared in 1982, digital audio became a commodity. It took many years before the same transformation began to work its magic on video. The step up from digital audio to digital video is a doozy, for in any segment of standard definition television there is about 250 times as much information as in the same-length segment of CD audio. The trick is to reduce the amount of information without significantly reducing the quality of the picture. As we have seen, the solution is digital compression.

In a sense, you employ compression in daily conversations. Picture yourself talking on the phone to a friend. You are describing the antics of a particularly striking bird outside your window. You might begin by depicting the scene and then mentioning the size, shape, and color of the bird. But when you begin to describe the bird's actions, you naturally don't repeat your description of the background scene or the bird. You take it for granted that your friend remembers this information, so you only describe the action — the part that changes. If you had to continually refresh your friend's memory of every detail, you would have very high phone bills. The problem with TV is that it has no memory — the picture has to be continually refreshed. It's as if the TV were saying, "There's a patch of grass and a small tree with a 4-inch green and black bird with a yellow beak sitting on a branch. Now there's a patch of grass and a small tree with a 4-inch green and black bird with a yellow beak hanging upside down on a branch. Now there's a patch of grass and a small tree with a 4-inch green and black bird with a yellow beak hanging upside down on a branch trying to eat some fruit," and so on, only in much more detail, redescribing the scene 30 times a second. In addition, a TV individually describes each piece of the picture even when they are all the same. It would be as if you had to say, "The bird has a black breast and a green head and a green back and green wing feathers and green tail feathers and..." (again, in much more meticulous detail) rather than simply saying, "The bird has a black breast, and the rest is green." This kind of conversational compression is second nature to us, but for computers to do the same thing requires complex algorithms. Coding only the changes in a scene is called *conditional replenishment*.

The simplest form of digital video compression takes advantage of *spatial redundancy*— areas of a single picture that are the same. Computer pictures are comprised of a grid of dots, each one a specified color, but many of them are the same color. Therefore, rather than storing, say, a hundred red dots, you store one red dot and a count of 100. This reduces the amount of information from 100 pieces to 3 pieces (a marker indicating a run of similar colored dots, the color, and the count) or even 2 pieces (if all information is stored as pairs of color and count), as shown in Figure 2.4. This is called *run-length encoding* (RLE). It is a form of *lossless* compression, meaning that the original picture can be reconstructed perfectly with no missing detail. Run-length encoding is great for *synthetic images* — simple, computer-generated images containing relatively few colors. However, this method does not work well for most *natural images* (e.g., photographs) because in the natural world there are continuous, subtle changes in color that thwart the RLE process.

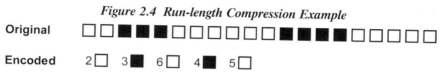

Figure 2.4 Run-length Compression Example

If you take the fundamental concept behind run-length encoding, which is to find redundancies by identifying correlations in the data, and then apply more sophisticated pattern search algorithms and efficient symbol replacement (e.g., replace a 10-byte pattern with an 8-bit symbol each time that pattern appears), you will have an even more effective version of lossless compression. The PNG image format, which is primarily used for graphics on Blu-ray Discs, supports *compression filters*, which provide different search algorithms for identifying and reducing redundant patterns in the image data. With these formats, even natural images can start to see at least some benefit (though it is still more effective for synthetic images where there tend to be more and larger repeating patterns).

To reduce picture information even more, *lossy* compression is required. This results in information being removed permanently. The trick is to only remove detail that will not be noticed. Many such compression techniques, known as *psychovisual* encoding systems, take advantage of a number of aspects of the human vision system —

- The eye is more sensitive to changes in brightness than changes in color.
- The eye is unable to perceive brightness levels above or below certain thresholds.
- The eye cannot distinguish minor changes in brightness or color. This perception is not linear, with certain ranges of brightness or color more important visually than others. For example, variegated shades of green such as leaves and plants in a forest are more easily discriminated than various shades of dark blue such as in the depths of a pool.
- Gentle gradations of brightness or color (such as a sunset blending gradually into a blue sky) are more important to the eye and more readily perceived than abrupt changes (such as pinstriped suits or confetti).

The human retina has three types of color photoreceptor cells, called *cones*.[9] Each is sensitive to different wavelengths of light that roughly correspond to the colors red, green, and

---

[9]*Rods*, another type of photoreceptor cell, are only useful in low-light environments to provide what is commonly called night vision.

Technology Primer

blue. Because the eye perceives color as a combination of these three stimuli, any color can be described as a combination of these primary colors.[10] Televisions work by using three electron beams to cause different phosphors on the face of the television tube to emit red, green, or blue light, abbreviated to RGB. Television cameras record images in RGB format, and computers generally store images in RGB format.

RGB values are a combination of brightness and color. Each triplet of numbers represents the intensity of each primary color. As just noted, however, the eye is more sensitive to brightness than to color. Therefore, if the RGB values are separated into a brightness component and a color component, the color information can be more heavily compressed. The brightness information is called *luminance* and is often denoted as Y'.[11] Luminance is essentially what you see when you watch a black-and-white TV. Luminance is the range of intensity from black (0 percent) through gray (50 percent) to white (100 percent). A logical assumption is that each RGB value would contribute one-third of the intensity information, but the eye is most sensitive to green, less sensitive to red, and least sensitive to blue, so a uniform average would yield a yellowish green image instead of a gray image.[12] Consequently, it is necessary to use a weighted sum corresponding to the spectral sensitivity of the eye, which is about 70 percent green, 20 percent red, and 10 percent blue (Figure 2.5).

The remaining color information is called *chrominance* (denoted as C), which is made up of *hue* (the proportion of color — the redness, orangeness, greenness, etc.), and *saturation* (the purity of the color, from pastel to vivid). For the purposes of compression and converting from RGB, it is easier to use *color difference* information rather than hue and saturation.

The color information is what's left after the luminance is removed. By subtracting the luminance value from each RGB value, three color difference signals are created: R-Y, G-Y, and B-Y. Only three stimulus values are needed, so only two color difference signals need to be included with the luminance signal. Since green is the largest component of luminance, it has the smallest difference signal (G makes up the largest part of Y, so G-Y results in the smallest values). The smaller the signal, the more it is subject to errors caused by noise, so B-Y and R-Y are the best choice for use as the color difference values. The green color infor-

---

[10]You may have learned that the primary "colors" are red, yellow, and blue. Technically, these are magenta, yellow, and cyan and usually refer to pigments rather than colors. A magenta ink absorbs green light, thus controlling the amount of green color perceived by the eye. Since white light is composed of equal amounts of all three colors, removing green leaves red and blue, which together form magenta. Likewise, yellow ink absorbs blue light, and cyan ink absorbs red light. Reflected light, such as that from a painting, is formed from the character of the illuminating light and the absorption of the pigments. Projected light, such as that from a television, is formed from the intensities of the three primary colors. Since video is projected, it deals with red, green, and blue colors.

[11]The use of Y for luminance comes from the XYZ color system defined by the Commission Internationale de L'Eclairage (CIE). The system uses three-dimensional space to represent colors, where the Y axis is luminance and X and Z axes represent color information.

[12]Luminance from RGB can be a difficult concept to grasp. It may help to think of colored filters. If you look through a red filter, you will see a monochromatic image composed of shades of red. The image would look the same through the red filter if it were changed to a different color, such as gray. Since the red filter only passes red light, anything that's pure blue or pure green won't be visible. To get a balanced image, you would use three filters, change the image from each one to gray, and average them together.

mation can be recreated by subtracting the two difference signals from the Y signal (roughly speaking). Different weightings are used to derive Y and color differences from RGB, such as YUV, YIQ, and $Y'C_bC_r$. DVD uses $Y'C_bC_r$ as its native storage format. (Details of the data variations are beyond the scope of this book.)

The sensitivity of the eye is not linear, and neither is the response of the phosphors used in television tubes. Therefore, video is usually represented with corresponding nonlinear values, and the terms *luma* and *chroma* are used. These are denoted with the prime symbols as Y' and C', as is the corresponding R'G'B'. (Details of nonlinear functions are also beyond the scope of this book.)

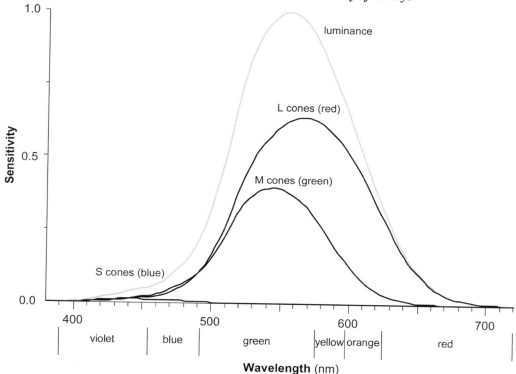

Figure 2.5 *Color and Luminance Sensitivity of the Eye*

Compressing Single Pictures

An understanding of the nuances of human perception led to the development of compression techniques that take advantage of certain characteristics. Just such a development is *JPEG compression*, which was produced by the *Joint Photographic Experts Group* and is now a worldwide standard. JPEG separately compresses Y, B-Y, and R-Y information, with more compression done on the latter two, to which the eye is less sensitive.

Technology Primer

To take advantage of another human vision characteristic — less sensitivity to complex detail — JPEG divides the image into small blocks and applies a *discrete cosine transform* (DCT) to the blocks, which is a mathematical function that changes spatial intensity values to spatial frequency values. This describes the block in terms of how much detail changes, and roughly arranges the values from lowest frequency (represented by large numbers) to highest frequency (represented by small numbers). For areas of smooth colors or low detail (low spatial frequency), the numbers will be large. For areas with varying colors and detail (high spatial frequency), most of the values will be close to zero. A DCT is an essentially *lossless transform*, meaning that an inverse DCT function can be performed on the resulting set of values to restore the original values. In practice, integer math and approximations are used, causing some loss at the DCT stage. Ironically, the numbers are bigger after the DCT transform. The solution is to *quantize* the DCT values so that they become smaller and repetitive.

Quantizing is a way of reducing information by grouping it into chunks. For example, if you had a set of numbers between 1 and 100, you could quantize them by 10. That is, you could divide them by 10 and round to the nearest integer. The numbers from 5 to 14 would all become 1s, the numbers from 15 to 24 would become 2s, and so on, with 1 representing 10, 2 representing 20, and so forth. Instead of individual numbers such as 8, 11, 12, 20, and 23, you end up with "3 numbers near 10" and "2 numbers near 20." Obviously, quantizing results in a loss of detail.

Quantizing the DCT values means that the result of the inverse DCT will not exactly reproduce the original intensity values, but the result is close and can be adjusted by varying the quantizing scale to make it finer or coarser. More importantly, as the DCT function includes a progressive weighting that puts bigger numbers near the top left corner and smaller numbers near the lower right corner, quantization and a special zigzag ordering result in runs of the same number, especially zero. This may sound familiar. Sure enough, the next step is to use run-length encoding to reduce the number of values that need to be stored. A variation of run-length coding is used, which stores a count of the number of zero values followed by the next nonzero value. The resulting numbers are used to look up symbols from a table. The symbol table was developed using Huffman coding to create shorter symbols for the most commonly appearing numbers. This is called *variable-length coding* (VLC). See Figures 2.6 and 2.8 for examples of DCT, quantization, and VLC.

The result of these transformation and manipulation steps is that the information that is thrown away is least perceptible. Since the eye is less sensitive to color than to brightness, transforming RGB values to luminance and chrominance values means that more chrominance data can be selectively thrown away. And since the eye is less sensitive to high-frequency color or brightness changes, the DCT and quantization process removes mostly the high-frequency information. JPEG compression can reduce picture data to about one-fifth the original size with almost no discernible difference and to about one-tenth the original size with only slight degradation.

Technology Primer

Figure 2.6 Block Transforms and Quantization

Encoding

Macroblock (luma)

Pixel values

134	142	145	131	114	122	131	130
129	143	134	130	135	144	134	118
123	117	118	111	97	109	130	143
129	116	112	116	120	126	130	118
118	127	141	138	138	148	141	125
125	129	119	127	143	149	145	136
131	126	128	142	141	135	126	116
131	140	146	154	133	118	124	124

DCT coefficients

+1037	-1	-6	+1	-12	+8	-4	-4
-16	+1	+28	-6	-14	0	+4	0
+19	+32	-7	-19	+2	-1	-4	-3
+29	-9	-14	+13	-10	-6	+1	0
+4	+14	-6	-13	-2	+7	+1	+2
-26	+2	+16	+2	+11	+6	+1	+1
-10	-11	+27	-18	+4	+1	0	0
-2	+1	+1	-19	-1	+6	+6	0

Quantized coefficients
$Q^{-}(DCT*16)/(QM*8)$

+130	0	0	0	-1	0	0	0
-2	0	+3	0	-1	0	0	0
+2	+3	0	-1	0	0	0	0
+3	-1	-1	+1	0	0	0	0
0	+1	0	-1	0	0	0	0
-2	0	+1	0	0	0	0	0
-1	-1	+2	-1	0	0	0	0
0	0	0	-1	0	0	0	0

Decoding

Reconstructed coefficients (dequantize)

+1040	0	0	0	-9	0	0	0
-12	0	+24	0	-10	0	0	0
+14	+24	0	-10	0	0	0	0
+24	-8	-9	+10	0	0	0	0
0	+9	0	-10	0	0	0	0
-19	0	+10	0	0	0	0	0
-9	-10	+21	-12	0	0	0	0
0	0	0	-14	0	0	0	0

Reconstructed values (IDCT)

136	141	138	125	119	125	132	134
136	137	133	130	134	139	133	121
121	125	122	112	107	117	130	138
125	123	119	117	121	128	127	122
123	129	135	139	140	139	136	131
129	125	124	130	139	144	141	136
129	130	134	139	140	135	127	120
132	138	144	141	131	122	122	127

Difference

-2	+1	+7	+6	-5	-3	-1	-4
-7	+6	+1	+0	+1	+5	+1	-3
+2	-8	-4	-1	-10	-8	0	+5
+4	-7	-7	-1	-1	-2	+3	-4
-5	-2	+6	-1	-2	+9	+5	-6
-4	+4	-5	-3	+4	+5	+4	0
+2	-4	-6	+3	+1	0	-1	-4
-1	+2	+2	+13	+2	-4	+1	-3

Quantization matrix

8	16	19	22	26	27	29	34
16	16	22	24	27	29	34	37
19	22	26	27	29	34	34	38
22	22	26	27	29	34	37	40
22	26	27	29	32	35	40	48
26	27	29	32	35	40	48	58
26	27	29	34	38	46	56	69
27	29	35	38	46	56	69	83

Technology Primer

Compressing Moving Pictures

Motion video adds a *temporal* dimension to the spatial dimension of single pictures. Another worldwide compression standard, *MPEG*, from the *Moving Pictures Expert Group*, was designed with this in mind. MPEG is similar to JPEG but also reduces redundancy between successive pictures of a moving sequence.

Just as your friend's memory allows you to describe things once and then only talk about what is changing, digital memory allows video to be compressed in a similar manner by first storing a single picture and then only storing the changes. For example, if the bird moves to another tree, you can tell your friend that the bird has moved without needing to describe the bird over again.

MPEG compression uses a similar technique called *motion estimation* or *motion-compensated prediction*. As motion video is a sequence of still pictures, many of which are very similar, and each picture can be compared with the pictures next to it, the MPEG encoding process breaks each picture into blocks, called *macroblocks*, and then hunts around in neighboring pictures for similar blocks. If a match is found, instead of storing the entire block, the system stores a much smaller *vector* describing how far the block moved (or did not move) between pictures. Vectors can be encoded in as little as one bit, so backgrounds and other elements that do not change over time are compressed most efficiently. Large groups of blocks that move together, such as large objects or the entire picture panning sideways, are also compressed efficiently.

MPEG uses three kinds of picture storage methods. *Intra* pictures are like JPEG pictures, in which the entire picture is compressed and stored with DCT quantization. This creates a reference frame from which successive pictures are built. These *I frames* also allow random access into a stream of video, and in practice occur about twice a second. *Predicted* pictures, or *P frames*, contain motion vectors describing the difference from the closest previous I frame or P frame. If the block has changed slightly in intensity or color (remember, frames are separated into three channels and compressed separately), then the difference (*error*) is also encoded. If something entirely new appears that does not match any previous blocks, such as a person walking into the scene, then a new block is stored in the same way as in an I frame. If the entire scene changes, as in a cut, the encoding system is usually smart enough to make a new I frame. The third storage method is a *bidirectional* picture, or *B frame*. The system looks both forward and backward to match blocks. In this way, if something new appears in a B frame, it can be matched to a block in the next I frame or P frame. Thus P and B frames are much smaller than I frames.

Experience has shown that two B frames between each I or P frame work well. A typical second of MPEG video at 30 frames per second looks like I B B P B B P B B P B B P B B I B B P B B P B B P B B (see Figure 2.7). B frames are more complex to create than P frames, requiring time-consuming searches in both the previous and subsequent I or P frame. For this reason, some realtime or low-cost MPEG encoders only create I and P frames. Likewise, I frames are easier to create than P frames, which require searches in the subsequent I or P frame. Therefore, the simplest encoders only create I frames. This is less efficient but may be necessary for very inexpensive realtime encoders that must process 30 or more frames a second.

Figure 2.7 Typical MPEG Picture Sequence

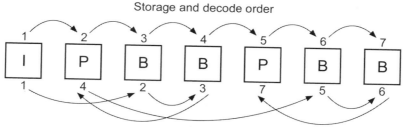

MPEG-2 encoding can be done in realtime (where the video stream enters and leaves the encoder at display speeds), but it is difficult to produce quality results, especially with *variable bit rate* (VBR) encoding. Variable bit rate allows varying numbers of bits to be allocated for each frame depending on the complexity of the frame. Less data is needed for simple scenes, while more data can be allocated for complex scenes. This results in a lower average data rate but provides room for data peaks to maintain quality. Encoding frequently is done with variable bit rate and is usually not done in realtime, so the encoder has plenty of time for macroblock matching, resulting in much better quality at lower data rates. Good encoders make one pass to analyze the video and determine the complexity of each frame, forcing I frames at scene changes and creating a compression profile for each frame. They then make a second pass to do the actual compression, varying quantization parameters to match the profiles. The human operator often tweaks minor details between the two passes. Many low-cost MPEG encoders, hardware or software, for personal computers use only I frames, especially when capturing video in realtime. This results in a simpler and cheaper encoder, since P and B frames require more computation and more memory to encode. Some of these systems can later reprocess the I frames to create P and B frames. MPEG also can encode still images as I frames. Still menus on a DVD, for example, are I frames.

The result of the encoding process is a set of data and instructions (see Figure 2.8). These are used by the decoder to recreate the video. The amount of compression (how coarse the quantizing steps are, how large a motion estimation error is allowed) determines how closely the reconstructed video resembles the original. MPEG decoding is *deterministic* — a given set of input data always should produce the same output data. Decoders that properly implement the complete MPEG decoding process will produce the same numerical picture even if they are built by different manufacturers.[13] This doesn't mean that all DVD players will produce the same video picture. Far from it, since many other factors are involved, such as conversion from digital to analog, connection type, cable quality, and display quality. Advanced decoders may include extra processing steps such as block filtering and edge enhancement. Also, many software MPEG decoders take shortcuts to achieve sufficient performance.

---

[13]Technically, the *inverse discrete cosine transform* (IDCT) stage of the decoding process is not strictly prescribed, and is allowed to introduce small statistical variances. This should never account for more than an occasional least significant bit of discrepancy between decoders.

Technology Primer

Software decoders may skip frames and use mathematical approximations rather than the complete but time-consuming transformations. This results in lower-quality video than from a fully compliant decoder.

Figure 2.8 MPEG Video Compression Example

DC	symbols
130	01 0

run codes (zeros, value)	symbols (from lookup table)
1,-2	0011 01
0,+2	1100
3,+3	0000 0001 1100 0
0,+3	0111 0
0,+3	0111 0
1,-1	0101
2,-1	0010 11
1,-1	0101
0,-1	101
0,-1	101
0,+1	100
0,-2	1101
0,-1	101
2,+1	0010 10
7,-1	0000 1001
0,+1	100
0,-1	101
2,+2	0000 1110
9,-1	1111 0001
1,-1	0101
EOB	0110

bytestream (14 bytes: 22 symbols)
46 E0 0E 1C E5 2D 6D 9B 4A 09
94 3B C5 58

compression
4.6:1 (64:14)

pixel values:

134	142	145	131	114	122	131	130
129	143	134	130	135	144	134	118
123	117	118	111	97	109	130	143
129	116	112	116	120	126	130	118
118	127	141	138	138	148	141	125
125	129	119	127	143	149	145	136
131	126	128	142	141	135	126	116
131	140	146	154	133	118	124	124

quantized coefficients:

+130	0	0	0	-1	0	0	0
-2	0	+3	0	-1	0	0	0
+2	+3	0	-1	0	0	0	0
+3	-1	-1	+1	0	0	0	0
0	+1	0	-1	0	0	0	0
-2	0	+1	0	0	0	0	0
-1	-1	+2	-1	0	0	0	0
0	0	0	-1	0	0	0	0

zig-zag scan sequence:

1	5	6	14	15	27	28
2	4	7	13	16	26	29	42
3	8	12	17	25	30	41	43
9	11	18	24	31	40	44	53
10	19	23	32	39	45	52	54
20	22	33	38	45	51	55	60
21	34	37	47	50	56	59	61
35	36	48	49	57	58	62	63

Encoders, on the other hand, can and do vary widely. The encoding process has the greatest effect on the final video quality. The MPEG standard prescribes a *syntax* defining what instructions can be included with the encoded data and how they are applied. This syntax is quite flexible and leaves considerable room for variation. The quality of the decoded video depends on how thoroughly the encoder examines the video and how cleverly it goes about applying the functions of MPEG to compress it. Video quality steadily improves as encoding techniques and equipment get better. The decoder chip in the player will not change, but the improvements in the encoded data will provide a better result. This can be likened to read-

Technology Primer

ing aloud from a book. The letters of the alphabet are like data organized according to the syntax of language. The person reading aloud from the book is similar to the decoder — the reader knows every letter and is familiar with the rules of pronunciation. The author is similar to the encoder — the writer applies the rules of spelling and usage to encode thoughts as written language. The better the author, the better the results. A poorly written book will come out sounding bad no matter who reads it, but a well-written book will produce eloquent spoken language.[14]

It should be recognized that random artifacts in video playback (aberrations that appear in different places or at different times when the same video is played over again) are not MPEG encoding artifacts. They may indicate a faulty decoder, errors in the signal, or something else independent of the MPEG encode-decode process. It is impossible for a fully compliant, properly functioning MPEG decoder core module to produce visually different results from the same encoded data stream, although in practice there are variations in implementation along with added video processing steps that make the output from every player or decoding software application look different.

MPEG (and most other compression techniques) are *asymmetric*, meaning that the encoding process does not take the same amount of time as the decoding process. It is more effective and efficient to use a complex and time-consuming encoding process because video generally is encoded only once before being decoded hundreds or millions of times. High-quality MPEG encoding systems can cost tens of thousands dollars, but since most of the work is done during encoding, decoder chips cost less than $20, and decoding can even be done in software.

Some analyses indicate that a typical video signal contains over 95 percent redundant information. By encoding the changes between frames, rather than re-encoding each frame, encoders can achieve amazing compression ratios (see Table 2.2).

Advanced Video Codecs

There are two new compression/decompression (codec) technologies that are part of the specification for Blu-ray Disc, and they are considered to be almost twice as efficient as MPEG-2. The first of these codecs, *SMPTE VC-1*, is a video codec defined by the *Society of Motion Picture Television Engineers* (SMPTE) as an "open standard" and is based on Microsoft's Windows Media Video 9 technology. Development for Windows Media began years after the creation and standardization of MPEG-2 and has the benefit of building on the lessons learned there. Windows Media (and now VC-1) have demonstrated dramatic improvements in compression efficiency over MPEG-2 and MPEG-1 codecs. For example, prior to adoption of the codec in either HD DVD or BD, Microsoft began promoting a proprietary Windows Media Video HD (WMV-HD) format that showcased Windows Media Video by putting high definition content onto standard red-laser DVDs. The content, at res-

---

[14] Obviously, it would sound better if read by James Earl Jones than by Ross Perot, but the analogy holds if you consider the vocal characteristics to be independent of the translation of words to sound. The brain of the reader is the decoder, the diction of the reader is the post-MPEG video processing, and the voice of the reader is the television.

Technology Primer

olutions of 1280 × 720 × 24 fps and 1920 × 1080 × 24 fps, was encoded at constant data rates from 6.0 to 12.0 Mbps for playback on Windows Media-capable PCs. Few personal computers at the time had the processing power needed to present the full-size HD video at a consistent frame rate, but the WMV-HD format did go a long way to establish Windows Media, and later VC-1, as a serious video codec.

Taking a more traditional development path, *MPEG-4 Advanced Video Codec* (AVC), also known as *ITU H.264 Part 10*, was developed by the Joint Video Team (JVT), a co-operation between MPEG (under the ISO) and the ITU-T.[15] Designed with the intention of improving coding efficiency of video by a factor of two or more, uses for H.264 span multimedia applications ranging from video on cell phones to high definition televisions, using a dramatic range of data rates and screen resolutions. In the evaluation of advanced codecs performed by both the DVD Forum and the Blu-ray Disc Founders group, MPEG-4 AVC quickly established itself as the leader and was the first advanced codec to be accepted into the Blu-ray format. The AVC codec includes in its toolset of compression algorithms a wide range of options, allowing AVC encoders to achieve high quality at lower data rates.

Both VC-1 and AVC push asymmetric encoding/decoding to extremes, with first generation implementations of the encoders requiring approximately 30 to 50 times more processing power than equivalent MPEG-2 encoders. Likewise, compared to MPEG-2, these advanced video codecs require significantly more processing power when decoding, but there the difference is perhaps only a factor of 4 to 1.

These advanced codecs achieve a 50 percent reduction in the size of a video stream. Table 2.2 shows some comparative data rates for the compression technologies.

Table 2.2 Compression Ratios for Disc Technologies

Native data	Native rate (kbps)	Compression	Average Compressed Rate (kbps)[a]	Ratio	Percent
Video					
720 × 576 × 25 fps × 12 bits	124,416	MPEG-2	5,400	23:1	96
720 × 576 × 24 fps × 12 bits	119,439	MPEG-2	5,200	23:1	96
720 × 480 × 30 fps × 12 bits	124,416	MPEG-2	5,400	23:1	96
720 × 480 × 24 fps × 12 bits	99,533	MPEG-2	4,400	23:1	96
1920 × 1080 × 30 fps × 12 bits	746,496	MPEG-2/AVC/VC1	25,000	30:1	97
1920 × 1080 × 30 fps × 12 bits	746,496	MPEG-2/AVC/VC1	18,500	40:1	98
1440 × 1080 × 30 fps × 12 bits	559,872	MPEG-2/AVC/VC1	11,200	50:1	98
1280 × 720 × 60 fps × 12 bits	663,552	MPEG-2/AVC/VC1	13,400	50:1	98
1920 × 1080 × 24 fps × 12 bits	597,197	MPEG-2/AVC/VC1	12,000	50:1	98

continues

[15]ITU stands for the International Telecommunications Union (the -T refers to the standardization Group. The Moving Pictures Experts Group (MPEG) falls under the International Organization for Standardization (ISO, whose name is not an acronym, but is derived from the Greek *isos*, meaning "equal"). Because the video codec was developed jointly, the resulting standard exists within both bodies. In the ITU-T, the codec is referred to as H.264. Under ISO/IEC, it is referred to as MPEG-4 Part 10 Advanced Video Codec (AVC).

Technology Primer

Table 2.2 Compression Ratios for Disc Technologies (continued)

Native data	Native rate (kbps)	Compression	Average Compressed Rate (kbps)[a]	Ratio	Percent
Video (continued)					
1440 × 1080 × 24 fps × 12 bits	447,898	MPEG-2/AVC/VC1	9,000	50:1	98
1280 × 720 × 24 fps × 12 bits	265,421	MPEG-2/AVC/VC1	5,300	50:1	98
720 × 480 × 30 fps × 12 bits	124,416	MPEG-2/AVC/VC1	2,700	46:1	98
720 × 576 × 25 fps × 12 bits	124,416	MPEG-2/AVC/VC1	2,700	46:1	98
720 × 480 × 24 fps × 12 bits	99,533	MPEG-2/AVC/VC1	2,200	45:1	98
Audio					
2 ch × 48 kHz × 16 bits	1,536	Dolby Digital 2.0	192	8:1	87
6 ch × 48 kHz × 16 bits	4,608	Dolby Digital 5.1	384	12:1	92
6 ch × 48 kHz × 16 bits	4,608	Dolby Digital 5.1	448	10:1	90
6 ch × 48 kHz × 24 bits	6,912	Dolby Digital 5.1[b]	448	15:1	94
6 ch × 48 kHz × 16 bits	4,608	DTS 5.1	768	6:1	83
6 ch × 48 kHz × 16 bits	4,608	DTS 5.1	1,536	3:1	67
6 ch × 96 kHz × 20 bits	11,520	MLP[c]	5,400	2:1	53
6 ch × 96 kHz × 24 bits	13,824	MLP[c]	7,600	2:1	45
2 ch × 48 kHz × 16 bits	1,536	Dolby Digital Plus[b]	192	8:1	88
6 ch × 48 kHz × 16 bits	4,608	Dolby Digital Plus[b]	448	10:1	90
8 ch × 48 kHz × 16 bits	6,144	Dolby Digital Plus[b]	896	7:1	85
8 ch × 48 kHz × 24 bits	9,216	Dolby Digital Plus[b,c]	896	10:1	90
2 ch × 192 kHz × 24 bits	9,216	DTS-HD[d]	4,900	2:1	47
6 ch × 48 kHz × 16 bits	4,608	DTS-HD[e]	1,717	3:1	63
6 ch × 48 kHz × 16 bits	4,608	DTS-HD[f]	2,044	2:1	56
6 ch × 48 kHz × 24 bits	6,912	DTS-HD[f]	3,954	2:1	43
6 ch × 96 kHz × 24 bits	13,824	DTS-HD[f]	8,198	2:1	41
6 ch × 48 kHz × 16 bits	4,608	Dolby TrueHD[g]	1,450	3:1	69
6 ch × 48 kHz × 24 bits	6,912	Dolby TrueHD[g]	3,440	2:1	50
6 ch × 96 kHz × 24 bits	13,824	Dolby TrueHD[g]	5,300	3:1	62

[a] MPEG-2 and MLP compressed data rates are an average of a typical variable bit rate.

[b] Dolby Digital Plus offers improved performance over a wider range of data rates and channel configurations compared to Dolby Digital. However, recommended data rates remain unchanged for these 2- and 6-channel configurations.

[c] As with Dolby Digital, source word lengths of 16, 20 and 24 bits all result in the same output rate.

continues

Table 2.2 Compression Ratios for Disc Technologies (continued)

ᵈLosslessly encoded with an embedded DTS 2.0 core compressed at 384 kbps. Note that the compressed rate of DTS-HD streams is heavily dependent on the source material.

ᵉLosslessly encoded with an embedded DTS 5.1 core compressed at 768 kbps. Note that the compressed rate of DTS-HD streams is heavily dependent on the source material.

ᶠLosslessly encoded with an embedded DTS 5.1 core compressed at 1,509 kbps. Note that the compressed rate of DTS-HD streams is heavily dependent on the source material.

ᵍNote that the compressed rate of Dolby TrueHD streams is heavily dependent on the source material. Values shown here represent movie source material, which can typically be more highly compressed than music.

Birds Revisited: Understanding Audio Compression

Audio takes up much less space than video, but uncompressed audio coupled with compressed video uses up a large percentage of the available bandwidth. Compressing the audio can result in a small loss of quality, but if the resulting space is used instead for video, it may improve the video quality. In essence, reducing both the audio and the video is usually most effective. Usually video is compressed more than audio, since the ear is more sensitive to detail loss than the eye.

Just as MPEG compression takes advantage of characteristics of the human eye, modern audio compression relies on detailed understanding of the human ear. This is called *psychoacoustic* or *perceptual coding*.

Picture, again, your telephone conversation with a friend. Imagine that your friend lives near an airport, so that when a plane takes off, your friend cannot hear you over the sound of the airplane. In a situation like this, you quickly learn to stop talking when a plane is taking off, since your friend won't hear you. The airplane has *masked* the sound of your voice. At the opposite end of the loudness spectrum from airplane noise is background noise, such as, a ticking clock. While you are speaking, your friend can't hear the clock, but if you stop, then the background noise is no longer masked.

The hairs in your inner ear are sensitive to sound pressure at different frequencies (pitches). When stimulated by a loud sound, they are incapable of sensing softer sounds at the same pitch. Because the hairs for similar frequencies are near each other, a stimulated audio receptor nerve will interfere with nearby receptors and cause them to be less sensitive. This is called *frequency masking*.

Human hearing ranges roughly from low frequencies of 20 Hz to high frequencies of 20,000 Hz (20 kHz). The ear is most sensitive to the frequency range from about 2 to 5 kHz, which corresponds to the range of the human voice. Because aural sensitivity varies in a nonlinear fashion, sounds at some frequencies mask more neighboring sounds than at other frequencies. Experiments have established certain *critical bands* of varying size that correspond to the masking function of human hearing (see Figure 2.9).

Technology Primer

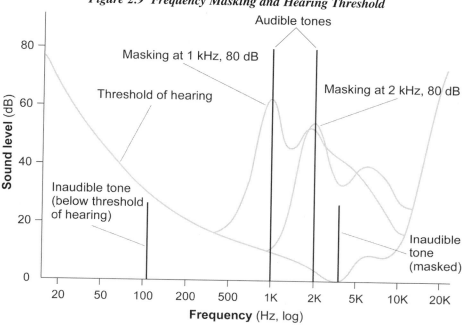

Figure 2.9 Frequency Masking and Hearing Threshold

Another characteristic of the human audio sensory system is that sounds cannot be sensed when they fall below a certain loudness (or amplitude). This sensitivity threshold is not linear. In other words, the threshold is at louder or softer points at different frequencies. The overall threshold varies a little from person to person — some people have better hearing than others. The threshold of hearing is adaptive; the ear can adjust its sensitivity in order to pick up soft sounds when not overloaded by loud sounds. This characteristic causes the effect of *temporal masking*, in which you are unable to hear soft sounds for up to 200 milliseconds after a loud sound and for 2 or 3 milliseconds before a loud sound.[16]

Perceptual Coding

Standard definition DVD uses three audio data reduction systems — Dolby Digital (AC-3) coding, MPEG audio coding, and DTS (Coherent Acoustics) coding. Blu-ray Disc uses two of those — Dolby Digital (AC-3), and DTS — and adds several audio codecs to the collection — Dolby Digital Plus, Dolby True HD, DTS-HD High Resolution, and DTS-HD Master Audio.[17] All use mathematical models of human hearing based on sensitivity thresh-

---

[16]How can masking work backward in time? The signal presented by the ear to the brain is a composite built up from stimuli received over a period of about 200 milliseconds. A loud noise effectively overrides a small portion of the earlier stimuli before it can be accumulated and sent to the brain.

[17]DTS-HD Master Audio is an optional audio codec for Blu-ray Disc. Blu-ray Disc can also use Linear PCM (Pulse Code Modulation) as an audio codec. As Linear PCM is regarded as uncompressed, we chose to not list it here, to avoid arguments about whether it does or does not reduce audio data. But we'll probably get an argument, anyways. Ah, well...

Technology Primer

olds, frequency masking, and temporal masking to remove sounds that you cannot hear. The resulting information is compressed to about one-third to one-twelfth the original size with little to no perceptible loss in quality (see Table 2.2).

Digital audio is *sampled* by taking snapshots of an analog signal thousands of times a second. Each sample is a number that represents the amplitude (strength) of the waveform at that instance in time. Perceptual audio compression takes a block of samples and divides them into frequency bands of equal or varying widths. Bands of different widths are designed to match the sensitivity ranges of the human ear. The intensity of sound in each band is analyzed to determine two things: (1) how much masking it causes in nearby frequencies and (2) how much noise the sound can mask within the band. Analyzing the masking of nearby bands means that the signal in bands that are completely masked can be ignored. Calculating how much noise can be masked in each band determines how much compression can be applied to the signal within the band. Compression uses *quantization*, which involves dividing and rounding, and this can create errors known as *quantization noise*. For example, the number 32 quantized by 10 gives 3.2 rounded to 3. When re-expanded, the number is reconstructed as 30, creating an error of 2. These errors can manifest themselves as audible noise. After masked sounds are ignored, remaining sounds are quantized as coarsely as possible so that quantization noise is either masked or is below the threshold of hearing. The technique of noise masking is related to *noise shaping* and is sometimes called *frequency-domain error confinement*.

Another technique of audio compression is to compare each block of samples with the preceding and following blocks to see if any can be ignored on account of temporal masking — soft sounds near loud sounds — and how much quantization noise will be temporally masked. This is sometimes called *temporal-domain error confinement*.

Digital audio compression also can take advantage of the redundancies and relationships between channels, especially when there are six or eight channels. A strong sound in one channel can mask weak sounds in other channels, information that is the same in more than one channel need only be stored once, and extra bandwidth can be temporarily allocated to deal with a complex signal in one channel by slightly sacrificing the sound of other channels.

Audio compression techniques result in a set of data that is processed in a specific way by the decoder. The form of the data is flexible, so improvements in the encoder can result in improved quality or efficiency without changing the decoder. As understanding of psychoacoustic models improves, perceptual encoding systems can be improved.

At minimal levels of compression, perceptual encoding removes only the imperceptible information and provides decoded audio that is virtually indistinguishable from the original. At higher levels of compression, and depending on the nature of the audio, *bit starvation* may produce identifiable effects of compression. These include a slightly harsh or gritty sound, poor reproduction of transients, loss of detail, and less pronounced separation and spaciousness.

Many listeners claim that DTS audio quality is better than Dolby Digital, but such claims are rarely based on accurate comparisons, and may be another form of perceptual influence — that of assuming higher bitrates means higher quality. Within the confines of any given

Technology Primer

codec, that is usually true, but it cannot be assumed when comparing codecs of differing efficiencies — something that has become abundantly clear when comparing MP3s to iTunes AAC recordings. DTS tracks are usually encoded at a reference volume level that is 4 dB higher, and most DTS soundtracks are mixed differently than their Dolby Digital counterparts, including different volume levels in the surround and LFE tracks. This makes it nigh impossible to compare them objectively, even using a disc that contains a soundtrack in both formats. The limited semi-scientific comparisons that have been done indicate that there is little perceptible difference between the two and that any difference may be noticeable only on high-end audio systems.

Linear Pulse Code Modulation (PCM)

Linear PCM is the "gold standard" against which all the other audio codecs are compared. This is because it represents uncompressed digital audio. Although, to say that Linear PCM is uncompressed is a bit of a misnomer since, in fact, the audio data it represents is based on quantized samples taken at regular periods across a given time interval, therefore some information is lost in the process, depending on the bit depth and sampling frequency compared to the original source. However, the 16-, 20- and 24-bit sample sizes and 48 to 192 kHz sampling frequencies are generally considered to capture all of the fidelity that the human ear is capable of hearing. So, we come back to the statement that it is uncompressed. In any case, it is enough to say that Linear PCM is the starting point from which all of the other audio codecs begin. Any divergence from the original set of Linear PCM samples would, therefore, be considered error, noise, or distortion.

So, if Linear PCM is the gold standard, then why not just use it exclusively? The simple answers are file size and bandwidth considerations. Large files generated by Linear PCM are a concern, but it is the data rate of the format that primarily impacts the decision to use it or some other audio format. Capturing Linear PCM audio involves taking a 16- to 24-bit sample of each audio channel at anywhere from 48,000 to 192,000 times each second. For stereo audio, that corresponds with 1.536 to 9.216 million bits per second (Mbps). For standard 5.1 surround sound audio, the numbers jump up to 4.608 to 27.648 Mbps. Given that the maximum tramsport stream data rate for Blu-ray playout is 48 Mbps, it is easy to see that with Linear PCM there may be insufficient capacity remaining for the video component.

Many of the first Blu-ray movie titles released in 2006 included Linear PCM surround sound audio tracks as a standard feature, although those discs had few other features since there was no space left for them on the single-layer 25 Gbyte discs used at that time.

Nonetheless, Linear PCM files are incredibly large, especially when compared to compressed counterparts from Dolby and DTS. For applications in which the highest audio quality is required, Linear PCM should be considered. However, for other applications in which audio is not the central concern and, perhaps, disc space or other features may be of greater importance, using one of the compressed audio options is a very attractive alternative.

Blu-ray Disc Demystified

Technology Primer

Dolby Digital (AC-3) Audio Coding

Dolby Digital audio compression (known as AC-3 in standards documents) provides for up to 5.1 channels of discrete audio.[18] One of the advantages of Dolby Digital is that it analyzes the audio signal to differentiate short, transient signals from long, continuous signals. Short sample blocks are then used for short sounds and long sample blocks are used for longer sounds. This results in smoother encoding without transient suppression and block boundary effects that can occur with fixed block sizes. For compatibility with existing audio/video systems, Dolby Digital decoders can downmix multichannel programs to ensure that all the channels are present in their proper proportions for mono, stereo, or Dolby Pro Logic reproduction.

Dolby Digital uses a frequency transform — somewhat like the DCT transform of JPEG and MPEG — and groups the resulting values into frequency bands of varying widths to match the critical bands of human hearing. Each transformed block is converted to a floating-point frequency representation that is allocated a varying number of bits from a common pool, according to the importance of the frequency band. The result is a *constant bitrate* (CBR) data stream.

Dolby Digital also includes *dynamic range* information so that different listening environments can be compensated for. Original audio mixes, such as movie sound tracks, which are designed for the wide dynamic range of a theater, can be encoded to maintain the clarity of the dialogue and to enable emphasis of soft passages when played at low volume in the home.

Dolby Digital was developed from the ground up as a multichannel coder designed to meet the diverse and often contradictory needs of consumer delivery. It also has a significant lead over other multichannel systems in both marketing and standards adoption. Millions of Dolby Digital decoders are in living rooms, and Dolby Digital was chosen for the U.S. DTV (digital television) standard and is being used for digital satellite systems and most other digital television systems.

Dolby Digital Plus (DD+)

Dolby Digital Plus, also known as Enhanced AC-3, is an extension of Dolby Digital that adds a number of features to the already popular codec. In particular, DD+ provides spectral coding techniques that help to improve audio reproduction at very low data rates (such as, for streaming applications), while adding a mechanism for supporting more discrete channels and higher data rates.

Unfortunately, Dolby Digital Plus comes with a few problems, as well. First, legacy A/V receivers that support Dolby Digital via digital connections (e.g., S/PDIF) will not natively support the decoding of Dolby Digital Plus. Fortunately, Dolby Digital Plus can address this by down converting to legacy Dolby Digital at a high data rate (640 kbps) to preserve as much of the DD+ quality as possible. If the Dolby Digital Plus stream to be converted contains more than 5.1 channels of audio, the channels beyond 5.1 are discarded during the conver-

---

[18]Dolby Digital can support up to 6.1 channels by matrixing an additional rear center channel into the left and the right surround channels, which may be reproduced with one or more rear center speakers.

Technology Primer

sion process, but due to the structure of the DD+ bitstream, the content from those channels is retained in the converted Dolby Digital stream by means of a downmix process that occurs in the Dolby Digital Pus encoder during the production process. However, only the 5.1 downmix of streams originally with 6.1 or 7.1 channels will be heard.

Second, most of the extra features of Dolby Digital Plus have the greatest effect at very low encoding rates rather than at the data rates one would typically expect on a Blu-ray disc. DD+ is able to retain its level of perceptual quality at lower data rates, but in the moderate and higher data rate ranges that are more consistent with what would be expected to be used on a BD disc, the difference in quality between AC-3 and DD+ is actually quite small. Even though DD+ can support data rates as high as 4.7 Mbps on BD, such high rates would offer little perceived improvement to the audio experience in a 5.1-channel surround environment.

Dolby TrueHD

For those who have followed DVD-Audio, Dolby TrueHD is a familiar technology under a different name. Dolby TrueHD is, essentially, the marketing name Dolby Laboratories uses for Meridian Lossless Packing (MLP), the lossless compression scheme developed by Meridian Audio, Ltd., but licensed by Dolby Labs. MLP was the codec of choice preferred by the recording industry for providing lossless audio compression on formats such as DVD-Audio. Capable of delivering the exact same quality as uncompressed Linear PCM audio but usually at much better compression rates, Dolby TrueHD provides an excellent alternative in cases where disc space or bandwidth are a concern.

One important advantage that losslessly compressed codecs, such as, Dolby TrueHD and DTS-HD Master Audio (defined in the following section) have over Linear PCM is the ability to include a lossless downmix in the stream. For example, an 8-channel Linear PCM stream contains eight discrete audio channels that are all expected to be heard by the listener. However, if your sound system only has six or even just two speakers, you will not be able to hear the full Linear PCM presentation. A Dolby TrueHD 8-channel stream, however, is required to also include both a 6-channel and a 2-channel downmix, each of which is losslessly encoded. This has almost no impact on the data rate and file size, but provides for a greatly improved listening experience no matter how many speakers you have.

One difficulty with Dolby TrueHD, however, is that it is a *variable bitrate* (VBR) coding scheme. That means the data rate of the stream fluctuates over time depending on the complexity of the audio being compressed. This can be a problem when the variable bitrate audio is combined with variable bitrate video, which can lead to unexpected multiplexing issues due to peak data rate overruns. In areas with intricate high frequency detail, which is difficult to encode, the data rate of a stream may reach levels as high as an uncompressed Linear PCM stream, while in other areas it may be dramatically lower. Although the Dolby TrueHD encoder uses a complex scheme to minimize peak data rates, some peaks may be unavoidable. The operator cannot simply specify data rate limits as one would do for variable bitrate video because doing so would require changing the audio data itself, making it no longer lossless. Because Dolby TrueHD is a lossless compression method (a key difference from VBR video encoding), it can only ensure a lossless encode by being allowed to raise the data rate

Technology Primer

to encode all of the audio perfectly. What's more, the operator cannot predict what the range of required data rates will be until *after* the stream has been encoded since it is entirely content dependent. This can turn the disc production process upside down if you do not know how much bandwidth will be available for your video until after you are done encoding all of your audio.

DTS and DTS-HD Audio Coding

Digital Theater Systems (DTS) Digital Surround uses the Coherent Acoustics differential subband perceptual audio transform coder, which is similar to the Dolby Digital audio coder. It uses polyphase filters to break the audio into subbands (usually 32) of varying bandwidths and then uses ADPCM (Adaptive Differential Pulse Code Modulation) to compress each subband. The ADPCM step is a linear predictive coding that "guesses" at the next value in the sequence and then encodes only the difference. The prediction coefficients are quantized based on psychoacoustic and transient analysis and then are variable-length coded using entropy tables.

DTS decoders include downmixing features using either preset downmix coefficients or custom coefficients embedded in the stream. Dynamic range control and other user data can be included in the stream to be used in post-decoder processes.

The DTS Coherent Acoustics format used for discs is different from the format used in theaters, which is Audio Processing Technology's apt-X, a straight ADPCM coder with no psychoacoustic modeling.

An addition to the statistically lossless group of codecs is the DTS-HD variable bitrate (VBR) codec. Built on the existing DTS foundation, the DTS-HD codec takes a backward compatible, hierarchical approach to implementing lossless compression. The encoded data is composed of a normal DTS core plus an additional extension to achieve lossless compression.

Both DTS and DTS-HD are built upon the same Coherent Acoustics technology, which defines a "core + extension" structure for audio coding. The core component is typically composed of either 768 kbps or 1.509 Mbps 5.1-channel 48 kHz encoded audio. For DTS audio, the coded audio is composed of a single *core substream* that contains the core audio data plus zero or one optional extension. For DTS-HD, however, at least one *extension substream* exists, which can contain one or two extension units. A legacy DTS encoder only pays attention to the core substream, while a new DTS-HD decoder will use both the core substream and the extension substream to provide an enhanced experience. The core substream is essentially equivalent to the DTS audio that has been used in the past on DVD titles. The new extension substream in DTS-HD is what allows it to transcend even that high quality experience, providing additional channels (up to 7.1), lossless compression, increased sampling rates (96 kHz, or 192 kHz for lossless encoding), and additional data rates (up to 24.5 Mbps).

The extensions supported by the legacy DTS codec include XCH and X96. The DTS-HD codec adds XXCH, XBR and XLL.

Technology Primer

XCH

The *Channel Extension,* XCH, provides an additional discrete mono channel for output as a single or dual rear center channel, and is marketed under the name "DTS-ES®." Many legacy DTS decoders and A/V receivers support this extension when found in the core substream.

X96

The *Sampling Frequency Extension*, X96, also known as "Core+96k," provides a method for extending the sampling frequency of DTS audio from the standard 48 kHz up to 96 kHz. An encoder using this extension will start by performing a standard DTS 48 kHz encode of the core data, then decode that and subtract it from the original to determine which residuals were lost in the process. The X96 extension data is then generated by encoding these residuals. As with XCH, many legacy decoders and A/V receivers support this extension under the name "DTS-96/24™" ("24" refers to the 24-bit audio data that is output).

XXCH

Similar to the XCH Channel Extension, XXCH is a newer version that supports a greater number of additional discrete channels. For Blu-ray, however, the number of additional channels is limited to two, extending the core 5.1 channel configuration to 7.1 channels. Only new decoders and A/V receivers that support DTS-HD directly will recognize this extension.

XBR

Like XXCH, the *High Bit-Rate Extension*, XBR, is an extension for DTS that provides greater audio quality by increasing the available encoded data rate via the extension substream. In short, an encoder implementing the XBR extension would do so in a similar manner to X96, above. First, the data is compressed using the core encoder to form the core substream. It is then immediately decoded and subtracted from the original audio data to calculate the residuals that were "missed" by the core encoder. These residuals are then encoded into the XBR extension in the extension substream, resulting in an overall increase in data rate. When the core and XBR extension are later decoded and recombined, the result is a more accurate reproduction of the original audio data. Like XXCH, only newer decoders and A/V receivers that directly support DTS-HD will benefit from the XBR extension.

XLL

The *Lossless Extension*, XLL, is perhaps the most exciting extension to the DTS codec. With the ability to perform lossless compression of up to eight channels of 24-bit audio at sampling frequencies of up to 192 kHz, *DTS-HD Master Audio* promises to deliver all of the quality of a Linear PCM stream at a fraction of the bandwidth and disc space. However, support for the XLL extension in Blu-ray players is only optional, and it requires a new A/V receiver that supports DTS-HD Master Audio to decode it, and then only if it is able to receive the extension substream from the player. Like the other variations, DTS-HD Master Audio includes a DTS core 5.1 channel component that can be decoded even by legacy devices.

Blu-ray Disc Demystified

Technology Primer

Like Dolby TrueHD, DTS-HD Master Audio is a *lossless* compression scheme, which means that every bit that is encoded from the original Linear PCM source audio will be recreated at the decoder with exact precision. However, it should also be noted that both compression methods have variable data rates that are content dependent. In other words, some content, such as 24-bit audio with a lot of noise in the least significant bits, can require significantly more bandwidth and disc space than other content. In extreme cases, this can cause a DTS-HD Master Audio or Dolby TrueHD stream to take up nearly as much room as an uncompressed Linear PCM stream, though this is quite rare. One recommended technique in such cases is to consider "bit shaving," in which the least significant one or two bits of the audio samples are removed (set to zero) prior to encoding. This often results in a stream that is far more efficient to encode, while having little or no impact on the audio quality of the stream. In any case, this is a technique that the mastering engineer ought to perform, as is seen fit. It is not something the encoder itself would do, since it would no longer be considered a lossless encode if this took place inside the encoder.

Speakers Everywhere

One of the key selling points for Blu-ray Disc is its support for discrete multichannel audio. BD players paired with A/V receivers are able to deliver a compelling experience — 5.1 and 7.1 channel surround sound with a quality equivalent to (or better than) CD audio. Interestingly, while a higher quality of video has usually been assumed to be the most compelling feature by the creators of DVD and BD, it seems that a higher quality audio experience may be an even more significant motivating factor.

Most consumers, it seems, are familiar with the common 5.1 channel speaker setup, with left (L), center (C), right (R), left surround (LS), and right surround (RS) speakers and low-frequency effects subwoofer (LFE) arranged around the room. Now, with Blu-ray Disc, new speaker configurations are appearing. For example, the 7.1-channel audio codecs allow for multiple different speaker arrangements, including the standard 5.1 plus either two additional "center surround" speakers, or two additional surround side speakers, or two additional surround rear speakers, or height speakers to provide altitude to the exprience. In the meantime, releases to the theaters rarely use a 7.1-channel mix, and many more standard practices must be defined before these channel configurations take on greater significance.

A Few Timely Words about Jitter

Jitter is one of the most confusing aspects of digital video and audio, and this confusion carries over to Blu-ray Disc. Even the experts disagree about its effects. Part of the problem is that jitter means many things, most of them quite technical. Modern episodes of *Star Trek* come closest to providing a comprehensive definition. Since it's not very dramatic to say, "Captain, we have detected jitter!" crew members instead say, "We have encountered a temporal anomaly!" In general terms, jitter is inconsistency over time.

Technology Primer

When most people speak of jitter, they mean *time jitter*, also called *phase noise*, which is a time-base error in a clock signal — deviation from the perfectly spaced intervals of a reference signal. Figure 2.10 compares the simplified square wave of a perfect digital signal to the same signal after being affected by factors such as poor-quality components or poorly designed components, mismatched impedance in cables, logic-level mismatches between *integrated circuits* (ICs), interference and fluctuations in power supply voltage, *radiofrequency* (RF) interference, and reflections in the signal path. The resulting signal contains aberrations such as phase shift, high-frequency noise, triangle waves, clipping, rounding, slow rise/fall, and ringing. The binary values of the signal are encoded in the transition from positive voltage to negative voltage, and vice versa. In the distorted signal, the transitions no longer occur at regularly spaced intervals.

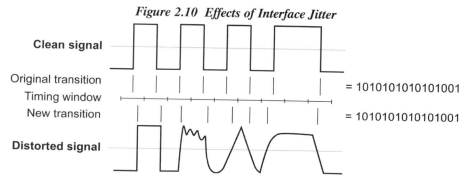

Figure 2.10 Effects of Interface Jitter

However, looking closely at Figure 2.10 reveals something interesting. Even though the second signal is misshapen to the point of displacing the transition points, the sequence of ones and zeros is still reconstructed correctly, since each transition is within the interval timing window. In other words, there is no data loss, and there is no error. The timing information is distorted, but it can be fixed. This is the key to understanding the difference between correctable and uncorrectable jitter. In the digital domain, jitter is almost always inconsequential. Minor phase errors are easily corrected by resynchronizing the data. Of course, large amounts of jitter can cause data errors, but most systems specify jitter tolerances at levels far below the error threshold.[19]

Jitter is an interface phenomenon — it only becomes a problem when moving from the analog to the digital world or from the digital world to the analog world. For example, jitter in the sampling clock of an analog-to-digital converter causes uneven spacing of the samples, which results in a distorted measurement of the waveform (Figure 2.11). On the other end of the chain, jitter in a digital-to-analog converter causes voltage levels to be generated at

---

[19]The AES/EBU standard for serial digital audio specifies a 163 ns clock rate with ±20 ns of jitter. The full 40 ns range is 24 percent of the unit interval. Testing has shown that correct data values are received with bandwidths as low as 400 kHz. Jitter in the recovered clock is reduced with wider bandwidths up to 5 MHz. The CD Orange Book specifies a maximum of 35 ns of jitter, but also recommends that total jitter in the readout system be less then 10 percent of the unit interval (that is, 23 ns out of 230 ns). The DVD-ROM specification states that jitter must be less than 8 percent of the channel bit clock period (8 percent of 38 ns comes to approximately 3 ns of jitter). In BD, layer 0 jitter must be less than 0.98 ns (6.5 percent of the 15 ns channel clock period) and layer 1 jitter must be less than 1.28 ns (8.5 percent).

Technology Primer

incorrect moments in time, resulting in audio waveform distortions such as spurious tones and added noise, producing what is often described as a "harsh" sound. Jitter that passes into an analog speaker signal degrades spatial image, ambience, and dynamic range. Actual data errors produce clicks or pops or periods of silence.

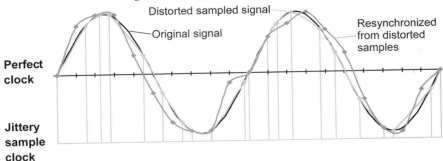

Figure 2.11 Effects of Sampling Jitter

In some cases there is nothing you can do about jitter (other than buy better equipment). In other cases, power conditioners and high-quality cables with good shielding reduce certain kinds of jitter. Before you do anything, however, it is important to understand the various types of jitter and which ones are worth worrying about. Many a shrewd marketer has capitalized on the fears of consumers worried about jitter and sonic quality, bestowing on the world such products as colored ink that supposedly reduces reflections from the edge of the disc, disc stabilizer rings that claim to reduce rotational variations, foil stickers alleged to produce "morphic resonance" to rebalance human perception, highly damped rubber feet or hardwood stabilizer cones for players, cryogenic treatments, disc polarizing devices, and other technological nostrums that are intimate descendents of *Dr. Feelgood's Amazing Curative Elixir*.

Basically, five types of jitter are relevant to DVD and BD[20] —

- **Oscillator jitter** Oscillating quartz crystals are used to generate clock signals for digital circuitry. The quality of the crystal and the purity of the voltage driving it determine the stability of the clock signal. Oscillator jitter is a factor in other types of jitter, since all clocks are "fuzzy" to some degree.

- **Sampling jitter** (recording jitter) This is the most critical type of jitter. When the analog signal is being digitized, instability in the clock results in the wrong samples being taken at the wrong time (see Figure 2.11). Reclocking at a later point can fix the time errors but not the amplitude sampling errors. There is nothing the consumer can do about jitter that happens at recording time or

---

[20] One particular phenomenon is incorrectly referred to as jitter. When optical drives perform digital audio extraction (DAE) from audio CDs, they can run into problems if the destination drive cannot keep up with the data flow. Most drives do not have block-accurate seeking, so they may miss or duplicate a small amount of data after a pause. These data errors cause clicks when the audio is played back. This is colloquially referred to as "jitter," and there are software packages that perform "jitter correction" by comparing successive read passes during DAE, but technically this is not jitter. It's a data error, not a phase error.

during production, because it becomes a permanent part of the recording. Sampling jitter also occurs when the analog signal from a player is sent to a digital processor (such as an AV receiver with DSP features or a video line multiplier). The quality of the DAC in the receiving equipment determines the amount of sampling jitter. Using a digital connection instead of an analog connection avoids the problem altogether.

- **Media jitter** *(pit jitter)*   This type of jitter is not critical. During disc replication, a laser beam is used to cut the pattern of pits in the glass master. Any jitter in the clock or physical vibration in the mechanism used to drive the laser will be transmitted to the master and thus to every disc that is molded from it. Variations in the physical replication process also can contribute to pits being longer or shorter than they should be. These variations are usually never large enough to cause data errors, and each disc is tested for data integrity at the end of the production line. Media jitter can be worse with recorded discs because they are subject to surface contamination, dust, and vibration during recording. Strange as it may seem, however, recorded media usually have cleaner and more accurate pit geometry than pressed media. However, with both pressed and recorded discs, the minor effects of jitter have no effect on the actual data.

- **Readout jitter** *(transport jitter)*   This type of jitter has little or no effect on the final signal. As the disc spins, phase-locked circuits monitor the modulations of the laser beam to maintain proper tracking, focus, and disc velocity. As these parameters are adjusted, the timing of the incoming signal fluctuates. Media jitter adds additional perturbations. Despite readout jitter, error correction circuitry verifies that the data is read correctly. Actual data errors are extremely rare. The data is buffered into RAM, where it is clocked out by an internal crystal. In theory, the rest of the system should be unaffected by readout jitter because an entirely new clock is used to regenerate the signal.

- **Interface jitter** *(data-link jitter)*   This type of jitter may be critical or it may be harmless, depending on the destination component. When data is transmitted to another device, it must be modulated onto an electrical or optical carrier signal. Many factors in the transmission path (such as cable quality) can induce random timing deviations in the interface signal (see Figure 2.10). There is also *signal-correlated jitter*, where the characteristics of the signal itself cause distortions. As a result of interface jitter, values are still correct (as long as the jitter is not severe enough to cause data errors), but they are received at the wrong time. If the receiving component is a digital recorder that simply stores the data, interface jitter has no effect. If the receiving component uses the signal directly to generate audio and does not sufficiently attenuate the jitter, it may cause audible distortion.[21] A partial solution is to use shorter cables or cables with more bandwidth and to properly match impedance.

---

[21]For example, a string of ones or a string of zeros may travel faster or slower than a varying sequence because the transmission characteristics of the cable are not uniform across the signal frequency range.

Technology Primer

To restate the key point, when the component that receives a signal is designed only to transfer or store the data, it need only recover the data, not the clock, so jitter below the error threshold has no effect. This is why digital copies can be made with no errors. However, when the component receiving a signal must reconstruct the analog waveform, then it must recover the clock as well as the data. In this case, the equipment should reduce jitter as much as possible before regenerating the signal. The problem is that there is a tradeoff between data accuracy and jitter reduction. Receiver circuitry designed to minimize data errors sacrifices jitter attenuation.[22] The ultimate solution is to decouple the data from the clock. Some manufacturers have approached this goal by putting the master clock in the DAC (which is probably the best place for it) and having it drive the servo mechanism and readout speed of the drive. RAM-buffered time-base correction in the receiver is another option. This *reclocks* incoming bits by letting them pile up in a line behind a digital gate that opens and closes to release them in a retimed sequence. This technique removes incoming jitter but introduces a delay in the signal. The accuracy of the gate determines how much new jitter is created.

The quality of digital interconnect cables makes a difference, but only to a point. The more bandwidth in the cable, the less jitter is introduced. Note that a digital audio cable is actually transmitting an analog electrical or optical signal. There is a digital to analog conversion step at the transmitter and an analog to digital conversion step at the receiver. That is the reason interface jitter can be a problem. However, the problem is less serious than when an analog interface cable is used because no resampling of analog signal values occurs.

Much ado is made about high-quality transports — disc readers that minimize jitter to improve audio and video quality. High-end systems often separate the transport unit from other units, which ironically introduces a new source of jitter in the interface cable. Jitter from the transport is a function of the oscillator, the internal circuitry, and the signal output transmitter. In theory, media jitter and transport jitter should be irrelevant but, in reality, the oscillator circuitry is often integrated into a larger chip, so leakage can occur between circuits. It is also possible for the servo motors to cause fluctuations in the power supply that affect the crystal oscillator, especially if they are working extra hard to read a suboptimal disc. Other factors such as instability in the oscillator crystal, temperature, and physical vibration may introduce jitter. A jitter-free receiver changes everything. If the receiver reclocks the signal, you can use the world's cheapest transport and get better quality than with an outrageously expensive, vacuum-sealed, hydraulically cushioned transport with a titanium-lead chassis. The reason better transports produce perceptibly better results is that most receivers do not reclock or otherwise sufficiently attenuate interface jitter. Even digital receivers with DSP (digital signal processing) circuitry usually operate directly on the bitstream without reclocking it.

Interface jitter affects all digital signals coming from a player — PCM audio, Dolby Digital, DTS, and so on. In order to stay in sync with the video, the receiver must lock the decoder to the clock in the incoming signal. Since the receiver depends on the timing information recovered from the incoming digital audio signal, it is susceptible to timing jitter.

---

[22]The jitter tolerance characteristic of a PLL (phase-locked loop) circuit is inversely proportional to its jitter attenuation characteristic. The more "slack" the circuit allows in signal transition timing, the more jitter passes through. This situation can be improved by using two PLLs to create a two-stage clock recovery circuit.

Technology Primer

There is an ongoing tug of war between engineers and critical listeners. The engineers claim to have produced a jitterless system, but golden ears hear a difference. After enough tests, the engineers discover that jitter is getting through somewhere or being added somewhere, and they go back to the drawing board. Eventually, the engineers will win the game. Until then, it's important to recognize that most sources of jitter have little or no perceptible effect on the audio or video.

Pegs and Holes: Understanding Aspect Ratios

The introduction of high definition television and widescreen displays has taken the subject of aspect ratios to an entirely new level. Yet, we are still, and will be for some time, dealing with standard definition televisions. This section describes the various cinematic and television system approaches to presentation on standard definition displays. The following sections address the issues of widescreen displays and their aspect ratio.

The standard television picture is a specific rectangular shape — a third again wider than it is high. This aspect ratio is designated as 4:3, or 4 units wide by 3 units high, also expressed as 1.33.[23] This rectangular shape is a fundamental part of the NTSC and PAL television systems — it cannot be changed without redefining the standards and reengineering the equipment.[24]

The problem is that movies are generally wider than television screens. Most movies are 1.85 (about 5.5:3). Extra wide movies in the Panavision or Cinemascope format are around 2.35 (about 7:3). The conundrum is how to fit a wide movie shape into a not-so-wide television shape (see Figure 2.12).

Fitting a movie into standard television is like the old puzzle of putting a square peg in a round hole, but in this case it is a rectangular peg and a rectangular hole. Consider a peg that is twice as wide as the hole (see Figure 2.13). There are, essentially, three ways that have been developed to get the peg in the hole —

1. Shrink the peg to half its original size or make the hole twice as big (see Figure 2.14).
2. Slice off part of the peg (see Figure 2.15).
3. Squeeze the peg from the sides until it is the same shape as the hole (see Figure 2.16).

---

[23]There's no special meaning to the numbers 4 and 3. They are simply the smallest whole numbers that can be used to represent the ratio of width to height. An aspect ratio of 12:9 is the same as 4:3. The ratio can be normalized to a height of 1, but the width becomes the repeating fraction 1.33333..., which is why the 4:3 notation is generally used. For comparison purposes it is useful to use the normalized format of 1.33:1 or 1.33 for short. On occasion, people will mention the "Academy Aperture" ratio of 1.37. In 1927, the Academy of Motion Picture Arts and Sciences officially chose 1.33 as the industry standard. In 1931 it was changed to 1.37 to allow room on the film for a soundtrack. From the 1920s to the early 1950s almost all films were shot at 1.37, providing a picture aspect ratio of 1.33.

[24]The next generation of television — known as HDTV, ATV, DTV, etc.— has a 1.78 (16:9) picture that is much wider than standard television. However, the new digital format is incompatible with the old standard recording and display equipment.

Blu-ray Disc Demystified

Technology Primer

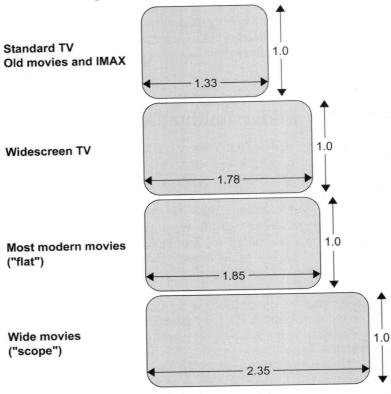

Figure 2.12 TV Shape vs. Movie Shape

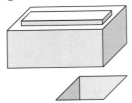

Figure 2.13 Peg and Hole

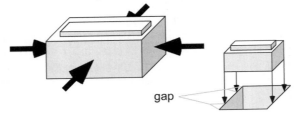

Figure 2.14 Shrink the Peg

Technology Primer

Figure 2.15 Slice the Peg

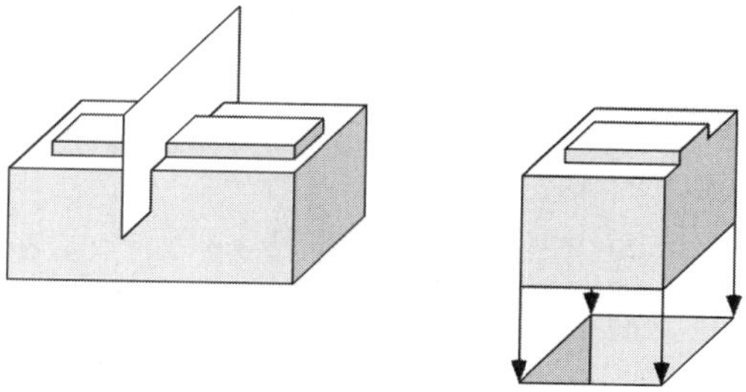

Figure 2.16 Squeeze the Peg

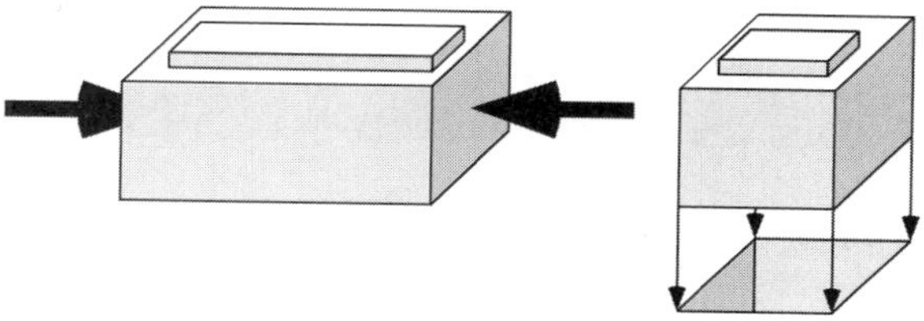

Now, think of the peg as a movie and the hole as a TV. The first two peg-and-hole solutions are used commonly to show movies on television. Quite often you will see horizontal black bars at the top and bottom of the picture. This means that the width of the movie shape has been matched to the width of the TV shape, leaving a gap at the top and the bottom. This is called *letterboxing*. It does not refer to postal pugilism but rather to the process of putting the movie in a black box with a hole the shape of a standard paper envelope. The black bars are called *mattes*.

At other times you might see the words, "This presentation has been formatted for television", at the beginning of a movie. This indicates that a *pan and scan* process has been used, where a TV-shaped window over the film image is panned from side to side, scanned up and down, or zoomed in and out (see Figures 2.17 and 2.18). This process is more complicated than just chopping off a little from each side; sometimes the important part of the picture is entirely or primarily on one side, and sometimes there is more picture on the film above or below what is shown in the theater, so the person who transfers the movie to video must determine for every scene how much of each side should be chopped off or how much additional picture from above or below should be included in order to preserve the action and story line. For the past 20 years or so, most films have been shot *flat*, sometimes called *soft matte*. The cinematographer has two rectangles in the viewfinder, one for 1.85 (or wider) and one for 4:3. He or she composes the shots to look good in the 1.85 rectangle while making sure that no crew, equipment, or raw set edges are visible above or below in the 4:3 area.

Technology Primer

Figure 2.17 Soft Matte Filming

Figure 2.18 Pan and Scan Transfer

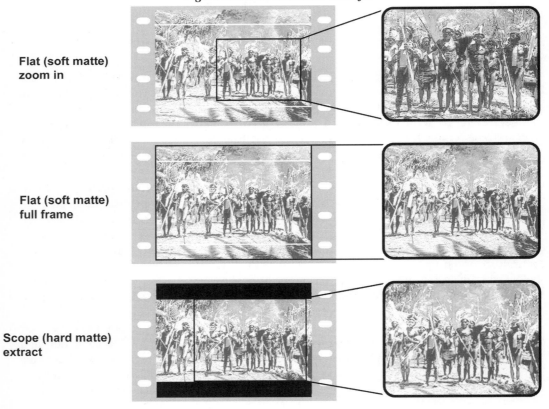

Flat (soft matte) zoom in

Flat (soft matte) full frame

Scope (hard matte) extract

Figure 2.19 The Anamorphic Process

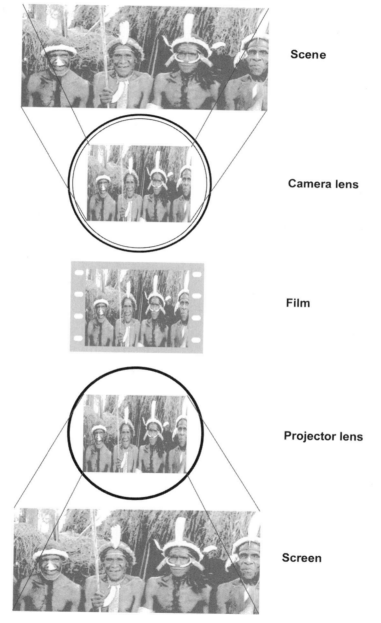

Technology Primer

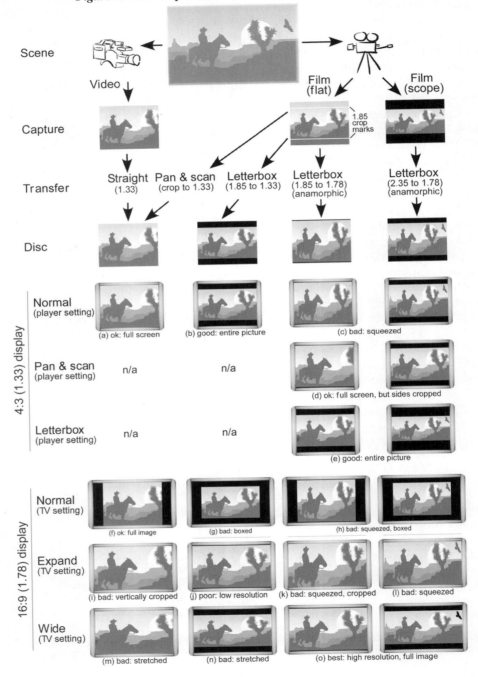

Figure 2.20a-o Aspect Ratios, Conversions, and Displays

Technology Primer

For presentation in the theater, a *theatrical matte* or *aperture plate* is used to mask off the top and bottom when the film is printed or projected. When the movie is transferred to video for 4:3 presentation, the full frame is available for the pan and scan (and zoom) process.[25] In many cases, the director of photography or the director approves the transfer to ensure that the integrity of the film is maintained. Full control over how the picture is reframed is very important. For example, when the mattes are removed, close-up shots become medium shots, and the frame may need to be zoomed in to recreate the intimacy of the original shot. In a sense, the film is being composed anew for the new aspect. The pan and scan process has the disadvantage of losing some of the original picture but is able to make the most of the 4:3 television screen and is able to enlarge the picture to compensate for the smaller size and lower resolution as compared with a theater screen.

There is another option that applies only to computer-generated films. The shots can be recomposed for 4:3 presentation by moving characters and objects closer together. The story, action, and sound all remain the same, but the movie is re-rendered to fit the dimensions of standard TVs.

The third peg-and-hole solution has been used for years to fit widescreen movies onto standard 35-mm film. As filmmakers tried to enhance the theater experience with ever wider screens, they needed some way to get the image on the film without requiring new wider film and new projectors in every theater. They came up with the *anamorphic* process, where the camera is fitted with an anamorphic lens that squeezes the picture horizontally, changing its shape so that it fits into a standard film frame. The projector is fitted with a lens that unsqueezes the image when it is projected (see Figure 2.19). It is as if the peg were accordion-shaped so that it can be squeezed into the square hole and then pop back into shape after it is removed. You may have seen this distortion effect at the end of a Western movie where John Wayne suddenly becomes tall and skinny so that the credits will fit between the edges of the television screen.

Figure 2.20 depicts the aspect ratio considerations when presenting images on standard television displays and widescreen display. Depending on the method of acquisition (1.33 or 1.85), the scene can be presented in a myriad of permutations.

Widescreen Displays

In recent years, we have seen a marked transition to widescreen and high definition displays. Display manufacturers have jumped on the February 2009 digital television switchover, and are producing widescreen displays in a diverse array of models and technologies. We have gone from a world dominated by *cathode ray tubes* (CRTs) to one in which it will soon be difficult to find a CRT model display. Although the CRT is fading from the landscape, it should not be dismissed too quickly.

Widescreen displays are quite flexible in the way they deal with different input formats. Widescreen displays present images in a 16:9 aspect ratio and, generally, this accommodates

---

[25]Contrast this to hard matte filming, where the top and bottom are physically — and permanently — blacked out to create a wide aspect ratio. Movies filmed with anamorphic lenses also have a permanently wide aspect ratio, with no extra picture at the top or bottom.

Technology Primer

high definition format presentations. They can also display 4:3 video with black bars on the sides — a kind of sideways letterbox that is sometimes called a *pillarbox* — and they also have display modes that enlarge the 4:3 video to fill their entire screen —

- *Wide* mode stretches the picture horizontally (see Figure 2.21). This is sometimes called *full* mode. This is the proper mode to use with anamorphic video, but it makes everything look short and fat when applied to 4:3 video. Some widescreen TVs have a *parabolic* or *panorama* version of wide mode, which uses nonlinear distortion to stretch the sides more and the center less, thus minimizing the apparent distortion. This mode should not be used with anamorphic DVD output or very strange fun-house-mirror effects will occur.

Figure 2.21 Wide (full) Mode on a Widescreen TV

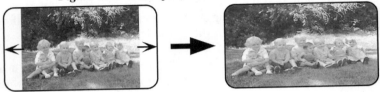

- *Expand* mode proportionally enlarges the picture to fill the width of the screen, thus losing the top and bottom (see Figure 2.22). This is sometimes called *theater* mode. Expand mode is for use with letterboxed video because it effectively removes the mattes. If used with standard 4:3 picture, this mode causes a Henry VIII "off with their heads" effect.

Figure 2.22 Expand (theater) Mode on Widescreen TV

When DVD contains widescreen video and is played in a DVD player, the different output modes of the player can be combined with different widescreen TV display modes to create a confusing array of options. Figure 2.20 shows how the different DVD output modes look on a regular TV and on a widescreen TV. Note that there is one "good" way to view widescreen video on a standard TV (Figure 2.20e) but that a very large TV or a widescreen TV is required to do it justice. Also note that there is only one good way to view widescreen video on a widescreen TV, and this is with widescreen (anamorphic) output to wide mode (Figure 2.20o).

Neither the *Wide* or the *Expand* modes should be used with Blu-ray players, as the player outputs a true widescreen picture. In cases where the disc contains standard definition 4:3 video (either 720 × 480 or 720 × 576) the player will center the image on the 16:9 display, or

Technology Primer

for anamorphic video (high definition 1440×1080 or standard definition) the player will expand the image to fit the 16:9 display.

Clearly, it is easy to display the wrong picture in the wrong way. In some cases, the equipment is smart enough to assist. The player can send a special signal embedded in the video blanking area or via the connector to the widescreen TV, but everything must be set up properly. If everything is working correctly and the TV is equipped to recognize widescreen signaling, it will automatically switch modes to match the format of the video.

Video comes out of a player in, basically, one of five ways —

1. Full frame 4:3 original
2. Pan and scan (widescreen original)
3. Full frame 16:9 original
3. Letterbox (widescreen original)
4. Anamorphic (widescreen original)

All five can be displayed on any TV, but the third, fourth and fifth modes are specifically intended for widescreen TVs. This may seem straightforward, but it becomes much more complicated. The problem is that very few movies are in 16:9 (1.78) format. Although, a great proportion of video is now being created in 16:9 format as 16:9 HD cameras are being more widely used.

Most movies are usually 1.85 or wider, although European movies are often 1.66. DVD and Blu-ray Disc support aspect ratios of 1.33 (4:3) and 1.78 (16:9) because they are the two most common television shapes. Movies that are a different shape must be made to fit, which brings us back to pegs and holes. In this case, the hole is the 16:9 shape. There are essentially four ways to fit a 1.85 or wider movie peg into a 1.78 hole —

- **Letterbox to 16:9** Perhaps the most common approach, when the movie is transferred from film, black mattes are added to box it into the 16:9 shape. These mattes become a permanent part of the picture. The position and thickness of the mattes depend on the shape of the original.

  - For a 1.85 movie, the mattes are relatively small. On a widescreen TV the mattes are hidden in the overscan area.[26]

  - For a 2.35 movie, the permanent mattes are much thicker. In this case, the picture has visible mattes no matter how it is displayed.

  - For a 1.66 movie, thin mattes are placed on the sides instead of at the top and bottom and may be hidden in the overscan area of a display.

- **Crop to 16:9** For 1.85 movies, slicing 2 percent from each side is sufficient and probably will not be noticeable. The same applies to 1.66 movies, except that about 3 percent is sliced off the top and off the bottom, but to fit a 2.35 movie

---

[26]Overscan refers to covering the edges of the picture area with a mask around the screen. Overscan was originally implemented to hide distortion at the edges. Television technology has improved to the point where overscan isn't usually necessary, but it is still used. Most televisions have an overscan of about 4 to 5 percent. Anyone producing video intended for television display must be mindful of overscan, making sure that nothing important is at the edge of the picture. It should be noted that when DVD-Video or Blu-ray Disc content is displayed on a computer there is no overscan and the entire picture is visible.

Technology Primer

requires slicing 12 percent from each side. This Procrustean approach throws away a quarter of the picture and is not likely to be a popular solution.

- **Pan and scan to 16:9** The standard pan and scan technique can be used with a 16:9 window (as opposed to a 4:3 window) when transferring from film to digital video. For 1.85 movies, the result is essentially the same as cropping and is hardly worth the extra work. For wider movies, pan and scan is more useful, but the original aspect ratio is lost, which goes against the spirit of a widescreen format. When going to the trouble of supporting a widescreen format, it seems silly to pan and scan inside it, but if the option is there, someone is bound to use it.

- **Open the soft matte to 16:9** When going from 1.85 to 16:9 (1.78), a small amount of picture from the top and bottom of the full-frame film area can be included. This stays close to the original aspect ratio without requiring a matte, and the extra picture will be hidden in the overscan area on a widescreen TV or when panned and scanned by the player. Even wider movies are usually shot full frame, so the soft matte area can be included in the transfer (see Figure 2.23).

Figure 2.23 Opening the Frame from 1.85 to 1.78

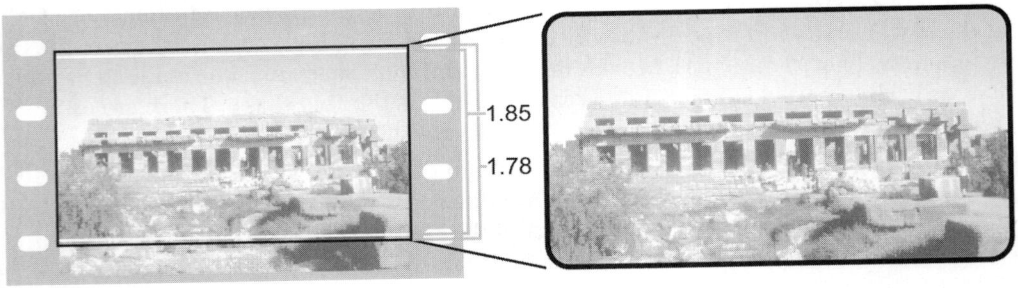

Why 16:9?

The 16:9 ratio has become the standard for widescreen. Most widescreen televisions are this shape, it's the aspect ratio used by almost all high-definition television standards, and it's the widescreen aspect ratio used by DVD and BD. You may be wondering why this ratio was chosen, since it does not match television, movies, computers, or any other format. But this is exactly the problem: there is no standard aspect ratio (see Figure 2.24).

Current display technology is limited to fixed physical sizes. A television picture tube must be built in a certain shape, while a flat-panel display screen must be made with a certain number of pixels. Even a video projection system is limited by electronics and optics to project a certain shape. These constraints will remain with us for decades until we progress to new technologies such as scanning lasers or amorphous holographic projectors. Until then, a single aspect ratio must be chosen for a given display. The cost of a television tube is based roughly on diagonal measurement (taking into account the glass bulb, the display surface, and the electron beam deflection circuitry), but the wider a tube is, the harder it is to maintain uniformity (consistent intensity across the display) and convergence (straight horizontal and vertical lines). Therefore, too wide a tube is not desirable.

Figure 2.24 Common Aspect Ratios

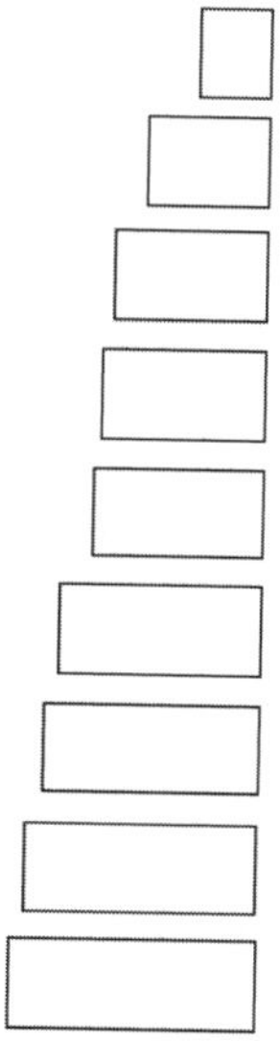

	7:9	2.3:3	0.77:1	Paper (8.5 x 11)
	12:9	4:3	1.33:1	NTSC / PAL television, full-frame DVD
	15:9	5:3	1.66:1	Photographs, many European movies
	16:9	5.3:3	1.78:1	Anamorphic DVD, HDTV, most widescreen TV
	16.7:9	5.6:3	1.85:1	Most movies
	19.9:9	6.6:3	2.21:1	Wide movies (70 mm)
	21.2:9	7.1:3	2.35:1	Wide movies (Panavision, Cinemascope)
	23.5:9	7.7:3	2.55:1	Wide movies (Cinemascope 55)
	24.3:9	8.1:3	2.7:1	Extra-wide movies (Ultra Panavision)

The 16:9 aspect ratio was chosen in part because it is an exact multiple of 4:3. Going from 4:3 to 16:9 merely entails adding one horizontal pixel for every three and going from 16:9 to 4:3 requires simply removing one pixel from every four.[27] This makes the scaling circuitry for letterbox and pan and scan functions much simpler and cheaper. It also makes the resulting picture cleaner.

The 16:9 aspect ratio is also a reasonable compromise between television and movies. It's very close to 1.85 and it is close to the mean of 1.33 and 2.35. That is, $4/3 \times 4/3 \times 4/3 \sim= 2.35$. Choosing a wider display aspect ratio, such as 2:1, would have made 2.35 movies look wonderful but would have required huge mattes on the side when showing as 4:3 video (as in Figure 2.20f, but even wider).

Admittedly, the extra space could be used for picture outside picture (POP, the converse of PIP), but it would be very expensive extra space. To make a 2:1 display the same height as

---

[27]In each case a weighted scaling function generally is used. For example, when going from 4 to 3 pixels, 3/4 of the first pixel is combined with 1/4 of the second to make the new first, 1/2 of the second is combined with 1/2 of the third to make the new second, and 1/4 of the third is combined with 3/4 of the fourth to make the new third. This kind of scaling causes the picture to become slightly softer but is generally preferable to the cheap alternative of simply throwing away every fourth line. Similar weighted averages can be used when going from 3 to 4.

Technology Primer

a 35-inch television (21 inches) requires a width of 42 inches, giving a diagonal measure of 47 inches. In other words, to keep the equivalent 4:3 image size of 35-inch television, you must get a 47-inch 2:1 television. Figure 2.25 shows additional widescreen display sizes required to maintain the same height of common television sizes.

Figure 2.25 Display Sizes at Equal Heights

Diagonal size (width x height)

1.33 (4:3)	1.78 (16:9)	2.0
27 (22x 16)	33 (29x 16)	36 (32x 16)
32 (26x 19)	39 (34x 19)	43 (38x 19)
34 (27x 20)	42 (36x 20)	46 (41x 20)
36 (29x 22)	44 (38x 22)	48 (43x 22)
42 (34x 25)	51 (45x 25)	56 (50x 25)
46 (37x 28)	56 (49x 28)	62 (55x 28)
50 (40x 30)	61 (53x 30)	67 (60x 30)
53 (42x 32)	65 (57x 32)	71 (64x 32)
60 (48x 36)	73 (64x 36)	80 (72x 36)
65 (52x 39)	80 (69x 39)	87 (78x 39)
84 (67x 50)	103 (90x 50)	113 (101x 50)
100 (80x 60)	122 (107x 60)	134 (120x 60)
120 (96x 72)	147 (128x 72)	161 (144x 72)

Figure 2.26 demonstrates the area of the display used when different image shapes are letterboxed to fit it (that is, the dimensions are equalized in the largest direction to make the smaller box fit inside the larger box).[28] The 1.33 row makes it clear how much smaller a letterboxed 2.35:1 movie is: only 57 percent of the screen is used for the picture. On the other hand, the 2:1 row makes it clear how much expensive screen space goes unused by a 4:3 video program or even a 1.85:1 movie. The two middle rows are quite similar, so the mathematical relationship of 16:9 to 4:3 gives it the edge.

In summary, the only way to support multiple aspect ratios without mattes would be to use a display that can physically change shape — a "mighty morphin' television." Since this is currently impossible (or, if possible, outrageously expensive as with motorized masks that change the shape of fancy home theater projection screens), 16:9 is the most reasonable compromise. That said, the designers of DVD and BD could have improved things slightly by allowing more than one anamorphic distortion ratio. If a 2.35 movie were stored using a 2.35 anamorphic squeeze, then 24 percent of the internal picture would not be wasted on the black mattes, and the player could automatically generate the mattes for either 4:3 or 16:9 displays. This was not done, probably because of the extra cost and complexity it would add

---

[28] If you wanted to get the most for your money when selecting a display aspect ratio, you would need to equalize the diagonal measurement of each display because the cost is roughly proportional to the diagonal size. This approach is sometimes used when comparing display aspect ratios and letterboxed images, and it has the effect of emphasizing the differences. The problem is that wider displays end up being shorter (for example, a 2:1 display normalized to the same diagonal as a 4:3 display would be 4.47:2.25, which is 12 percent wider but 25 percent shorter). In reality, no one would be happy with a new widescreen TV that was shorter than their existing TV. Therefore, it is expected that a widescreen TV will have a larger diagonal measurement and will cost more.

to the player. The limited set of aspect ratios presently supported by the MPEG-2 standard (4:3, 16:9, and 2.21:1) also may have had something to do with it.

Figure 2.26 Relative Display Sizes for Letterbox Display

Display shape	Image shape		
	1.33 (4:3)	1.85	2.35
(4:3) 1.33	100%	72%	57%
(16:9) 1.78	75%	96%	76%
1.85	72%	100%	79%
2.0	67%	92%	85%

The Transfer Tango

Most Blu-ray Disc video is in native widescreen resolution. However, the Blu-ray specification does allow for 1440×1080 anamorphic high definition video and standard definition anamorphic video. So, the option still remains to transfer a 1.85 movie to a 4:3 aspect ratio instead of 16:9, thus creating a pillarbox presentation when viewing on widescreen. At first glance, there may seem to be no advantage in doing this because 1.85 movies are so close to 16:9 (1.78). It seems simpler to do a 16:9 transfer and let the player create a letterbox or pan and scan version, when viewing on a 4:3 display. However, there are disadvantages to having the player automatically format a widescreen movie for 4:3 display — the vertical resolution suffers by 25 percent, the letterbox mattes are visible on movies wider than 1.85, and the player is limited to lateral motion. In addition, many movie people are averse to what they consider as surrendering creative control to the player. Therefore, almost every pan and scan transfer is done in the studio and not enabled in the player. During the transfer from film to video, the engineer has the freedom to use the full frame or to zoom in for closer shots, which is especially handy when a microphone or a piece of the set is visible at the edge of the shot.

Many directors are violently opposed to pan and scan disfigurement of their films. Director Sydney Pollack sued a Danish television station for airing a pan and scan version of his *Three Days of the Condor*, which was filmed in 2.35 Cinemascope. Pollack felt strongly that the pan and scan version infringed his artistic copyright. He believed that, "The director's job is to tell the film story, and the basis for doing this is to choose what the audience is supposed to see, and not just generally but exactly what they are to see." Some directors, such as Stanley Kubrick, accept only the original aspect ratio. Others, such as James Cameron, who closely supervise the transfer process from full-frame film, feel that the director is responsible for making the pan and scan transfer a viable option by recomposing the movie to make the most of the 4:3 TV shape.

Technology Primer

About two-thirds of widescreen movies are filmed using a 1.85 aspect ratio. When a 1.85 film is transferred directly to full-frame 4:3 by including the extra picture at the top and bottom, the actual size of the images on the TV are the same as for a letterbox version. In other words, letterboxing only covers the part of the picture that also was covered in the theater.

A pan and scan transfer to 4:3 makes the "I didn't pay good money for my 30-inch TV just to watch black bars" crowd happy, but a letterbox transfer to 16:9 is still needed to appease the videophiles who demand the theatrical aspect ratio and to keep the "I paid good money for my widescreen TV" crowd from revolting. As widescreen TVs replace traditional TVs, 4:3 transfers will become less common, and even letterboxed 4:3 transfers will become more appreciated.

More on Interlaced vs. Progressive Scanning

One of the biggest problems facing early television designers was displaying images fast enough to achieve a smooth illusion of motion. Early video hardware was simply not fast enough to provide the required flicker fusion frequency of around 50 or 60 frames per second. The ingenious expedient solution was to cut the amount of information in half by alternately transmitting every other line of the picture (see Figure 2.27). The engineers counted on the persistence of the phosphors in the television tube to make the two pictures blur into one. This approach to image presentation is continued with the high definition format of 1080 lines, first the 540 odd lines are sent and displayed, followed by the 540 even lines. This is called *interlaced scanning*. Each half of a frame is called a *field*.

Figure 2.27 Interlaced Scan and Progressive Scan

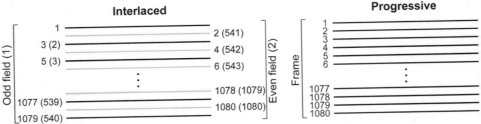

The alternative approach, *progressive scanning*, displays every line of a complete frame in one sweep. Progressive scanning requires twice the frequency in order to achieve the same refresh rate. Progressive scan monitors have generally been used for computers since their introduction. And progressive scanning is also used for HDTV.

Progressive scan provides a superior picture, overcoming many disadvantages of interlaced scanning. In interlaced scanning, small details, especially thin horizontal lines, appear only in every other field. This causes a disturbing flicker effect, which you can see when someone on TV is wearing stripes. The flicker problem is especially noticeable when computer video signals are converted and displayed on a standard TV. In addition to flicker, line crawl occurs when vertical motion matches the scanning rate. Interlaced scanning also caus-

Technology Primer

es problems when the picture is paused. If objects in the video are moving, they end up in a different position in each field. When two fields are shown together in a freeze-frame, the picture appears to shake.

Film transfer processes are now capable of producing 24 frames per second (fps) masters that are recorded to server and hard drive arrays. (Thie film may have been shot or telecined at 23.976, but we will use 24 fps for both cases.) By eliminating a videotape recording from the transfer process, the native film rate is preserved and allows the encoding process to smoothly move to an encoded video stream that can be presented at the 24 frames per second rate. Some widescreen displays are capable of reproducing the 24p fps imagery which, in virtually all instances, requires an HDMI connection between player and display. See **Chapter 7, Players**, for more information about HDMI.

The 24 fps transfer process precludes the tedious and error-fraught method of generating a transfer master that accommodates display at the 59.94 frame rate of standard televisions. That, however, does not exempt the likely display of the encoded footage at that ol' bugaboo rate of 60 video fields per second. (Yes, we've again taken the low road and used "60" when it's really 59.94.)

Further, for legacy films that had previously been transferred to videotape, a transfer method was instituted that took the film running at 24 progressive frames per second (fps) and displayed it at the rate of 60 video fields per second. This is done with a film *telecine* machine using a process called *2-3 pulldown*, where one film frame is shown as two fields, and the following film frame is shown as three fields (see Figure 2.28).

This 2-3 pulldown pattern results in pairs of 24-per-second film frames converted to 60-per-second TV fields ($[2+3] \times 12 = 60$). Unfortunately, this causes side effects. One is that film frame display times alternate between 2/60 of a second and 3/60 of a second, causing a *motion judder* artifact — a jerkiness that may be visible when the camera pans slowly. Another side effect is that two of every five television frames contain fields derived from two different film frames, which does not cause problems during normal playback but can cause problems when pausing or playing in slow motion. A minor problem is that video actually plays at 59.94 Hz, so the film runs 0.1 percent slow and the audio must be adjusted to match. Displaying film at the rate of 50 video fields per second is simpler and usually is achieved by showing each film frame as two fields and playing it 4 percent faster.[30] This is sometimes called *2-2 pulldown*, and is used extensively in PAL countries, where 25 fps is the norm.

When an encoder processes video made with 2-3 pulldown, it would be inefficient to encode the extra fields, so the encoder recognizes and removes the duplicate fields. This is called *inverse telecine* (no, it's not called 3–2 pushup). But, when the goal is to present 24p material on a 59.94 display, the encoder adds flags in the data stream that indicate which fields to show when and which fields to repeat when. The encoder sets the repeat_first_field flag on every fifth field, which instructs the decoder to repeat the field, thus recreating the 2–3 pulldown sequence needed to display the video on an interlaced TV. In other words, the decoder in the player performs 2-3 pulldown, but only by following the instructions in the data stream.

---

[30]Since the video is sped up 4 percent when played, the audio must be adjusted before it is encoded. In many cases the audio speedup causes a semitone pitch shift that the average viewer will not notice. A better solution is to digitally shift the pitch back to the proper level during the speedup process.

Blu-ray Disc Demystified

Technology Primer

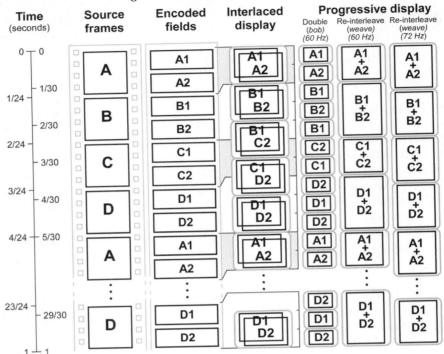

Figure 2.28 Converting Film to Video

New Display Technologies

In recent years, there has been a dramatic expansion in the area of new display technologies. We have gone from a world dominated by *cathode ray tubes* (CRTs) to one in which it will soon be difficult to find one. With advances in these new display technologies, it is important to understand how Blu-ray and DVD are affected.

If you go down to your local consumer electronics store, you will see that standard definition televisions have all but disappeared, and there is a wide array of widescreen high definition displays available. Most standard televisions are of the CRT genre and although there are a small number of CRT widescreen televisions still available, most widescreen displays are either plasma or LCD. The leading reason for this differentiation is size. It is very difficult to manufacture a glass tube CRT at very large sizes. Given that the CRT is a type of vacuum tube technology, the tube itself must be able to structurally withstand the pressure of containing a vacuum without imploding. As the tube grows larger, this becomes more difficult without any sort of internal support. Additionally, it's a lot of glass, making it expensive and, more importantly, HEAVY.

Does that mean CRT is going away? Almost, but not completely. It's true that a 500+ pound CRT television will have a difficult time competing against the convenience of a relatively light, wall-mounted plasma screen or LCD. Even in situations where taking up a large

Technology Primer

space is not a problem, the convenience of a lightweight DLP system may outweigh (pun intended) the CRT heavyweight. However, it will be a shame to lose CRTs. For one thing, there has been over 60 years of research applied to developing the color display technologies that CRT uses, including phosphors, screen masks and color dot distribution to name a few. However, as the new technologies mature, considerable improvements have been made that are proving to make the good ol' CRTs relics of a bygone era.

Types of New Displays

Digital light processing (DLP) This type of display is based on the *digital micromirror device* (DMD), a silicon chip whose surface is covered by a matrix of tiny mirrors. Through electrical signals, these mirrors can toggle from an "off" position to an "on" position and back. Basically, each micromirror represents a pixel; by reflecting a light source off the mirror, one can project a black and white image. Flickering the mirror at variable high rates creates the impression of varying amounts of light, thereby displaying a grayscale. Passing the image through a rotating color wheel that is properly synchronized with the mirror flashes generates a color display. This is the basis for most consumer rear-projection DLP televisions, which can offer a bright, crisp picture. However, depending on the color wheels used and the sophistication of the display logic, DLP displays tend to have difficulty reproducing some of the darker color ranges. Likewise, being a rear-projection device, there may be some *light bleed*, which can wash out the image, making black areas appear gray. Some DLP displays may exhibit color shift problems, exaggerated sharpness, and excessive brightness. The latter two characteristics, in particular, can overemphasize encoding artifacts in video.

Plasma display device panel (PDP) Plasma displays work by applying a charge to a small gas-filled cell. The gas becomes ionized and, in turn, interacts with a phosphor on the surface of the cell, which glows a given color. Because plasma displays work with phosphors, they are able to utilize much of the research that has gone into phosphor research for CRTs and reproduce more natural colors. However, it also means that they may *burn in* quickly, just like a CRT, leaving a permanent after-image where something remained on screen for a length of time. For example, channel logos that occupy a corner of the screen can begin to burn in after a period of time. Plasma displays can consume large amounts of power compared to other types of displays, and tend to suffer from bleed-over effects in which the plasma from one cell bleeds over to other cells.

Liquid-crystal display (LCD) Liquid-crystal displays operate by sandwiching color filters and polarizing filters on either side of a liquid-crystal cell matrix, and placing a white backlight behind the sandwich. The backlight remains lit while the display is active. The liquid crystal in each cell, by default, stays "curled" up such that it blocks the polarized illumination from the backlight. When a charge is applied to a cell, the crystal "unwinds," allowing the illumination from the backlight to escape through the color filter. One of the most noticeable problems with LCD is the limited viewing angle, although with the latest updates to these technologies, the viewing angle has been extended dramatically (up to as much as 178 degrees). Increased viewing angle tends to lead to increases in *light bleed* — white light that escapes to wash out the picture and raise black levels. In addition, the "unwinding" process

Blu-ray Disc Demystified

2-55

Technology Primer

takes time at a molecular level. As a result, it takes time to refresh the display. An LCD, with a refresh time of more than 33 ms will blur 30 fps video. A refresh time of 16.7 ms is needed to maintain the 59.94 fields per second rate of de-interlaced video. Newer LCDs with very low pixel latency can handle even 120 Hz display rates (8.3 ms). LCDs do not suffer from burn in.

Liquid crystal on silicon (LCoS) Similar to DLP rear projection displays, this type of television operates by projecting colored light through a translucent matrix of liquid crystals, in which each cell can be individually made opaque, transparent or somewhere in between. Unfortunately, because LCoS uses liquid crystal, these displays tend to suffer from the same performance issues as LCD and often require three separate light sources (or one light source with three pathways through three separate color filters) in order to provide full color reproduction.

With the rapid acceptance of these display technologies and with new ones emerging, such as, *organic light-emitting diodes* (OLED), *high dynamic range* (HDR), D-ILA (*Digital Direct Drive Image Light Amplifier*) , and SXRD (*Silicon X-tal Reflective Display*), the content producer's job becomes a bit more complicated. It used to be that a producer only had to worry about NTSC and PAL video formats. Now they have a half dozen new display technologies that bring their own problems to the reproduction mix.

HD Discs Meet HD Video Meet HD Television

As the Blu-ray Disc format enters the marketplace, it joins the collection of already introduced HD video formats. In the broadcasting arena, there is an ongoing battle between the high-definition video formats — 720p (1280 × 720 resolution at 59.94 progressive frames per second) and 1080i (1920 × 1080 resolution at 29.97 interlaced frames per second). Television stations, depending on their affiliations, are broadcasting in one or the other HD flavor. Yet, there is a massive push for technology to catch up with the desire to reproduce imagery in the 1080/24p format. The problem is that broadcast television is limited in its high-definition data rate to that of 19 Mbps. The advanced video codecs, VC-1 and AVC, can do a passable job of squeezing 1080/24p video to under 19 Mbps, but broadcast formats limited to MPEG-2 cannot manage it at acceptable quality. In fact, broadcasters are going in a totally different direction and are splitting their available bandwidth to distribute secondary digital channels that use some of that 19 Mbps. The result is that a television station's HD broadcast channel may be going on the air at a rate of 12 to 14 Mbps or even less.

Even more challenging to the consumer is understanding and accepting what their widescreen display is doing to reproduce both the television station signals and their Blu-ray Disc player output. For the forseeable future, a Blu-ray Disc in a state-of-the-art Blu-ray player, playing on a 1920 × 1080 HD display, will be the best way to view high-definition video because Blu-ray can use up to 40 Mbps for the video stream. Yet, even that is a result of an encode/decode process that concludes with a data stream consisting of 8-bit video in a 4:2:0 colorspace.

It is only recently that displays are being produced and marketed as capable of reproducing the elusive 1080/24p image format, with a native resolution of 1080 lines and progressive

Technology Primer

scan. A considerable number of high-definition displays only have a native resolution of approximately 1280 × 720. Some are slightly smaller (e.g., 1024 × 768 for some DLPs) while others may be higher (e.g., 1368 × 848 for some plasma displays). However, there is still a significant number of displays (rear projection or CRTs) that prefer 1080i.

The good news is that EICTA[31] developed HD 1080p logos that help consumers identify displays with a native resolution of, at least, 1920 × 1080. The "Ready" logos refer to displays without built-in HD TV tuners. Note that the HD logos only require 720 lines. The DTV and HDTV logos from the US Consumer Electronics Association (CEA) are not helpful in this case, because they do not distinguish 1080p from 720p (Figure 2.29).

Finally, an overriding concern is that a display may state that it supports the 720p or the 1080i or the 1080p video format, but they rarely indicate what scaling, frame rate conversion, deinterlacing, or other signal processing these formats may undergo prior to being displayed. Finding a good match between video format and display device may be just a matter of luck for some time to come. Education and understanding of these constantly evolving technologies is the only good course, albeit a never-ending one. Ah, well...

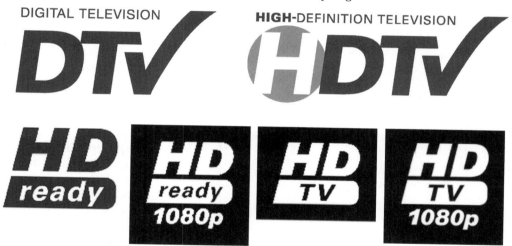

Figure 2.29 DTV and HD Ready Logos

---

[31] EICTA was formed in 1999 as the European Information and Communications Technology Industry Association. In 2001 it merged with EACEM (European Association of Consumer Electronics Manufacturers). EICTA represents more than 10,000 European enterprises.

Chapter 3
Features

The features available with Blu-ray Disc™ technology are considerably more advanced than the features introduced with standard DVD. This chapter describes the breadth of general feature approaches that may be incorporated in a BD title.

Feature playability is also dependent on the capabilities of the player being utilized for display. The Blu-ray Disc Association established a stepped introduction of players, referred to as *Profiles*, that allowed manufacturers to roll out players and feature capabilities in stages. Please see **Chapter 7, Players**.

Building on DVD Features

The creators of DVD realized that to succeed, DVD had to be more than just a roomier CD or a more convenient laserdisc. Hollywood had started the ball rolling by requesting a digital video consumer standard that would hold a full-length feature film, had better picture quality than existing high-end consumer video plus widescreen aspect ratio support, contained multiple versions of a program with parental control, supported high-quality surround-formatted audio with soundtracks for at least three languages, and had built-in content protection. Next, the computer industry added their requirements of a single format for computers and video entertainment with a common cross-platform file system, high performance for both movies and computer data, compatibility with CDs and CD-ROMs, compatible recordable and rewritable versions, no mandatory caddy or cartridge, and high data capacity with reliability equal to or better than CD-ROM. Hollywood agreed that they too wanted a content protection system, with the added requirement that the then new disc standard had to include a locking system to control release across different geographic regions of the world.

Following those initial development steps, the designers threw in a few more features, multiple camera angles and graphic overlays for subtitling or karaoke, and DVD was born. Unlike CD, where the computer data format was cobbled on top of the digital music format, the digital data storage system of DVD-ROM is the base format for the standard DVD. DVD-Video is built on top of DVD-ROM, using a specific set of file types and data types. From a purely storage application point, a DVD-ROM may contain digital data in almost any conceivable format, as long as a computer or other device can make use of it. Whereas, DVD-Video requires simple and inexpensive video players and the format capabilities and features are strictly defined.

DVD-Audio was very late out of the specifications gate and remains the nearly forgotten red-headed stepchild of the DVD format family. The plan was to create a separate DVD-Audio format based on input from the music industry. Their requirements included copyright identification and content protection, compatibility with DVD-ROM and DVD-Video, CD

Features

playback including an optional hybrid DVD/CD format, navigation with random access similar to DVD-Video but also usable on players without an attached video display, a slideshow feature and, of course, superior sound quality. Unfortunately, DVD-Audio is a classic case of too little, too late. It did not dramatically improve on CD and it required a special DVD-Audio player because it was developed after DVD-Video was released. DVD-Audio was either not quite a success or not quite a failure, depending on how you look at it. Only a few hundred DVD-Audio discs were made. Music on standard DVD-Video discs sold much better. The creators of Blu-ray wisely chose to make the Blu-ray audio-only capability an integral part of the format from the beginning. Profile 3 audio-only players use a subset of the HDMV format without video and with simplified controls to play audio from many standard Blu-ray Discs.

Now that DVD has become firmly entrenched in the marketplace it is incumbent on would-be-successor technologies, such as Blu-ray, to incorporate innovative features and compelling advances in order to have a hope of success. The Hollywood studios and major production houses continue to be driving forces in establishing the functions and features of media technologies that are heavily oriented toward entertainment (see Table 3.1). Their motives are not solely altruistic, of course, with the overarching condition being content security. With the early breaking of DVD's CSS mechanism, the powers that be in content ownership demanded that systems be developed to prevent the unauthorized illegal duplication of data. Towards that end the criteria for new media technologies remains heavily slanted in favor of security restrictions (see *Chapter 4*).

Table 3.1 Hollywood Studio Requirements for Next-Generation Formats - October, 2003

Proper launch timing	No cartridge
Best quality	Play CDs and DVDs
Up to 1920 × 1080 video	Advanced content protection
Uncompressed/compressed multichannel audio	Prevent casual copying
Multistream audio mixing in player	Deter professional piracy
Adequate capacity	Protect all outputs
2-hours or more for a movie	Watermarking
Bonus content, often in SD	Regions
Advanced features	Internet connection
Single format for PC and CE, including games	Interactivity (Network connection)

The Makeover from DVD to Blu-ray Disc

As most everyone is familiar with DVD and will inevitably compare it to Blu-ray, the following sections present the features that were introduced with DVD-Video and describe how these features were adapted for Blu-ray. The term *player* here also applies to software players on computers, as well as other devices such as video game consoles that have the ability to play DVD-Video or Blu-ray Discs.

High-Quality Digital Video and Audio

More than 95 percent of Hollywood movies are shorter than 2 hours and 15 minutes, so 135 minutes was chosen as the goal for DVD. Uncompressed, this much video would require more than 255 gigabytes[1] of storage capacity. DVD exclusively uses MPEG-2 compression to fit digital video onto a disc. The length of a movie that can fit on a standard DVD depends almost entirely on how many audio tracks are available and how heavily the video is compressed. Other factors do come into play, such as, the frame rate of the source video (24, 25, or 30 frames per second), the quality of the original images (soft video is easier to compress than sharp video or video with a lot of film grain, and clean video is easier to compress than noisy video), and the complexity (slow, simple scenes are easier to compress than scenes with rapid motion, frequent changes, and intricate detail). In any case, the average Hollywood movie easily fits on a standard-definition DVD.

Blu-ray adds two more video compression systems, AVC and VC-1. All three provide video in widescreen, high-definition as well as standard-definition. Blu-ray also provides the ability to decode and play a second stream of video in a picture in picture (PIP) window. The PIP window can be scaled to smaller sizes and can be made partly transparent to reveal the video underneath. Early Profile 1 Blu-ray players did not have the PIP feature.

Widescreen Movies

Television and movies shared the same rectangular shape until the early 1950s when movies began to get wider. Television stayed unchanged until recently. With the advent of HDTV, widescreen became the norm, and it is becoming harder to find "narrowscreen" TVs in the old 4:3 shape. DVD supported widescreen with anamorphic video, which was squeezed into a 4:3 shape to fit on the disc in standard TV format, then expanded back out by widescreen displays.

Blu-ray uses 16:9 widescreen high-definition video formats along with older 4:3 standard-definition video. Blu-ray players generally windowbox the narrow video for widescreen display. Blu-ray players can letterbox widescreen video when connected to older 4:3 televisions, and some can squeeze the video anamorphically to be compatible with older widescreen televisions. Aspect ratios are discussed further in *Chapter 2*.

Multiple Surround Audio Tracks

The DVD-Video format provides for up to eight soundtracks to support multiple languages or supplemental audio. Each of these tracks may include surround sound with 5.1 channels of discrete audio[2]. Audio may be encoded using *Dolby Digital* (AC-3), *DTS Digital Surround*, or MPEG-2 audio. The 5.1-channel digital tracks can be downmixed by the player for compatibility with legacy stereo systems and *Dolby Pro Logic* audio systems. An option is also available for better-than-CD-quality PCM audio. Almost all DVD players include digital audio connections for high-quality output.

---

[1] Digital studio masters for standard-definition video generally use 4:2:2 10-bit sampling, which at 270 Mbps eats up over 32 megabytes every second. High-definition video is even more bit hungry. A 4:4:4 12-bit master of 1080p24 video devours 213 megabytes per second. (See Table 2.1.)

[2] Discrete means that each channel is stored and reproduced separately rather than being mixed together (as in Dolby Surround) or simulated. The ".1" refers to a low-frequency effects (LFE) channel that adds bass impact to the presentation.

Features

The usefulness of multiple audio tracks was discovered when digital audio was added to laserdiscs, leaving the analog tracks free. Visionary publishers, such as *Criterion*, used the analog tracks to include audio commentary from directors and actors, musical sound tracks without lyrics, foreign language audio dubs, and other fascinating or obscure audio tidbits. Most DVD players allow the user to select a preferred language so that the appropriate menus, language track, and subtitle track are selected automatically, when available. In many cases, the selection also determines the language used for the player's on-screen display.

For DVD-Audio, quality was significantly improved by doubling the PCM sampling rates of DVD-Video to 96 kHz, improving support for multichannel PCM audio tracks and audio downmixing, and using MLP lossless compression to increase disc playing times. A single-layer DVD-Audio disc can play 74 minutes of super-fidelity multichannel audio or over seven hours of CD-quality stereo audio.

Blu-ray ups the ante with *Dolby Digital Plus* and *DTS-HD High Resolution Audio*, which provide higher sampling rates and up to 7.1 channels of surround audio. Higher sampling rates than DVD can be used with uncompressed audio, as well as with lossless audio compression using *Dolby TrueHD* or *DTS-HD Master Audio*. A dual-layer Blu-ray disc can hold over ten hours of highest quality uncompressed stereo (192/24 PCM) or over 200 hours of 5.1-channel Dolby Digital.

Blu-ray adds sound effects that can be incorporated with the main audio, such as, *click* sounds for menus or *pop* sounds when popup information appears during movie playback. Audio mixers are built into Blu-ray players so that a secondary audio stream can be overlaid on the main audio. This opens up a large variety of options, such as, adding a director commentary or fan commentary, swapping in different audio soundtracks, providing comedic "heckling" of scenes in a movie, creating a censor's "bleep" sounds to block certain words in the dialog, karaoke track mixing, and more.

Subpictures and Subtitles

The primary video on a standard DVD can be supplemented with one of 32 subpicture tracks for subtitles, captions, and more. Unlike existing closed captioning or teletext systems, DVD subpictures are graphics that can fill the screen. The graphics can appear anywhere on the screen and can create text in any alphabet or symbology. Subpictures can be Klingon characters, karaoke song lyrics, *Monday Night Football*-style motion diagrams (telestrator diagrams), pointers and arrows, highlights and overlays, and much more. Subpictures are limited to a few colors at a time, but the graphics and colors can change with every frame, which means subpictures may be used for simple animation and special effects. Some DVDs use subpictures to show silhouettes of the people speaking on a commentary track, in the style of the *Mystery Science Theater 3000* TV show. Being able to see the gestures and finger pointing of the commentators enhances the audio commentary.

A transparency effect can be used to dim areas of the picture or make picture areas stand out. This technique can be used to great effect for educational video and documentaries. Other options include covering parts of a picture for quizzes, drawing circles or arrows, and even creating overlay graphics to simulate a camcorder, night-vision goggles, or a jet fighter cockpit.

Blu-ray extends this concept in subtle but important ways. It implements two simultaneous graphic overlays with significantly more color and transparency options than DVD (256 colors and transparencies instead of DVD's 4 colors and 16 transparencies). This means that animated overlays, menus, popups over video, and the like, are smoother, more detailed, and more vivid. A powerful computer language can be used to display text, graphics, and animation over the video.

Blu-ray also adds up to 32 text subtitles. DVD was severely limited because subtitle text had to be pre-rendered as graphic subpictures, a complicated and error-prone process. Once on the disc they were immutable. Subtitles in Blu-ray can be taken directly from text sources in almost any language and displayed in various fonts, colors, and positions on the screen. Players and discs can provide settings for the viewer to choose display preferences for subtitle text.

Musical Slideshows

The DVD-Audio format added a slideshow feature for showing pictures while audio plays. The pictures may appear automatically at preselected points in the program, or the viewer can choose to browse them at will, independent of the audio. Although the DVD-Video format also supports programmed slideshows, it does not support browsable pictures that do not interrupt the audio. This feature is handy for displaying lyrics of songs as they play. Blu-ray carries on the tradition of browsable slideshows with uninterrupted music.

Karaoke

DVD includes special karaoke audio modes to play music without vocals or to add vocal backup tracks, along with information for identifying the music and singer, such as, male vocalist, female soloist, chorus, and so on. Few DVD players support the karaoke features. DVD's subtitle feature is also essential for karaoke, providing up to 32 different sets of lyrics in any language, complete with bouncing balls or word-by-word (or ideogram-by-ideogram) highlighting.

Blu-ray does not need a special karaoke mode or karaoke player features because even the most basic Blu-ray player has everything needed to control audio mixing and show animated on-screen lyrics.

Different Camera Angles

One of the most innovative features of DVD is the option to view a scene from different angles. A movie or television event may be filmed with multiple cameras, which can then be authored so that the viewer can switch, at will, between up to nine different viewpoints.

In essence, camera angles are multiple simultaneous video streams. As you watch the DVD, you can select one of the nine video tracks just as you can select one of the eight audio tracks. The storytelling opportunities are intriguing, and many creative uses have appeared on DVD. A movie about a love triangle can be watched from the point of view of each main character. A murder mystery ends with multiple solutions. A nature scene can be played at different times of the day, different seasons of the year, or different points in time. Music videos include shots of each performer, enabling viewers to focus on their favorite band

Features

member or to pick up instrumental techniques. Classic sports videos are designed so that armchair quarterbacks have control over camera angles and instant replay shots. Exercise videos allow viewers to choose their preferred viewpoints. Instructional videos provide close-ups, detail shots, and picture insets containing supplemental information.

The options are endless, and the multiangle approach has become a new tool for film-making and video production. The disadvantage of this feature is that each angle requires additional footage to be created and stored on the disc. A program with three angles always available can only be one third as long if it has to fit in the same amount of space as a single-angle program. Blu-ray multiangle video is essentially the same as DVD, although there is more room on the disc for multiangle programs.

Multistory Seamless Branching

A major drawback of almost every video format that came before DVD, including laserdiscs, Video CD, and even computer-based video formats, such as *QuickTime*, is that jumping to another part of the video caused a break in play. DVD-Video overcame this drawback and achieved completely seamless branching. For example, a DVD may contain additional "director's cut" scenes for a movie but can jump over them without a break to recreate the original theatrical version.

The multistory branching feature creates endless possibilities of the mix-and-match variety. At the start of a movie, the viewer could choose to see the extended director's cut, with alternate ending number four, and the punk rock club scene rather than the jazz club scene, and the player would jump around the disc, unobtrusively stitching the selected scenes together. Of course, this requires significant additional work by the director or producer, and most mass-market releases skip this option, leaving it to small, independent producers with more creative energy.

Blu-ray provides the same multi-story branching feature as DVD. Because Blu-ray players have relatively larger data buffers there are fewer restrictions on seamless branching, and even the switch between layers can be seamless, whereas on DVD players it caused the video to freeze for a moment.

Parental Lock

DVD includes parental management features to block playback of an entire disc or to control which parts of the disc or which versions of a multi-rated movie can be viewed. A player can be set to a specific parental level using a password protected onscreen menu. If a disc with a rating above the selected level is put in the player, it will not play. Each program on the disc can have a different rating.

A DVD may be designed so that it plays a different version of the movie depending on the parental level that has been set in the player. By taking advantage of the branching feature, scenes can automatically be skipped over or substituted during playback, usually without a visible pause or break. For example, a PG-rated scene can be substituted for an R-rated scene, along with dialog containing less profanity. This requires that the disc be carefully authored with alternate scenes and branch points that do not cause interruptions or discontinuities in the soundtrack.

Sadly, fewer than one percent of DVDs use the multi-rating feature. Most of the ones that do, ironically, use it to include unrated scenes rather than provide a more family-friendly version. Hollywood studios are not convinced that the demand merits the extra work involved, which may include shooting extra footage, recording extra audio, editing new sequences, creating seamless branch points, synchronizing the soundtrack across jumps, submitting new versions for MPAA rating, dealing with players that do not implement parental branching properly, having video store chains refuse to carry discs with unrated content, and much more. The few discs that have multirated content do not have standard package labeling or other ways to be easily identified.

Another option is to use a software player or a specialized DVD player that can read a playlist or a use a filter telling it where to skip scenes or mute the audio.[3] Playlists can be used to retrofit thousands of DVD movies that have been produced without parental control features.

Blu-ray implements parental control in a more simple and flexible way. The player has a parental level that can be set using a password protected onscreen menu but, otherwise, the player does not check the disc or control playback. Commands on the disc can check the parental level of the player and respond accordingly — stop playback, choose different paths through a movie, substitute a different audio track, play alternative versions, and so forth.

Menus

To provide access to all the content on a disc, the DVD-Video format includes on-screen menus. Menus are used to select from among multiple programs, choose scenes (chapters) within a program, choose audio tracks and subtitles, navigate through multilevel or interactive programs, and more. Playback can stop at any point for interaction with the viewer, or selectable "hot spots" can be visibly or invisibly placed over live video.

For example, a DVD may have a main menu from which you can choose to watch the movie, view a trailer, watch a "making of" featurette, or peruse production stills. Another menu may also be available from which you can choose to hear the regular soundtrack, foreign language sound track, or director's commentary. Selecting a "bonus features" option from the main menu may bring up another menu with options such as cast and crew bios, script pages, storyboards, and outtakes.

Menus were a new and unfamiliar feature to many people when DVD was introduced.[4] A decade later, when Blu-ray rolled out, DVD menus had become comfortably familiar yet, in some ways, frustratingly primitive. (See *Chapter 10, Interactivity*, for more on limitations of modal menus.) There was an expectation that Blu-ray would take menus out of the stone age, which it did.

Blu-ray menus can popup over the video at any time, allowing the viewer to make selections in the middle of a program without interrupting the program flow. Blu-ray also implements multipage menus so that submenus appear quickly and without interruption in the

---

[3] *ClearPlay* is the primary company providing this capability.

[4] One newcomer to DVD complained that the picture and sound were great, but the discs only showed shortened versions of the movies over and over. It turned out that the hapless novice thought that the intro video loop in the menu was all there was.

Features

video and audio. Sound effects can be added, which provide useful feedback when on-screen buttons are selected. Beyond built-in menu support, the Blu-ray programming language can be used to construct just about any type of interactive user interface that can be imagined, limited only by the buttons on the remote control.

Interactivity

DVD can be even more interestingly interactive if the creator of the disc takes advantage of the rudimentary command language that is built into all DVD players. DVD-Video can be programmed for simple games, quizzes, branching adventures, and so on. DVD brought a level of personal control to video programs that simply did not exist before. While it is debatable just how much control the average couch potato craves, directing the path and form of a visual experience is certainly an appealing option to some. Diverse adventurous souls in the creative community embraced the new genre of nonlinear cinema afforded by DVD.

For example, some music video DVDs provided an editing environment where the viewer could choose music, scenes, and so on, to create their own custom version. Instructional videos followed lessons with comprehension quizzes. If a wrong answer was selected, a remedial segment could be played to further explain the topic.

Again, Blu-ray upped the ante with new and exciting levels of user interactivity. While continuing to offer "standard content" similar to DVD with the HDMV (high-definition movie) mode, Blu-ray added BD-J, a specially tailored version of *Java*, the powerful computer programming language. BD-J puts Blu-ray interactivity into an echelon far beyond DVD. Features previously seen only on computers, Websites, and game consoles can now be part of the experience. Sophisticated applications can be created that go well beyond the simple menus and games that standard navigation offers. For instance, imagine an application that allows the viewer to choose from a pre-selected set of stillframes from a movie, generate a coloring book version of that frame, color it with painting tools, and e-mail the new image to a friend's Blu-ray player or upload it to a computer for printing. Or, a complete arcade-style video game played with the buttons on the remote control. More in **Chapter 10, Interactivity**.

Customization

As noted previously, players may be customized with a parental lock. Other options can be set on a Blu-ray player to customize the viewing experience. Most Blu-ray players can be set for the preferred audio language, text and subtitle language, menu language (for disc menus and, sometimes, for the player's own menus), subtitle style, and country code. For example, if you are studying French, you can set your preferences to watch movies with French dialog and English subtitles. If these are available on the disc, they will be selected automatically.

Instant Access

Consumer surveys have indicated that one of the most appealing features of DVD and Blu-ray is that there is never a need to rewind or to fast forward. Convenience and time can be inestimably important to consumers — just consider our penchant for microwave ovens, electric pencil sharpeners, electric windows, and escalators. A DVD or a Blu-ray player can obligingly jump to any part of a disc -program, chapter, or time position — in less than a second.

Features

Trick Play

In addition to near-instantaneous search, most DVD and Blu-ray players include features such as freeze-frame, frame-by-frame advance, slow motion, double-speed play, and high-speed scan. Most players scan backward at high speed, but due to the nature of video compression, many cannot play at normal speed in reverse or step a frame at a time in reverse.

Access Restrictions

The author of a DVD or Blu-ray disc can choose to restrict user operations such as fast forward, chapter search, and menu access. Almost every button on the remote control can be blocked at any point on the disc. This is not always a benefit to the viewer (when locked into the FBI warning or "coming attractions" at the beginning of a disc), but it is helpful in complicated discs to keep button-happy viewers from going to the wrong place at the wrong time.

Durability

Unlike tape, optical discs are impervious to magnetic fields. A disc left on a speaker or placed too close to a motor will remain unharmed. Discs are also less sensitive to extremes of heat and cold. Because they are read by a laser that never touches the surface, the discs will never wear out — even your favorite one that you play six times a week or the kids' favorite one that they play six times a day. Discs are susceptible to scratching, but their sophisticated error-correction technology can recover from minor damage. Blu-ray includes a special *hard coat* that helps protect the surface from scratches.

Playback on PCs

DVDs can include more than what is playable on a standard player. Computer software, such as, screen savers, games, and interactive enhancements can be included on a disc. Computers, video game consoles, even cable settop boxes can play DVDs, sometimes with enhanced DVD features. A single disc might contain a movie, a video game based on the movie, an annotated screenplay with "hot" links to related scenes or storyboards, as well as the searchable text of the screenplay's novelization complete with illustrations and hyperlinks. The disc might contain links to Internet Websites with fan discussion forums, related merchandise, and special promotions.

The quantum leap in power and features on Blu-ray players enables them to do almost everything that formerly could be done only when a DVD was played in a computer. Many of the features that were pioneered to enhance DVDs when played in computers were later incorporated into the Blu-ray format. Of course Blu-ray discs can also be played in computers equipped with Blu-ray drives.

Network Connection

Blu-ray players that implement the Profile 2, *BD-Live*, feature set can be hooked up to the Internet, directly or through a wired or wireless home network. Extra content can be delivered via the network connection, such as, updated information about the cast and crew including previews of their latest films, additional language subtitles, games, a new director's

Features

commentary, popup notes for music videos, or dubbed audio tracks that were not ready when the disc was released. Discs can provide for e-commerce transactions, allowing the viewer to browse a collection of products and perform a purchase online. Network support does not go so far as to implement general Web browsing capabilities, but it does provide for streaming video and audio, uploading and downloading files, and communicating with devices, such as, cell phones, computers, networked printers, and even other Blu-ray players.

The world of DVD was circumscribed — cozy but cramped. Content creators soon bumped into every wall and learned the limits of the technology. There was room for creativity, but only within the compartment delineated by the format specifications. Blu-ray blows the walls away, especially with connected players. Blu-ray players are specialized yet incredibly powerful media processing systems cleverly disguised as innocuous plastic rectangles that sit next to a TV and connect to the Internet. They do not pretend to be computers or Internet appliances — failures such as WebTV and other browser-equipped settop boxes have proven that few people want to surf the Internet from their couch — and that is both the challenge and the opportunity: how to craft interactive, connected experiences that appeal to someone putting their feet up on the sofa to watch a movie.

The possibilities for creative use of BD-J programs connected to the Internet and to other devices are boundless. In addition to straightforward downloading of video, audio, graphics, text, and interactive programs from content provider servers, online Web pages can change the way a disc works in players, driven by updates from individual users or entire online communities. Information flows the other way as well, so that user interactions with the discs in their players can bubble up to Web sites. Enclaves of enthusiastic and inventive content developers scattered across the world are germinating ideas, from sensible to harebrained —

- Multiplayer games between BD players or even between BD players, computers, and mobile phones
- Movie blogs to share with other BD player owners or upload to the Internet
- Live trivia games
- Multiplayer games between BD players or even between BD players, computers, and mobile phones
- Online "leaderboards" showing high scores, number of Easter eggs found, or other achievements that bring coveted status amongst a group of likeminded users
- Online chats for fans simultaneously watching a movie (*Rocky Horror Picture Show* with virtual toast and rice?)
- Shareable video mashups
- A feature to upload your face to be superimposed on actors or turned into an animated character
- The option to record your voice to add to a scene, perhaps morphed ala Chipmunks
- Online and in-player progress tracking and charting for exercise videos
- Home education discs that connect students together in a virtual classroom
- Progressive story threads, where each person contributes a snippet of text or voice or video to an ever-changing narrative

Features

- Face recognition search for when you know you have seen that actor before but you can't remember where
- Cross promotion, affiliate marketing, and contextual selling (buying something from a popup e-commerce window on the player)[5]
- "Movie mail" between players, including personalized postcards and favorite scene lists
- Automatic "what I'm watching now" notifications on micro-blog sites, such as, *Twitter*
- Using a mobile phone or PDA as a "super remote" to control the Blu-ray player
- Uploading desktop pictures, "skins," songs, ringtones, games, and more from the disc in the player to computers, mobile phones, music players, and other devices
- Two-screen viewing — synchronizing related content (screenplays, trivia, historical notes, lyrics, and the like) to laptops or WiFi-enabled mobile phones
- and the list goes on and on and on...

Some of these ideas will plop into the cyberpond and deservedly sink to the bottom without a ripple, but others could generate huge waves of interest and cultivate new genres of media interaction and social networking. A lot of people thought *YouTube*, *Twitter*, *MySpace*, and the like, were faddish notions that would quickly fade, but look what has happened.

Persistent Storage and Local Storage

Unlike DVD players, which are required to perform self-administered amnesia when the disc is taken out, Blu-ray players have persistent storage, memory space set aside for each disc to store a small amount of information such as game high scores, last viewed position, and viewer preferences.

Players also have local storage (which technically includes persistent storage) for holding content downloaded via the network connection or transferred from disc for later playback. Local storage may be a small, fixed amount built into the player, or it may be expandable to many megabytes or gigabytes with an internal hard drive, external memory card, or external hard disk drive.

Authorized Copying and Authorized Recording

As touched on elsewhere in this book, Blu-ray technology emerged in the middle of the slow transition from physical to virtual delivery of content. Around the time Blu-ray was introduced, various approaches were being attempted to allow consumers to legally copy

---

[5]One brainstorm that has been around since the first days of WebDVD is the notion that you'd say, "Look at that shirt she's wearing, I want one just like it," and buy it on the spot. It's never been pulled off, and probably never will be, but, on the other hand, there's huge potential (good or bad, depending on your viewpoint) in promoting a sequel to viewers who just finished watching the first movie, offering the soundtrack CD or downloadable music files from a film or concert, advertising action figures while the credits roll, and so on. Consider the art of Amazon.com's "We have suggestions for you..." and how it can be applied after viewers have been deeply engaged with a video.

Features

their DVDs to recordable discs, hard disks, portable devices, and the like, and conversely to legally download and burn commercial content to recordable DVDs using CSS and other protection systems. Lessons learned from these initiatives made their way into the *Advanced Access Content System* (AACS) so that similar features could be available for Blu-ray.

There are essentially two paths through the convergence zone between physical and digital — yin and yang bridges, if you will. One is from physical to virtual, where a disc or other physical medium is copied or translated into digital form for transmission to other devices. This is *managed copying* or *authorized copying*. The bridge in the other direction goes from virtual to physical, where digitally distributed content is put onto physical media, such as, a disc or a memory card. This is called *managed recording*, *managed burning*, or *authorized recording*.

Authorized Copying

As digital technology has increased the demand for content to play on more than one device, some content owners have espoused the concept of "buy once, play anywhere." For this to work there must be a way to represent a customer's purchased access rights to a particular set of content. This can be done with an online "digital locker" or with a physical token such as a disc or printed serial number.[6] Based on the representation of ownership, an authorized copy can be made into an approved content protection or *digital rights management* (DRM) technology with permission from the content provider, who has full control to set the price, number of copies, allowed formats, viewing period, and so on.

There are essentially four models for copying content based on physical media. The first two models are supported by the *managed copy* feature of AACS, which is sometimes called *mandatory managed copy*, since at least one copy of almost every disc must be permitted.

Duplication. An equivalent copy is made from the original disc to a similar recordable disc. For example, a BD-ROM movie disc is copied to a BD-R disc.

Conversion. A copy is made from the original disc and adapted in one or more ways to fit the destination — new physical medium, new video or audio encoding, new application format, new picture size (resolution), or new content protection system. For example, a Blu-ray Disc is converted to *Microsoft Windows Media* format (video and audio are transcoded, menus and interactivity are discarded), at a lower resolution for display on portable media players, and stored on a flash memory card with Windows Media DRM file encryption.

Digital Copy. The original disc contains an alternative version of the content or a subset of the content in different form, intended to be copied from the disc. This is also called *second session* or *digital file*. For example, a Blu-ray Disc includes a low-resolution version of the video encoded in H.264 format and protected with *Apple FairPlay* DRM so that it can be copied using *iTunes* software and played on a compatible device such as a *Mac*, *iPod*, or *iPhone*.

E-copy. The original disc is used as the token of ownership to authorize the

---

[6]It is possible to combine multiple access rights representation mechanisms, such as, by registering a purchased Blu-ray movie in an online system that also allows purchasing of music downloads, streaming videos, and mobile phone ringtones. The system — or other allied services if it acts as a clearinghouse — can then grant various access rights to each owned content item.

download of the same content or related content over the Internet. For example, a Blu-ray Disc is inserted into a computer drive to unlock the download of the music soundtrack as MP3 files that can be played on a computer or portable music player or burned to CD.

Most authorized copy approaches incorporate the following elements —

- A unique identifier (ID) on the disc, either in a) electronic format such as serial number in the BCA or in a file on the disc, which the copy system can read directly, or b) in printed form on the disc or packaging, which must be entered by the user into the copy system.
- An authorization server to verify the unique ID (and to store it in a central database so that attempts to use the same ID again can be blocked or handled appropriately)
- An optional transaction process for the consumer to pay, watch an advertisement, buy something else, join a mailing list, take a survey, or engage in some other transaction before being allowed to make the copy.
- A license server that issues the license and usage rights associated with the copy. The license server is usually part of the DRM used to protect the copy.
- The actual process of copying, converting, or downloading the content.

AACS approves various *managed copy output technologies* (MCOT), including copying to a recordable Blu-ray disc using AACS, copying to a recordable DVD using protection systems such as CPRM or VCPS, copying to memory cards using a protection system such as CPRM or *Sony MagicGate*, and copying to digital files using a protection system such as *Windows Media* DRM.

Authorized Recording

Given the hundreds of millions of Blu-ray players, DVD players, and CD players that play discs, and the millions of portable devices that play video from memory cards, plus the recognition that it will take another century or so before every media device on the planet is connected to a vast electronic delivery network, there is clearly a need to put electronically delivered content onto physical media to bridge the gap to the player. BDs and DVDs are the ultimate *sneakernet* for video.[7] And, there's the tangibility factor. At some level, humans will never get past the ingrained notion that you do not really own something unless you can hold it in your hands (and show it off to your friends and neighbors).

The fundamental idea of authorized recording is that protected content, usually delivered in a digital rights management (DRM) system, can be transferred to portable physical media

---

[7]Sneakernet refers to the fact that sometimes it is easier, faster, or cheaper to simply copy data to storage media such as disc or memory card, walk to the destination, and copy it over. In other words, in the OSI seven-layer model, the physical layer that is usually implemented as Ethernet or WiFi is replaced with feet in sneakers. For devices not connected to a network, this is typically the only choice. Well, other than RFC 1149, specifying avian carriers, which was actually implemented on April 28, 2001 in Bergen, Norway using carrier pigeons. In a sense, *Netflix* operates the biggest sneakernet in the world, shipping approximately 2 million DVDs per day. This equates to 510 petabytes per month, which is astonishing considering that the total estimated Internet traffic in the U.S. in 2007 was 750 to 1250 petabytes per month (ignoring business traffic it was around 500 to 825). Sources: *Netflix, University of Minnesota Internet Traffic Studies* (MINTS).

Features

and bound to the media with an encryption method that is tied to that particular piece of media. A related concept is a *secure move*, where the content is transferred between devices or media in a way that makes the original unplayable, at least while the copy exists.

There are basically two types of authorized secure recording —

Transmitted. Content marked as *copyable with restrictions* is broadcast to multiple users over the air, via cable, over the Internet, or so on, or is moved or copied from another device or medium. In other words, the content is not free to copy, but certain types of copying are pre-authorized. Usually this means a *copy once flag* is attached to the content. In most cases, *generational copies* — copies of the first copy — are not allowed. Therefore, the pre-authorization stipulates that the recording can only be made into a content protection system that does not allow additional copies to be made.

Transacted. Content is delivered to a single user based on a purchase or other transaction that gives the user the right to make one or more copies. *Transacted recording* is often used for content that may not sell enough discs to make it worth the expense of manufacturing, distribution, warehousing, and returns. Instead, the content is sold online or in retail —taking negligible shelf space — and is recorded on demand at relatively low cost. This enables physical formats such as Blu-ray and DVD to move much farther into *The Long Tail*.[8] Other portable media such as memory cards can be used, but their relatively high price compared to optical media makes them impractical for permanent collections, so we'll focus on discs. Transacted copying falls into two broad categories —

Manufacturing On Demand (MOD). MOD is used in two environments. In *factory MOD*, merchants use centralized fulfillment centers where content is stored in digital form and recorded to disc after a customer places an order, which is then shipped. This is a form of zero-inventory manufacturing that can be more cost effective than traditional inventory distribution. The second environment is *retail MOD*, also called *retail managed recording* (RMR). Retail stores or other locations such as airports, train stations, coffee shops, fast food outlets, truck stops, and even city streets, can be outfitted with stand-alone kiosks or with touch-screen panels connected to behind-the-counter media recording stations. Customers browse through hundreds to tens of thousands of titles (more than could ever fit on a retail shelf), choose the one they want, and have it recorded, labeled, and packaged in minutes while they shop or wait for their prescription, their plane, or their train.

---

[8]Chris Anderson, editor of *Wired Magazine*, injected the concept of *The Long Tail* into mainstream consciousness in an October, 2004 article. He argued that technology can bring about low enough cost and wide enough availability to a large enough market that it becomes profitable to sell hard-to-find items. That is, providers with low-volume sales of a lot of content may do as well as providers who have a few top-selling hits.

Features

Electronic sell-through (EST).[9] Also called *consumer download and burn* (CDB) or *network download*. Customers use a computer or consumer electronics device, or even a mobile phone, to browse titles available from an online *content service provider* (CSP). Their purchases are downloaded to their home system and burned to disc without them ever needing to leave the house.

A key difference between transmitted recording (or secure recording) and transacted recording (or managed recording) is that with secure recording the copy is pre-authorized. That is, for content that is appropriately marked, the user is able to record it with an approved technology, such as AACS, often without being aware that a content protection system is engaged. In contrast, managed recording requires authorization from the content owner or content service provider, usually associated with a transaction by the user. Often an Internet connection to an authorization server is required, and in the case of AACS, the unique media identifier of the target recordable disc must be sent to the authorization server to cryptographically bind the content to the disc.

All of the above are accomplished on Blu-ray Discs through provisions of AACS. AACS *Recordable Video* is for recording protected video from transmission sources. AACS *Network Download* mode is used for managed recording of AACS *Prepared Video* for MOD and EST. AACS Prepared Video is also used for managed copying from a pre-recorded Blu-ray Disc to a recordable Blu-ray Disc. Prepared Video content has the necessary certificates and hash tables to be encrypted using AACS.

When this book was finished in fall 2008, the AACS managed copy and managed recording features for Blu-ray were finally, after delays that had stretched over more than two years, ready to be put in place and launched in the second half of 2009. See **Chapter 4** for more details regarding AACS.

---

[9]In general terms, EST refers to any method of selling a permanent electronic copy, as opposed to a transient electronic copy that is streamed or has a limited viewing period or limited number of views. EST includes downloads of electronic files for playback on PCs and mobile devices. Therefore, the term EST should be carefully placed in context when referring only to copies that are recorded to portable media.

Chapter 4
Content Protection, Licensing, and Patents

For the Hollywood studios to support Blu-ray Disc™, they had to be assured that high-definition content would be protected and that it would not be possible to make a digital copy of the content. To achieve that goal, a number of protection schemes have been adopted by Blu-ray. These schemes can be segregated into four major categories — conditional access, protected distribution, protected transmission, and protected storage (see Figure 4.1).

Figure 4.1 Copy Protection

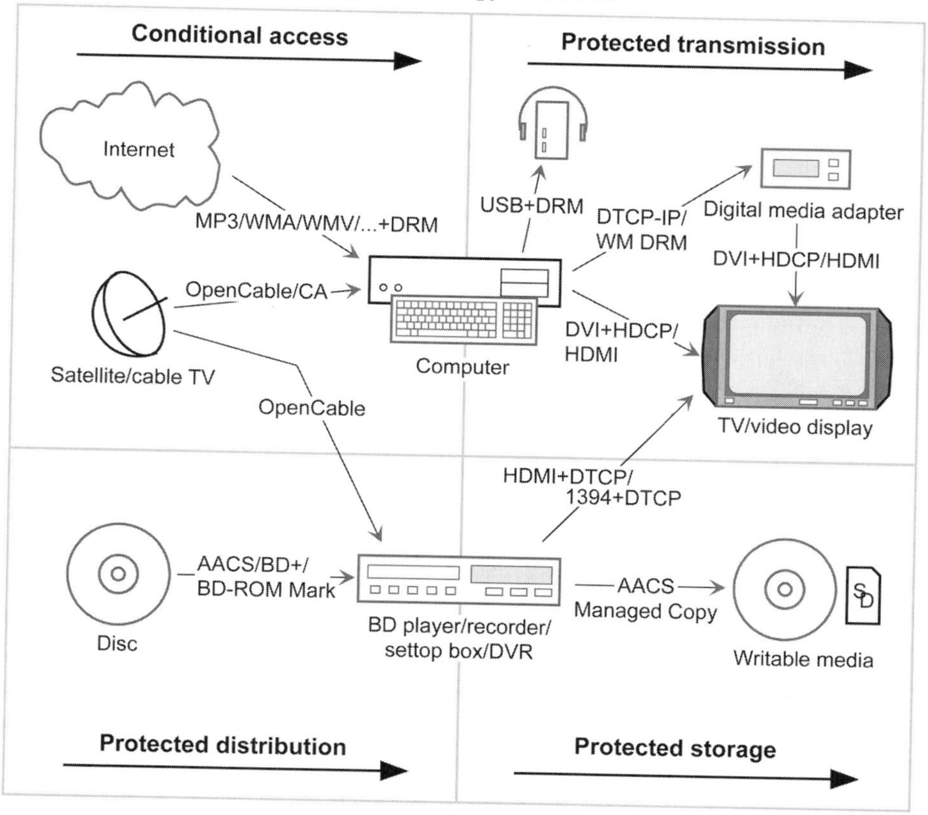

Content Protection, Licensing, and Patents

Conditional access. This method primarily applies to broadcast, cable, or Internet delivery, where unprotected distribution channels are used. Conditional access ensures that only the users who are allowed to receive or view the content, actually do.

Protected distribution. To maintain content security during the delivery to the enduser, the distribution method must be protected. For Blu-ray, a newly developed protection scheme called AACS (Advanced Access Content System) provides this security.

Protected transmission. Once the originally protected content has been decoded, most likely into baseband form, it is transmitted to either a video display or an audio receiver. In order to continue protection of the content from unauthorized copies, the transmission needs to be protected, as well. Even though there are a number of protection mechanisms for digital transmissions, a so-called *analog hole* exists for analog transmissions of high-definition video. In fact, during the rush to get the first high-definition televisions (HDTVs) into the market, they were released with unprotected HD analog inputs. Obviously, this would allow any recording device to easily make a high quality copy of the content.

Protected storage. To complete a full content protection ecosystem, it is important to provide means to properly protect copies when they are authorized. Broadcast content may carry "copy once" constraints that only allow it to be copied using an approved content protection system. Fair use doctrine historically suggests allowing one copy of a work for backup purposes. AACS follows these guidelines, and furthermore provides an entirely new protection system called *Managed Copy*. This system allows the consumer to create additional copies of a disc's content, but the content owner manages whether the copy is either free of charge or only possible for an additional fee.

Implementing Content Protection

Even though the pursuit of total content protection is an exciting exercise, realistically, it is in vain because it likely would make the disc impossible to play and would ultimately result in an unsatisfying user experience. However, many engineering hours have been and will be spent on improving content protection systems, to not only prevent digital copies but also to ensure degradation of the quality of analog copies. Otherwise, an analog copy would only be slightly degraded from the original digital source, and could very well be used as the source for creating multiple additional digital copies. Some of the AACS measures to prevent duplication are discussed later in this chapter.

An additional security measure is *watermarking*. Watermarking can either be used for preventing playback of unauthorized copies, which requires the playback devices to recognize watermarked content, or it can help to trace back to the origin of an unauthorized copy, thus allowing for appropriate action to be taken against the perpetrator.

In order for a content protection system to be effective, it requires three areas to work hand in hand — technology, business, and legislation. The technology piece clearly defines the specific protection method of encryption or watermarking to prevent digital or analog copies. The business aspect is usually addressed by license agreements requiring manufacturers to only release compliant devices and for content owners to produce reasonably priced

Content Protection, Licensing, and Patents

content. And, in order to ensure the effectiveness of the entire system, legal enforcement is necessary.

Content owners, manufacturers, and consumers have different expectations as to what they want out of a content protection system, and all their needs must be balanced to make content protection a success.

The requirements of content owners are —
- Maintain the highest quality possible of the content
- Protection against unauthorized use
- Robustness and tamper-resistance
- Renewable system (recovery after breach)
- Applicable to all forms of distribution or media
- Suitable for implementation on CE devices and PCs
- Low cost

The requirements of system manufacturers are —
- No effect on normal use of a system
- Low resource requirements
- Voluntary implementation
- Low cost

The requirements of consumers are —
- No effect on normal use of a system
- Maintain the highest quality possible of the content
- Fair-use copying
- No additional cost
- No limitations on playback equipment or environment
- No artificial barriers (inconvenience) to access the content

As some of the requirements may conflict at times, the resulting content protection system ends up being a compromise. However, in the case of the content protection systems for Blu-ray, each group seems fairly happy with the outcome. As a result, manufacturers are busy building players, studios are creatively producing content, and consumers are buying players and discs.

Although the Blu-ray Disc Association (BDA) defines which content protection systems are mandatory and which are optional for Blu-ray Discs, the technical specifications of these technologies are not defined by the BDA. Instead, separate groups under separate licenses establish them. However, the BDA, through its Content Protection Group (CPG), provides a framework for coordination with the industry groups. The CPG reviews the various content protection technologies and submits them to the BDA for approval. Once approved, the various groups within the BDA may amend the necessary specifications to incorporate support for the required content protection technologies.

There are a multitude of content protection systems in place covering encryption, watermarking, playback control, protection of analog and digital outputs, et cetera. Figure 4.2 pro-

Content Protection, Licensing, and Patents

vides an overview of the content protection ecosystem, describing the various pieces. The applicable technologies for Blu-ray are described in detail in the following sections of this chapter.

Figure 4.2 Content Protection Systems

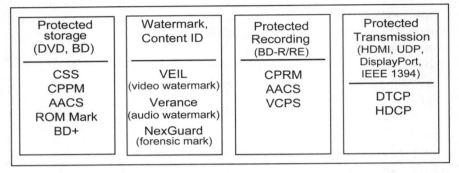

Advanced Access Content System (AACS)

The mandatory and, thus, primary copy protection system for Blu-ray is called AACS, Advanced Access Content System. As optical discs have been around for quite some time, it is only natural that AACS is based on its predecessors with improvements from lessons learned in the past. The copy protection system used for DVD is called Content Scrambling System (CSS). The CSS technology uses *Title Keys* that are stored in sector headers of a disc and *Disc Keys* that are stored in the control area of a disc. Even though both disc areas were considered to be very secure, as they cannot be directly accessed from a DVD-ROM drive, it took less than two years after format launch for CSS to be hacked. A security flaw in a software player exposed some of the keys, which was enough to reverse engineer the entire CSS system and disable the protection mechanism on all players.

The successor format to CSS is called Content Protection for Prerecorded Media (CPPM) and includes a number of improvements over CSS that added robustness and reliability. For instance, CPPM does not use Title and Disc Keys anymore, replacing them with identifiers that are stored in the disc lead-in area. This area does not exist on recordable media, making it impossible to create a duplicate copy of the disc, even if it could be decoded. This method was first used for protecting DVD-Audio discs.

Yet, Hollywood was trying to find a content protection system strong enough for their valuable high-definition content. Eight companies joined forces — IBM, Intel, Microsoft, Matsushita, Sony, Toshiba, Walt Disney, and Warner Bros. — to develop the AACS protection scheme. The resulting system promises a seamless, advanced, robust, and renewable content protection system. But it does not stop with the protection of optical discs. New business models, such as electronic distribution to home media servers or portable devices were also considered during the development process. In this way, the same content security could

Content Protection, Licensing, and Patents

be applied to multiple destinations providing a futureproof and manageable system for content owners.

Technical Overview

The underlying encryption technology used for AACS is the Advanced Encryption Standard (AES) with 128-bit keys. For key management, the *Media Key Block* (MKB) technology is used, which was first introduced for CPPM and proved to be very robust. The AACS licensing entity provides a set of *Device Keys* for each compliant device and a Media Key Block to a replicator for inclusion on each disc. Using the MKB on a disc in combination with a specific set of Device Keys, each compliant device is able to calculate a *Media Key*. This Media Key is then used to decrypt the *Title Key*, which then is used to decrypt the audio/video content on the disc. This encryption chain adds additional layers of security.

An additional security measure for AACS is *revocation*. The MKB on each disc includes a *Host Revocation List* (HRL) for software player applications and a *Drive Revocation List* (CRL) for ROM drives. The most up-to-date version of each revocation list is permanently stored in non-volatile memory of a device. In this way, disc playback in a computer will only be successful if neither of the two components is on a revocation list and if both components can verify each other as compliant, uncompromised devices. This revocation mechanism allows for future discs to contain updated Media Key Blocks with new revocation lists that disable the disc playback on compromised devices.

There is one drawback to device revocation, however. In order to revoke a device, the corresponding Device Key needs to be known. As Device Keys could be shared across a manufacturer's entire line of player models, it is a very difficult task to identify which individual device was compromised. A starting point to address this issue is the *Sequence Keys* process (Figure 4.3).

Figure 4.3 Sequence Key Example

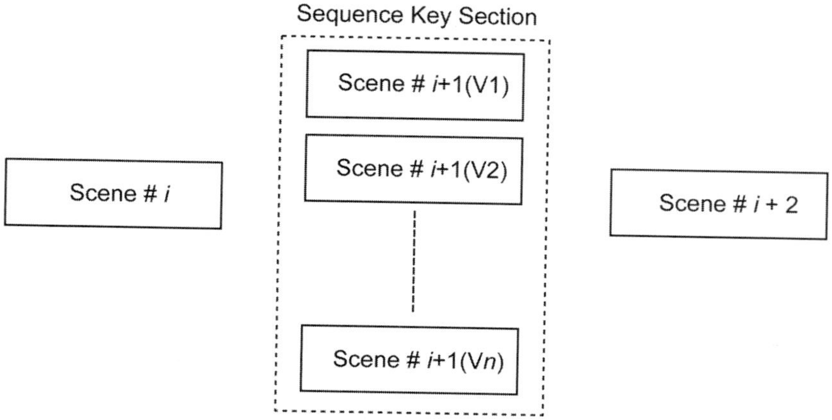

Content Protection, Licensing, and Patents

The approach is to divide a movie into multiple sections — up to six Sequence Key Blocks (SKB). Each SKB can contain up to 256 PlayLists, with each PlayList containing a set of PlayItems for Sequence Key segments and non-Sequence Key segments. Each segment contains multiple versions of the same scene, but with differing forensic marks embedded. Each individual player calculates a particular path through the movie, thereby allowing investigators to narrow down which particular player was used to pirate the movie. However, as the number of segments is limited, an individual player cannot be singled out by using only one disc. Instead, it will take a few discs to identify an individual player.

To address the threats from computer playback environments, yet another security measure was introduced by AACS — a proactive software renewal process. The ultimate aim of this method is to prevent hacks similar to the breaking of CSS. This proactive software renewal approach mandates that software players have to authenticate themselves on a regular basis to ensure that they are not compromised. If the authentication fails, the player is disabled. Further, an enhanced drive authentication has been introduced that enables new usage scenarios, such as, content distribution to home media servers or network connectivity for e-commerce applications.

Authentication and Decryption

The AACS decryption process is essentially a method for the management and distribution of cryptographic keys. In fact, there is not just one key but there are multiple keys necessary to encrypt AACS protected content.

Basically, the decryption process of the playback device starts by processing the most recent Media Key Block (MKB), either off the disc or from within the drive. Calculating this MKB with the Device Keys stored inside the player generates the Media Key. The Media Key is then cryptographically calculated with the Volume ID from the disc to create a *Key Variant Unit*. In combination with the Unit Key Files stored on the disc, this Key Variant Unit is used to decrypt the Title Key. And, finally, the Title Key is used to decrypt the content from the disc. These rather complicated processes are in place to properly secure the Title Key, since it is the only information needed to decrypt the content. By implementing such encryption chains, a level of added security can be achieved.

For communication between a computer drive and a software player application, there is an additional authentication process needed to verify the integrity of both components. The process is divided into two stages. Authentication between the host computer (a software player) and the drive unit ensures that neither component is listed on the respective Revocation List and both can be authenticated as being compliant with the specification. The result of this authentication process is a shared *Bus Key*. This Bus Key is used for the second stage of the process — *Bus encryption*. Because only these components have the Bus key, the data being sent between the components cannot be compromised, and any other unauthorized component is not able to make sense of the data. Finally, the drive calculates a *Read Data Key* that is encrypted by the Bus Key and sent over to the host for decryption. Once decrypted, the host is ready to receive actual data from the disc. Now, whenever the drive is reading data off the disc, is encrypted using this Read Data Key, because only that specific software player understands how to decrypt the data. Figure 4.4 outlines the AACS Authentication Process, including the Bus encryption.

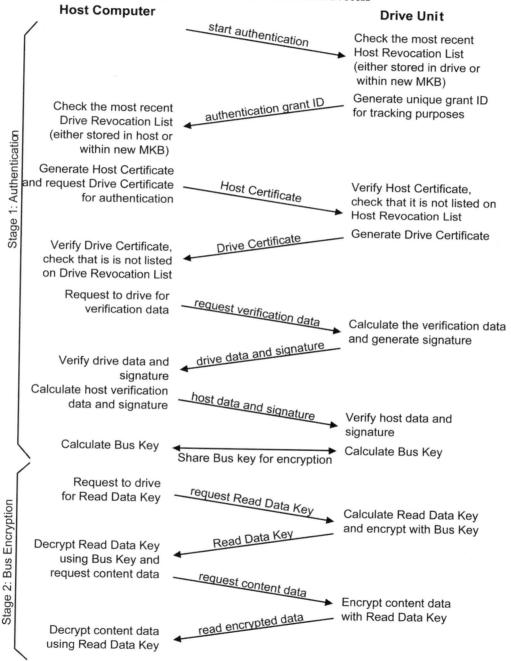

Figure 4.4 AACS Authentication Process

Content Protection, Licensing, and Patents

AACS Managed Copy

In addition to conventional disc-based security measures, new business models for electronic distribution were also considered during the AACS development process. A procedure termed AACS Managed Copy can authorize a consumer to move content from the disc to a home media server or a portable device. This also bridges the gap to new business models such as electronic distribution of content. The Blu-ray Disc not only contains the high-definition version of a movie but may also contain information pertinent to making an authorized copy of the disc. The Blu-ray Disc may also provide links to additional or related content that is available for download. The content owner fully controls, on a title-by-title basis, what is being offered and under what conditions. And, since the Internet drives all of this, the offers and rules can change at any time, keeping the system highly flexible.

Technically, an AACS protected disc provides a link to a specific Managed Copy Server (MCS), which authenticates a specific disc predicated on an embedded *Prerecorded Media Serial Number* (PMSN). This PMSN is a unique identifier stored in the *Burst Cutting Area* (BCA) for each individual disc. Each player contains a Managed Copy Machine (MCM) that is used to establish secure communication and transaction with the Managed Copy Server.

Once the MCS has presented the offers that are available for a given title, it is up to the customer to select an option — copy, or download, or other — and the transaction and authentication process between the MCM and MCS can then take place. As soon as the transaction is finished, either for a fee or free of charge based on the rules defined by the content owner, the copy process begins.

BD-ROM Mark

In addition to AACS, there is a second mandatory content protection system for replicated Blu-ray Discs that uses a physical perturbation of the disc called the BD-ROM Mark. The concept of physical marks has been around since the days of Super Audio CDs and it proved very effective back then. The Blu-ray Disc Association picked up on the concept of a physical mark and instituted its use in combination with AACS. The physical mark contains information that is required for AACS processing. In detail, a BD-ROM Mark contains a 128-bit key as payload on the disc as the Volume ID. As previously noted, a Volume ID is essential in the decryption process of a disc. Without this information, the content on the disc cannot be decrypted successfully. The BD-ROM Mark does not prevent a copy from being made, but it blocks playback of the copy if it is recorded onto another disc.

In addition to the added security provided by the BD-ROM Mark, the physical mark can also be used for tracking purposes in the event that pirated discs are discovered. This tracking function is made possible by the BD-ROM manufacturing process. When the Volume ID is calculated, the manufactured disc is irrevocably tied to the mastering facility. Hence, the place where the counterfeit discs were replicated can be identified in this way.

All prerecorded Blu-ray Discs (other than BD-5 and BD-9) must contain this mark, otherwise they cannot be played back. As BD Recorders are not able to reproduce this mark, it

Content Protection, Licensing, and Patents

makes it impossible to create bit-to-bit copies of a disc. And, tampering with the mark is very difficult as well given that only licensed BD players are able to read the BD-ROM Mark, and the mark is never transmitted outside the player. This process creates a very closed environment, thereby making this a formidable solution in the fight against piracy, even on a professional level.

For disc replication, a special piece of hardware called the *BD-ROM Mark Inserter* is required to encode and insert the Volume ID key, and to format the BD-ROM Mark properly. The Inserter is only available to licensed BD-ROM manufacturers under strict security obligations to ensure integrity of the process. Further, both the mastering facilities and the player manufacturers have to be licensed, thereby providing a substantial security threshold for protecting disc content, in addition to AACS.

BD+

The technology that is now called BD+ was initially developed by Cryptography Research, Inc. under the name *Self-Protecting Digital Content* (SPDC). Subsequently, it was acquired by Macrovision and is licensed by BD+ Technologies, LLC. The technology does exactly what is implied by the original name — it protects digital content on a disc and provides for countermeasures that actively react to attacks and to fix security holes, hence the digital content protects itself against attacks. The implementation is based on a dedicated *Security Virtual Machine* (SVM) inside each player. This SVM can be compared to the Java VM used for BD-J, but the SVM is much smaller. BD+ is mandatory for Blu-ray players and PC player software, but is optional for Blu-ray Discs. The only purpose of the SVM is to load security code (also called content code) from the Blu-ray Disc and run this code while the disc is playing. Essentially, the content code may perform all sorts of operations in a very secure environment. The main goal of BD+ is to detect security problems inside the player and react to correct the problem. Since the content code can also interact with the HDMV and BD-J layer of the disc, for example, by passing parameters back and forth, the behavior of the disc can be altered when a security problem is detected (e.g., show a warning screen). BD+ also provides additional features, such as, countermeasures, Media Transform, and Forensic Marking.

If security problems are found inside a player, countermeasures are intended to reverse them. This task requires intimate knowledge of the player software, so these countermeasures are typically developed by the player manufacturer in order to fix the security leaks identified. Countermeasures allow problems to be fixed on an individual basis, player by player, rather than impacting an entire series of players.

Media Transform allows the content code to execute transformations to the media on the disc (e.g., the video). The concept is to store heavily distorted video on the disc, which would not be viewable in its native form following AACS decryption. The BD+ content code takes charge of correcting the video by applying a Fix Up Table (FUT) during playback. In this way, a player cannot be easily hacked to circumvent the content code, as the video material would not be viewable. As the video is being manipulated during playback, this provides an excel-

Content Protection, Licensing, and Patents

lent opportunity for an additional security feature — *Forensic Marking*. It is even possible to individualize the forensic mark by inserting the player serial number into the payload. This allows identifying the specific player that was used for piracy, and allows for a countermeasure to be written for that specific player in order to fix the security problem.

Although Figure 4.5 outlines how the three major content protection systems for Blu-ray Disc — AACS, BD-ROM Mark, and BD+ — work together, there are a lot of additional possibilities for BD+, at its core, it provides a secure environment to run code and perform myriad operations.

Figure 4.5 Overview of BD-ROM Content Protection Systems

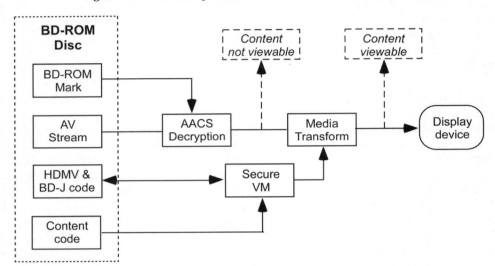

Content Protection on Recordable Discs

Because a large emphasis was put on the consideration of new business models during the development process of the AACS specification, it does not come as a surprise that content on recordable discs and other electronic distribution channels can also be protected using AACS. The method to be used for such things is called *AACS Protection for Prepared Video*. The technology is very similar to that used for prerecorded media (i.e. ROM discs), but on recordable discs it allows content protection for other business models, such as —

- **Electronic Sell-through** (EST)
  EST, also known as download and burn, describes the content typically distributed through the Internet that can be stored on a computer, or burnt onto a recordable disc.
- **Manufacturing on Demand** (MOD)
  This service provides a distribution channel that is particularly suited for niche content that may create discs on-demand. Typical scenarios

Content Protection, Licensing, and Patents

include individual disc replication as needed (once an individual order is received), or a kiosk model, where discs are created in stores based on a customer's order. This allows for a large amount of content (especially deep library content) to be prepared and stored on large servers, so that discs can be created when an order is received.

- **Managed Copy to Optical Discs**

    AACS allows endusers to make an authorized copy of a disc under the control of the content provider — Managed Copy. To make a copy to another optical disc, AACS provides the means to properly secure the copy.

Generally, prepared video follows the same criteria as AACS on prerecorded media, but it does enable different business models and distribution channels by gaining independence from physical storage formats. For example, it provides a robust protection for content delivered through EST or Managed Copy. It also enables the content owners to fully control the EST channels and allows for validating prepared video discs that were created with content owner approval.

From a technical perspective, there are a number of steps involved in creating a recordable disc with AACS security. To begin with, there are at least four different entities involved in the process —

❶ License Preparer

Entity preparing the content and issuing the license

❷ Prepared Video Authorization Server (PVAS)

Entity that hosts the key pairs, and licenses. The PVAS also generates the security tokens to be included on the recordable disc. The PVAS can host the content itself, or hosting can also be provided by another entity.

❸ Recorder

Provides information about the recordable media and ultimately records the data onto the disc

❹ AACS LA

As with any AACS process, the AACS Licensing Agency must sign the content certificate for each prepared video.

Although there are a number of entities involved, the technical workflow is rather simple. It starts with the License Preparer and the preparation of the content by applying the AACS technology. The License Preparer also has a *Public Key* of the PVAS and uses this key to generate a content certificate for the specific piece of prepared video content. AACS LA will have to sign this content certificate, which will subsequently be returned to the PVAS. From this point on, the content is available for distribution. If a copy of that content is made on a recordable disc, the content needs to be bound to the media. Every individual disc has a unique Media ID that the recorder sends to the PVAS along with a *Binding Nonce*, which is used to create the *Prepared Video Token* (PVT). The PVAS generates a PVT based on the

Content Protection, Licensing, and Patents

Content Certificate, the Media ID, and the Binding Nonce. This marries the content to an individual disc. Once that is complete, the prepared video (the content), the Content Certificate for the content, and the associated PVT are sent to the recorder. As every licensed Blu-ray player is required to validate the Content Certificate and the PVT prior to playback of the disc, this process assures the integrity of each individual AACS protected recordable disc.

The decryption process for recordable discs is akin to that for prerecorded discs. Yet, there are a few differences. Because recordable discs do not have a BD-ROM Mark, the Volume ID cannot be stored on the disc in the typical fashion. However, as the Prepared Video Token is calculated based on the Volume ID and the Content Certificate ID, the playback device can easily extract the requisite information from the PVT for the decryption process.

Watermarking

Watermarking is a process that embeds information within the content yet remains transparent to the enduser, thus making it difficult to be removed or to be easily changed. If a watermark is present, certain digital video frames and/or segments of audio are permanently marked with noise that is supposed to be undetectable by human eyes or ears. In practice, there is considerable debate about how undetectable watermarks really are. To reduce the impact of watermarks on the user experience, watermark intensity can be adjusted based on the content itself. As an example, for a very dynamic piece of content, the amount of watermarking could be reduced to decrease the possible impact, whereas for a more static scene, the amount could be increased. The watermark is essentially a digital signature that can be recognized by recording and playback equipment. Regardless of the digital or analog transformation of the content, the watermark signature remains embedded. Watermarking is not really protecting the content, but it identifies the status of the content.

Using watermarking in conjunction with a content protection system usually carries *Content Management Information* (CMI) data. Whenever the content is played back, a compliant device recognizes the CMI carried in the watermark and abides by its constraints. Obviously, this only works if the device implements the watermark detection. To encourage device manufacturers to do so, a requirement is included and defined in other encryption technology license agreements. If a manufacturer wants to get the keys and the secrets to play encrypted content, they must sign a license agreement which mandates the implementation of the watermark detection. AACS exemplifies such a scenario in which a manufacturer has to implement a specific watermarking technology in order to get access to the keys necessary for encryption. In this case, all Blu-ray Disc players must implement the circuitry to recognize the watermarking technology developed by Verance (and originally used by DVD-Audio).

Another way to use watermarks is for forensic purposes. Forensic marking is used to trace the path of legitimate content. If unauthorized copies of the content are created, the forensic mark will help identify the source of the piracy. As an example, the forensic mark could help identify which source tape was used to create the pirated copy. In the case of Blu-ray,

Content Protection, Licensing, and Patents

Thomson's NexGuard technology is closely integrated with the BD+ protection system to provide a more active solution. NexGuard embeds invisible information into the video output. Then, if content were to be pirated, it could be traced back to the player used, which would allow future titles to contain updated BD+ security code to fix the security hole.

Digital Transmission Content Protection (DTCP)

To preserve the high quality of digital video, it is imperative that the digital signal be transmitted all the way to the receiving end — the digital television or digital video recorder. But, due to content protection concerns, this has not always been the case (i.e., as with some DVD players). Instead, the player converted the pristine digital signal to an analog signal and the receiver would convert back to digital, resulting in a loss of quality.

Digital Transmission Content Protection (DTCP) provides a solution for the security concerns when content is being sent via digital connections. Developed by Intel, Sony, Hitachi, Matsushita, and Toshiba (also referred to as *5C*), DTCP primarily focuses on IEEE 1394 (aka FireWire™) and Internet protocol (IP) connected devices, but it can also be applied to other transmission protocols. The technology defines an authentication protocol for digitally connected devices, such as DVD players, digital TVs, and digital recorders. With DTCP, connected devices establish a secure transmission channel by exchanging keys and authentication certificates. Once established, the player sends an encrypted audio/video signal to the receiving device, which decrypts the signal. Other connected devices, if present, that are not authorized to receive the signal cannot hijack it because they do not have the decryption keys. If the content on a disc is not protected, it doesn't need to be encrypted during transmission.

An additional bonus is the "renewability" of the security. New content (new discs or new broadcasts) and new devices can carry updated key blocks and revocation lists identifying compromised devices. There is a difference between the privileges of playback-only devices and recorders. If a device is only able to reproduce audio and video, it is allowed to receive data through the DTCP connection, as long as it can authenticate itself for being a playback-only device. However, digital recording devices can only receive data that either is unprotected or is marked as copyable. In the latter case, if the source is marked as "copy once", the recorder must change this flag to "do not copy" or "no more copies". Since DTCP only impacts the digital connection between playback and receiving devices, there are no changes needed for the content itself.

High-bandwidth Digital Content Protection (HDCP)

During the search for a digital replacement of the analog VGA connection standard, the Digital Display Working Group (DDWG) was formed in 1998. The mission was to create a universal digital interface standard between computers and displays. The founding members of DDWG included Silicon Image, Intel, Compaq, Fujitsu, Hewlett-Packard, IBM, and NEC. Progress was made rather quickly with the release of the *Digital Visual Interface* (DVI) specification in April 1999.

Content Protection, Licensing, and Patents

The DVI specification is based on the panelLink technology originally developed by Silicon Image. DVI quickly gained wide acceptance in the industry due to its low-cost, high-speed digital connection for video displays. DVI supports all HDTV resolutions, in fact, it supports up to 1600×1200 (UXGA) resolution and even higher resolutions are possible with dual links. DVI exceeds by far the bitrate limitation of contemporary IEEE 1394 implementations. IEEE 1394 supports bitrates up to 40 Mbps, which would be enough for most compressed MPEG videos, whereas DVI supports up to 4.95 Gbps, which is more than enough for high-definition video of all sorts.

Alas, in its initial configuration, DVI was missing a security component. But Intel quickly proposed a solution — *High-bandwidth Digital Content Protection* (HDCP). HDCP can be used with multiple digital video monitor interfaces, such as DVI, HDMI (High Definition Multimedia Interface), and DisplayPort. See **Chapter 7** for more information on HDMI.

HDCP contains three components — authentication with key exchange, encryption, and revocation. HDCP ciphers and circuitry are implemented both in transmitters and in receivers, to ensure a secure end-to-end connection. Each transmitter (e.g., a player or a computer) and each receiver (e.g., a display or a recorder) has programmable read-only memory (PROM) circuitry that is used to store secret device-specific keys provided by the HDCP licensing entity. The authentication between devices is a cryptographic process verifying that the receiving device is indeed authorized to receive.

It is not mandatory that manufacturers include HDCP circuitry in their devices. Further, because HDCP was introduced after the first DVI-compliant monitors were released, there is a possibility that a device with HDCP could detect that it is connected to a non-HDCP-capable display. When this situation occurs, the transmitting device will lower the image quality of the protected content before transmission.

Each HDCP-compliant device has an array of 40 keys that are 56 bits each and a corresponding 40-bit binary *Key-selection Vector* (KSV), provided by the HDCP licensing authority. The transmitter sends the KSV with a random 64-bit value that initiates the authentication process. The receiver returns its own KSV, which is then checked by the transmitter to ensure that the key has not been revoked. Both devices individually calculate a shared value from the exchanged data that will be equal if both devices have valid keys. Because both ends of the transmission share the same value, it can be used to encrypt the protected data in the transmitter and to decrypt the data in the receiver. To ensure that the connection is still secure, a reauthentication occurs every few seconds. This avoids scenarios in which an authorized device is unplugged and an unauthorized device is connected instead. An unauthorized device will not be able to decrypt the stream and the content will appear as random noise.

Summary of Protection Schemes

Contrary to DVD, where all content protection systems were optional for the producer of the disc, there are mandatory content protection implementations for Blu-ray — AACS and BD-ROM Mark. Obviously, since BD-ROM Mark contains hidden keys that are necessary for AACS decryption, if one is mandatory both have to be. Once the AACS licensing agree-

Content Protection, Licensing, and Patents

ments become finalized it may become mandatory to embed a Verance audio watermark on the disc, as well.

The other protection schemes described above are optional for the disc producer. It is a choice for the producer whether or not BD+ protection is added to a disc, with or without forensic marks. As for the transmission protection systems, DTCP and HDCP, the player or computer handles them, not the disc developer.

The Analog Sunset

There are no reliable protection mechanisms for analog transmissions of high-definition video content — the so-called "analog hole". This presents a serious problem since there are many HD-ready displays in the marketplace that do not have digital connection technology. These HD-ready devices can only display HD content delivered via analog inputs, which may allow for easier copying of high-definition content. Obviously, this is not a satisfying scenario for the Hollywood industry that is trying to protect their high quality content.

As much as Hollywood would like to dismiss the fact that there are a number of analog HD displays in the market and to simply not release unprotected content into the analog realm, a compromise had to be found in order to avoid dissatified customers. As a result, an agreement was instituted that will result in the so-called "analog sunset", or termination of non-digital high-definition connectivity. This analog sunset takes place in two stages. The first stage begins with all products manufactured and sold after December 31, 2010. They will be allowed to only transmit high-definition content through encrypted digital outputs. Content that may be passed through analog connections can only be in standard-definition interlace mode, such as, composite, s-video, or 480i component. The second stage of this analog sunset requires that products manufactured and sold after December 31, 2013 are not allowed to pass any content through analog outputs.

In the meantime, the AACS specifications provide mechanisms to control the behavior of the protected content. There are two flags that can be applied to disc behavior — the *Image Constraint Token* (ICT) and the *Digital Only Token* (DOT). The ICT restricts the resolution of the content to be passed through analog outputs to not exceed 520,000 pixels per frame. This translates to a resolution of 960×540 pixels for a 16:9 aspect ratio. There are no further restrictions on post-processing algorithms, such as scaling, line doubling, or other video processing techniques that can be applied within the display. AACS also defines which devices are allowed to pass content through their analog outputs prior to the first sunset date of December 31, 2010, as outlined in Table 4.1.

The other flag, the Digital Only Token, prevents the transmission of decrypted AACS content through analog outputs. When the DOT flag is set, only protected digital outputs are allowed to transmit the content. Furthermore, digital interfaces are required to be protected by DTCP, HDCP, or by the Windows Media Digital Rights Management (WMDRM) process from Microsoft™. DOT was not implemented in the AACS interim agreement.

Content Protection, Licensing, and Patents

Table 4.1 Allowed Analog Output Devices for AACS, prior to December 31, 2010

Item	Conditions
Computer monitors	Image Constraint Token (ICT) must be set
Component Video Outputs – Standard definition and high definition	Image Constraint Token (ICT) must be set Macrovision APS1[a] setting has to be applied CGMS-A[b] has to be applied with RC[c] optional
Composite Video Outputs – Standard definition only	Macrovision APS1[a] setting has to be applied CGMS-A[b] has to be applied with RC[c] optional

[a] Macrovision provides three different options of their analog protection system (APS). APS 1 is the most common and widely supported implementation that was also used for DVD-Video.

[b] CGMS-A is the analog implementation of the copy generation management system and conveys general information over analog video interfaces –

    00: Unlimited copies allowed

    01: One generation of copies already been made, no more copying

    10: One generation of copies may be made

    11: no copies may be made

[c] As an optional part of CGMS-A, the Redistribution Control (RC) provides information to control unauthorized consumer redistribution.

Ramifications of Content Protection

Ever since digital content was released in a high quality, such as that for DVD, the studios have become more and more wary of the risks surrounding the protection of digital content. It did not take long to blame the creation of casual copies for billions of dollars in lost revenues. As a result, most of the efforts in content protection technologies were focused on stopping easy consumer copies, and hence to "keep honest people honest". It has always been very clear that stopping professional pirates or even determined consumers would be a very hard task. Video pirates have a lot of professional equipment at their disposal and in their possession that make it easy to create bit-to-bit copies of discs, or to generate high-quality masters for mass replication.

Given this history, and the fact that the protection systems for DVD did not last very long, it does not come as a surprise that a great deal of effort has been put into the renewability of the content protection systems for Blu-ray. The revocation systems within AACS, and the additional renewable protection mechanisms through BD+ are two examples of this improved security. Additional watermarking opportunities provide further enhancements for Blu-ray that allow a forensic analysis of where the pirated content originates. But it does not stop at the protection from casual copying, for there are also attempts to stop professional pirates with the BD-ROM Mark technology. It is probably a false hope to expect these technologies to entirely stop digital copies of the content, and it has already been proven that such copies can be created but, at least, they provide a means to close security holes and pro-

Content Protection, Licensing, and Patents

tect future titles with improved security. As a result, a continuous cat-and-mouse game between studios and pirates is guaranteed.

Besides the technological aspects to prevent illegal copies, the movie studios have started addressing the serious issue of piracy on the legal level. Dating back to 1994, they started promoting the idea of making it illegal to defeat technical content protection measures, which resulted in the World Intellectual Property Organization (WIPO) Copyright Treaty, the WIPO Performances and Phonograms Treaty (December 1996), and the compliant US Digital Millennium Copyright Act (DMCA) in October 1998. With this legislation, processes and devices that circumvent content protection are now illegal in the United States and many other countries.

Although the major studios consider content protection one of the most important issues, many smaller, independent studios do not necessarily share this point of view. Many of the smaller studios or commercial content producers may not mind if their content is being copied. In fact, some of them may even encourage copying, as it provides an additional audience. Although it was at the content producer's liberty to decide whether or not the disc was protected for DVD-Video, this situation changes for Blu-ray. AACS and BD-ROM Mark are mandatory features that everybody has to adhere to. As a result, smaller content producers now have to sign various agreements. Each of the agreements are associated with certain license fees to release content in the BD format. That obviously raises the bar for some producers to get involved in this format, and may even prevent some of them from releasing content at all. Whether or not this situation will prevent Blu-ray from gaining the market acceptance that DVD did is a heavily discussed topic on numerous levels. It remains to be seen whether the content protection obligations take priority or the necessity of mandatory protection for certain types of content will change in the future to address this problem.

Region Playback Control [RPC]

Control over geographic distribution of packaged media content became an important issue for motion picture studios with the introduction of DVD-Video. Region playback control (RPC) provides a tool to manage the release of home video content in different regions of the world. As an example, a movie may already have been released on video in the United States when it is just beginning to play in theatres in Europe. However, release windows have gotten shorter and technology has made it easier to debut movies worldwide, so this exclusivity is less common nowadays. A more typical scenario is exclusive local content distribution agreements. For example, the studio distributing a movie in the United States may not be the same studio with the rights to distribute the movie in other parts of the world. Region playback control is designed to support this business model.

The original implementation of region control for DVD-Video defined a set of seven region codes to allow or prevent playback of a disc in each region (plus a special region for airplanes, hotels, and the like). Each DVD player is set to the code of the region where it is sold. Each DVD contains flags to define which regions where it will play. A disc can allow any combination of regions. The player is only allowed to play a disc when its own region

Content Protection, Licensing, and Patents

code matches the corresponding flag in the disc. Region codes on DVD players, including PCs, are mandatory, but region codes on discs are optional. Discs without region codes play in any country, although 525/60 (NTSC) discs require a player that can convert to 625/50 (PAL) output and vice versa.

Blu-ray takes a similar path but specifies only three regions — A, B, and C (see Figure 4.6 and Table 4.2, also see Table B.9 for a complete list of countries).

Figure 4.6 Blu-ray Disc Regions

Table 4.2 Blu-ray Disc Regions

Region	Countries
Region A	Americas, Korea, Japan, Southeast Asia
Region B	Europe, Middle East, Africa, Australia, New Zealand
Region C	Russia, India, People's Republic of China, Rest of World

The most remarkable change in the constellation of countries for each Blu-ray region is that the United States, Japan, and South Korea (countries producing a significant portion of players), are all part of the same region, whereas with DVD-Video they were in separate regions. As with DVD, each Blu-ray player must be permanently coded for a single region. Each Blu-ray disc can be set to play in one, two, or all three regions. Region playback control is intended to apply only to pre-recorded (BD-ROM) discs — the license prohibits region-checking code on recordable discs. Logos on the players and discs indicate their region settings (see Figure 12.7). Of course Blu-ray players are also set to one of the DVD regions for DVD region playback control.

Another key difference for Blu-ray is that the disc itself, not the player, must manage region playback control with HDMV or BD-J code that checks the player setting and takes

Content Protection, Licensing, and Patents

action, such as limiting access to some of the content, playing different content, putting up a warning, or stopping all playback. Player manufacturers are explicitly obligated (by the Blu-ray format license) to stop playback if the disc requests it.

Implementation of region playback control on a computer is more difficult than on a standalone player since two components, a drive and a software player, have to work together, yet the components are sold independently and worldwide. The initial DVD computer implementations mostly had ROM drives with no built-in region code, so the software player application or the operating system was responsible for maintaining the region code. After 2000, all DVD drive manufacturers were required to produce drives that maintain the region code in their firmware. The user has the chance to change the region code up to five times. Having the region managed by the drive hardware rather than the software added more security to the system.

For Blu-ray the burden shifted back to the computer operating system and/or player application software, which must allow the user to change the region up to five times but otherwise must maintain the region code setting (and the count of changes) across uninstall/reinstall cycles and must be designed to prevent users from tampering with the software to alter the region code or defeat the feature.

It didn't take long for workarounds to appear with both DVD and Blu-ray. As soon as the first DVD players were released, people figured out how the region coding could be reset to "region 0"[1]— by a switch on a circuit board, a combination of keypresses on the remote control, replacing the firmware with a "modded" chip, or some other innovative way to defeat the system. There are dozens of legitimate vendor outlets, especially on the Internet, where code-free DVD players can be purchased. In response to code-free players, some studios designed discs with specific program code to query the region setting of the player. If a player was set to more than one region, the disc would refuse to play. Not surprisingly, this endeavor did not last long either — the next multi-region products were released with switchable region codes.

The same cat and mouse game is being played with Blu-ray regions. This time the Blu-ray license includes specific conditions stating that players cannot include switches, buttons, circuit-board jumpers or traces, or player functions to change the region or defeat the region playback control function. This covers software players as well. Nevertheless, by the end of 2007 multi-region players (with modified firmware) were available. In addition, hacking instructions were soon posted on the Internet, giving details on how to reset the region change count in computer Blu-ray player software applications, allowing unlimited region changes.

Given the ability of BD-J code to examine the state of the player, especially in conjunction with BD+, it is inevitable that content owners who care about region control will make discs that attempt to recognize multi-region players. The funny thing is most of the Hollywood studios don't care about region control the way they used to. Business models and release patterns have changed since DVD came out. A look at the list of Blu-ray Discs maintained at blurayregioncodes.com (766 discs as of September 2008) showed that only 30 percent

---

[1]Region 0 is a common but misleading term for players that support all regions. There is no region 0 defined in the DVD specification, this instead refers to the fact that no single region is set inside the player.

Content Protection, Licensing, and Patents

had region codes. In fact, the only major studios that had more region-coded discs than region-free discs were Fox (95 percent region coded, representing one third of the entire list of region-coded discs), Disney (including Miramax and Buena Vista, at 56 percent) and MGM (77 percent, although ironically MGM is controlled by Sony, yet only 24 percent of Sony Picture titles were region coded). Notably, every Paramount and Universal release was all-region, and all of the 133 Warner titles were all-region, apart from a handful of New Line titles.

Region codes can't be changed on BD-ROM discs, but they can be removed from a copy. Software utilities are available on the Internet to copy discs and strip out region coding. Some even claim to remove BD-based region detection, which requires analyzing and modifying the BD-J code.

There are arguments that region codes are an illegal restraint of trade, although no legal cases have been made. An important point is that while player manufacturers are obligated by their Blu-ray format license to implement region playback control, there are no laws requiring it. More importantly, circumventing region playback control is, in practically every case, perfectly legal for consumers, retailers, and others.[2] In fact, countries such as Australia, New Zealand, and Switzerland have legal provisions that allow circumvention of regional coding or access controls, and the European Commission stated in 2007 that it was allowed. Only Japan, the United States, and South Korea have bodies of law that could arguably make it illegal to circumvent region coding.

For the typical consumer, none of this really matters. Players and discs bought from local stores or from domestic Internet retailers will work perfectly together. Region mismatch is primarily an issue for those who buy imported discs from other regions or who move from one region to another. But again, the good news for those who may be adversely affected by region coding is that it's not nearly as prevalent on Blu-ray discs as it was on DVDs.

Licensing

No single company "owns" Blu-ray. The official format specification was developed initially by a founding group of nine companies — Hitachi, LG, Panasonic, Pioneer, Philips, Samsung, Sharp, Sony, and Thomson. Multi-company groups within the Blu-ray Disc Association worked on different parts of the BD specifications. Although people around the world have contributed to the Blu-ray format, it is not an international standard in the formal sense. It is an open specification, in that anyone can sign the proper agreements to get a copy of the technical documents, but use of the format and related logo is strictly controlled and comes with obligations for testing and certification as well as implementing selected content protection systems and region playback control. The Blu-ray format specification books are only available from the Blu-ray Disc Association after signing an information agreement, which includes confidentiality obligations, and paying a $2500 fee for the first book, $1500 for the second book, and $1000 for each additional book. These fees must be paid again each year or the books must be returned.

---

[2]The authors of this book are not lawyers, and they don't even play them on television. So anyone relying on this book for legal advice will not get much more than poetic justice.

Content Protection, Licensing, and Patents

Anyone who manufactures Blu-ray Disc products (media, players, PC drives, or PC software) must sign a format and logo license agreement (FLLA). There are different agreements for each format (ROM, R, RE, and AVCREC), each with a one-time fee of $15,000 for a five-year term, although the maximum for combined fees is $30,000. The agreements include content protection obligations (CPO) that mandate the use of certain content protection systems, as shown in Table 4.3. The BDA may choose to add a new content protection system at any time, in which case manufacturers have 24 months to implement the new system. Those who simply want to use the Blu-ray Disc logo for promotion of products may do so after signing the free Logo License Agreement (LLA).

Table 4.3 Format and Logo License Content Protection Obligations

Product	Content Protection System		
	AACS	BD-ROM Mark[a]	BD+
BD-ROM media (video, including Hybrid)	Mandatory	Mandatory	Optional
BD-ROM media (data and games)	N/A	Optional	N/A
BD player (standalone or game console)	Mandatory	Mandatory	Mandatory
BD PC drive (not AACS capable)	N/A	Mandatory	N/A
BD PC playback drive (AACS capable)	Mandatory	Mandatory	N/A
BD -R/RE PC recording drive (not AACS capable)	Mandatory	N/A	N/A
BD player PC application software	Mandatory	N/A	Mandatory
BD-R/RE recorder PC application software	Mandatory	N/A	N/A
AVCREC player (including PC software application)	Mandatory	Not applicable to AVCREC	Not applicable to AVCREC
AVCREC recording application software (AACS capable)	Mandatory	Not applicable to AVCREC	Not applicable to AVCREC

[a]BD-ROM Mark is not required for BD content on red-laser media (BD5 and BD9).

Content providers must sign a separate content participant agreement (CPA) and pay $3000 per year per format (BD-ROM, BD-RE, BD-R, and AVCREC, BD-Live) in order to release content on Blu-ray discs. There is a "light" version of the agreement that omits third-party beneficiary rights — the ability to participate in a lawsuit against a Blu-ray licensee if they breach the terms of the agreement, which basically boils down to the licensee doing something that allows the content provider's product to be illegally copied.

The format and logo license does not convey any patent royalties, which are claimed by many companies, some of whom have banded together and pooled their patents to make licensing easier, but there is still a bewildering array of companies holding out their hands for their slice of royalties (see Table 4.4). MPEG LA formed a one-stop-shopping pool of all essential Blu-ray technology patents, including interactivity and downloaded content, but as of late 2008 they had not yet finalized the licensing program, and not all relevant patent holders were necessarily on board. The Blu-ray Disc Association provides comprehensive format and logo licensing and content protection system licensing information on their Website (www.blu-raydisc.info), but no patent licensing information.

Content Protection, Licensing, and Patents

Table 4.4 BD and Related Format and Patent Licensing

Licensing Entity	License	Cost	Who Pays
BDA	BD format and logo	$15,000 per format (max. $30,000, $1,000 credit for each format book fee already paid)	Player/recorder and drive manufacturers, software player developers, disc replicators, recordable media manufacturers
	BD-ROM format and logo	$4,000 for 5 years for "Commercial Audiovisual Content" license	Authoring houses
	BD logo	Free	Anyone promoting BD
	BD logo and region logos (content participant agreement)	$3,000 per year or $500 per year for "light" version	Content providers
	BD-Live logo and online certificates	$1,000 per key	Content providers, authoring houses, disc replicators
AACS LA	AACS format and patents	$25,000 per year for first product category, $5,000 for each additional up to max. $40,000	Player/recorder and drive manufacturers, software player developers, recordable media manufacturers
		$5,000 per year	Component resellers
		$15,000 per year	Disc replicators
		$3,000 one time for basic content provider (less than x300,000 discs per year) or $15,000 per year for volume content provider	Content providers
		$40,000 per year	Major content providers ("Content Participants")
		$0.04 per disc or $12,000 to $2,500,000 per year depending on volume, $800 per order (up to 100 MKBs)	Disc replicators or content providers
		$0.02 per disc, $1,000 per order (1 MKB per glass master)	Recordable media manufacturers
		$0.10 per unit ($0.08 if ECDSA private key is used), $1,000 per order (up to 50,000 device keys)	Player/recorder manufacturers (use Type A device keys)
		$3,500.00 for up to 100,000 units per year, $12,000 for up to 1,000,000 units per year, $30,000.00 for up to 10,000,000 units per year (cap of $30,000 per year), $1,000 per order (up to 10 keys)	Software player developers (use Type C shared device keys for proactive renewal)
		$0.02 per unit, $1,000 per order	Drive manufacturers

continues

Content Protection, Licensing, and Patents

Table 4.4 BD and Related Format and Patent Licensing (continued)

Licensing Entity	License	Cost	Who Pays
AACS LA		$0.02 per unit (cap of $2,000 per year), $1,000 per order	Recorder manufacturers or recording software developers with non-updatable MKBs
		$500 per title (content certificate), $800 per order	Content producers
BD+ Licensing, LLC	BD+ (SPDC) format and patents	$20,000 per year	Player manufacturers, game console manufacturers, software player developers
MPEG LA	BD technology patent pool (including portions of GEM)	Not established as of Oct 2008, estimated to be $2 to $4 per device	Player and drive manufacturers, software player developers, disc replicators, recordable media manufacturers
	MPEG-2 patents	$0.03 per disc ($0.01 per disc if 12 minutes or less of video)	Disc replicators
		$2.50 per decoder and/or encoder	Player manufacturers, software player developers
	AVC/H.264 patents	$0.02 per disc or 2% of sale price (no fee if 12 minutes or less of video)	Disc replicators
		Fees per decoder and/or encoder based on units per year: no fee for up to 100,000, $0.20 up to 5,000,000, $0.10 for more than 5,000,000; annual cap of $5 million in 2009 and 2010;	Player manufacturers, software player developers
	VC-1 patents	$0.02 per disc or 2% of sale price (no fee if 12 minutes or less of video)	Disc replicators
		Fees per decoder and/or encoder based on units per year: no fee for up to 100,000, $0.20 up to 5,000,000, $0.10 for more than 5,000,000; annual cap of $5 million ($8 for computer OS) in 2006-2012	Player manufacturers, software player developers
Dolby	Dolby Digital and Dolby TrueHD/MLP patents	$0.66 per 2-channel decoder, $0.71 per 2-channel decoder + 2-channel encoder; $1.03 per TrueHD decoder	Player manufacturers, software player developers
DTS	DTS and DTS-HD Audio patents	$11 to $2 per DTS 5.1 decoder or $17 to $3 per DTS HD or DTS 6.1 decoder depending on volume; discounts for non-retail PC applications	Player manufacturers, software player developers

continues

Content Protection, Licensing, and Patents

Table 4.4 BD and Related Format and Patent Licensing (continued)

Licensing Entity	License	Cost	Who Pays
Via Licensing	MHP (GEM) patents	One-time fee of $15,000 ($1000 for small companies), $1.75 per product	Player manufacturers, software player developers
Digital Content Protection, LLC (Intel)	HDCP format and patents	$15,000 per year; $1000 for 10,000 device keys ($0.10 per key), $2500 for 100,000 keys ($0.025 per key), $5000 for 1,000,000 keys ($0.005 per key)	Player manufacturers, display manufacturers
5C	DTCP format and patents	$14,000 (small) or $18,000 (large) per year; $0.06 (small) or $.05 (large) per device (for key), $200 per order	Player manufacturers, display manufacturers
HDMI Licensing, LLC	HDMI format	$10,000 per year (less for volumes of 10,000 units or less); $0.05 per product ($0.04 per product if HDCP implemented)	Manufacturers of players, displays, receivers, cables, and related hardware
Verance	Verance Copy Management Systems (audio watermark) format and patents	$0.04 per disc	Disc replicators or content providers
	Watermark detector	$10,000 to $300,000 per year depending on unit volume	Player manufacturers, software player developers
		$50 per watermarked track	Production houses
DVD FLLC	DVD format and logo	$10,000 per year per format	Player/recorder and drive manufacturers, software player developers, disc replicators, recordable media manufacturers
6C	DVD patents	Greater of 4% of price ($8 max.) or $3 per unit	Player manufacturers (DVD-Video or DVD-Audio), ROM drive manufacturers
		Greater of 4% of price or $6 per unit	Recorder manufacturers, recording drive manufacturers
		Greater of 4% of price or $0.50 per unit	DVD "decoder" manufacturers
		Greater of 4% of price or $0.75 per unit	DVD "encoder" manufacturers
3C (by Philips)	DVD and optical disc patents	$3.50 per DVD-ROM/Video unit; $2.50 per DVD-Audio unit	Player manufacturers
	DVD-R/-RW/ +R/ +RW patents (incl. HP for +R/+RW)	$8 per DVD-R/-RW/+R/+RW unit	Drive manufacturers (recording)

continues

Content Protection, Licensing, and Patents

Table 4.4 BD and Related Format and Patent Licensing (continued)

Licensing Entity	License	Cost	Who Pays
Philips	VCPS	$5,000 one time, $0.05 per device, $220 per order of keys	Player/recorder manufacturers, drive manufacturers
	CD patents	2% of price	Player manufacturers, drive manufacturers
	CD recordable patents (incl. Sony, Yamaha, Ricoh)	$5 per unit	Recorder manufacturer
	Video CD patents (incl. Sony, France Telecom, IRT, JVC, Panasonic)	$0.75 per unit	Player manufacturers, software player developers
	Dolby Digital patents (incl. IRT and France Telecom)	$0.60 per unit (3 or more channels)	Player manufacturers
Thomson	DVD patents	~$1 per player/drive	Player manufacturers
Discovision	Optical disc patents	~$0.20 per disc	Disc replicators
DVD CCA	CSS	$15,000 annually per license category	Player manufacturers, software player developers, disc replicators, large content developers
4C	CPRM, CPPM	$12,000 per year per product category ($30,000 annual cap); CE devices: $0.05 per unit for player or $0.10 per unit for recorder/player; PC software keys: $10,000 for 1M units, $25,000 for 10M units	Player/recorder manufacturers, software player developers
		$0.06 per DVD-Audio disc	Disc replicators
		$0.02 per CPRM recordable disc	Recordable media manufacturers
Macrovision	Macrovision ACP	$50,000 initial charge and $25,000 yearly renewal or $15,000/$10,000 for quantities up to 15,000 per year or $125,000 one-time perpetual	Player manufacturers, PC or graphics card manufacturers
		$0.04-$0.10 per disc	Content providers

The licensor of AACS encryption technology is AACS LA, LLC. Dolby and DTS license their decoders and encoders at varying prices depending on format and number of channels. Philips (in joint licensing on behalf of France Telecom and IRT) also charges per player and per disc for patents underlying Dolby Digital. MPEG LA, LLC represents most MPEG-2 patent holders, with licenses per player and per disc. MPEG LA also licenses patent portfo-

Content Protection, Licensing, and Patents

lios for AVC/H.264 and VC-1. Via Licensing (a subsidiary of Dolby) covers patents for DVB-MHP (which may apply to GEM as part of BD-J).[3] Sun Microsystems charges per-unit royalties for its reference implementation of Java and the technology compatibility kits (TCKs) needed to validate BD-J implementations. Alternative Java implementations are available from other suppliers such as IBM. AACS charges per disc and also per title for encryption keys. BD+ Licensing LLC, not surprisingly, handles BD+ licensing. BD-ROM Mark licensing can be obtained from Panasonic, Philips, or Sony. Patent royalties may be owed to Discovision Associates, which owns about 1300 optical disc patents going back to laserdisc days, but many are expiring.

Essentially all BD players can play DVDs, which requires format and logo licenses from the DVD Format and Logo Licensing Corporation (FLLC) and patent licenses from a Hitachi, Panasonic, Mitsubishi, Samsung, Sanyo, Sharp, Toshiba, Victor (JVC), and Warner pool (known as 6C for the original 6 companies), from an LG, Philips, Pioneer, and Sony pool (administered by Philips but known as 3C since there were originally three companies), and from Thomson (jokingly referred to as the "1C pool").[4] DVD disc royalties are paid by the replicator. Some BD recorders (video recorders or PC writers) can also record to DVD-R/RW, DVD+R/RW, and DVD-RAM, which have associated format and patent licenses. Philips and Sony charge per-disc royalties for DVD+R/RW. Royalties for DVD+R/RW devices are covered by 3C fees. Implementation of the DVD-RAM specification incurs no royalties so long as no patented technologies are used. Many DVD players are also Video CD (VCD) players. Philips licenses the Video CD format and patents (which also cover Super Video CD) on behalf of itself, France Telecom, IRT, JVC, Panasonic, and Sony.

All the pools give licensees the choice to negotiate directly with each patent holder rather than with the agent for the pool. This allows for cross-licensing programs, where companies agree to license patent rights to each other at little or no cost instead of each paying money to the other. This is good for companies with cross licenses but hard on companies without any patents, since they have to pay full royalty fees. One bit of overall good news is that many of the newer technology license programs include caps, where payments stop after a few million dollars.

---

[3]Via Licensing originally started a pool for AVC/H.264 licensing but abandoned it. Some of the patent holders moved to the MPEG LA pool.

[4]IBM originally held about 250 DVD patents, but sold them to Mitsubishi in August 2005.

Chapter 5
Physical Disc Formats

CDs and DVDs are marvels of modern technology, storing data as dots smaller than a human hair whizzing past a laser at hundreds of kilometers per hour. Blu-ray Discs™ (BD) are even more complicated to produce because of their thinner data layer and smaller pit structure chararacteristics. More care needs to be taken to properly replicate the pits and to bond the substrates together.

Table 5.1 lists the physical characteristics of the Blu-ray Discs. Most of these characteristics are shared by all the physical formats (read-only, writable, 12 centimeter and 8 centimeter). Blu-ray discs are similar to DVDs and HD DVDs but with different substrate thicknesses (see Figure 5.1). The data substrate, often called the *cover layer*, is 0.1 mm thick, and the other substrate is 1.1 mm thick. The thin cover layer allows the laser to converge more sharply[1] without undue aberration from disc tilt[2]. This leads to smaller spot sizes and thus higher data density compared to DVD or HD DVD. Minor variations in pit length on the disc result in three slightly different capacities (see Table 5.2).

Table 5.1 Physical Characteristics of BD

Thickness	1.2 mm (±0.3) (two bonded substrates)
Substrate thickness	0.1 (SL) or 0.75 mm (DL) plus 1.1 mm (±0.003)a
Spacing Layer thickness	0.025 mm (±0.005)
Hard coat thickness	2 μm
Diameter	120 (±0.3) or 80 mm
Center hole diameter	15 mm (+0.10/-0.00)
Clamping area diameter	23.0 to 33.0 mm
Burst cutting area (BCA) diameter	42 (+0.0/-0.6) to 44.4 (+0.4/-0.0) mm
Lead-in diameter	44.0 to 44.4 mm (+0.4/-0)
Data diameter (12-cm disc)	48.0 (+0, -0.2) to 116.2 mm
Data diameter (8-cm disc)	48.0 (+0/-0.2) to 76.2 mm

continues

---

[1] Convergence performance of an objective lens is expressed as a unitless number called numerical aperture (NA), calculated as the sine of the half angle of convergence. The diameter of converging light is inversely proportional to the NA. That is, $D = \lambda / NA$, where λ = wavelength (405 nm, in this case). The larger the NA, the smaller the convergence.

[2] Think of looking at reeds in a pond. If you look straight down, the reeds look normal, but if you look at an angle, the reeds seem fractured at the point they penetrate the surface of the water because of light refraction. Likewise, a laser beam passing through the plastic material of a disc is distorted if it doesn't enter at a perfectly perpendicular angle, and the thicker the plastic the more severe the aberration.

Blu-ray Disc Demystified

Physical Disc Formats

Table 5.1 Physical Characteristics of BD (continued)

Lead-out outer diameter (12-cm disc)	117 mm
Lead-out outer diameter (8-cm disc)	77 mm
Mass (12-cm disc)	12 to 17 g
Mass (8-cm disc)	5 to 8 g
Readout wavelength	405 nm (±5)
Read power	0.35 (±0.1) mW (SL), 0.70 (±0.1) mW (DL)
Polarization	Circular
Numerical aperture (objective lens)	0.85 (±0.01)
Wavefront aberration (ideal substrate)	≤0.033 λ rms
Relative intensity noise (RIN) of laser	≤-125 dB/Hz
Beam diameter	0.58 μm
Optical spot diameter	0.11 μm
Reflectivity	35 to 70% (SL), 12 to 28% (DL)
Refractive index	1.45 to 1.70
Birefringence	<0.030 μm
Radial runout (disc)	≤0.050 mm (SL), ≤0.075 mm (DL), peak to peak
Radial runout (track)	-
Axial runout (disc)	≤0.3 mm (≤0.2 mm for 80-cm disc)
System radial tilt margin (angular deviation, α)	±0.60°
System tangential tilt margin (angular deviation, α)	±0.30°
Disc radial tilt margin (α)	<1.60°, peak to peak
Disc tangential tilt margin (α)	<0.60°, peak to peak
Asymmetry	-0.10 to 0.15

continues

Physical Disc Formats

Table 5.1 Physical Characteristics of BD (continued)

Track spiral (outer layer)	Clockwise (hub to edge)
Track spiral (inner layer)	Counterclockwise (edge to hub)
Track pitch	0.32 μm (±0.003)
Pit length	0.149 to 0.695 μm (2T to 8T)
Data bit length (avg.)	0.11175 μm
Channel bit length (avg.)	0.745 μm
Jitter (of channel clock period)	<6.5%(DL or L0) or 8.5%(L1)[b]
Channel bitrate	66.000 Mbps
Channel clock period	15 ns
Correctable burst error	7 mm
Symbol error rate	$<2\times10^{-4}$
Maximum local defects	100 μm (air bubble), 150 μm (black spot)
Rotation	Counterclockwise to readout surface
Rotational velocity (CLV)[c]	
Scanning velocity[c]	5.280 m/s (23.3 GB) 4.917 m/s (25.0 GB) 4.554 m/s (27.0 GB)
Storage temperature	-10 to 55°C (14 to 131°F), ≤15°C (59°F)/hour change
Storage humidity	5 to 90% relative, 1 to 30 g/m³ absolute
Operating temperature	5 to 55°C (41 to 131°F), ≤15°C (59°F)/hour change
Operating humidity	3 to 90% relative, 0.5 to 30 g/m³ absolute; ≤10%/hour change

[a]SL = single layer; DL = dual layer.

[b]L0 = layer 0; L1 = layer 1

[c]Reference value for a single-speed drive.

Blu-ray Disc Demystified

Physical Disc Formats

Figure 5.1 BD Disc Structure

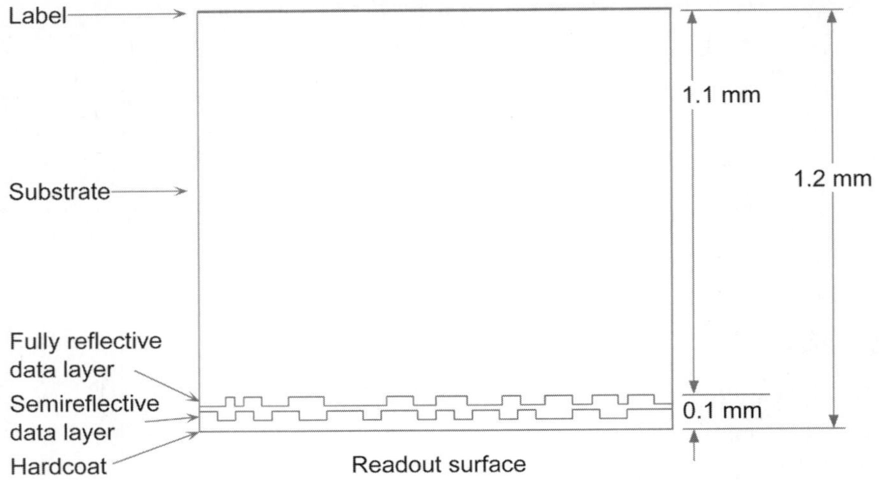

Label
Substrate
Fully reflective data layer
Semireflective data layer
Hardcoat
Readout surface

1.1 mm
1.2 mm
0.1 mm

Table 5.2 Blu-ray Disc Capacities

Name	Size	Sides and Layers	Billions of Bytes (10^9)	Gigabytes (2^{30})
BD-25	12 cm	1 side, 1 layer	25.025	23.306
BD-27 (future)			27.020	25.164
BD-23 (obsolete)			23.304	21.704
BD-50	12 cm	1 side, 2 layers	50.050	46.613
BD-54 (future)			54.040	50.328
BD-46 (obsolete)			46.609	43.408
BD-8	8 cm	1 side, 1 layer	7.791	7.256
BD-16	8 cm	1 side, 2 layers	15.582	14.512

Since the laser does not read through the backing substrate it does not have to be clear. It can even be made from non-plastic material such as paper or corn starch. It is theoretically possible to make a two-sided, three substrate BD with a 1.0-mm middle substrate sandwiched between two 0.1-mm cover layers, but this is not covered in the format specification.

A standard feature of BD is a 2-nanometer hardcoat at the readout surface to protect the disc from fingerprints and scratches. Think of it as transparent Teflon. Hard coatings can be applied to DVD, as well.

Blu-ray discs can optionally use a cartridge to protect the disc (see Figure 5.2). There are two types of cartridge — sealed (from which the disc cannot be removed) and open (allowing the disc to be removed and replaced.) The original BD-RE 1.0 format used a sealed cartridge, but since hardcoat technology was adopted, the BDA recommends that cartridges should not be used for any of the BD formats except in special situations.

Figure 5.2 BD Disc Cartridges

Open cartridge

Sealed cartridge

BD-ROM Mastering

Different technologies are used for creating BD masters. One technology is *phase-transition metal* (PTM) mastering that uses a 405-nm laser to create a pattern in inorganic photo-resist material similar to that used for phase-transition recordable DVD-RW and BD-RE. The PTM layers are placed on a silicon wafer instead of a glass plate. One advantage to PTM is that the pickup head can monitor the creation of pit patterns as they are created so that adjustments can be made in real time. Another approach is to modify a laser beam recorder (LBR) to use *liquid immersion*, where a stream of liquid is injected in front of the recording head and retrieved behind it. The interface created by the liquid in contact with the recording lens and the master surface allows a deep-UV laser to create a sufficiently small spot size for BD-ROM. A third approach uses an electron beam recorder (EBR) to expose a photo-resist layer. Electron beam recording can take hours to cut a master, but researchers are making progress on enhanced photo-resists material to speed up the process to make it more commercially viable. Regardless of the approach, the resulting masters are used to create stampers for replication.

BD-ROM Composition and Production

BD-ROM discs are created, complete with tracks of pits holding data, by a stamping process. There are two basic approaches to producing Blu-ray discs: 1) transfer a data pattern to a 0.1 mm layer (typically by molding a PMMA base, injecting polycarbonate, then peeling off the PMMA) and bonding it to a 1.1 mm substrate, and 2) injection-molding a pattern directly on top of a 1.1 mm substrate, then adding a cover layer. The latter method, which is more popular, includes approaches to forming the cover layer such as 1) spin-coating the cover with an ultraviolet (UV) resin, 2) bonding a pre-made cover sheet with a UV resin, and 3) bonding a pre-made cover sheet with pressure sensitive adhesive (PSA) (see Figure 5.3). The resin coating process spins the disc to spread the resin from the hub to the edge, which makes it difficult to keep the thickness uniform. The two sheet coating processes require tight tolerances when manufacturing the sheets and performing the bonding.

Physical Disc Formats

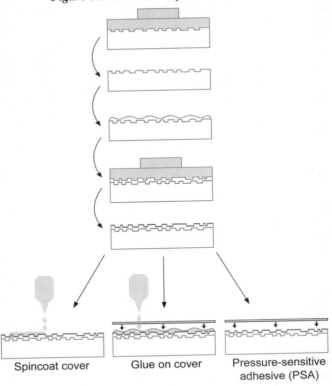

Figure 5.3 BD Dual-layer Construction

BD-RE Composition

Blu-ray rewritable discs have a structure similar to Blu-ray read-only (BD-ROM) discs (see Table 5.3). They are produced using one or two phase-change recording layers with a high-frequency modulated groove to provide addressing information and speed control for the recorder. Wobble period is approximately 5 μm, with embedded addresses in *address in pregroove* (ADIP) units of 56 wobbles. ADIP information is stored using *minimum-shift-keying* (MSK) modulation and *saw-tooth wobble* (STW). Data is written in the groove rather than on the land between.

BD-R Composition

Blu-ray write-once discs are similar to Blu-ray rewritable (BD-RE) discs, although the recording layers can be formulated with organic dye, inorganic alloys such as Si/Cu, or write-once phase-change material (see Table 5.4). Data can be written in the groove or on the groove (on the land between).

Physical Disc Formats

Table 5.3 BD-RE Characteristics

Recording material	Phase change
Transmission stack thickness	0.095 to 0.105 mm (±0.002) (TS0), 0.070 to 0.080 mm (±0.002) (TS1)
Recording method	In groove
Write power (1x)	<7 mW (SL), <12 mW (DL)
Address tracking	High-frequency modulated wobbled groove with addresses
Wobble frequency (1x)	956.522 kHz
Channel bits per wobble	69
Nominal wobble length	5.1405 μm
Storage temperature	-10 to 55°C (14 to 131°F), ≤ 15°C (59°F)/h change
Storage humidity	5 to 90% relative, 1 to 30 g/m³ absolute, ≤ 10%/h change
Operating temperature	5 to 55°C (41 to 131°F)
Operating humidity	3 to 90% relative, 0.5 to 30 g/m³ absolute

Table 5.4 BD-R Characteristics

Recording material	Organic dye, inorganic alloy, or phase change
Transmission stack thickness	0.095 to 0.105 mm (±0.002) (TS0), 0.070 to 0.080 mm (±0.002) (TS1)
Recording method	On groove or in groove
Write power (1x)	<6 mW
Address tracking	High-frequency modulated wobbled groove with addresses
Wobble frequency (1x)	956.522 kHz
Channel bits per wobble	69
Nominal wobble length	5.1405 μm
Storage temperature	-10 to 55°C (14 to 131°F), ≤ 15°C (59°F)/h change
Storage humidity	5 to 90% relative, 1 to 30 g/m³ absolute, ≤ 10%/h change
Operating temperature	5 to 55°C (41 to 131°F)
Operating humidity	3 to 90% relative, 0.5 to 30 g/m³ absolute

BD Error Correction

Error correction is the same for all BD formats. Data is recorded in 64K partitions, called clusters, each containing 32 data frames with 2048 bytes of user data each. 64KB clusters are protected by two error-correction mechanisms. The first is a *long-distance code* (LDC) using Reed-Solomon (RS) in a (248, 216, 33) structure. The second error-correction mechanism multiplexes the data with a *burst indicator subcode* (BIS) using (62, 30, 33) Reed-Solomon codewords. BIS includes addressing information and application-dependent control data information (18 bytes per data frame). BIS pinpoints long burst errors that can then be removed to improve LDC error correction.

Physical Disc Formats

A data frame holds 2048 bytes of user data and 4 bytes of simple error-detection code (EDC) for a total of 2052 bytes. Each data frame is scrambled to spread the bits around. Then, 32 data frames are combined into a data block with 216 rows of 304 columns. Each column is one byte. A data block is extended into an LDC block by appending the LDC codes for the data block as 32 rows of 304 columns. The LDC block is internally interleaved and shifted to improve burst error correction, resulting in an LDC cluster of 152 columns and 496 rows.

The 64KB physical cluster is divided into 16 *address units* (AU). The 4-byte address unit numbers are derived from the physical sector numbers and together with 1 byte of flags, 4 bytes of error correction, and user control data, they make up the data used for the BIS, which goes through a RS (62, 30, 33) coding and is arranged into a BIS cluster of 496 rows by 3 columns.

The LDC cluster is split into four groups of 38 columns, and each of the three columns from the BIS cluster is inserted between them, forming an ECC cluster. An additional column of frame sync bits is added at the beginning of the ECC cluster and DC control bits are inserted to form a *recording frame* of 496 rows by 155 columns, also called a physical cluster.

BD Data Modulation

Data in a recording frame, except for the sync bits, are modulated using the 1-7PP technique, an RLL (1, 7) code (2T through 8T run lengths) where the first P stands for *parity preserve* and the second P stands for *prohibit repeated minimum transition run length* (RMTR). Parity preserve means that the parity of the data stream matches the parity of the modulated stream. That is, if the number of ones in the selected chunk of data is even, then the number of ones in the modulated bits is even, and the same if the number of ones is odd. This is an efficient way to control the low-frequency content of the recorded signal. Prohibit RMTR limits the number of consecutive minimum run lengths (2T) to six, which avoids low signal levels and improves readout performance. A modulation conversion table is used to map sequences of data bits to modulation bits, which are then converted to a *non-return to zero inverted* (NRZI) channel bit stream that is recorded on the disc.

BD-R/RE Recording

The smallest unit for recording data is a *recording unit block* (RUB) consisting of 2,760 channel bits of *run-in*, followed by a 64KB physical cluster, followed by 1,104 channel bits of *run-out*. A continuously written sequence of one or more RUBs is terminated with a *guard_3* field of 504 channel bits. The run-in and run-out provide buffers so that clusters can be randomly written and rewritten, and to allow for *start position shift* (SPS), which randomly shifts the start position of each recording sequence by up to 128 channel bits before or 127 channel bits after the nominal start position to help the recording material last longer through multiple overwrites of the same data.

The lead-in at the beginning of the disc contains a pre-embossed section, called the *permanent information and control data* (PIC) zone, followed by a rewritable section. The PIC

Physical Disc Formats

zone holds general information about the disc and includes a special section called the *emergency brake* (EB). Up to 62 emergency brake fields can be defined, which specify a drive manufacturer, drive model, and firmware version, with associated special handling procedures to avoid damage to the drive or disc. The rewritable section is used for optimum power control (OPC) tasks and to store information about the recorded data on the disc, including defect management information, physical access control, and drive-specific information. Discs must be initialized before use, which largely consists of creating a defect list, if any.

BD-R/RE Defect Management

Defect management during recording can be handled by the file system or by the drive. If the drive manages defects, it maps defective clusters using linear replacement and a single defect list into one or two optional spare areas per layer. The inner spare area of layer 0 is at the inner side of the data zone and has a fixed size of 4,096 physical clusters (256 megabytes). The outer spare area of layer 0 and the inner spare area of layer 1 each have a variable size of 0 to 16,384 clusters (1,024 megabytes). The outer spare area of layer 1 has a variable size of 0 to 8,192 clusters (512 megabytes). The total spare areas represent about 5 percent of the storage capacity of the disc. If the drive detects a defective physical cluster it may replace the cluster, mark it for future replacement, mark it as defective without replacement, or ignore the error.

Phase-Change Recording

Phase-change recording technology, used for BD-R and BD-RE, as well as other optical recording technologies such as CD-RW, DVD-RW, DVD-RAM, DVD+RW, depends on changes in reflectivity between the amorphous state and the crystalline state of special alloys. When the alloy is heated by a laser at low power (*bias*) to reach a temperature around 200°C (400°F), it melts and crystallizes into a state of high reflectivity. When the alloy is heated by the laser at high power (*peak*) to a temperature between 500 and 700°C (900 and 1300°F), it melts and then cools rapidly to an amorphous state in which the randomized atom placement causes low reflectivity. As the disc rotates under the laser, it writes marks with high-power pulses and "erases" between the marks with low-power pulses. The marks can then be read with a much lower power setting to sense the difference in reflectivity (see Figure 5.4).

Figure 5.4 Phase-Change Recording

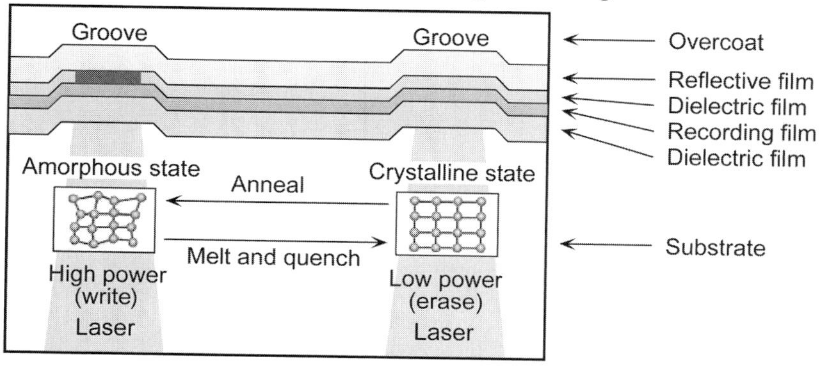

Physical Disc Formats

Phase-change discs are made of a polymer substrate to which a recording layer and a protective lacquer are applied. The recording layer is a sandwich of four thin films: the lower dielectric film, the recording film, the upper dielectric film, and the reflective film. The upper and lower dielectric films rapidly draw heat away from the recording layer to create the 10-second supercooling effect needed to keep it from returning to a crystalline state. The recording film is composed of an alloy such as germanium, antimony, and tellurium (Ge-Sb-Te) or indium, silver, antimony, and tellurium (In-Ag-Sb-Te). The dielectric layers are made of a material such as zinc sulfide and silicon dioxide (ZnS-SiO2). The reflective film is usually aluminum or gold.

Burst Cutting Area (BCA)

A section near the hub of the disc, called the *burst cutting area* (BCA), optionally can be used for individualizing discs, usually during the manufacturing process. The BCA applies to all BD formats — BD-ROM, BD-R, and BD-RE. A strong laser is used to cut a series of low-reflectance stripes, somewhat like a barcode, to store simple information such as ID codes or serial numbers. The BCA sits in the ring from 21.0 to 22.2 mm from the center of the disc (Figure 5.5). The BCA is only written on layer 0.

Figure 5.5 Burst Cutting Area

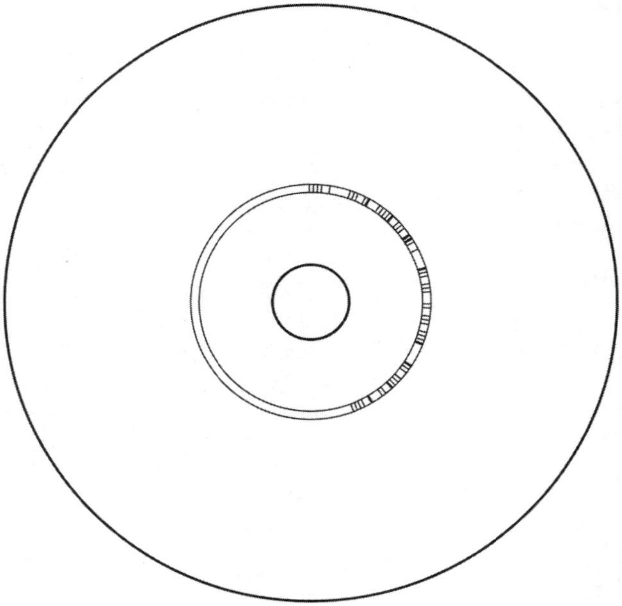

The BCA holds up to 64 bytes of information, broken into four 16-byte chunks (Data Units). The BCA can be read by the same laser pickup head that reads the disc. BCA information can be used for inventory purposes or by storage systems, such as, disc jukeboxes, to quickly identify discs. The BCA is used on rewritable discs to store information about the disc so the drive can very quickly determine the type of disc. One Data Unit of the BCA on recordable discs therefore contains information about the disc type (recordable or rewritable), size (120mm or 80mm), specification version, structure (single or dual layer), and channel bit length (80, 74.5, or 69nm).

The BCA is also used in AACS recording to uniquely encrypt the data on a recordable disc so that it can be decrypted only with the key stored in the BCA. Manufacturers of AACS-compatible recordable media write a unique serial number in the BCA of each disc. (See *Chapter 4* for more on AACS recording.)

Source Identification Codes (SID)

Two different marks, usually legible strings of letters and numbers, are placed near the hub of a Blu-ray Disc to identify the facility where the disc is made. The *Mastering Code* is written by the mastering recorder along with the pit structure that is transferred to the stamper. The *Mold Code* is placed on the back of the mold used in the stamping process. The marks then appear on all discs manufactured with that combination of stamper and mold.

Hybrid Discs

The Blu-ray format includes *Hybrid* discs, which are a combination of layers in BD, DVD, and CD format. More specifically, the following formats are accounted for — BD-ROM, BD-R, BD-RE, DVD-ROM, DVD-R, DVD+R, DVD-RW, DVD+RW, CD-ROM, CD-R, and CD-RW. *Intra-hybrid* discs use two mixed BD layers, such as, a BD-ROM layer and a BD-R layer. *Inter-Hybrid* discs use a BD layer and either a CD or a DVD layer. All Hybrid discs are single sided, which means they have a normal label side and, in all cases, there is one semitransparent BD layer so that the layer behind it (CD, DVD, or BD) can be read or recorded.

Unfortunately, as of the BD 1.01 (December 2005) specification, most of the interesting combinations are not allowed, but they could be supported in the future. Only the two combinations of BD-ROM + CD-ROM and BD-ROM + DVD-ROM are allowed.

Media Storage and Longevity

BDs can be stored and used in a surprisingly wide range of environments (see Tables 5.1, 5.2, and 5.3, for recommended temperature and humidity ranges). You can keep them in your refrigerator or in your hot water heater, although neither of these storage methods is a recommended alternative to a sturdy shelf or cabinet. A cool, dry storage environment is best for long-term data protection. If a disc has been kept in an environment significantly different from the operating environment, it should be conditioned in the operating environment for at least two hours before use.

BDs are quite stable, and if treated well, they usually will last longer than the person who

Physical Disc Formats

owns them. Estimating the lifetime of a storage medium is a complex process that relies on simulated aging and statistical extrapolations. Based on accelerated aging tests and past experience with optical media, the consensus is that replicated discs will last from 50 to 300 years.

Organic-dye-based discs, such as, some types of BD-R discs, are expected to last from 20 to 250 years, about as long as CD-R discs[3]. Shelf life before recording is about 10 years. The primary factor in the lifespan of these BD-R discs is aging of the organic dye material, which can change its absorbance properties. Long-term storage of this type of BD-Rs should be in a relatively dark environment, because the dyes are photosensitive, especially to blue or UV light (both of which are present in sunlight). Cyanine media are more susceptible than phthalocyanine media. Anecdotal reports have circulated of CD-R and DVD-R discs "wearing out" after being played for long periods of time, ostensibly because of alteration in the dye caused by reading it with a laser, but there are counter-reports of discs that play in kiosks 24 hours a day for years without a hitch.

The phase-change format (BD-RE and some BD-R) discs are expected to last from 25 to 100 years. The primary factor in the lifespan of these discs is the chemical tendency of phase-change alloys to separate and aggregate, thus reducing their ability to hold state. This also limits the number of times a disc can be rewritten.

In all cases, longevity can be reduced by materials of poor quality or a shoddy manufacturing process. Pressed BDs of inferior quality may deteriorate within a few years, and cheap recordable BDs may produce errors during recording or become unreadable soon after recording.

For comparison, magnetic media (tapes and disks) last 10 to 30 years; high-quality, acid-neutral paper can last a hundred years or longer; and archival-quality microfilm is projected to last 300 years or more.

Remember that computer storage media often become technically obsolete within 20 to 30 years, long before they physically deteriorate. In other words, long before the discs become nonviable, it may become difficult or impossible to find the equipment to read them.

BDs may be subject to "laser rot" — oxidation of the reflective layer. Media types such as dual-layer discs and BD-Rs that use gold for the reflective layer are not susceptible to oxidation because gold is a stable element.

Handling and Storage

Handle discs only by the hub or outer edge. Do not touch the shiny surface with your popcorn-greasy fingers. Store the disc in a protective case when not in use. Do not bend the disc when taking it out of the case, and be careful not to scratch the disc when placing it in the case or in the player tray. Make certain that the disc is seated properly in the player tray before you close it.

Keep discs away from radiators/heaters, hot equipment surfaces, direct sunlight (near a window or in a car during hot weather), pets, small children, and other destructive forces.

---

[3]Kodak officially states that its CD-R media will last 100 years if stored properly. Testing by Kodak engineers indicates that 95 percent of the discs should last 217 years.

Physical Disc Formats

Magnetic fields have no effect on DVDs or BDs. The format specifications recommend that discs be stored at a temperature between 20 and 50°C (4 and 122°F) with less than 15°C (59°F) variation per hour at a relative humidity of 5 to 90 percent.

Coloring the outside edge of a BD with a green marker (or any other color) makes no difference in video or audio quality. Data is read based on pit interference at one-quarter of the laser wavelength, a distance of less than 102 nanometers. A bit of dye that, on average, is more than 4 million times farther away is not going to affect anything.

Care and Feeding of Discs

Because BDs are read by a laser, they are resistant — to a point — to fingerprints, dust, smudges, and scratches. However, surface contaminants and scratches can cause data errors. On a video player, the effect of data errors ranges from minor video artifacts to frame skipping to complete unplayability. Therefore, it is a good idea to take care of your discs. In general, treat them the same way as you would a CD.

Your player cannot be harmed by a scratched or dirty disc unless there are globs of nasty substances on it that might actually hit the lens. Still, it is best to keep your discs clean, which also will keep the inside of your player clean. Never attempt to play a cracked disc because it could shatter and damage the player. It probably does not hurt to leave a disc in the player (even if it is paused and still spinning), but leaving the player running unattended for long periods of time is not advisable.

In general, there is no need to clean the lens on your player, since the air moved by the rotating disc keeps it clean. However, if you commonly use a lens cleaning disc in your CD player, you may want to do the same with your BD players. It is best to only use a cleaning disc designed for BD players because there are minor differences in lens positioning.

There is no need for periodic alignment of the pickup head. Sometimes the laser can drift out of alignment, especially after rough handling of the player, but this is not a regular maintenance item.

Cleaning and Repairing Discs

If you notice problems when playing a disc, you may be able to correct them with a simple cleaning. Do not use strong cleaners, abrasives, solvents, or acids. With a soft, lint-free cloth, wipe gently in only a radial direction (in a straight line between the hub and the rim). Since the data is arranged circularly on the disc, the microscratches you create when cleaning the disc will cross more error correction blocks and be less likely to cause unrecoverable errors. Do not use canned or compressed air, which can be very cold from rapid expansion and may stress the disc thermally.

For stubborn dirt or gummy adhesive, use water, water with mild soap, or isopropyl alcohol. As a last resort, try peanut oil. Let it sit for about a minute before wiping it off. Commercial products are available to clean discs, and they provide some protection from dust, fingerprints, and scratches. Cleaning products labeled for use on CDs work as well as those that say they are for DVDs and BDs.

Physical Disc Formats

If you continue to have problems after cleaning the disc, you may need to attempt to repair one or more scratches. Sometimes even hairline scratches can cause errors if they happen to cover an entire ECC block. Examine the disc, keeping in mind that the laser reads from the bottom. There are essentially two methods of repairing scratches: (1) fill or coat the scratch with an optical material or (2) polish down the scratch. Many commercial products do one or both of these, or you may wish to buy polishing compounds or toothpaste and do it yourself. The trick is to polish out the scratch without causing new ones. A mess of small polishing scratches can cause more damage than a big scratch. As with cleaning, polish only in the radial direction.

Libraries, rental shops, and other venues that need to clean many discs may want to invest in a commercial polishing machine that can restore a disc to pristine condition, even after an amazing amount of abuse. Keep in mind, though, that the data layer on a BD is only one-tenth as deep as on a CD and one-sixth of a DVD so the BD cannot be repolished too many times.

Improvement over DVD

The storage capacity of a single-layer DVD is seven times higher than that of a CD. BD further improves on DVD capacity by reducing the laser spot size with smaller wavelength (2.6× increase) and a more tightly focused beam (2× increase), resulting in more than five times the capacity.

At the reference 1× linear velocity of 4.917 m/s, Blu-ray discs spin about 1.4 times faster than DVDs but, because of the higher data density, the data rate is 35.965 Mbps, over three times faster than DVD's 11.08 Mbps and over 29 times faster than CD's 1.23 Mbps. However, to handle the high data rate requirements of video and audio, discs are played at a minimum of 1.5 times reference velocity, giving a data rate of 53.947 Mbps, about 4.8 times faster than DVD.

Chapter 6
Application Details

With the exception of disc diameter, the physical format of a Blu-ray Disc™ is significantly different from that of a standard DVD. With 25 GB for a single layer, and 50 GB for a dual layer disc, a Blu-ray Disc (BD) holds more than five times the data of a DVD. There are format variations that provide applications for pre-recorded discs (BD-ROM), recordable discs (BD-R), and re-writable discs (BD-RE).

When DVD was introduced, it took a few months after the release of DVD-ROM discs for DVD recorders to appear in the marketplace, but the progression was reversed for Blu-ray. BD Recorders for the BD-RE re-writable format were introduced in the Japanese market starting in 2003, significantly earlier than the first BD-ROM discs, which shipped in 2006. The initial BD recorder applications targeted recording and playback of Japanese digital broadcast signals, and an update of the BD-RE specification was required in order to ensure compatibility with the subsequent BD-ROM application formats and other video recording formats.

The Blu-ray Disc Association (BDA) controls all of the BD formats and applications and ensures that they are all compatible. In order to do that objectively, the BDA is composed of a Board of Directors along with different committees, as described in Figure 6.1.

Figure 6.1 Blu-ray Disc Association Organizational Structure

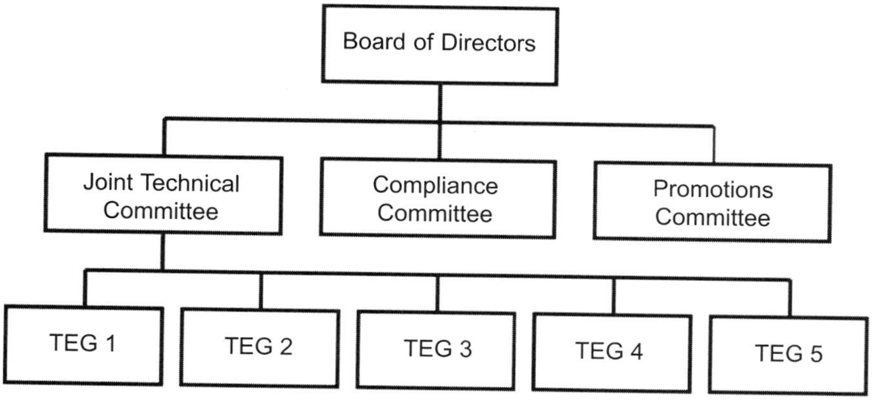

```
                    Board of Directors
                           │
        ┌──────────────────┼──────────────────┐
   Joint Technical    Compliance         Promotions
     Committee        Committee          Committee
         │
   ┌─────┬─────┬─────┬─────┐
 TEG 1 TEG 2 TEG 3 TEG 4 TEG 5
```

The Joint Technical Committee (JTC) is responsible for the development of the technical specifications and is subdivided into five Technical Expert Groups (TEGs) that cover the various aspects of the format (Table 6.1).

Application Details

Table 6.1 Blu-ray Disc Association Technical Expert Groups

Group	Responsibility
TEG 1	Physical specification for BD-RE (rewritable)
TEG 2	AV application
TEG 3	Physical specification for BD-ROM (pre-recorded)
TEG 4	Physical specification for BD-R (recordable)
TEG 5	File system and command set

To ensure that all products released into the market have been properly tested and are compliant with the technical specifications, the Compliance Committee is tasked with creating test procedures for each product category. The third committee is the Promotions Committee, which is responsible for marketing the format in various geographical locations around the globe.

All of these committees report to the BDA's Board of Directors (BoD), which consists of representatives from all key contributor companies. The BoD sets the overall strategy, discusses major issues and approves key decisions.

Beyond the division into three physical formats (BD-ROM, BD-R, and BD-RE), the specification for each format is further broken down into three parts —

- the physical specifications defining various parameters of the disc,
- the file system requirements, and
- the application formats.

Though some of the specifications are continually being updated, table 6.2 lists the most recent versions of each specification book (see Figure B.3 for a diagram of relationships).

Table 6.2 Blu-ray Disc Association Specification Books

Book	Physical	File System	Application
BD-RE (rewritable)	Part 1 (Ver. 2.12), Nov. 2007	Part 2 (Ver. 2.11), Oct. 2007	Part 3 (Ver. 3.02), Oct. 2007
BD-R (recordable)	Part 1 (Ver. 1.22), Nov. 2007	Part 2 (Ver. 1.11), Oct. 2007	
BD-ROM	Part 1 (Ver. 1.32), Nov. 2007	Part 2 (Ver. 1.21), Oct. 2007	Part 3 (Ver. 2.2), Dec. 2007
AVCREC (BDAV on recordable DVD)	DVD-R/-RW -RAM/+R/+RW Parts 1	BD-R/RE Part 2	AVCREC Part 3 (Ver. 1.0)

With every new technology comes a number of acronyms that can be very intimidating and confusing; naturally, it is no different with Blu-ray Disc. In fact, as the format evolved, about a dozen acronyms with "BD" in the name were used, making it very difficult to stay on track with what is being talked about. To help with some of these acronyms, table 6.3 deciphers the BD alphabet soup.

Application Details

Table 6.3 Guide to BD Alphabet Soup

Acronym	Full Name, Meaning
BD	Blu-ray Disc, the overall format family
BDF	Blu-ray Disc Founders group, original BD member companies
BDA	Blu-ray Disc Association, the larger association open to new members
BD-RE	Blu-ray Disc Rewritable, initial physical disc format created by the BDF
BD-R	Blu-ray Disc Recordable, record-once version of Blu-ray discs
BD-ROM	Blu-ray Disc Read Only Memory, read-only version of Blu-ray discs
BD-FS	Blu-ray Disc File System, original file system for BD-RE 1.0 (not PC compatible)
BDAV	Blu-ray Disc Audio Visual, application format for BD-RE and BD-R discs
BDMV	Blu-ray Disc Movie, content storage and playback format for BD-ROM (uses BDAV streams)
HDMV	HD Movie mode, declarative navigation environment for BDMV files
BD-J	Blu-ray Java, procedural software environment for BDMV files
BoD	Blu-ray Board of Directors, top-level voting group in the BDA

BDAV

The Blu-ray format for recordable discs is called BDAV (Blu-ray Disc Audio Visual). Its primary purpose is the recording of digital broadcast content in high-definition on recordable and rewritable Blu-ray Discs (BD-R, BD-RE). Unlike DVD, the BD recordable format was introduced prior to the introduction of the BD-ROM pre-recorded format. With that in mind, it is understandable that Blu-ray uses MPEG-2 Transport Streams (TS) technology for BDAV rather than the MPEG-2 Program Streams (PS) technology that is used for DVD. The advantage of using transport streams is that much smaller packet sizes (188-bytes) are capable of multiplexing multiple channels as well as providing additional information, such as, electronic program guides (EPG). This makes transport streams far more suitable for broadcast applications. Given the BDAV format's main purpose as a recording format for digital broadcast, it provides the means to extract, record, and play partial transport streams that were composed from multiplexed full transport streams (digital broadcasts) containing several channels of content.

In addition to the primary intent of recording digital broadcasts, the BDAV format supports the recording of self-encoded streams (e.g., an analog standard-definition video input) and recording of DV video streams (e.g., a camcorder). To generate a self-encoded stream, a built-in MPEG-2 video encoder encodes the incoming video, multiplexes it with the incoming audio, and creates a valid transport stream similar to digital broadcast. Whereas, in a DV stream, the video already exists in an MPEG-2 format and can be directly recorded without further encoding. In that way, no quality degradation would occur.

Application Details

Organizational Structure

To allow the user to seamlessly, and non-destructively, edit the recorded stream data, BDAV provides an organizational structure that separates the physically recorded stream from a logical playlist. Additionally, the player provides a user interface for assembling and editing of the recorded content. Figure 6.2 outlines this organizational structure.

Figure 6.2 Structural Organization of BDAV Content

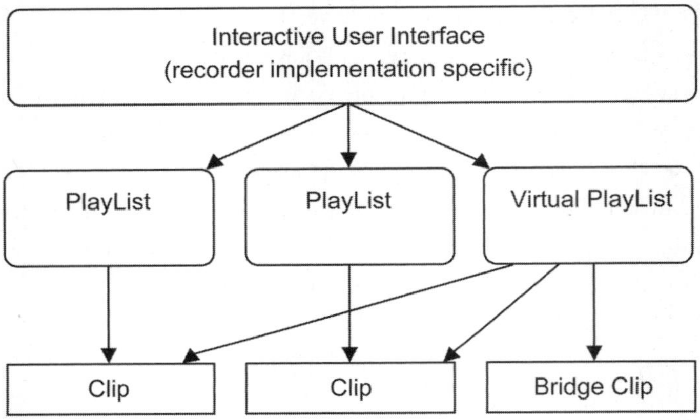

```
          ┌─────────────────────────────────────┐
          │    Interactive User Interface       │
          │  (recorder implementation specific) │
          └─────────────────────────────────────┘
              │           │           │
              ▼           ▼           ▼
        ┌─────────┐ ┌─────────┐ ┌──────────────┐
        │PlayList │ │PlayList │ │Virtual PlayList│
        └─────────┘ └─────────┘ └──────────────┘
              │           │           │
              ▼           ▼           ▼
          ┌──────┐    ┌──────┐   ┌───────────┐
          │ Clip │    │ Clip │   │Bridge Clip│
          └──────┘    └──────┘   └───────────┘
```

Interactive User Interface

Because the user wants to be in control of the navigation and the behavior of the recorded disc, each individual player implementation provides a user interface for establishing that control.

Playlist

There are two types of playlists defined in the BDAV format — a *real playlist* and a *virtual playlist*. A real playlist has a one-to-one correspondence to the clips on the disc. The enduser can also create a virtual playlist as part of editing the recorded content. Each playlist is composed of one or more *playitems*. This allows a virtual playlist to refer to multiple portions of existing clips on the disc, each portion being defined as a separate playitem. In the end, multiple playitems can be tied together into a single virtual playlist for seamless playback.

Clip

Each clip contains two components — the physical *Clip AV stream data* and a corresponding *Clip Information* file. There is also a playlist created for each recorded clip, containing a single playitem. When the content is being edited, and a virtual playlist is created, the physical stream data of a clip is not changed. Instead, the virtual playlist refers to the desired portion of each clip to allow seamless playback. In order to fully support the seamless playback from one playitem to another playitem in the virtual playlist, it is sometimes necessary to create a *bridge clip*, which requires the re-recording of a small portion of recorded clips.

BDMV

To allow interoperability between the recordable BDAV format and the pre-recorded Blu-ray Disc Movie (BDMV) format, the BDAV organizational structure was subsequently adopted as the basis for the organization of BDMV. For pre-recorded content, however, the control of the navigation and user experience is not handled by the player implementation's user interface. Instead, the content producer is in control, which results in a more sophisticated application layer for BDMV content.

The BDMV format is used for pre-recorded discs and is built on the same key technologies as the BDAV recording format. On the lowest level, an MPEG-2 transport stream is used together with corresponding clip information and playlist files as described for BDAV.

The biggest difference between the formats is the necessity of a more enriched and sophisticated user interface for BDMV discs. The reason is simply that the user interface for a recordable disc is created and driven by the recorder itself, which guarantees a similar interface for each disc, regardless of the content being recorded. While this is desirable for recordable discs to keep the user environment recognizable, it would be less than desirable for pre-recorded discs. Instead, the content owner wants to create a unique user interface for each disc to match the look and feel of the content.

The **BDMV** directory represents the root level of the pre-recorded disc and contains all of the necessary files for a player to render the audio/video presentation material and for the user to navigate through the logic of the disc (Figure 6.3).

First of all, the `index.bdmv` and the `MovieObject.bdmv` files identify the basic organization of the disc, listing all titles with the corresponding Movie Objects.

Secondly, a number of sub-directories exist containing the actual data to be rendered. As mentioned before, the **PLAYLIST**, **CLIPINF**, and **STREAM** directories are very similar to the BDAV disc structure in the sense that they contain the actual AV stream data with the corresponding clip information and playlists.

In addition, the **AUXDATA** directory is used to hold auxiliary data, such as, any OpenType fonts or any sound effects that will be used on the disc. The **META** directory contains optional metadata that may be used for Disc Library or Search features in certain Blu-ray players. If the disc is programmed with BD-J features, the **BDJO** and **JAR** folders hold the *BD-J Objects* (BDJO) and *Java Archive* (JAR) elements, respectively. And, to ensure that not everything is lost in case some critical files are damaged on the disc, the **BACKUP** directory contains a copy of those critical files (e.g., playlists, clip information, and others).

Application Details

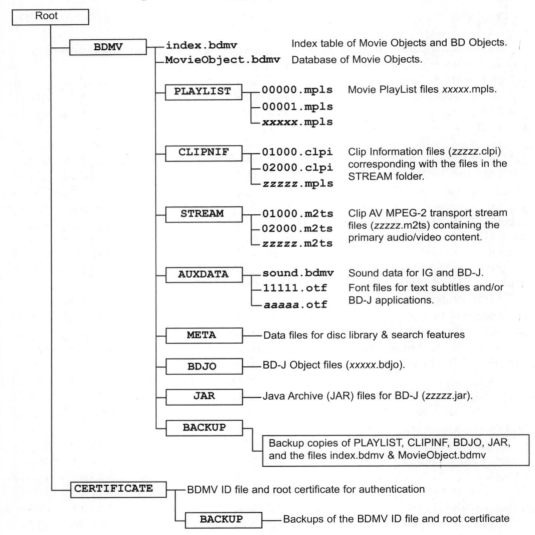

Figure 6.3 Directory Structure of a BDMV Disc

Application Details

Presentation Data

Even though the clip AV data is stored in the same format as BDAV content — in MPEG-2 transport streams — BDMV allows a greater variety of audio and video codecs and other associated parameters for BDMV discs. This ultimately affords content producers more flexibility in choosing the right presentation environment for their content. Table 6.4 outlines some general characteristics of the BDMV format.

Table 6.4 General Characteristics of BDMV Presentation Data

Disc capacity	25 Gbytes / 50 Gbytes
Video codecs	MPEG-4 AVC: HP@4.1/4L and MP@4.1/4/3.2/3.1/3L SMPTE VC-1: AP@L3 and AP@L2 MPEG-2: MP@ML and MP@H/H1440L profiles
Picture size	1920×1080, 1440×1080, 1280×720, 720×576, 720×480
Display aspect ratio	4:3 or 16:9
Frame rate	24 / 23.976 fps (film) 29.97 / 59.94 fps (NTSC) 25 / 50 fps (mandatory in PAL regions only)
Audio codecs	LPCM Dolby Digital DTS DTS-HD (core + extension[a]) Dolby Digital Plus[a] Dolby TrueHD[a]
Audio channel configuration	1.0 up to 7.1 channels[b]
Audio sampling frequency	48 kHz, 96 kHz and 192 kHz[a]
Subtitles / graphics	8-bit Interactive Graphics (IG) stream for menus 8-bit Presentation Graphics (PG) stream for subtitles HDMV Text Subtitle stream 32 bit RGBA graphics for BD-J menus
Stream structure	MPEG-2 System Transport Stream
Max. data rate	48 Mbps
Max. number of video streams	9
Max. number of audio streams	32
Max. number of subtitle streams	32

[a]Support for these features is optional in the player.

[b]A wide range of possible channel configurations exists, including mono, stereo, 5.1, and 7.1, among others.

Similar to the content presentation method of DVD, the various elementary streams (video, audio, subtitle) for Blu-ray have to be multiplexed together to create a contiguous stream that allows for seamless playback within the player. The reason for this is the fact that a decoder will only read small chunks of data at a time in order to render the output in a timely manner. As a result, the various elementary streams are broken into smaller packets and stitched together in a *multiplexing* process, as depicted in Figure 6.4.

Application Details

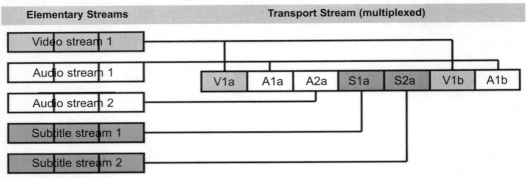

Figure 6.4 Multiplexing Process

Paths and Subpaths

The Blu-ray format introduces a new concept with the definition of a *path* and a *subpath*. A main path typically represents the primary audio/video content on a disc, for instance, a movie or other feature video with its corresponding audio and subtitles, while a subpath allows for supplementary content. As an example, secondary audio or video would be defined as a subpath. Similarly, certain types of interactive graphics (IG) streams, text-based subtitles, or audio for browsable slideshows could be defined as subpaths, as well.

Subpaths may also be used for picture in picture (PIP) presentations. For this feature, there are three different PIP video stream scenarios —

- in-mux synchronized,
- out-of-mux synchronized, and
- out-of-mux asynchronized.

The in-mux synchronized variety is a secondary video stream that is multiplexed in the main program stream with the feature video content. It will always be synchronous with the main feature and the position and size of the secondary video are determined and fixed during the disc content authoring.

For a more flexible solution, an out-of-mux video stream can be used. This allows a file to be played back from the local storage of a player (e.g., after downloading it from the network or copying from the disc). A typical out-of-mux feature would be a director's commentary that is made available over the network after the disc is released. In such a case, the out-of-mux synchronized picture in picture approach would be used, as the timing of the commentary with the main feature is essential. However, there are scenarios in which the synchronicity to the main feature is not important, in which case an out-of-mux asynchronous stream approach can be used to allow the user to define when to play the picture in picture video stream.

BD-ROM Application Types

Blu-ray offers two modes for BD-ROM application programming — *HD Movie* (HDMV) and Blu-ray Disc Java (BD-J).

The more basic mode is the *HD Movie* (HDMV) mode, which can be described as a declarative execution environment. This mode is similar to that employed for DVD-Video, with very little support for dynamic control of the presentation. HDMV authoring requires the content author to anticipate all possible scenarios and create all necessary graphics and menu compositions ahead of time, along with the logic programming for determining which menu to show when. However, HDMV authoring still provides a lot more capabilities than DVD authoring. The advantage of a declarative environment is that it is more straightforward to create and to verify the content, and to comply with the specifications, which ensures a more consistent performance across the variety of player implementations.

In order to achieve a more flexible and dynamic user experience, the BD-Java (BD-J) mode can be used. BD-J is considered a procedural environment where programming software is written to handle different playback scenarios. BD-J is based on a number of other standards, such as, DVB's[1] Multimedia Home Platform (MHP) standard, and provides additional application programming interfaces (APIs) specifically necessary for Blu-ray. Essentially, BD-J programming code is a multimedia software application executed in a Java virtual machine (JVM) that generates graphics and animations, interacts with the user, and controls the media playback from the disc or from other storage. Because Java is a software programming language, there is no declarative portion within the code. Even though this allows the greatest flexibility and the most control over the content, it is not a trivial process. In fact, the use of a software programming language for the development of rich multimedia applications defines a rather huge paradigm shift within the industry. A comparison between HDMV and BD-J mode is outlined in Table 6.5.

Table 6.5 Comparison of HDMV and BD-J Features

Feature	HDMV	BD-J
Application types	Declarative	Procedural
Programming logic	Simple	Complex
General purpose memory size	16 KB[a]	9 MB
Font buffer size for text rendering	4 MB	4 MB
Maximum sound effects buffer size	2 MB	5 MB[b] / 6.5 MB[c]
Maximum graphics buffer size	16 MB	45.5 MB[b] / 61.5 MB[c]
Graphics bit depth	8 bpp (Indexed)[d]	32 bpp
Text-based subtitle rendering	Yes	Yes
Popup and multipage menus	Yes	Yes

continues

---

[1] The Digital Video Broadcasting Project (DVB) is a European industry consortium established in 1993 with the aim to establish a pan-European platform for digital terrestrial television. The first Multimedia Home Platform (MHP) specifications were released by the DVB in June 2000.

Application Details

Table 6.5 Comparison of HDMV and BD-J Features (continued)

Feature	HDMV	BD-J
Alpha-blended graphics	Yes	Yes
Overlapping alpha-blended graphics	No	Yes
Image scaling	No	Yes
Frame-accurate animation	Yes	Yes
Resize primary video	No	Yes
Secondary video	Yes[e]	Yes[e]
Local storage access	No	Read/Write
Network access	No	Yes[e]
Applications (e.g. menus) may persist across titles	No	Yes
Applications (e.g. menus) may persist across discs	No	Yes

[a]In the form of 4,096 32-bit general purpose registers (GPRs).

[b]Profile 1 players

[c]Profile 2 players

[d]HDMV graphics use an 8-bit (256-color) palette of Y'CbCr colors with 8-bit alpha.

[e]Only available on Profile 1.1 and Profile 2 players.

Presentation Planes

In order to present the different media types, a so-called plane model is used by both HDMV and BD-J. It basically uses five independent planes to render the various media types as depicted in Figure 6.5.

Figure 6.5 Blu-ray Disc Presentation Planes

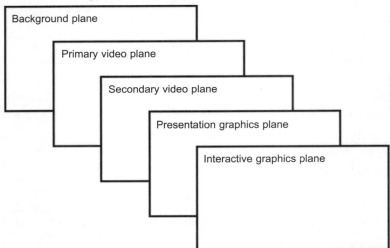

Application Details

These five planes are (from back to front) the *background plane*, the *primary video plane*, the *secondary video plane*, the *presentation graphics plane*, and an *interactive graphics (IG) plane*. The presentation plane is used for subtitles and graphics. However, there are subtle differences between HDMV and BD-J modes in the way these planes are used. For instance, the background plane can only be used in BD-J. And the IG plane will be populated by different applications — either HDMV interactive graphics, or BD-J graphics, both with their own limitations.

It is important to understand this plane model or the final presentation of the disc may not turn out as expected. As an example, if the background plane is supposed to be used to display a graphic, the video planes cannot be rendered in full screen HD resolution, as they would fully cover the background. Hence, in order to use the background plane, the primary video needs to be scaled to a smaller image size.

Also, in the event where both primary and secondary video are displayed, the BD specification requires one of the video planes to be full-screen HD resolution. As a result, to display the background plane only the primary video can be used. And, since it is not possible to switch the order of planes, whenever the secondary video is scaled up to full HD resolution, it will fully cover the primary video plane. Thus, applications where the secondary video plays in full-screen with the primary video in a PIP window are not possible.

Similarly, with the interactive graphics plane sitting on top of the subtitle plane, this means that graphics displayed in the subtitle plane may be covered by other on-screen graphics in the IG plane. Bottomline, during the design of a Blu-ray Disc, the position of the various planes should be carefully considered to ensure the final presentation turns out as expected.

Organizational Structure

With the BDMV format being based on the BDAV format, it does not come as a surprise that the organizational structure is very similar, as well. However, the higher levels of the BDMV format allow for much more sophistication. Further, this means that a disc can jump back and forth between the modes, HDMV and BD-J, to allow content interaction and to enable more flexibility. Figure 6.6 depicts the structural organization of a BDMV disc.

Figure 6.6 Structural Organization of BDMV Content

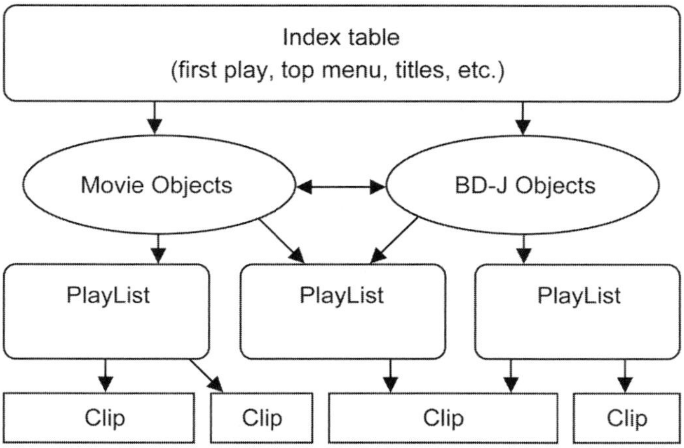

Application Details

Index Table

The *index table* is the starting point of the disc — the highest level. It lists all the titles on the disc with the corresponding Movie or BD-J Objects. The index table can be referenced whenever a title or a menu is to be executed. Hence, it is called during initial playback of the disc, during title search, and with menu call operations.

Movie and BD-J Objects

While *movie objects* refer to HDMV objects, *BD-J objects* (BDJOs) refer to Java applications. A movie object contains commands that can launch a playlist or execute other code to navigate through the disc. Similarly, the BD-J object can also launch a playlist, and refers to specific Java Archives (JARs) that contain the programming code to control the disc playback. The index table binds each movie and BD-J object to a specific title. Note that a number of titles can share the same BDJO.

Playlist

Similar to BDAV, a *playlist* consists of one or multiple playitems. Each playitem refers to a physical stream file, or a so-called clip. The in and out points of the playitem do not necessarily have to coincide with the start and end time of the clip. In other words, the playitem can define a portion of a given clip. In the end, a playlist can refer to a main path and additional sub-paths and will be composed of one or multiple playitems.

Clip

While the playlist is a logical description, a *clip* describes a physical AV stream file and its associated clip information data. Besides the actual MPEG-2 Transport Stream, as defined in the BDAV specification, the clip information describes additional time map and other relevant information of a transport stream (e.g., valid entry points into the stream).

Navigation Data

With an understanding of the organizational structure of the disc, it is important for the content author to identify the tools available to navigate through this structure. This can best be described as the next layer, on top of the presentation data — the *Navigation data*. It is a collection of information that allows the content author to access the presentation data and navigate the disc. Leaving the much more complex and flexible BD-J application programming interfaces (APIs) aside, the navigation data presents a more fundamental toolset, which can be used by the BD-J application as well.

The navigation commands are generally grouped into three categories — **branch commands**, **compare commands**, and **set commands**. The branch commands provide the means to link different pieces of the disc and actually physically navigate around the disc. Compare commands allow the content author to compare various values and parameters to make further decisions. And, the set commands allow for mathematical operations that enable the content author to examine the current state of the player or disc, set streams, buttons, or other elements, and calculate the next steps for navigation. Table 6.6 outlines the navigation commands available to content authors.

Application Details

Table 6.6 Navigation Commands

Branch Commands

Go To Group — Commands for controlling the program flow in a command sequence

Nop	No operation (do nothing)
GoTo	Jump to a specified command in the sequence (e.g. Movie Object)
Break	Stop executing commands in the command sequence

Jump Group

Jump Object	Start a specified Movie Object
Jump Title	Start a specified Title
Call Object	Stop current playback and start specified new Movie Object
Call Title	Stop current playback and start specified new Title
Resume	Resume at location where playback was suspended by Call Object or Call Title command

Play Group

Play PL	Start playback of a specified Playlist
Play PLatPI	Start playback of a specified Playlist at a specified PlayItem
Play PLatMK	Start playback of a specified Playlist at a specified Playlist Mark
Terminate PL	Stop playback of current Playlist
Link PI	Start at specified PlayItem number within the same Playlist
Link MK	Start at specified PlayItem Mark within the same Playlist

Compare Commands

BC	Compare bitwise (logical and)
EQ	Test if equal
NE	Test if not equal
GE	Test if greater than or equal
GT	Test if greater than
LE	Test if less than or equal
LT	Test if less than

Set Commands

SetSystem Group — Commands for setting system parameters

SetStream	Set audio, angle, subtitle, or interactive graphics stream
SetNVTimer	Set navigation countdown timer
SetButton Page	Set Button ID and Page ID
EnableButton	Set Button to "enabled" state
DisableButton	Set Button to "disabled" state
SetSecondaryStream	Set secondary video, audio, and PiP subtitle stream
PopUpMenu Off	Turn Popup menu off
Still On	Pause playback
Still Off	Resume playback after pause

continues

Application Details

Table 6.6 Navigation Commands (continued)

Set Commands (continued)

Set group — Commands for setting and manipulating values in general purpose parameters

Move	Set GPR value (from a constant, GPR, or PSR)
Swap	Exchange the values in two GPRs
Add	Add a value (constant or GPR) to a GPR
Sub	Subtract a value (constant or GPR) from a GPR
Mul	Multiply a GPR by a value (constant or GPR)
Div	Divide a GPR by a value (constant or GPR)
Mod	The remainder of dividing a GPR by a value (constant or GPR)
Rnd	Set the GPR to a random number between 1 and a value (constant or GPR)
And	The bitwise product of a GPR and a value (constant, GPR, PSR)
Or	The bitwise sum of a GPR and a value (constant, GPR, PSR)
Xor	Exclusive or a GPR and a value (constant, GPR, PSR)
Bit Set	Set a specific bit in a GPR
Bit Clear	Clear a specific bit in a GPR
Shift Left	Zeros are shifted in low order bit in GPR
Shift Right	Zeros are shifted in high order bit in GPR

The Blu-ray player has a total of 128 *Player Status Registers* (PSRs). These are 32-bit registers that hold information about the current state of the player, such as, the current audio language (PSR16), current chapter number (PSR5), and the current angle (PSR3). This enables content authors to use commands that track the state of the disc in order to influence the behavior. As an example, if a user navigates to a scene selection page on a popup menu, simply reading PSR5 and setting the button based on this parameter could highlight the chapter that is currently playing. PSRs are read-only registers and cannot be directly influenced. However, they can be set by special commands defined in the SetSystem group, such as, the SetNVTimer command which would influence PSR9 (Timer). Similarly, the SetStream command could influence PSR1 (Primary Audio), PSR2 (Subtitles), or PSR3 (Angle), depending on which streams are being set. Even though the total of 128 PSRs are available only about two-dozen of them are actually useful for the content author. The others are used for backup purposes, or are reserved for other uses (table 6.7).

Table 6.7 Player Status Registers (PSRs)

Number	Register Name	Access
0	Interactive Graphics stream number	read/write
1	Primary audio stream number	read/write
2	PG TextST and PiP PG TextST stream numbers	read/write
3	Angle number	read/write
4	Title number	read/write
5	Chapter number	read/write

continues

Application Details

Table 6.7 Player Status Registers (PSRs) (continued)

Number	Register Name	Access
6	PlayList ID	read/write
7	PlayItem ID	read/write
8	Presentation Time	read-only
9	Timer	read-only
10	Selected Button ID	read/write
11	Menu Page ID	read/write
12	TextST User Style number	read/write
13	Parental Level	read-only
14	Secondary Audio and Secondary Video stream numbers	read/write
15	Audio Capability	read-only
16	Audio Language	read-only
17	PG and TextST language	read-only
18	Menu Language	read-only
19	Country	read-only
20	Region	read-only
21 to 28	Reserved	
29	Video Capability	read-only
30	TextST Capability	read-only
31	Player Profile and Version	read-only
32 to 35	Reserved	
36	Backup PSR4 (for resume)	read/write
37	Backup PSR5	read/write
38	Backup PSR6	read/write
39	Backup PSR7	read/write
40	Backup PSR8	read-only
41	Reserved	
42	Backup PSR10	read/write
43	Backup PSR11	read/write
44	Backup PSR12	read/write
45 to 47	Reserved	
48 to 61	TextST capability for each language	read-only
62 to 95	Reserved	
96 to 111	Reserved for BD system use	
112 to 127	Reserved	

Application Details

In addition to the Player Status Registers, there are also *General Purpose Registers* (GPRs). GPRs are useful for on-disc programs to keep scores, track what sections of the disc have already been viewed, and to store user responses, and so on. There is a total of 4096 GPRs available, each of them able to hold a 32-bit unsigned integer. In comparison with DVD, where only 16 registers (called *General Parameters*) are available, this is a huge improvement and provides much better capabilities for mathematical calculations and tracking of the disc.

User Interaction

The enduser is in the driver's seat when it comes to navigating the disc using a remote control or other available interface device, such as a keyboard. The user decides when to change audio or subtitle streams, when to jump to a different piece of content, and any other playback decision. There are generally two different ways for this decision to be made, either by selecting the appropriate option on the on-screen menu page or by making a decision on-the-fly from any part of the disc (menu or video) using keys on the remote control or other device.

User Operations

Although the general control of the disc playback is based on user interaction, there is a way for content authors to monitor these user operations. Furthermore, content authors can mask user operations to prohibit use by the enduser. In the case where a user operation (UO) is masked, the player module will still register the push of the respective button on the remote control or keyboard, but it will be ignored. For instance, in order for content providers to ensure that all trailers and warning cards at the beginning of the disc are seen, the fast-forward and skip functions may be masked to disable them. However, the same buttons should work during other parts of the disc, such as, during the main feature. To accomplish this, the user operations can be controlled on a granular level, either for an entire PlayList, a PlayItem, or a page of the interactive display (e.g., menu page).

There are three different categories of user operations — *Title Control User Operations*, will be used to change from one title to another during playback; *Playback Control User Operations*, as one might guess, control the playback engine, such as, forward and backward play, or changing streams; and, *Interactive User Operations*, which controls the navigation through the interactive graphics and enables or disables functions such as selecting or activating buttons.

Title Control User Operations and Playback Control User Operations are controllable from either the HDMV mode or the BD-J mode. However, the third category, Interactive User Operations, is only available from HDMV menus. Table 6.8 outlines the various user operations.

One additional thing to note is that not all user operations have to be supported by all players. Some user operations are optional functions for player implementations and may or may not be available on a player's remote control.

Application Details

Table 6.8 User Operations (UO)

Title Control UO	Player Support
Menu call	Mandatory
Title search	Mandatory
Resume	Mandatory
Play FirstPlayback title	Mandatory
Stop	Mandatory

Interactive UO	Player Support
Move up selected button	Mandatory
Move down selected button	Mandatory
Move left selected button	Mandatory
Move right selected button	Mandatory
Select button	Optional
Activate button	Mandatory
Select button and activate	Optional
Popup on	Mandatory
Popup off	Mandatory

Playback Control UO	Player Support
Chapter search	Optional
Time search	Optional
Skip to next point	Mandatory
Skip back to previous point	Mandatory
Pause on	Optional
Pause off	Optional
Still off	Mandatory
Forward play (speed)	Optional
Backward play (speed)	Optional
Primary audio stream number change	Mandatory
Angle number change	Mandatory
PG textST enable / disable	Mandatory
PG textST stream number change	Mandatory
Secondary video enable/disable	Optional
Secondary video stream number change	Optional
Secondary audio enable /disable	Optional
Secondary audio stream number change	Optional
PIP PG textST stream number change	Optional
Text subtitle style change	Optional

Key Events

Similar to the Interactive User Operations for HDMV menus, there are controls available for BD-J menus. Because BD-J is using a programming language instead of a simple command set, the remote control or keyboard buttons will be tracked via Virtual Key (VK) events. Each VK is mapped to a button on the remote control or keyboard. For BD-J menus, event listeners must be implemented in the code to listen for the virtual keys. If the user pushes a button on the remote control, the event listener for the VK registers the event and an appropriate action is followed as defined in the programming. There are four groups of Virtual Keys defined in the specification — *Playback Key Events* handle playback control issues; *Numeric Key Events* register entries on the numeric keypad; *Interactive Key Events* handle navigation, and there are four *Color Key Events* (Table 6.9). The color keys may be used as additional buttons if needed for an application. For example, a BD-J game may need a couple of additional buttons on the remote control to ease the navigation tasks. The color keys are inherited from settop boxes as used for digital TV receivers, and they are always red, green, yellow, and blue. Unfortunately, the order of the keys is not specified and, thus, varies from country to country. Further, the color VKs are numbered from zero to three and it is recommended to query the order of the color keys before representing them on-screen.

Application Details

Table 6.9 Virtual Key (VK) Events

Playback Key Events	Interactive Key Events
VK_PLAY	VK_LEFT
VK_STOP	VK_UP
VK_STILL_OFF	VK_RIGHT
VK_TRACK_NEXT	VK_DOWN
VK_TRACK_PREV	VK_ENTER
VK_FAST_FWD	VK_POPUP_MENU
VK_REWIND	**Numeric Key Events**
VK_PAUSE	VK_0
VK_SECONDARY_VIDEO_ENABLE_DISABLE	VK_1
VK_SECONDARY_AUDIO_ENABLE_DISABLE	VK_2
VK_PG_TEXST_ENABLE_DISABLE	VK_3
	VK_4
Color Key Events	VK_5
VK_COLORED_KEY_0	VK_6
VK_COLORED_KEY_1	VK_7
VK_COLORED_KEY_2	VK_8
VK_COLORED_KEY_3	VK_9

Video Formats

The Blu-ray specification supports three high-definition video codecs — MPEG-2, SMPTE VC-1, and MPEG-4 AVC. That triples the amount of supported video codecs compared to DVD. Although the MPEG-2 codec is very well understood, given its legacy from DVD, the situation is different with VC-1 and AVC. Both of these codecs are considered more advanced and can achieve better compression ratios. In other words, the algorithms used by such advanced codecs are much more complex, but as a result, a better video quality can be achieved with much lower data rates.

In addition to the various video codecs, there are also a number of resolutions and frame rates to choose from. The full high-definition resolution (also known as *Full HD*) of 1920 × 1080 pixels is supported, but Blu-ray also allows for smaller resolutions to be used, such as, 1280 × 720, 720 × 576 (PAL), and 720 × 480 (NTSC). This begs the question of how these different resolutions and, particularly, their aspect ratios, will be rendered on the various screens available (see figure 6.7 and 6.8)? For instance, how is 4:3 content rendered on a 16:9 widescreen display, or vice versa? Ideally, the aspect ratio of the source content should be respected, otherwise the picture will look distorted. As an example, if the source content is 4:3 and played back on a 16:9 widescreen display, it will be stretched, making the actors look "fat". Since that is certainly not desired, the setting on the display should be changed to a *pillarbox* presentation by adding black bars on the left and right of the picture which remains in the original aspect ratio.

Application Details

Figure 6.7 Display Formats of 4:3 Aspect Ratio Content

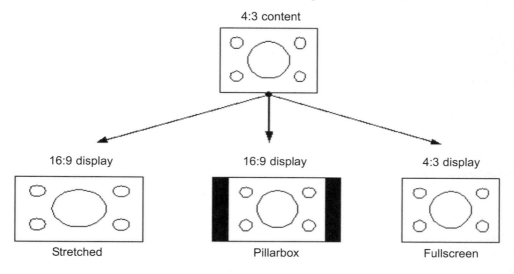

Figure 6.8 Display Formats of 16:9 Aspect Ratio Content

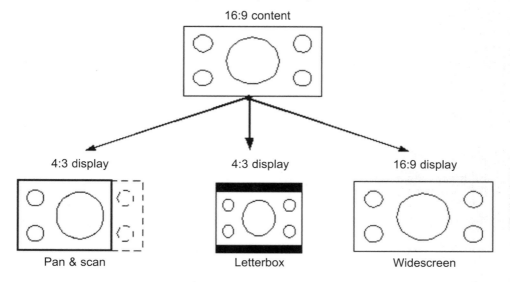

Application Details

Each of the video resolutions allow for different kinds of frame rates as detailed in Table 6.10. The typical Hollywood movie is shot with a film camera, which natively produces 24 frames per second. This film frame rate is translated to 23.976 frames per second to accommodate NTSC television sets. Content can also be acquired through other video acquisition methods, which results in different frame rates. For instance, some high-definition video cameras can mimic film and record at 24 frames per second, or even 29.97 frames per second in interlaced mode.

Table 6.10 Supported Resolutions and Frame Rates for Primary Video on Blu-ray Disc

Resolution	Frame Rates
1920 × 1080	29.97i, 25i, 24p, 23.976p
1280 × 720	59.94p, 50p, 24p, 23.976p
720 × 576	25i
720 × 480	29.97i

There are two modes in which video can be rendered to the screen — *interlaced* and *progressive*. A small letter "p" after the frame rate (e.g., 24p) indicates progressive mode, while a small letter "i" (e.g., 29.97i) indicates interlaced mode. In interlaced mode, a frame is composed of two fields, which display successively. All odd-number lines (e.g., 1, 3, 5, ...) will be rendered to the screen as the first field, and the even line numbers (e.g. ,2, 4, 6, ...) will be rendered as the second field to compose the full frame image. This is traditionally used for legacy television displays. However, the trend for home entertainment is to become more and more like home theatres, so a native film resolution using progressive frames is desired. This means that rather than splitting the frame into two fields, an entire frame is rendered at once. Even though computer displays already use progressive display for a long time, only recent introductions of next-generation television sets, such as, LCD or Plasma displays, or others, allow the same frame rate film viewing experience. Figure 6.9 describes the difference between interlaced and progressive mode.

Figure 6.9 Interlaced versus Progressive Display Modes

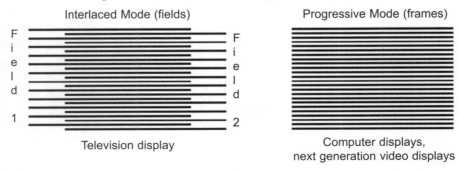

Application Details

Given the increased resolutions of high-definition content, the bitrates have to be adjusted, as well. However, they certainly did not adjust scale in a linear manner. In other words, with a high-definition resolution that is six times greater than standard definition, the maximum bitrates for video did not increase by a factor of six. Instead they only increased by a factor of four. For DVD, the maximum video bitrate is 9.8 Mbps with a combined multiplexed bitrate (including all audio, video, and subtitle streams) of 10.08 Mbps. For Blu-ray, the maximum video bitrate is 40 Mbps and the maximum multiplexed bitrate is 48 Mbps. The reason for the larger delta between video and multiplexed bitrate is to allow enough room for additional audio streams, especially lossless audio streams.

So now, the million-dollar question is how can the larger amount of pixels be encoded with fewer bits while keeping the quality equal or even better? The answer is rather simple: through advances in compression technologies. Although both VC-1 and AVC still use a block-based motion compensation and spatial transform scheme similar to MPEG-2, there are several enhancements in the two new codecs. Figure 6.21, presented at the end of this chapter, shows a generalized block diagram of an MPEG-2 encoder.

VC-1 provides adaptive block-size transforms, improved motion compensation, and a simple de-blocking filter, among others tools. Figure 6.22, also presented at the end of this chapter, shows a block diagram of a VC-1 encoder.

AVC provides even more advanced features and enables a lot of flexibility in the encoding process to adapt to the complexity of the scene. As an example, the configuration of blocks and slices, with a slice being defined as a contiguous group of multiple blocks, is flexible. So are the multi-picture inter-picture predictions that are able to provide a higher quality image. AVC also provides an integer-based transform algorithm to eliminate reconstruction errors during decode, a strong de-blocking filter, and greater precision for motion compensation. However, the biggest improvement is achieved through the *Context-Adaptive Binary Arithmetic Coding* (CABAC) approach, which allows a more efficient entropy coding of the syntax elements, and results in very big quality improvements. Figure 6.23 shows a generalized block diagram of an MPEG-4 AVC encoder, and is presented at the end of this chapter.

One thing to consider for both of the advanced codecs is that the increased efficiency does not come for free. The penalty to pay is related to processing times. While an entire film using MPEG-2 can be encoded in about the time it takes to play the film, this may not be possible with VC-1 or AVC. The processing requirements for both advanced codecs are much higher than for MPEG-2 resulting in much longer encoding times. One way to address this problem is the use of so-called *render farms*. Many professional users take advantage of the method of multiple computers connected via a network to share resources for encoding tasks to significantly reduce processing times. Similarly, the advanced codecs also require more expensive decoding circuitry in the players.

Advanced Video Applications

Besides the standard way of playing a video stream, such as a linear presentation of the main feature, there are additional advanced video applications that allow a much richer user experience. As with DVD, it is possible to have multiple camera angles or seamless multi-story presentations (different contextual paths through the movie). However, the advances

Application Details

do not stop there, Blu-ray also incorporates new features, such as, picture in picture (allowing two parallel video streams to be presented at the same time), with or without luma keying applied during playback.

Camera Angles

Blu-ray allows up to nine camera angles. This means that multiple video streams are interleaved and the viewer has the opportunity to switch between the angles by using the *Angle* key on the remote control. Additionally, the program code on the disc can determine and automatically effect an angle change, as well. One thing to keep in mind for multiangle features on a disc is the fact that all angles need to be in sync and, consequently, have the same duration. A typical application for synchronous multiple camera angles is a concert disc, which allow the user to switch to various camera angles, each angle following a different member of the band. The enduser can essentially "direct" their own version of the concert recording. Another application would be to allow multiple languages of the same scene. In such a scenario, it is very likely that the program code makes the decision on which angle to play based on the language selection of the user.

When multiple camera angles are used, they are multiplexed as interleave units inside the AV stream. This is to support the buffer management of the disc, as it can automatically jump from one interleaved unit for an angle to the next, ignoring units for other angles (Figure 6.10).

Figure 6.10 Camera Angles Example

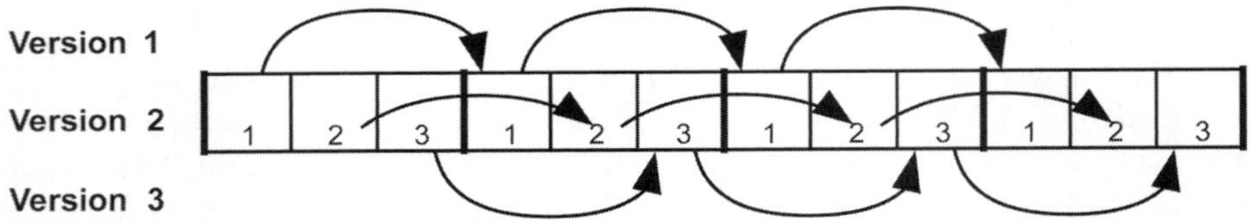

Adding a camera angle has an impact on the data rates and bit budget of the disc; having more angles will reduce the overall running time of the disc, as well as limit the maximum quality of the material. For example, a single-angle program of 5-minute duration encoded at 20 Mbps would take up 750 megabytes of the disc. If the program had two angles with the same duration, both could still be encoded at 20 Mbps. However, the program would take up twice the amount of data on the disc — 1,500 megabytes. This obviously reduces the possible overall playback time of the disc. In addition to that, there are further limitations on the maximum video bitrate to be used. For a single-angle program, the maximum video bitrate is 40 Mbps while the maximum bitrate for the multiplexed stream (TS_Recording_Rate) is 48 Mbps. The limitations for multiangle are imposed on the multiplexed stream, which will ultimately impact the maximum video bitrate. As an example, using the maximum rate for the multiplexed stream and subtracting the data rates used by all audio streams, the author can calculate the maximum video bitrate for each angle. Table 6.11 outlines the impact on the maximum multiplexed data rates (TS_Recording_Rate) depending on the number of angles being used.

Application Details

Table 6.11 Recommended Data Rate Limitations for Camera Angles

Number of Camera Angles	Maximum Data Rate (TS_Recording_Rate)
2	28.61
3	22.89
4	22.89
5	19.07
6	19.07
7	17.17
8	17.17
9	15.26

Seamless Multistory

The term, *seamless multistory*, refers to a video presentation that allows different paths through the disc content, hence allowing for multiple story lines. It is realized by jumping from place to place on the disc without any break or pause in the video (see Figure 6.11). This feature may be used for providing a viewer with alternate endings, or a director's cut, or similar applications.

Figure 6.11 Seamless Playback Example

From a conceptual point of view, the way seamless multistory is implemented in Blu-ray is rather straightforward. Different playlists define the multiple stories and the necessary video pieces are tied together for the desired presentation. Although two pieces of video that are to be connected seamlessly do not have to be contiguous on the disc, the jump distance between them is restricted. This means that any two video pieces that will be connected seamlessly should be very close together on the disc, so that the pickup head of the drive can jump to the next position without causing any breaks or pauses of the video. The higher the data rate of the program stream, the smaller the jump distance needs to be, because the pickup head has to start reading data sooner to avoid running out of data in the buffer.

Picture in Picture

Blu-ray supports showing a secondary video image called, *picture in picture* (PIP). This requires a second video decoder in the player and adds some complexity to the disc. While the initial Blu-ray players (Profile 1.0) did not have to support picture in picture as a mandatory feature (which allowed for faster market introduction of players without the added complexity of a secondary video decoder), all current players that are compliant with either Profile 1.1 or Profile 2, must support PIP as a mandatory feature.

Application Details

There are some restrictions on the resolutions, frame rates, and combinations of codecs for PIP streams (Tables 6.12 and 6.13). The content author can also determine the position of the picture in picture window and the position can change during playback. In order to do this, each picture in picture stream has positioning and scaling metadata associated with the secondary video file. For in-mux synchronous PIP streams, this metadata cannot be changed on-the-fly and it has to be established during the multiplexing stage.

Table 6.12 Allowed Combinations of Primary and Secondary Video Codecs

Primary Video	Secondary Video		
	MPEG-2	MPEG-4 AVC	SMPTE VC-1
MPEG-2	Allowed	Allowed	Allowed
MPEG-4 AVC	Prohibited	Allowed	Prohibited
SMPTE VC-1	Prohibited	Prohibited	Allowed

Table 6.13 Allowed Combinations of Primary and Secondary Video Formats

Primary Video		Secondary Video	
Resolution	Frame Rate	Resolution	Frame Rate
1920 × 1080 or 1440 × 1080	29.97i	1920 × 1080[a] 1440 × 1080 720 × 480	29.97i
	25i	1920 × 1080[a] 1440 × 1080 720 × 480	29.97i
	23.976p	1920 × 1080[a] 1440 × 1080 720 × 480	23.976p
	24p	1920 × 1080[a] 1440 × 1080 720 × 480	24p
1280 × 720	59.94p	1280 × 720 720 × 480	59.94p 29.97p
	50p	1280 × 720 720 × 480	50p 25p
	23.976p	1280 × 720 720 × 480	23.976p
	24p	1280 × 720 720 × 480	24p
720 × 480	29.97i	720 × 480	29.97i
720 × 576	25i	720 × 576	25i

[a]Support of 1920 x 1080 resolution for secondary video streams is optional for players.

There are three kinds of picture in picture scenarios possible — *in-mux synchronous* PIP, *out-of-mux synchronous* PIP, and *out-of-mux asynchronous* PIP.

The in-mux synchronous PIP is typically used for director's commentary or similar applications. Both video streams are multiplexed together into a single transport stream and play from the disc. Using this scenario, the primary and secondary videos are locked to the same time clock when presented on the screen.

The two out-of-mux PIP options do not require the secondary video to be multiplexed together with the primary video. And, in these scenarios, the secondary video does not have to live on the same disc. The PIP stream could be made available as a download to a BD-Live (Profile 2) player. A typical scenario would require the player to use a network connection to download the PIP stream to the local storage in the player. Once the download is completed, an application on the disc would recognize the stream and use the *Virtual File System* (VFS) to bind the new stream into the disc so it becomes available for the viewer (more information on VFS is presented later in this chapter). The fact that the stream is not on the disc will be transparent to the viewer.

With the out-of-mux synchronous approach, the application is identical to the in-mux scenario where the context of the secondary video is tied to the context of the primary video. Hence, they both have to be tied to the same clock. However, the out-of-mux scenario allows a commentary to be made available after the disc has been finished. For instance, with release windows for home entertainment getting shorter all the time, it may not be possible to record a director's commentary in time for the production of the disc. By relying on the out-of-mux option, this commentary can be produced after the disc shipped and then made available through a network download by the player (given that this option is authored into the initial disc).

In the out-of-mux asynchronous case, the out-of-mux PIP stream is not tied to the primary video. Instead, user interaction can determine when the PIP stream is played. Imagine a question and answer component on a disc: during playback of the feature, the viewer has the option to ask various questions to the creators of the movie. The answers will be displayed as a PIP commentary stream. Since it is up to the viewer when a question is asked, the PIP stream cannot be synchronized to the primary video. With the asynchronous PIP, such an application scenario is possible. However, as in the out-of-mux synchronous case, the data cannot be played off the disc. Instead, it must be stored in the local storage area of the player.

Luma Keying for Picture in Picture

An additional feature for Blu-ray is the option for luminance (luma) keys that can be applied to PIP streams, in certain cases. The luma keying is accomplished by defining an upper value for the luminance (Y) channel in the PIP metadata. All the pixels with a luminance value from 0 to the upper value are rendered transparent. This blends the two video streams seamlessly together. However, the luma key can only be applied if the secondary video is not scaled up to full screen. In the event where the secondary video is presented in full screen, the keying cannot be applied, hence all pixels are displayed fully opaque. To avoid such a scenario, it is recommended to prohibit the scaling of the secondary video when luma keying is used.

Blu-ray Disc Demystified

Application Details

Audio Formats

Considering that the number of supported video formats for Blu-ray increased compared to DVD, the supported audio format situation is also dramatically different from DVD! While DVD supports four audio codecs (PCM, MPEG, Dolby Digital, and DTS), Blu-ray supports seven primary audio codecs — PCM, Dolby Digital, Dolby Digital Plus, Dolby TrueHD, DTS, DTS-HD High Resolution, and DTS-HD Master Audio (Table 6.14). Why is there a need for three Dolby formats and three DTS formats? The difference can be found in the details. While Dolby Digital and DTS are the legacy formats from DVD, the other audio codecs are format extensions with improved quality.

Table 6.14 Supported Primary Audio Formats for Blu-ray

	Lossless	Mandatory	Sampling Frequency	Bits per Sample	Maximum Data Rate (Mbps)	Maximum Channels
Linear PCM[a]	Yes	Yes	48/96/192 kHz	16/20/24 bits	27.6	8
Dolby Digital	No	Yes	48 kHz	Compressed	0.640	5.1
Dolby Digital Plus	No	No	48 kHz	Compressed	1.7[b]	7.1
Dolby TrueHD[a]	Yes	Yes	48/96/192 kHz	Compressed	18.64[c]	8
DTS	No	Yes	48 kHz	Compressed	1.509	5.1
DTS-HD High Resolution	No	Yes	48/96 kHz	Compressed	6[d]	7.1
DTS-HD Master Audio[a]	Yes	No	48/96/192 kHz	Compressed	24.5[d]	7.1

[a]Sampling frequency of 192 kHz is only allowed for 2, 4, and 6-channel audio that is losslessly encoded.

[b]Dolby Digital Plus is only supported for more than 5.1 channels. The initial 5.1 channels are encoded with the core Dolby Digital Substream - channel 7 and 8 (if available) are encoded in Dolby Digital Plus.

[c]It is mandatory for a player to play back Dolby TrueHD. However, only the core Dolby Digital portion with up to 640 kbps has to be supported. The lossless portion of the Dolby TrueHD stream is optional for a player and may or may not be supported.

[d]DTS-HD streams consist of a core DTS portion and an "HD" portion. Players only need to be able to decode the DTS core of 1.509 Mbps. Supporting the HD portion of the stream is optional for a player and may or may not be supported.

Dolby Digital Plus (DD+) is an extension of the legacy Dolby Digital format and provides two additional channels, up to 7.1 channels. DD+ is based on the core Dolby Digital stream and only the additional two channels are encoded using the DD+ extension while the core 5.1 streams are still encoded with the legacy Dolby Digital Substream.

The situation is similar with DTS. The DTS-HD High Resolution and the DTS-HD Master Audio formats are extensions to the legacy DTS format. They provide up to 7.1 channels with higher bitrates, but still embodies the legacy DTS stream for backwards compatibility (legacy DTS only supported 5.1 channels).

Even though this may already seem a little complicated, let's add some more confusion.

Application Details

Why would we need two more audio formats — Dolby TrueHD and DTS-HD Master Audio? Don't we already have the capability for higher data rates and additional channels? The short answer is "Yes", but that's not enough. The two additional audio formats are lossless formats, whereas the Dolby Digital, DD+, DTS, and DTS High Resolution audio are lossy formats. The difference is that lossy audio codecs are much more efficient than lossless codecs. In other words, they eliminate a lot of redundant information in the audio source data to achieve lower bitrates and use less disc space (see **Chapter 2** for more on perceptual encoding). Lossless streams provide the highest possible quality, but can require much higher bitrates. Since lossless codecs are focused on delivering the best possible quality, they have to allocate as many bits as needed and use a variable bitrate. This drastically changes the production workflow compared to lossy codecs which use a constant bitrate. With a constant bitrate, the content author is aware of how much bitrate is necessary for the audio streams and, as a result, can easily allocate the maximum video bitrate to be used.

Having the audio stream as a variable bitrate changes this situation. The audio encoding has to be finished before the maximum bitrate used for the lossless audio stream can be determined. After that is done, the available maximum bitrate for the video can be calculated. Given that the video encoding historically happens before the audio encoding is finished, the production workflow has to adapt, taking this new requirement into consideration. As one can imagine, with the production workflow getting more complicated due to the variable bitrate of the lossless audio codecs, the audio decoders are getting more complicated, as well. Figures 6.24 and 6.25, presented at the end of this chapter, show block diagrams of the two lossless audio decoders.

There is a caveat with the support of the seven primary audio formats — it is not mandatory for players to support the full number of channels and sampling resolutions (Table 6.14). As some of the audio features are optional in the specification, it is up to the player manufacturers to decide whether or not they want to support them. The idea behind this approach is to allow manufacturers to produce differently-featured products and to differentiate themselves in the marketplace. For instance, a portable player may not need to support eight audio channels or lossless audio. By not having to support certain optional formats, the product can be produced easier and cheaper. The content producer should be aware that in the case where optional features are used, some players may not support them and an appropriate behavior should be built into the disc.

Multichannel Configuration

With all the additional channels beyond stereo, what does that mean for the living room? Where should the additional two speakers in a 7.1 audio configuration be positioned? Although there is the possibility for different positioning options, a very typical scenario is depicted in Figure 6.12. Besides the center speaker, the right and left speaker should be positioned at an angle of 30 degree, and the left surround and right surround speaker should be positioned at an angle of 150 degrees, respectively. While this is the same configuration as a traditional 5.1 positioning, including the position of the LFE (low frequency effects or subwoofer speaker), the additional two speakers would be positioned as left and right surround side speakers, angled at 90 degree to the center speaker.

Application Details

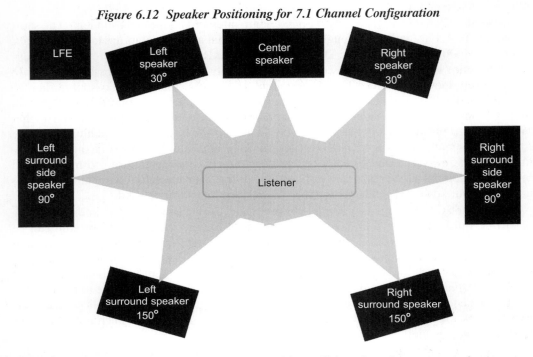

Figure 6.12 Speaker Positioning for 7.1 Channel Configuration

Now, if you believe that eight channels should be sufficient for a home entertainment system, think again. Japan's national broadcaster, NHK, has already started working on systems of the future supporting up to 22.2 audio channels. Will that, finally, be the maximum number of audio channels? Probably not!

Secondary Audio

Besides the primary audio stream, Blu-ray also supports a secondary audio stream. This allows discs to contain audio streams to be played in parallel and to be mixed with the primary stream. A typical application for a secondary audio stream is an audio commentary. While watching and listening to the main feature the secondary audio commentary can be mixed in. To allow this to happen, each Blu-ray player is required to have two audio decoders, as well as a *panner/mixer* that can mix the audio channels from the secondary audio stream with those of the primary audio stream.

The secondary audio stream can come from different sources. It can be contained on the disc and synchronized to the movie, which would be a typical scenario for an on-disc audio commentary. Or, it could also be made available as a download from a network source. In such a scenario, the audio would be stored in the local storage of the player, yet could still be synchronized to the movie. This would allow the audio commentary to be recorded later. Note that the secondary audio does not have to be synchronized to the primary audio. An example of an asynchronous secondary audio application might be a game where the user

Application Details

can select certain buttons on the screen that drive an audio response. This response would not be synchronized to the main feature audio as the user determines at a random point in time when the audio is played. The asynchronous secondary audio scenario also requires the files to come from local storage, not from the disc.

Blu-ray supports two secondary audio codecs — Dolby Digital Plus (DD+) and DTS-HD LBR (low bitrate) (Table 6.15). As the secondary audio streams are limited to bitrates of up to 256 kbps, both DD+ and DTS-HD are special low bitrate versions of the codecs that only consist of an extension substream. Besides the pure audio data, secondary audio streams must include additional mixing metadata. This allows the content producer to define the mixing parameters rather than leaving it to the player to make creative decisions.

Table 6.15 Supported Secondary Audio Formats for Blu-ray

	Mandatory	Sampling Frequency	Bits per Sample	Maximum Data Rate	Maximum Channels
Dolby Digital Plus	Yes	48 kHz	Compressed	256 kbps	5.1
DTS-HD LBR	Yes	48 kHz	Compressed	256 kbps	5.1

Subtitles

Blu-ray supports two types of subtitles — graphics-based and text-based. The graphics based subtitles are called *Presentation Graphics* (PG) and are very similar to subpictures used with DVD. Whereas, text-based subtitles, referred to as *TextST*, are a new method uniquie to Blu-ray that can provide a more dynamic way of rendering subtitles, albeit with some advantages and some disadvantages.

Presentation Graphics

While the graphics-based subtitles for DVD only allowed for images with four colors (two-bit images) to be used for subtitles, Blu-ray had to add some improvements. Otherwise, the subtitles would look awful when rendered to high definition television screens. So, the first improvement was to increase the allowed bit-depth for the graphics to 8 bits, or 256 colors. This allows much smoother image rendering since it is possible to use more anti-aliasing.

Looking at the Presentation Graphics (PG) decoding model, it should be mentioned that there are three different kinds of segments described — *graphics object segment*, *composition segment*, and *palette segment*. A segment defines a specific timestamp at which the element should be decoded into the buffer. At that point in time, the graphics processor decodes the RLE-compressed bitmap image into an uncompressed 8-bit graphic to be stored in the *object buffer*. Once the image has been decoded, it can be used by one or by multiple graphics displays, as described in the composition segment. The next step has the *graphics controller* mapping the information from the composition segment and the palette segment to the image data in the object buffer. The graphics controller is essentially responsible for compositing the image data onto the player's graphics plane, following the descriptions in the composition and palette segments. In other words, the controller applies the cropping, color, and transparency information to the uncompressed image in order to render the final full-color

Application Details

and full-transparency graphic on top of the video background. As mentioned earlier, each palette segment can contain 256 entries defining color and transparency combinations to be mapped against the 8-bit indexed source image.

As one might imagine, some basic graphics effects are possible by manipulating the color and transparency values in the palette. In addition to that, Blu-ray supports other graphics effects such as scrolls, wipes, cuts, and fades. Even though all of these effects may be useful for advanced subtitle presentations, such as, karaoke streams, they could also be used for other graphical displays over video, such as, storyboards that are timed to the video. Bottomline, any graphical elements that can be displayed with sufficient quality using only 8-bit graphics can take advantage of the Presentation Graphics system model.

Text-based Subtitles

Rather than using pre-rendered graphics subtitles, text-based subtitles provide a more dynamic way for displaying subtitle data. Text and associated style information are stored and used to render subtitles to the screen. As the text-based subtitles are implemented as out-of-mux streams, they do not take away any bitrate capacity from the AV streams. This implementation allows text-based subtitle streams to be present on a disc without impacting the audio or video quality. It also provides a perfect solution for downloadable subtitles as the files are very small and they are easy to imbed as an out of-mux stream. While it is the content authors' choice whether to use text-based or graphics-based subtitles, it should be considered that only one stream can be displayed at any given point in time. It is not possible to display both graphics and text-based subtitles simultaneously.

Each text subtitle segment can contain up to two text regions, and each region, besides having different text content, can be presented with a different style. However, both regions are tied to the same timing, which is frame-accurately synchronized to the AV presentation. The text region defines a size and position within the graphics plane and can have a unique background color. Within the text region, a text box defines the size and position of the text to be displayed. This framework provides flexibility for a user to define how much border will be shown outside the text. Each text box also has attributes that can be defined, such as, text flow, alignment, and line spacing. And, obviously, the various attributes such as font type, style, size and color can be set for each region, as well. In addition to defining the text attributes for a given text box, they can also be set individually for each character as in-line styles.

So having listed all the flexibility and bandwidth advantages of text-based subtitles, what are the trade-offs? Well, there are two disadvantages. First, it should be recognized that each player has a different font renderer and although the content author is able to define the fonts, they may look different on each player. But, besides this artistic challenge, there is a more fundamental issue called *font licensing*. Which font will be used should be considered very carefully. Although some fonts are available for free, most of them are not. And, in order to include them on a Blu-ray Disc, each font needs to be licensed, which adds some administrative and financial obstacles, especially when considering international releases. One way around font licenses is to design a unique font. Although this may seem a little bit over the top, it is actually not too difficult to implement, could add something unique to the disc and certainly avoids getting surprised by font royalty payments.

Application Details

Parental Management

One of the requirements from the Hollywood studios is to have a mechanism in place to control the levels of parental control for their movies. The idea is to allow a user to set a certain parental level on a player and prevent the playback of movies above that level. For example, when there are under-aged children at home, the Blu-ray player can be set to only allow playing movies appropriate to their age, unless the parent enters a specific password to unlock the playback of more mature content when the parents are watching themselves.

With Blu-ray, PSR 13 (Player Status Register 13) provides this kind of parental management. The user defines the parental level for a given player in the player settings menu, which sets the corresponding value in the register. Whenever content is played back, commands on the disc can check for the value of this PSR. In case the value is too low to allow playback, a warning card with an explanatory message can be displayed on-screen. In case the value is high enough to permit playback, the disc presentation would continue without problem.

PSR13 can contain an integer value from '0' to '255' and represents the youngest age for which the content is permitted. The value '0' is the most restricted content and value '255' signals that the parental control is switched off. As an example, Table 6.16 outlines how this PSR could be mapped against the MPAA rating system used in North America.

Table 6.16 Correlation of PSR 13 Settings to MPAA Ratings

Player Status Register 13	MPAA Rating
0	G
8	PG
13	PG-13
17	R
18	NC-17
255	Switched off

Because different countries have different rating systems, it is common that a parental management system is also bound to the country code setting inside the player. This setting is also done within the settings menu of the player and defines which country the player is used in, which is stored in PSR19. In order to support parental levels for multiple countries, the value of PSR19 should be checked or determined before checking for PSR13. By combining both results, the rating system for the respective country can be applied. Note that, as with DVD, this parental control system is, of course, predicated on the film content owners assigning practical ratings to their movies and for that information to be stored on the discs and properly read by the players.

Metadata

Metadata is defined as data describing data and, in the context of Blu-ray, metadata is used to describe the data on the disc. This can be very useful for certain applications, such as disc search. For example, if each disc included a metadata description that could be stored

Application Details

inside a player, a user could very easily start searching for titles with their favorite actor, director, or genre. To allow such a disc library search, the Blu-ray implementation of the application metadata is split into two parts — a metadata description file and associated thumbnails.

The metadata description file contains information about the name of the disc and, on a more granular level, the content of each title, each chapter, and track name, the various languages, and so on. All, or some, of this information can be provided by the content author and becomes available for a disc search. The way this is implemented is that once a disc is inserted into a player, the metadata file is copied to the local storage of the player. As discs get inserted into the player, a disc library starts building up. However, how much of a disc search is possible within a given player depends on the specific player implementation.

Thumbnails are an additional feature that provides a visual link to the content. They are similar to the jacket pictures in DVD. For instance, while the metadata of a specific disc is displayed on the screen, it could also show an associated thumbnail, such as, the disc cover. In addition to a thumbnail for the disc, the metadata descriptors can contain thumbnails for each title. Basically, all the thumbnails are stored as separate files with references inside the metadata information file. In this way, it is easy for content authors to decide whether to use thumbnails for a disc or title search. It is also possible to store multiple resolutions of thumbnails to accommodate different display resolutions.

HDMV Details

The HD Movie (HDMV) disc authoring model can be thought of as the more basic level of menu programming. It can be compared to standard DVD authoring, while providing a number of new functionalities and advanced concepts that can enrich the user experience. The terminology for HDMV is also very different from DVD. However, in terms of the authoring process, HDMV and DVD are, indeed, very similar in the sense that each requires a dedicated piece of software to format the disc. And, it is up to that software to determine whether it provides a low level access to all functions and parameters or whether it abstracts them to a higher level for easier authoring.

Popup Menus

One of the new concepts introduced on Blu-ray discs is *popup menus*, which is one application of the *Interactive Graphics* (IG) framework. Actually, this is not really new as settop boxes for TV's have been using this feature for quite a while but, compared to having to navigate to full-screen menus on DVD, popup menus can provide a much less disruptive user experience. Rather than leaving the feature presentation to jump to a full-screen menu page in order to select a different audio stream, this can now be done while the feature keeps playing, using menus that only cover a portion of the video presentation.

However, the good ol' full-screen menus in a dedicated area (referred to as *menu space*) are not completely lost. Blu-ray differentiates between a Top menu and a popup menu. The Top menu remains a designated area on the disc, similar to DVD, where full-screen menus can be placed and are separate from the video titles. This differentiation also means that, depending on the player implementation, there will most likely be two different menu buttons on a remote control — "Title Menu" and "Popup Menu".

As the name suggests, the popup menu appears whenever the user pushes the appropriate button on the remote control. At that point in time, the video keeps playing while a menu selection is displayed over the video and the user can navigate through various menu pages to select different audio or subtitle streams, browse different chapters or other available content on the disc. Once the user is done, by pushing the same remote control button again or through a simple timer, the popup menu disappears.

Multipage Menus

Menus may be composed of multiple pages, with a page being defined as a composite interactive display. As there can only be one page displayed at a time, all menu elements need to be contained therein. This includes all buttons with the various states — normal, selected, activated — and all other displayed graphics. As Figure 6.13 depicts, the first page may show the main menu in its starting layout with all submenus collapsed. Page 2 would show the main menu including the submenu items for the audio settings, page 3 would show the submenu items for the scene selection, and so on. This approach is flexible and scalable, but requires the necessary programming logic to make sure that the appropriate page is shown at any given time.

Figure 6.13 HDMV Multipage Menu Example

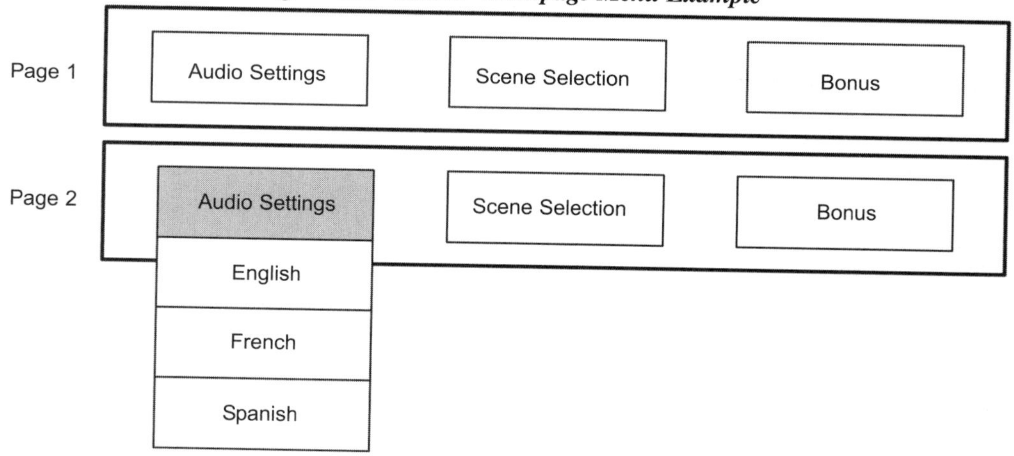

| Page 1 | Audio Settings | Scene Selection | Bonus |
| Page 2 | Audio Settings / English / French / Spanish | Scene Selection | Bonus |

In addition to displaying multiple static pages, HDMV menus also offer the ability for effects and animations. Each page can have an in- and out-effect associated with it that plays prior to loading the page or right before transitioning the page into the inactive state. A common application of this would be to animate the main menu bar onto the screen when the menus are turned on and to animate them off the screen when the menus are turned off. Similar to presentation graphics (subtitles), the effects supported are scrolls, wipes, cuts, fades, and color changes.

Application Details

While the page effects certainly provide a nice visual aid for the menu presentation, animations are also possible for buttons. A sequence of images can be displayed for each button state (normal, selected, activated). Imagine having static images for each button in the normal state, and then the selected button would spin or run some other animation to signal the selected state to the viewer. There would not be any doubt about which button is currently selected.

HDMV Graphics Limitations

All of the new fancy menu effects are interesting, but there are limitations. There is a limit of 16MB for all graphical elements used in the total of the menu pages of the HDMV menu set. While this sounds plentiful, it really is not when you look into the details. For instance, in order to render menu animations, a separate page has to be created for every frame displayed. Where a main menu bar is animated in ten frames onto the screen, ten menu pages would need to be rendered. And, although the pages would all be using the same source graphics, the amount of memory taken up would be a multiple of that, as each of the source elements needs to be decoded separately into the graphics buffer for each menu page. This restriction makes it very challenging to use animations within the HDMV menu framework.

On a more fundamental level, as Interactive Graphics follow the same principles as Presentation Graphics, they are limited to 256 color/transparency combinations. While each menu page can have its own color palette associated with it, in practice this is not very common due to the fact that menu source elements will likely be shared between various menu pages, and differing color palettes may result in color shifts when switching pages. When using HDMV graphics may not result in a satisfying viewing experience, there is always the option to go with BD-J menus, as described later in this chapter.

Interactive Audio

In addition to providing for button animations, Blu-ray Disc accommodates an option for button sounds or *interactive audio*. Whenever a user interaction occurs, an audio snippet can be played, as well. Interactive audio can be associated with each selected and each activated button state. When a button is selected, a sound effect is heard and, when the button is activated, a different sound effect is heard. This can be used in many creative ways: just envision using this feature to create a version of menus for the visually impaired. And in the context of button sounds for popup menus, the audio will be mixed with any other currently playing audio streams.

The format for interactive audio is LPCM, which may contain up to two channels with a sampling frequency of 48kHz and 16 bits per sample. All button sounds are stored in a `sound.bdmv` file, which can contain up to 128 individual pieces but cannot exceed a total aggregated file size of 2MB.

Interactive audio can also be used by BD-J applications. In fact, the support of interactive audio in HDMV mode is optional for players but it is mandatory for BD-J applications. Fortunately, most players do support interactive audio for either scenario.

Browsable Slideshow

Another feature introduced with Blu-ray is the browsable slideshow. This feature basically decouples a contiguous audio presentation from the images being viewed. As a result, it is possible to present a number of still images or video clips while audio continues to play. The user is able to switch back and forth between the slides without interrupting the audio. With DVD, such an application feature was desired, but was only possible with DVD-Audio players. Now, using the path and subpath concept of Blu-ray, this scenario is available. The images are stored in the main path for the user to navigate while the audio is stored in a subpath. The user navigation within the main path will not interfere with the audio play back from the subpath.

BD-Java (BD-J) Details

The most sophisticated method of programming a Blu-ray Disc takes advantage of the Java programming language, and is called *BD-Java* (BD-J). This allows a lot of flexibility for the content author at the core of the system, as there are no restrictions introduced by specific authoring software[2]. This is pure software development and can be best compared to writing a software application for a PC. This, obviously, requires a much higher level of commitment and a very different skill set than what regular DVD authors would normally bring to the table.

The benefit of a full-fledged programming language is the level of flexibility it provides to programmers. Other than your own imagination and the capabilities of the playback hardware, there are no further constraints. A second benefit is the reusability of the software code. Java code can be developed in an abstract way where the code portions are separated from the creative or content portions. This way it is easy to reuse the same code for multiple discs, as only the specific content (graphics, animations, text, et cetera) would have to change. Another benefit is extensibility and scalability. As the code interacts with the application and the player, it is possible to scale an application based on the player's capabilities. This means that a more powerful player may show more advanced menu animations while other players may not. BD-J guarantees a great user experience on every player without having to code only for the lowest common denominator. And, because players will most likely get better and faster as time goes on, writing extensible code allows for new features to be introduced further down the line. In other words, there may be ideas that cannot be implemented in today's applications due to player limitations, such as slow processors or minimal loca storage capacity. However, this situation may change over time as player processors get faster and hare larger internal and external hard drives, at which point such application features would become possible.

---

[2]There are software applications available with the aim of simplifying the BD-J development process by automatically generating BD-J code. Such applications typically provide a Graphical User Interface (GUI), which means it doesn't require a Java programmer to build a BD-J application. On the flip side of that, most of these applications have their own set of limitations and it may not always be possible to implement the features and functions desired.

Application Details

System Overview

BD-J applications interact with the user to control the player. The system model, depicted in Figure 6.14, shows four components that communicate with each other. On the lowest level, the *Resources* represent the capabilities of the player, such as decoding, memory management, and the player control. The *BD-J module* includes the *Java Virtual Machine* (JVM) on which the applications run and the *Application Manager*, which determines what application runs. The BD-J module also translates the *Application Programming Interface* (API) calls to access the necessary resources. The *BD-J API* component is a collection of functions and objects to communicate with the player. The BD-J applications use the API to load files, access the network, control the disc playback, any any other function that requires the player to perform a task. The top level of the system model is the *BD-J application*, which uses a so-called *Xlet*, a term inherited from JavaTV™. It is very similar to an applet, which is a lightweight application commonly used for web applications. In this context, Xlets follow the same lifecycle definitions with multiple states — loaded, paused, activated, and destroyed.

Figure 6.14 BD-J System Overview

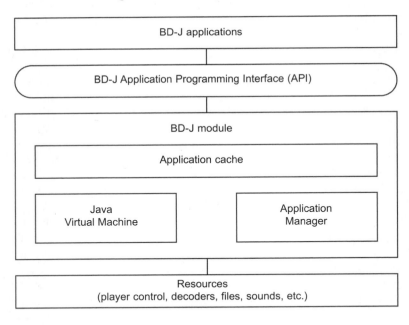

Resources and BD-J Module

There are a number of different resources that can be accessed by the BD-J application, such as graphics, sounds, JARs (Java Archives), and the AV structure on or off the disc. In order to access and manage these resources, the BD-J module is used. The BD-J module provides an abstraction layer between the player and the disc application. It uses the *appli-*

Application Details

cation cache, which is best described as a pre-load buffer for the application. All files (JARs, and other files or directories) necessary to properly run the application will be loaded into this cache. Once they are loaded into the cache, the application manager can take control. The manager can load or unload individual applications and manage their lifecycles, accordingly. Besides the application cache, there are other memory buffers holding additional resources that can be accessed by the BD-J module. Table 6.17 outlines the various buffers and their respective sizes, which vary between the different player profiles.

Table 6.17 BD-J Memory Overview

Memory	Description	Profile 1	Profile 2
Application cache	Used to load JAR files and directories	4MB	4MB
Font buffer	Used to load fonts to be rendered by the BD-J application	4MB	4MB
Image and audio memory[a]	Images and audio files used by an application will be decoded into this memory. Audio files can only take a portion of this memory.	45.5MB (5MB can be used for audio)	61.5MB (6.5MB can be used for audio)

[a]Button sounds are stored in a `sound.bdmv` file. The memory for images and audio does not include the `sound.bdmv` file (up to 2MB) but, instead, provides additional space for interactive audio.

Now, a question may come to mind of why the different buffers are necessary. In a real-life scenario, the JAR files may contain a lot of class files (Java code) and a bunch of compressed images (PNG, JPEG, et cetera.). These JAR files will then be loaded into the application cache. However, in order for the application module to use the images inside the JARs, they have to be decompressed. This would happen next and, as a result, all uncompressed images will be available in the image memory. Then the application can be executed and displayed on the screen.

It is very similar with the font buffer. In order for an application to use a font, it has to be copied into a buffer to be accessible. Because some fonts, particularly Asian fonts with many complex characters, may take up a lot of space, it may not be possible to render the font in a timely manner. By using the buffer, however, fast interaction is possible. In other words, fonts can be generated on-the-fly as the characters are pre-loaded in the buffer.

BD-J Application Programming Interface (API)

A programmer interacts with the player by using Application Programming Interfaces (API). There are API functions for pretty much anything you can imagine — mathematical operations, file I/O, playback control, graphics display and animations, text rendering, network access, local storage management, and so on. However, the API set for BD-J programmng is a bit more complicated and requires some background on how the BD-J specification came about.

It is very common for a format specification to reference other specifications. For instance, the DVD specification references various audio and video specifications such as MPEG, Dolby, DTS, and others. However, from a programming perspective, DVD has always been very self-contained in the sense that everything you needed to know was defined in one doc-

Application Details

ument — the *DVD Specification*. While it is generally the same for HDMV, it is a little different for BD-J. The BD-J specification leverages a multitude of other specifications as depicted in Figure 6.15.

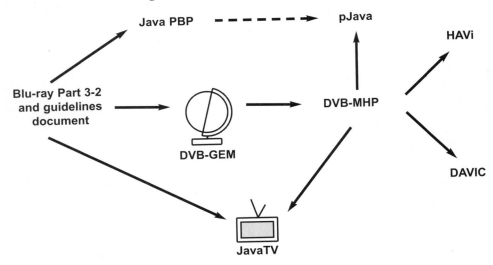

Figure 6.15 BD-J Specification Structure

For starters, BD-J uses the *Personal Basis Profile* (PBP) of Java, which was developed after PersonalJava™ (pJava), which was started in 1997, as a more lightweight programming environment for applications running on interactive TV devices or something similar. Although the PBP specification is a great start, it does not address broadcast specific functions, such as, scaling and positioning video, switching audio, video, or subtitle channels, and so on. The good thing is that other standards subsequently started developing on top of pJava and, later, PBP to address such items. For instance, the *JavaTV* specifications, which include the *Java Media Framework* (JMF), were developed to introduce media APIs for streaming applications on personal computers. Later on, when this was not enough, the GEM (*Globally Executable MHP*) specifications introduced more functionality by leveraging even more specifications such as MHP (Multimedia Home Platform), which references *HAVi* (Home Audio Video Interoperability) and *DAVIC* (Digital Audio Video Council). And since BD-J also requires some custom APIs that were not required for anything other than Blu-ray, the BD-J specifications contain additional APIs.

With that background in mind, it can be a little exhausting to find the necessary API for a specific function. Table 6.20 attempts to provide an overview of the various packages contained in each set of specifications — PBP, JavaTV, GEM, MHP, and BD-J specifications. The table is by no means an exhaustive list but, hopefully, it helps as a starting point. Please keep in mind that an MHP *broadcaster* becomes a BD content provider, a *terminal* or *receiver* becomes a BD player, and a *service* becomes a title.

Application Details

Table 6.18 BD-J API Overview

API	Function
Personal Basis Profile (PBP)	
java.awt	Basic APIs for drawing 2D images to the screen.
java.io	Read and write data streams.
java.net	Network operations such as opening sockets, and supporting protocols such as HTTP or HTTPS.
java.lang	Fundamental data types required for Java programming including threads and exceptions
java.math	Arithmetical operations.
java.util	Utility classes for operations such as random number generators, JAR file reader and writer, and others.
java.security	Cryptographic operations that may be used to securely send data over network ports.
java.text	Text and number formatting, which can be used to internationalize an application.
JavaTV	
javax.tv.xlet	Communication between Xlets and the application manager, primarily Xlet lifecycle management.
javax.tv.service	Access to a service, which is historically a channel in the broadcast environment. For BD-J, it translates to a title.
javax.tv.locator	Access to a resource by using a URI (Unique Resource Identifier).
javax.media & javax.tv.media	Media player control to change audio streams, pause the video, or other operations.
GEM (Globally Executable MHP) / MHP (Multimedia Home Platform)	
org.havi.ui	Screen resolution and other graphical user interface parameters.
org.davic.resources	Resource notification to manage conflicts between shared resources across multiple Xlets or other parts of BD-J.
org.dvb.application	List and launch applications from another Xlet.
org.dvb.dsmcc	Digital storage media command and control (DSMCC) to mount a filesystem contained inside a JAR.
org.dvb.ui	Overlaps to some extent with java.awt, but contains the important FontFactory class for loading fonts.
org.dvb.event	Event registration for remote control keypresses. This enables an application to listen for the remote control events and act upon them.
org.dvb.io.ixc	Inter Xlet Communication (IXC), which allows multiple Xlets to communicate with each other and handle priorities.
org.dvb.io.persistent	Access to local storage of a player.
org.dvb.media	Extensions to the JMF (Java Media Framework) to manage streams and playback functions.

continues

Application Details

Table 6.18 BD-J API Overview (continued)

API	Function
BD-ROM Specification, Part 3-2 (BD-J Specification)	
org.bluray.media	Blu-ray specific JMF controls for picture in picture, camera angles, or other BD player functions
org.bluray.ti	Information about the disc. "ti" stands for title information, which includes PlayList, PlayItems, metadata and other important information.
org.bluray.vfs	Virtual File System (VFS) access.
org.bluray.ui	Graphics animations synchronized to AV playback.
org.bluray.storage	Information about the local storage capacity and free space.
org.bluray.system	Access to the Player Status Registers (PSR) to communicate with the HDMV layer.

As one may imagine, with multiple specifications referencing each other, some functionalities can be implemented in multiple ways using different APIs. This creates some complexity for player implementation. As an example, a player manufacturer needs to support multiple ways of loading images, making the testing procedures more complex. As a result, it is recommended to properly test every disc in as many players as possible to ensure that the APIs used are properly implemented in the player and do not cause any compatibility or performance problems.

BD-J Menus

With BD-J positioned as the advanced programming method for Blu-ray, it can, of course, at minimum mimic the menu functionality of HDMV. Full-screen and popup menus, as described in the HDMV part of this chapter, can also be achieved using BD-J. However, the way these menus are implemented in BD-J differs very much, and provides additional capabilities. First of all, the way that graphics are used is very different. BD-J menus do not follow the page model that HDMV uses. Instead, BD-J provides a minimum graphics buffer of 45.5MB for Profile 1 or 61.5MB for Profile 2 players, as noted in Table 6.17. All graphical elements to be used by the BD-J menus are decoded into the image and audio buffer[3]. This means that a decoded image can be used multiple times without taking up additional space. This model benefits more advanced animations. Graphical elements can be moved across the screen as much as the author would like them to, and the animations can even be dynamically created by the Java code, without impacting the graphics buffer.

---

[3]Smart applications can also manage the loading and unloading of images as needed to deal with the buffer restrictions, if necessary.

Application Details

Blu-ray supports three image formats — JPEG, PNG, and GIF. The JPEG format is commonly used for a variety of Internet applications as the files are small in size. However, the disadvantage of the JPEG format is that it does not support transparencies. When multiple graphical objects are overlapping and must be composited together before being displayed, transparency information is necessary. For such a case, the *Portable Network Graphics* (PNG) format is typically used as it supports alpha blending. The third graphics format supported by Blu-ray is GIF but, in reality it is not used very often as it only supports one transparency color. As a result, JPEG and PNG are most commonly used in real-life Blu-ray applications, with the former best used for background and opaque images with no mattes, and the latter used for matted or transparent graphics.

Graphics Drawing

For compositing images, there are different graphics drawing methods available for BD-J. Depending on the application scenario, an image can be drawn to the screen in different ways. First, there is *source mode* (SRC), which takes an image the way it is and renders it to the screen, which is also called *direct draw*. This mode does not take other graphical images into account which may be overlapped. Instead, source mode simply renders the image that is on top without blending any overlapped graphical areas. If the image is semi-transparent it will render as such and allow the background video plane to be seen through the image. But, other rendered graphic elements on the graphics plane will not be blended. Figure 6.16 is an example of source mode, where the rectangle cannot be seen through the ellipse.

Figure 6.16 Graphics Drawing in Source Mode

(Courtesy of the United States Capitol Historical Society and HeritageSeries, LLC)

Application Details

The second graphics-drawing mode is called *source-over mode* (SRC_OVER). This mode takes transparency values into account when compositing overlapped graphical areas. This method is typically used for graphics animations, and takes multiple overlaid images and blends them together. Where graphics are semi-transparent, this mode shows the background video through the transparent areas of the rendered images. Figure 6.17 is an example of the source-over (SRC_OVER) drawing mode, where both the rectangle and the background video can be seen through the ellipse.

Figure 6.17 Graphics Drawing in Source-over Mode

(Courtesy of the United States Capitol Historical Society and HeritageSeries, LLC)

The reason for having both modes is to allow for different creative scenarios. However, there is a slight performance hit when using the source-over mode. Since the images have to be composited together, additional processing power is required inside the player. As a result, the source-over drawing speed is slower than using source mode and this should be considered during any application development.

Graphics Animations

The third graphics drawing mode for BD-J is called *frame-accurate animation* (FAA). It comes in two variants — *image frame-accurate animation* and *synchronized frame accurate animation* (SFAA). The idea of this drawing mode is to have a tighter connection of the graphics animation to the video presentation in the background. The FAA graphics are locked frame-by-frame to the video playback, which means that graphics will only appear at

Application Details

the timecode desired. This feature provides a great potential for creative applications. However, it, too, does not come for free.

Looking at all three drawing methods — SRC, SRC_OVER, and FAA, the player performance numbers vary. The fastest drawing method is the SRC mode as it draws the image directly to the screen without applying any alpha blending. The recommended minimum performance for the SRC drawing mode is 15 megapixels per second. Since the SRC_OVER and FAA modes require additional processing for operations, such as applying alpha blending or timing, the minimum performance is lower at 10 megapixels per second. Such implications should be kept in mind while designing applications to ensure that the animations will perform as expected on all players.

BD-J Text

One of the main benefits for using BD-J menus is they allow a lot of flexibility. An area where this becomes particularly apparent is for displaying dynamically-generated text. This provides a huge contrast to DVD or to HDMV menus where all the text has to be pre-composed and cannot be dynamically created during playback. For example, for a DVD game, an individual graphic for every possible score has to be created and stored on the disc. This limitation is history when using BD-J where the scores can be dynamically generated by the Java code and displayed on the screen. Additionally, for example, it is possible for users to enter their name to personalize the game or to allow multi-player scenarios.

There are two ways that BD-J text can be implemented on a Blu-ray Disc. One approach is to use fonts in a manner similar to text subtitles. The BD-J application loads an OTF font and then can render any text string to the screen. Of course, this has the same drawbacks as the text subtitles, which means each player uses a slightly different font renderer, making the final result potentially look different on each player. However, to display certain information, such as, game scores, user names, or runtimes of videos, this is a great method to use.

A second, but a little bit more complicated, way is to implement bitmap fonts. Basically, there is one large image file (often called a *character strip*) containing all the letters and numbers in the alphabet with an associated coordinates list that describes the cropping parameters for each letter. However, it should be noted that bitmap fonts are not natively supported by BD-J but, instead, must be implemented in the application. But, by using this method, any text string can be assembled on the screen using the bitmap font. This allows the creative designers to define how the font will look and does not leave it up to the player for interpretation. The amount of additional graphics memory necessary to accommodate this may be negligible. Although this method works great for western languages, applying it to Asian languages is much more difficult considering the larger amount of characters available. However, both options should be considered to provide a more dynamic user experience.

Advanced BD-J Features

Being able to create interactive, dynamic, and animated menus is a great benefit of using BD-J, but there are additional features that are much more advanced and provide a lot of creative potential that is waiting to be unleashed. Some of these features are not possible

Blu-ray Disc Demystified

Application Details

with the Profile 1.0 players that were first released to the market. Other features are only possible with Profile 2, which is optional for players. However, it is worth noting these features, as they are truly innovative and very interesting in certain applications.

Multi-disc Sets

Before getting too far into the territory of the newer generation players that are using Profile 1.1 or 2, let's focus on advanced features that are possible with any player. One of those features is a multi-disc set. Imagine a Blu-ray boxed set with multiple discs holding all episodes to a TV series. Wouldn't it be great if the entire boxed set could be enjoyed without interrupting the user experience? In other words, once all the episodes on the first disc are finished, the menu would ask for the next disc to be inserted and, while the disc change happens, the menus remain on-screen with some animations continuing to run. Once the new disc is ready, playback automatically resumes with the next episode without having to press a single button. All this is possible with BD-J by using *disc-unbound* applications.

A BD-J application can be bound to either a title or a disc allowing the following three scenarios —

- First, the application can be *disc and title bound*, which means that the application is only active while a specific title is being played. In the event where the title is changed, for instance to jump to a bonus piece, the application is terminated.
- The second scenario is *disc bound and title unbound*, which allows the application to continue running while a title change occurs. This allows a more integrated disc experience, as the main menu may always be present regardless of the title being played back.
- However, in the first two scenarios, the application is terminated whenever the disc is ejected. This is where scenario three comes into play, the *disc-unbound* application. This scenario allows a disc to be ejected and another disc to be inserted without terminating the application. During the disc change, the application continues to run, provided that both discs have the proper permissions set. This multi-disc feature provides excellent creative opportunities to improve the user experience.

Programmable Audio Mixing

As one may expect, BD-J also provides capabilities for interactive audio just like HDMV does. Yet, since BD-J is a programming language it allows a lot more flexibility. Basically, a Blu-ray Disc can play multiple different audio streams simultaneously, such as, main audio for the primary feature track, secondary audio for a commentary or PIP track, and effect audio for button sounds or other effects. Because the player only allows one audio stream to be output, this requires the player to mix the streams together. An audio mixer inside the player performs the mixing function, as depicted in Figure 6.18.

Application Details

Figure 6.18 Audio Mixer Components

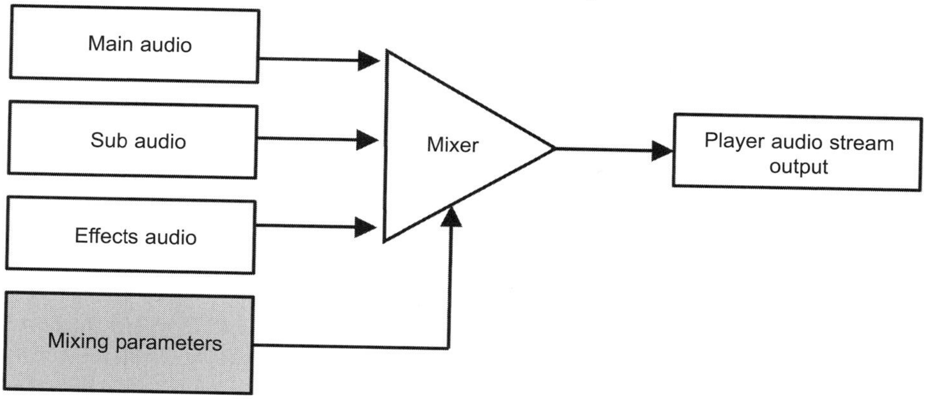

```
Main audio  ──►
Sub audio   ──►  Mixer  ──►  Player audio stream output
Effects audio ──►
Mixing parameters ──►
```

The audio mixer accepts mixing parameters that determine how the final output will sound. And since it is important for an application to control these parameters, BD-J allows a content developer to have full control over the audio mixing happening inside the player. It is even possible to programmatically change the audio mixing parameters to make it even more flexible or to potentially pass the control to the enduser via an on-screen tool.

Local Storage

The local storage feature of Blu-ray allows a player to have onboard memory to store additional data for various purposes. Each Blu-ray player has two different local storage areas — an *Application Data Area* (ADA, sometimes also referred to as APDA, and known in MHP as *persistent storage*) and a *Binding Unit Data Area* (BUDA).

The Application Data Area resides inside the player and is guaranteed to only be 64KB in size. This area is typically used to store small amounts of information such as user preferences, progress information, or other small data files. The Binding Unit Data Area can either reside inside or outside the player, or both. The intention for BUDA is to be able to store larger files — AV streams, subtitles, applications, or other types of files. Even if a player already has a large amount of built-in local storage, it may have the option for additional storage that can be added externally, for instance, through the use of a USB port.

While the ADA support has been available inside players from the very beginning (starting with the Profile 1.0 players), support of BUDA was only introduced with Profile 1.1 players. The minimum requirement for a Profile 1.1 player level is to support 256MB of BUDA storage, while a Profile 2 player has to support a minimum of 1GB. The reasoning for this should be rather obvious. If a player is network-connected, it is more likely that additional storage space for downloaded material will be required.

However, the BD requirements state that as long as a player provides the capability to add external storage, the minimum requirement is fulfilled. In other words, as long as a player

Application Details

comes with a USB port to add external storage, it does not need to ship with the required 256MB or 1GB itself. And, in reality, a lot of players indeed follow this direction and are only shipping with a USB port and the 64KB of ADA storage, leaving it up to the consumer to add the additional storage.

Virtual File System

Once content is available in the Binding Unit Data Area of local storage, either through network downloads or via a copy from a physical disc, a disc application can take advantage of it. This is accomplished by the creation of a *Virtual File System* (VFS). The VFS binds the files in local storage securely to the file system on the disc and creates a new "virtual package". None of the files on the disc are actually replaced but instead the pointers to the files are changed to the ones in local storage. Additionally, new files can be added to the file system to extend the capabilities of the user. For instance, a disc can be sold with three trailers during the intro and five audio languages for the main feature. Using the VFS, at a later point in time the initial three trailers can be replaced by newer trailers and the number of audio streams can be increased.

To do all this, a *manifest file* is needed that describes which files will be used from local storage and which files are to be used from the disc in order to create the ultimate user experience. Then, a new file system is created that maps all of the streams and makes viewing the disc a seamless experience. The viewer will not be able to tell whether the content is playing off the disc or from local storage.

Figure 6.26, presented at the end of this chapter, depicts how such a virtual package might be mapped.

BD-Live Functionality

Profile 2 players are assumed to either have a WiFi or an Ethernet connection that allows them to connect to a network or to the Internet. If a player provides this capability, a BD-J application can talk to the network using sockets, HTTP, and HTTPS connections. In the event where a secure connection is necessary (for example, to enable e-commerce applications), the *Secure Socket Layer* (SSL), formally known as the *Transport Security Layer* (TSL), can be used. In such a case, the BD-J application would open an HTTPS connection and use the specific SSL APIs provided by BD-J. This allows the same protection that is common to other kinds of web applications.

Uploading and Downloading Content

Before a BD-Live application gets into action, it should be considered that just because a player has the capability to connect to the Internet, does not mean that the player is actually hooked up. One of the first steps for any BD-Live application should be to check whether or not a network connection is indeed available. Once this is confirmed, any of the above mentioned connection types — sockets, HTTP, or HTTPS — can be used to transfer data either from a server to the player, or from the player to a server.

Application Details

Typical download applications include pulling new AV streams from the Internet, such as the most recent trailers, additional audio or subtitle streams, bonus videos, or new BD-J applications. Network connectivity provides the perfect mechanism to update BD-J applications of a disc to either enable more functions or to fix bugs that were encountered after the disc was released. In fact, a great way to start implementing BD-Live on discs is by adding a so-called *bootstrap functionality*. This means that a disc, after insertion into the player, will automatically try to connect to a network server to check for available updates. If an update is available, the user can decide if it should be downloaded or not. This bootstrap functionality allows additional features or fixes to be added without having to plan for them in the beginning. It's basically getting a foot in the door to add more functionality of any kind in the future. And if it is not necessary to add anything in the future, then it doesn't have to happen.

The counterpart of downloading features is uploading features. Typical scenarios for this are to allow users to upload information and share it with others on the network. This could be high-scores for games, bookmarks to their favorite scenes, or other types of user-specific comments for a given title. The possibilities provided by uploading or downloading functionalities are great, and combining them both makes it an even stronger proposition. A disc could include chat functionalities, as known from other computer environments, that allow users to talk about their favorite movie with other enthusiasts. Or, network games are also possible, where multiple online players can compete against each other. These kinds of scenarios all require data to be transferred both ways and need an application to be flexible enough to render instant information dynamically to the screen. In order to do this, though, the use of fonts or bitmap fonts is necessary.

Progressive PlayList

A great feature introduced by BD-Live is called *Progressive PlayList*. The intent is to allow AV streams delivered from the network to start playing before the entire stream is downloaded. A user does not have to wait until the entire stream is downloaded before it can be watched. The playback can begin as soon as a minimum amount of data is transferred into a local buffer, while the rest of the stream continues to download in the background. This kind of functionality is very common to Internet downloads that provide pseudo-streaming experiences using progressive downloads.

In Blu-ray, this is implemented by using a *virtual asset* (a Progressive PlayList) that is similar in structure to a regular PlayList in the sense that it contains multiple *PlayItems* and references AV stream data. However, it differs in the way that it allows playback to start before all assets are available. Because a PlayList is divided into multiple PlayItems, the playback can start once the first PlayItem is successfully downloaded. While the playback starts, the remaining PlayItems continue downloading in the background. As additional PlayItems become available, they are enabled for playback. It only depends on the download speed to determine whether or not the playback will be seamless. If the download speed is too slow, the playback will pause at the end of a PlayItem until the next item is available, at which point the playback continues.

Application Details

Application Signing

If a BD-J application is attempting to perform an operation that is beyond a self-contained "sandbox", meaning it needs to interact with other parts of the disc or with other applications — jumping to titles, starting and stopping other applications, reading and writing to local storage, using the virtual file system, or accessing the network — an authentication process is necessary. This authentication process is called *application signing* and the permissions of every application are defined in the *Permission Request File* (PRF). The PRF contains information about which area of the local storage can be accessed, or which network ports the player may connect to. The application signing is not a content protection mechanism but instead provides an authentication procedure to ensure that the content on the disc has not been altered and is indeed allowed to access other content located in local storage or on a network server.

The way this works is that each application is signed and a PRF, as well as the corresponding certificate, is attached to it. The certificate is authenticated against a disc root certificate through a certificate chain. The disc root certificate is authenticated through AACS. This allows scenarios where different parties may create individual parts of the disc. In this case, each party signs their own application and creates their own PRF, and the authoring studio putting it all together incorporates all of the applications with the signatures and PRFs to create the disc with their own root certificate.

As the next step, a SHA-1 hash[4] of the root certificate is generated that is incorporated into the master AACS List as part of the encryption process. As a result, whenever a disc is inserted into a player, the hash of the root certificate contained on the disc is calculated by the player and compared against the hash in the AACS list. If the content has not been modified, the hash values will match and the application will load. Figure 6.19 outlines the workflow scenario for application signing including multiple application developers and the use of content from different locations — disc, local storage, and the network.

The *root certificate* is also an important piece when it comes to managing the local storage in a player. As one could imagine, having a lot of discs accessing the same storage area can make it difficult to manage that space and will most likely come to naming conflicts if the space is not managed properly. As such, Blu-ray implemented a directory structure that allows sharing the common storage space. The idea is to have multiple layers of directories that differentiate between studios (content owners), producers (authoring facilities), and discs.

The root certificate is used to differentiate between studios. Each studio should have their own root certificate that it carefully manages, and each studio should issue leaf certificates to every authoring vendor they are using. Hence, every disc created by Studio A will be based on the same root certificate, regardless of the authoring studio. As a second security step, the

---

[4]SHA is shorthand for "Secure Hash Algorithm". A SHA-1 hash is a cryptographic signature for a text or a data file. It is sometimes called a "digest". A hash is not encryption, as it cannot be decrypted back to the original file content. It is a one-way cryptographic function. It is a process that may be used to validate passwords, authenticate "handshakes", verify digital signatures of files, et cetera.

Application Details

Organization ID is used to differentiate between different authoring vendors, or producers. Alternatively, the Organization ID could also be used as a differentiator between different product categories of a given studio. For example, children's discs could use a different Organization ID than mature content discs. The third security level is the *Disc ID*, which differentiates between different discs by a given studio and authoring facility.

This kind of structure helps to manage the content of discs and keeps them organized. It should be mentioned that a disc is only allowed to access the content of its own tree. In other words, a disc can access the data stored in its own Disc ID directory, its Organization ID directory, and the root certificate directory. This allows for sharing assets across multiple discs of the same category. As an example, if multiple children's discs of a given studio should share the latest trailers, it probably makes sense to store them in the area of the Organization ID, where each disc underneath can access them. On the other hand, if the trailers were stored in the disc-specific area of the local storage, another disc would not be able to access the data, unless specific credentials allowed for it.

Figure 6.19 Workflow for Application Signing

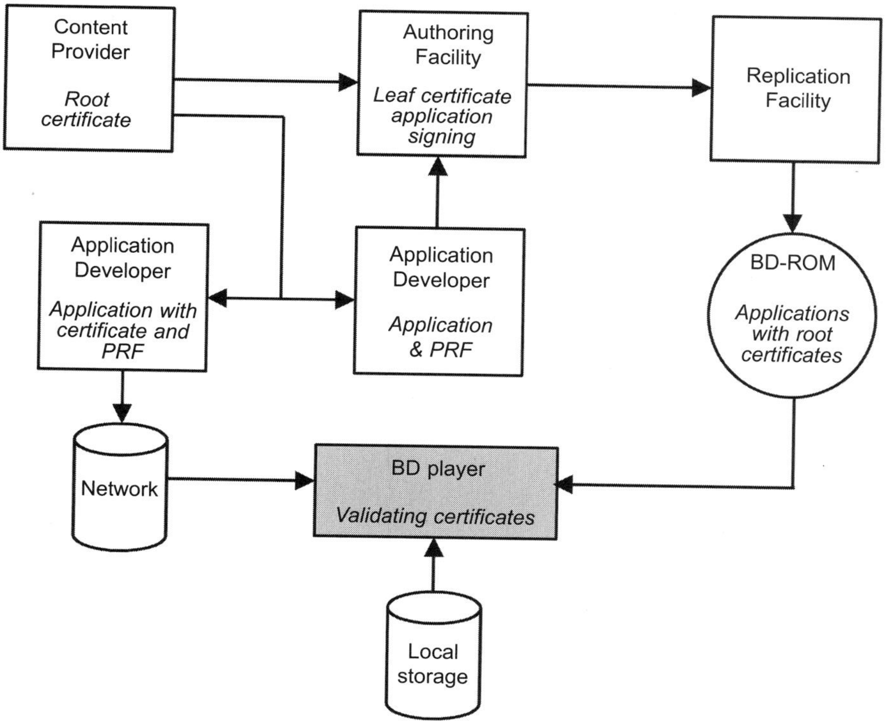

Blu-ray Disc Demystified

Application Details

Credential Process

When one disc needs to access data stored in the disc-specific area of another disc, the *credential process* is a method implemented in BD-J to allows for that. Credentials grant permission to applications to read or write data in other areas of the local storage. The way this works is illustrated in Figure 6.20.

Figure 6.20 Credential Creation Process

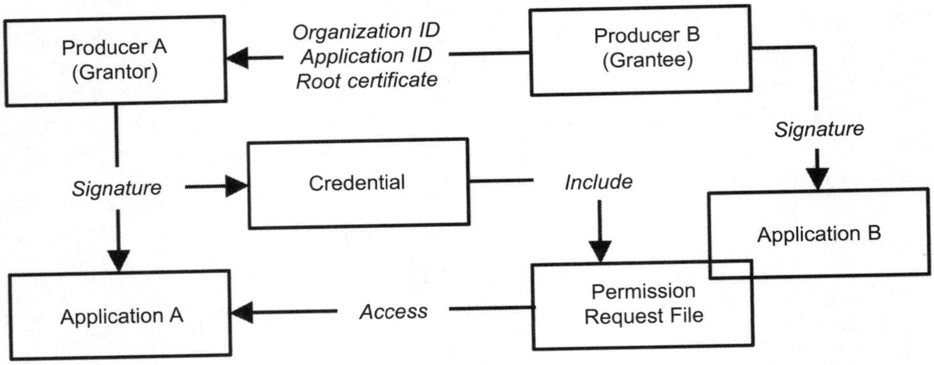

Basically, one producer (Producer A) grants permission to another producer (Producer B) to access Producer A's application. Thus, Producer B can access the disc-specific area of the application from Producer A to share data. In order to do so, a certain information exchange is necessary. First, Producer B (grantee) has to provide the Organization ID, Application ID, and root certificate of Application B to Producer A (grantor). Then, Producer A will create a credential and provide it to Producer B along with its own signature that was used for the initial Application A. As the last step, the credential is implemented in Application B's permission request file (PRF), and Producer B can then access Application A's data in the local storage area.

This all may seem a little bit confusing, but the process provides a lot of flexibility in managing BD-J applications and data stored in local storage areas. However, to keep it all organized, it is recommended to properly plan a disc beforehand to make sure that the data stored in local storage is available to all discs that need to access it. And, bearing the complexity of the process in mind, it is also recommended to thoroughly test the workflows before implementing it on discs. In that way, a desired functionality and user experience can be guaranteed.

Application Details

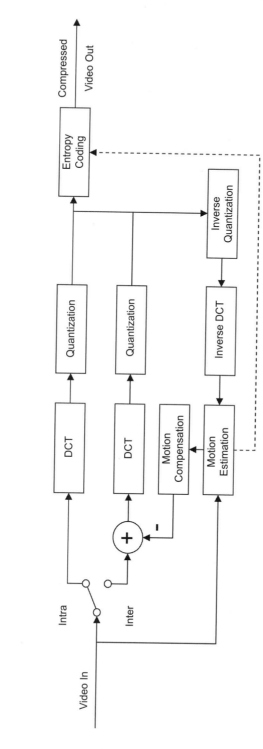

Figure 6.21 Generalized Block Diagram of an MPEG-2 Encoder

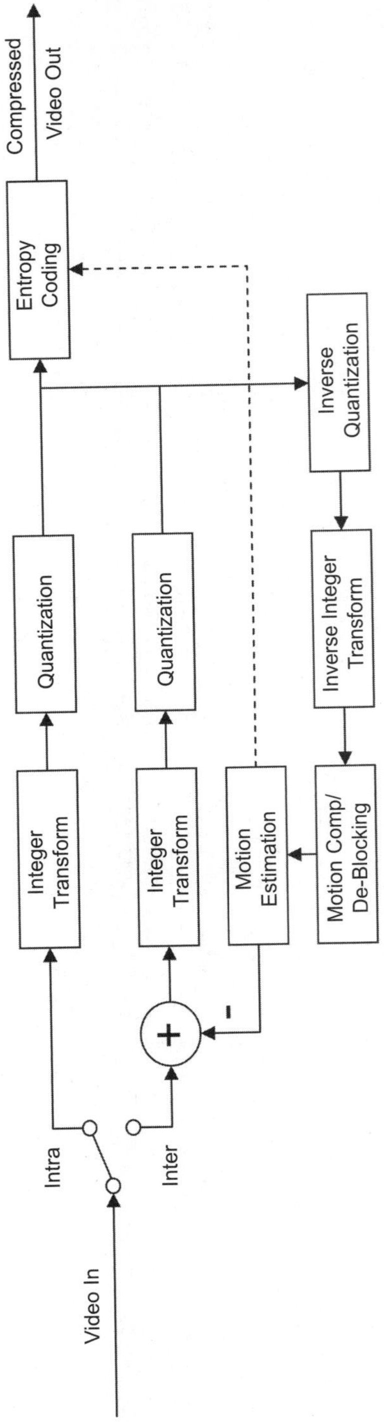

Application Details

Figure 6.22 Generalized Block Diagram of a VC-1 Encoder

Application Details

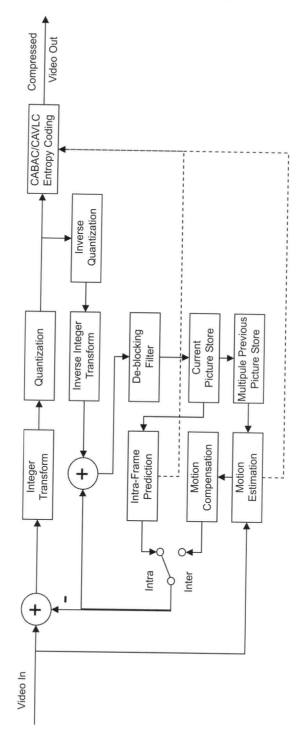

Figure 6.23 Generalized Block Diagram of an AVC Encoder

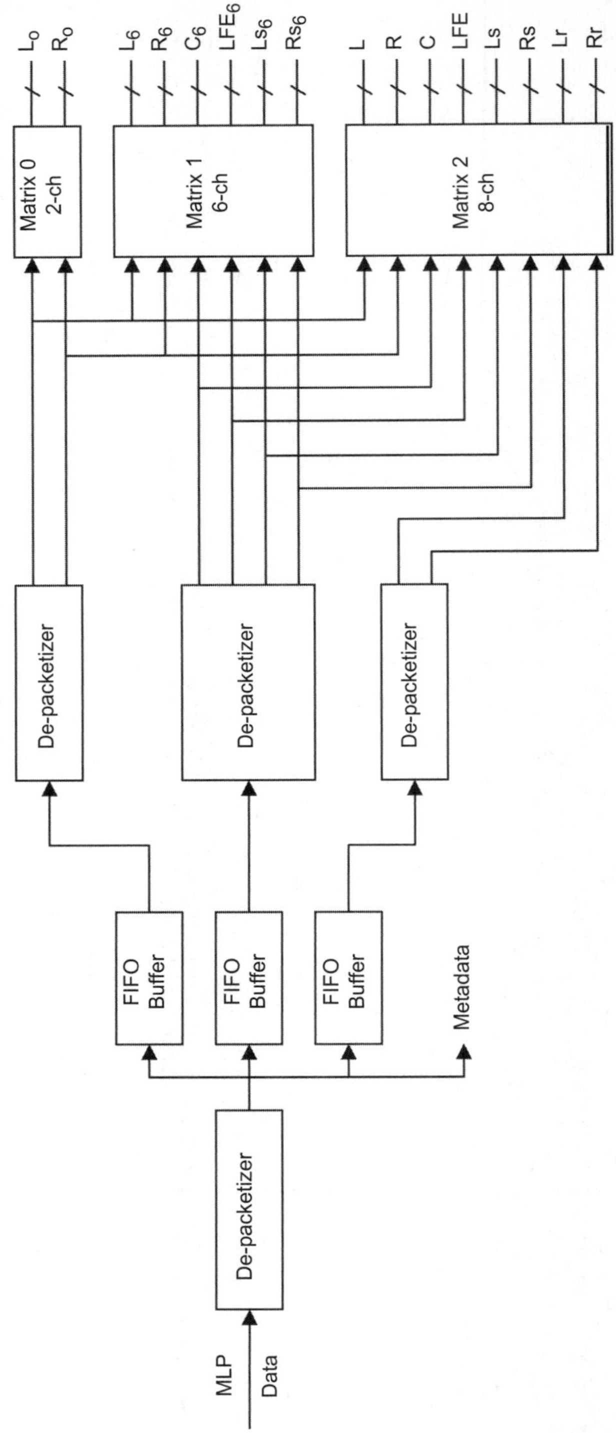

Application Details

Figure 6.24 Block Diagram of a Dolby TrueHD Decoder

Application Details

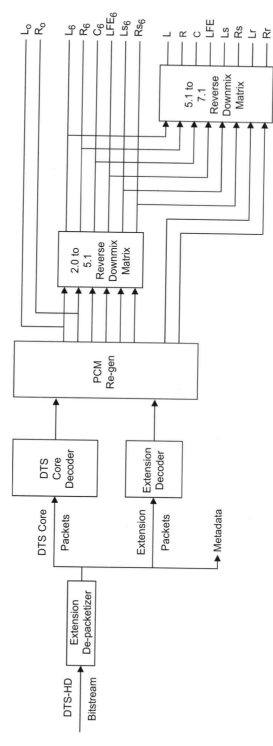

Figure 6.25 Block Diagram of a DTS-HD Lossless Decoder

Application Details

Figure 6.26 Example of Virtual Package

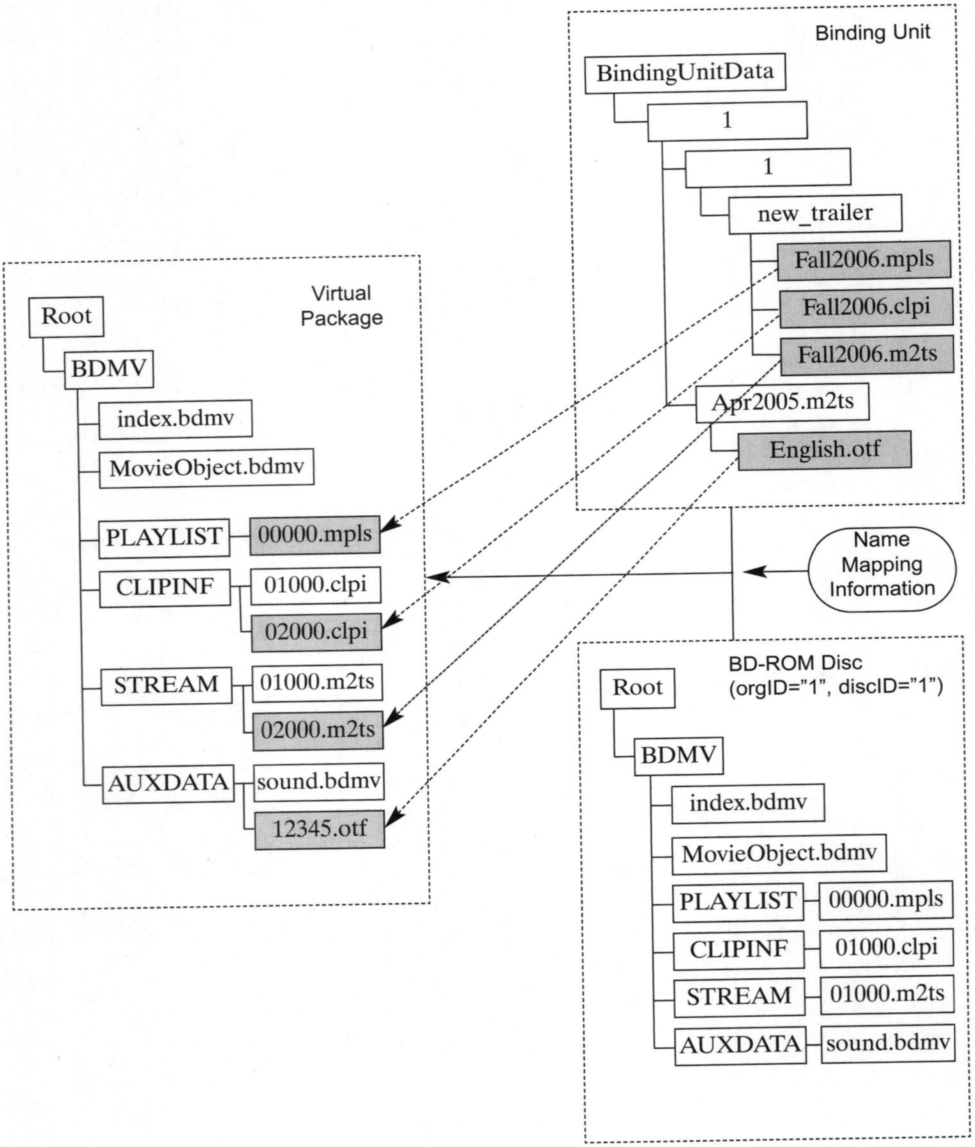

Chapter 7
Players

To consumers, the world of Blu-ray Disc™ players is, likely, a very confusing world. From the *Profile* method of player capabilities adopted by the Blu-ray Disc Association (BDA) to the gaming flexibility of the *Sony PlayStation 3*™, consumers face a daunting task when trying to make a player selection. Table 7.1 presents the Profile levels defined by the BDA, reflecting the stages of player development.

Table 7.1 Blu-ray Disc Player Profiles and Features

	BD-Video			BD-Audio	BDMV Recording
Profile Number and Name	Profile 1 "Initial Standard" "Grace Period"	Profile 1.1[a] "BonusView" "Final Standard"	Profile 2 "BD-Live"	Profile 3 "Audio Only"	Profile 4 "Recording"
Feature summary	BD playback (HDMV and BD-J)	PIP, Persistent Storage (VFS), increased JVM performance	Network capability	Subset for audio playback without video	Real-time recording and editing (RREF)
Available	Start of format	November 2008	March 2008	TBD in 2009	TBD
Formats	HDMV & BD-J	HDMV & BD-J	HDMV & BD-J	HDMV	HDMV
SD & HD Video[b]	Mandatory	Mandatory	Mandatory	No	AVC only[e]
Secondary video	Optional	Mandatory	Mandatory	No	No
Browsable Slideshow	Mandatory	Mandatory	Mandatory	No	No
Closed Captions	Optional	Optional	Optional	No	No
Audio[c]	Mandatory	Mandatory	Mandatory	Mandatory	DD only[f]
Secondary Audio	Optional	Mandatory	Mandatory	No	No
Audio Mixing	Optional	Mandatory	Mandatory	No	No
192 kHz LPCM audio	Optional	Optional	Optional	Optional	Optional
Persistent Storage	64 KB	64 KB	64 KB	64 KB	
Local Storage[d]	None	256 MB	1 GB	?	
Virtual File System	Optional	Mandatory	Mandatory	No	N/A
Internet Access	No	No	Mandatory	No	No
Metadata Search	Optional	Optional	Optional	Optional	No

[a]Profile 1.1 replaced Profile 1, which became officially obsolete November 1, 2007.

[b]AVC/H.264, VC-1, MPEG-2, text subtitles, presentation graphics, interactive graphics.

[c]Dolby Digital, Dolby TrueHD, DTS, DTS-HD, 48 and 96 kHz LPCM

[d]Total of built-in and external "potential" storage combined, which means the player can ship with no storage if it has the capability for the user to add sufficient storage. This requirement refers only to the BUDA.

[e]No text subtitles, no angles, no shuffle or random playback, no interactive graphics other than in menu (FirstPlayback and TopMenu titles).

[f]Only 8 streams of Dolby Digital (DD).

Blu-ray Disc Demystified

Players

Player technology has advanced to a level where the device components may match or exceed the components found in a typical desktop computer. Further, it ostensibly "won" the high-definition format war, Blu-ray faces competition from an unlikely source — DVD players that can upconvert the video output to high-definition displays. Added to the challenge of choosing a player, there is the confusing complexity confronted when trying to make a decision about those very high-definition displays.

Early adopters of HD displays are now caught in the technological gulch of having a great-looking display but the display is simply not equipped to present the best possible image from a Blu-ray Disc player. The early model widescreen televisions may not have the latest in connection security with HDMI (High-Definition Multimedia Interface) capability. If a Blu-ray Disc is authored to require that a secure HDMI connection exist between the player and the display, the player will only allow a less than full-resolution image to be passed to the display when there is no HDMI security. And, lurking in the wings is the soon-to-emerge (slow-to-emerge?) DisplayPort™ connection technology that replaces DVI and VGA connections for PCs while respecting HDMI connections to HDTVs (whew!). Those videophiles who bought into the high definition arena before HDMI are just plain out of luck (unless they take their chance with a reverse adapter, HDMI to RGB).

Conversely, early Blu-ray adopters most likely do not have a player that is capable of being upgraded to take advantage of the latest Blu-ray Profile features. When Blu-ray players were introduced they were only required to meet the initial feature set specified for a Profile 1 player. Further, the early players may not be capable of accepting a feature upgrade as their components are simply too simple (see *Chapter 8, Myths*).

Yet, it is not only the endusers who are flummoxed by the myriad player choices. Disc authors are caught in a never-neverland of what player is capable of what features and whether they should program for what the player is supposed to be able to execute according to the latest format specifications or should they program for the lowest common denominator in player capability. Their dilemma is exacerbated by the fact that the ballyhooed Profile 2, aka *BD-Live*, player level is merely an optional feature set, not mandatory. Player manufacturers are not required to accommodate network interactivity.

Players exhibit a range of characteristics. The player Profiles define which functions are mandatory and which functions may be optional. It is up to the player manufacturer to decide which optional features, if any, may be implemented in a player. Subsequent player generations need to fully support all previous features to ensure backwards compatibility of discs.

Blu-ray Disc supports JPEG, PNG, and GIF graphics format files. Each of these formats has its own advantages and disadvantages — for example, only PNG files support multiple transparency levels. Here, again, the decoding for each of these image types is implemented differently in each player depending on the player components, which will cause varying performance results from one player to the next. All of the players are quickest when decoding JPG images, followed by PNG images, and slowest when decoding GIF files.

While the first generation players do not support the now current level of features, this can change over time by either upgrading the player firmware and/or releasing successive generations of players. And, at the same time, newer discs are becoming available that take advantage of the enhanced feature capabilities.

For a disc author, different Profiles mean that their latest applications can possibly use more sophisticated features but, since first-generation players are still on the market, new applications have to work with the early players, too. There are ways for disc applications to query a player and determine which features are presently implemented. Knowing the level of the feature implementation allows the disc author to enable or disable specific features on different players. In this way, disc features can be scaled based on the sophistication of the player. More enhanced applications can be developed for more sophisticated players, while basic features would still be available on all players.

Profile 1.1, aka *Bonus View*, requires advances in the Java VM (virtual machine)™ performance characteristics of players. Yet, the player manufacturers can adopt different methods to achieve the desired performance results, thus creating another performance variable among players. And, disc authors who create applications where a lot of computational operations are required may end up with extremely different playback performances on devices.

As the market matures, player prices will reflect their feature capabilities. Although we have passed the Profile 1.1 threshold, manufacturers are not blindly altruistic and can be expected to continue producing players that meet the Profile 1.1 standard, not exceed it. Which may be welcomed by those users who merely wish to watch a great looking HD movie but have no desire to chat with other viewers while doing so.

High-Definition Multimedia Interface (HDMI)

The struggle to protect intellectual property is a major cause of the difficulties being faced by player and by display manufacturers. The desire to view the best possible image generates the demand for high-resolution, high-definition displays. But the need to protect the content from illicit duplication has created the unforgiving connection standards that have been developed, which are in addition to the AACS and BD+ systems for protecting the disc content. A chief concern is that an HD display could be used as a conduit for recording the image being presented, or that a player could have its output siphoned off to a recording device. So a standard was needed that protected the literal connection between the image source and the image display.

Expanding on the HDCP (High-bandwidth Digital Content Protection) process that was developed for the Digital Video Interface (DVI), manufacturers are instituting the High-Definition Multimedia Interface (HDMI) procedures and connection standards.

HDMI supports television and computer video formats, plus multichannel audio configurations, via a single connection cable. First promulgated in 2002 as HDMI version 1.0, the technology has matured to version 1.3. Each successive iteration provided for ever-increasing data throughput, growing from 4.9 Gbps with version 1.0 to 10.2 Gbps with version 1.3. With each new version of standards, devices adopted the new data rate. Unfortunately, this has lead to the situation where legacy devices that only supported a lower data rate are being connected to newer devices that recognize and/or output the higher rates. This mismatch of data capability may create the condition where the HDMI "handshake" between the devices cannot take place. What occurs subsequent to this connection refusal is the transmission of a less than optimum image format with an attendant diminution of video and audio quality.

Players

The backwards compatibility of the HDMI 1.3 connection should allow for devices with the earlier HDMI 1.1 and 1.2 versions to be recognized. These earlier versions support all of the mandatory audio and video formats of the Blu-ray Disc format. Once again, though, here is where confusion may be introduced in the connection equation. When multiple devices can be connected to a display, the display may not have the capability to automatically recognize the signal being sent to it by a Blu-ray player and does not reset for the newly-connected source. The display settings need to be manually adjusted before a proper connection can be sanctioned. The user is then faced with the task of accurately re-setting both devices in order to establish the best connection.

These security technologies add a great deal of complexity and, alas, confusion to the format, for the content creators and for the endusers. But, without the protection standards that have been developed for content and for transmission, the movie studios and content producers would not participate in the high-definition presentation arena. So, the difficulty of working with these protection schemes has become something of a necessary evil.

DisplayPort Interface Standard

The DisplayPort 1.1 interface standard has been established by VESA (Video Electronics Standards Association) (OMG, yet another committee!!!). Primarily expected to replace analog VGA and digital DVI connections currently in use for computer hardware, DisplayPort adds audio capacity, improved color signal passthrough, and enhanced screen resolution capabilities to a single cable. DisplayPort offers an open standard alternative to HDMI. As such, there is no inherent royalty cost to manufacturers, whereas HDMI technology is licensed by HDMI LLC and the cost for HDMI licensing raises device costs to consumers (see the *Licensing* section of *Chapter 4*).

DisplayPort (DP) is fully compatible with HDMI 1.3, and consumers will be able to purchase HDMI-to-DisplayPort adaptors and/or cables. DP supports HDCP and DPCP (DisplayPort Content Protection)(doh!) thus satisfying the content mavens at the Hollywood studios and allaying their Digital Rights Management (DRM) concerns.

There are minor differences between DP and HDMI. DP 1.1 supports a maximum bandwidth of 10.8 Gbit/s while HDMI supports 10.2 Gbit/s. The connector for DisplayPort is similar to that for USB and provides an optional latching mechanism for surer connections. There is an expected future improvement for DisplayPort that will accommodate multiple displays in a daisy-chain configuration.

But, the bottom line difference between DisplayPort and HDMI is economics. The PC industry is supporting DP because they do not have to pay royalties for using it, while a few companies in the consumer electronics industry developed HDMI and require licensing agreements for using the HDMI interface.

Fortunately, the ramp-up for DisplayPort incorporation with computers and other devices is taking longer than initially projected thereby also delaying any confusing impact on the marketplace.

Oh, did we mention that the DP connector is smaller than the VGA and the DVI connectors that it replaces?

Upscaling DVD

When we first saw the term, *upscaling DVD*, another phrase came to mind — *silk purse*. But, some of the biggest players in the player arena are deadly serious about providing this feature. What everyone must recognize is that ***NO DVD-VIDEO PLAYER IS CAPABLE OF PLAYING NATIVE HIGH-DEFINITION FORMAT VIDEO BECAUSE DVD-VIDEO IS ONLY STANDARD DEFINITION!***[1]

In the summer of 2008, Toshiba raised the confusion of digitally upscaling standard-definition DVD to new heights. After losing the HD format war, Toshiba sought to reassert itself in the optical disc arena by introducing Toshiba XDE™ (eXtended Detail Enhancement) technology. This latest attempt at infusing a "near HD"[2] quality presentation when watching a standard DVD will likely only prolong the general malaise that is affecting the marketplace when it comes to buying into the latest HD technologies.

Electronics manufacturers have hit on a way to provide purchasers of high-definition televisions with a display size matching video output from a standard DVD-Video player, with the additional requirement that a digitally secure connection must exist between the devices. Through the use of specially designed integrated circuit chipsets, the DVD video output may be scaled to match the desired display setting. These chipsets use various techniques to achieve the higher display rates, such as line doubling, pixel interpolation or pixel repetition, enhanced edge detail, deeper color with higher contrast, and some chipsets provide better pictures than other chipsets.

A standard DVD contains MPEG-2 video encoded as either 525/60 (NTSC) video at 720 pixels by 480 lines or 625/50 (PAL) format video of 720 pixels by 576 lines, and the digital upscaling recomposes the video to match the 720p, 1080i, or 1080p presentation settings of the display.

The advantage of having a DVD player perform the upscaling function is that the data remains in the digital domain. Previously, the DVD video would be converted to analog and sent to the display as composite, s-video, or component video. The display would then perform an analog-to-digital conversion prior to presenting the rescaled imagery on the screen.

An improved viewing capability may be most noticeable when watching a digitally upscaled output of a progressive-scan DVD player using the HDMI or DVI link. In fact, the digitally upscaled output is not viewable on any of the player's analog outputs.

The prices of these upscaling DVD players are considerably below that of a Blu-ray Disc player or a PlayStation 3 game console. Giving extended, if not new, life to existing DVD collections with an upscaling player may prove very appealing to a significant portion of the market. But, it is a somewhat surprising case of manufacturers advocating "good enough quality" as a strategy, which will likely further contribute to the delay in consumer adoption of true high-definition.

---

[1] There are a few DVD players that are capable of playing Windows Media High-Definition Video (WMV HD) and HD DivX files from DVD-5 or DVD-9 discs.

[2] Saying that "upscaled SD" is "near HD" is akin to saying the Moon is "near" the Earth. It is true, from a cosmic perspective, but when you are gazing up at the Moon while still on Earth it certainly does not look "near".

Player Types

Generally, there are three types of Blu-ray Disc players —
- CE (consumer electronics) standalone, or settop, players,
- software players in computer systems with a BD drive, and
- Game consoles, such as the Sony PlayStation 3, that have a BD player.

The three player varieties have very different capabilities in terms of processing power, memory/storage, and graphics support. Settop players are targeted for the living room with the main purpose of watching movies, where interactivity provides a heightened user experience and offers new viewing features. However, a settop player is not designed to be a gaming device, while a game console can provide excellent graphics with a lot of fast-paced interactivity, transporting the game player to a different world. Yet, a game console requires more processing and graphics power which means higher cost.

The second playback environment, software players in computer systems or media centers, provides a lot of processing power and memory but may not have the graphics power of a game console. A computer system provides a more powerful playback environment than a settop player, but the settop is generally targeted to only playback movies.

The mandatory playback capabilities for players are —
- Video decoder circuitry for MPEG-2, MPEG-4 AVC/H.264, and SMPTE VC-1
  - Encoded frame rates of 23.976, 24, 29.97, and 59.94 (25 and 50 are optional)
- Audio reproduction capability for —
  - Dolby® Digital (AC-3), at least two channels
  - DTS Digital Surround®, at least two channels
  - Linear PCM (LPCM) at 48 and 96 kHz, at least two channels

There are three optional audio formats for Blu-ray Disc —
- Dolby® Digital Plus, up to 7.1 channels
- Dolby® TrueHD, up to 7.1 channels
- DTS-HD Audio™, up to 7.1 channels

Customers should arm themselves with a list of criteria matched to their viewing needs. Criteria for evaluating player performance spans a diverse list of player characteristics such as, but not limited to, the following —
- Disc loading time
- Disc access time to playlists or switching applications
- Graphic image loading
- Audio format options
- Persistent storage capacity and speed
- CPU performance

Judging player performance is extremely subjective, and the enduser should recognize that their expected primary player usage should be the major contributor to their player decision. Gamers expect a level of player performance that is very different from that of a passive occupant of a favorite couch or armchair.

Blu-ray Disc Recorders

As the Blu-ray Disc format matures, disc recorders are becoming available for the format[3]. These recordable drives are available for installation in laptop or desktop computers, and may be used in standalone HD disc recorders, too.

BD recordable drives for computers have been introduced worldwide. These are available in both the internal and the external drive flavors, using recordable and re-recordable (BD-R/BD-RE) media. Additionally, both single and dual layer discs can be utilized for recording. Manufacturers are working on even higher capacity discs, but it is unlikely that a disc recorder of greater than dual-layer capacity will be marketed to the general consumer.

It is difficult to pin down the minimum requirements for a computer that incorporates a high definition optical drive, whether a laptop or a desktop. The data rate and throughput demands that a 1080p image at 24 frames per second puts on a computer are massive and relentless. Couple those demands with the typical operating environment of a computer with its power needs, display needs, media player needs, application needs, and that's a whole lotta needs!

Thus the computing power required for HD OD (high definition optical disc) play, let alone record, remains an area for experimentation and imagination. As a reference, here are the specifications for the Sony VAIO™ desktop, model VGT-LT39U —

- Intel® Core™ 2 Duo Processor T9300 @ 2.50 GHz
- Two 750 GB hard drives (which, by the way, is 1.5 terabytes!)
- 4 GB PC2-5300 DDR2, 667 MHz SDRAM
- NVIDIA® GeForce® 8400M GT graphics card with 256 MB video RAM
- 22" WSXGA display, 1680 × 1050
- Blu-ray Disc Read/Write (BD-R/-RE/-ROM,DVD±RW/±R DL/RAM, CD-R/-RW)

(Interestingly, the display does not natively reproduce 1080 lines. Ah, well...)

Player Connections

The BD format is designed around some of the latest advances in digital audio and video, yet the players are also designed to work with HD TVs, SD TVs, and video systems of all varieties, as well as audio systems from monaural to surround. As a result, the back of a player can have a confusing diversity of connectors producing a potpourri of signals.

One of the greatest difficulties for player manufacturers is determining how best to connect legacy equipment. For example, if you are one of the video enthusiasts who purchased an "HD Ready" television when that meant the TV did not contain an HD tuner and only supported analog component ($Y'P_bP_r$) high-definition inputs, you most likely will not be able to enjoy the best resolution images from a BD player. The consumer device manufacturers

---

[3]Sony launched a Blu-ray Disc recorder in 2003 in Japan, the BDZ-S77, to allow personal home recording of HD broadcasts, but it recorded on cartridge-type, version 1 media and would not play feature films.

Players

want to protect their customers while the content industry wants to prevent the high-definition players from having (unprotected) high-definition analog video outputs. The DVI, HDMI, and DisplayPort connection standards, in conjunction with disc authoring restrictions, can prevent the transmission of digital content to unprotected displays. For those without a secure connection capability, their display may only be able to show a lower-resolution video stream (see *The Analog Sunset* section in *Chapter 4*).

Another of the legacy connection issues that will be confronted by endusers has to do with digital audio signals. There are millions of A/V receivers on the market that have built-in decoders for Dolby Digital and/or DTS audio. Regrettably, none of these receivers directly support Dolby Digital Plus, DTS-HD, or Dolby TrueHD decoding, and very few provide a mechanism for delivering up to eight channels of uncompressed audio. In fact, most continue to sport S/PDIF or Toslink connectors, which are only able to carry up to two channels of uncompressed audio data. This is particularly problematic when a BD player is added to the setup because the BD disc formats include audio mixing capabilities that allow commentaries and sound effects to be mixed in with the feature audio content during playback. This may pose a problem, because in order to mix audio from different encoded sources, it must all be decoded in the player and mixed in the uncompressed audio domain. If any of the audio sources has more than two channels, the mixed result may exceed the capacity of a standard digital audio connection.

Several solutions are being effectuated for this audio bandwidth problem, among them an audio encoder embedded in the player that would recompress the high-bandwidth audio data into, for example, a Dolby Digital 5.1 stream for delivery to the legacy decoder. Other solutions implement a bypass option, in which the user can choose to bypass audio mixing and send the primary Dolby Digital or DTS-encoded audio stream to his or her external A/V receiver. Still other options require new A/V receivers to support HDMI or other high bandwidth methods for delivering up to eight channels of uncompressed digital audio.

BD players will provide a selection of output signals with a variety of connections —

- *Analog stereo audio*. This standard two-channel audio signal can include Dolby Surround encoding.
- *Digital audio*. This raw digital signal can connect to an external digital-to-analog converter or to a digital audio decoder in an A/V receiver. There are two signal interface formats for digital audio — S/PDIF and Toslink. Both formats can carry *pulse-code modulated* (PCM) audio, multichannel Dolby Digital (AC-3) encoded audio, and multichannel DTS encoded audio.
- *Composite baseband video*. This is the standard video signal for connecting to a TV with direct video inputs or an audio-visual (A/V) receiver. But, this does not carry the HDCP protected signal circuitry.
- *S-video (Y/C)*. This is a higher-quality video signal in which the luminance and chrominance portions of the signal travel on separate wires within a bundle or wrapper. As with composite video, this connection type does not provide secure content protection.

- *Six-, seven-, or eight-channel analog surround*. These audio signals from the internal audio decoder connect to a multichannel amplifier or a "Dolby-Digital-ready" (AC-3-ready) receiver.
- *Radio-frequency (RF) audio/video*. These older-style combined audio and video signals are modulated onto a VHF RF-carrier for connecting to the antenna leads of a TV tuner or to a TV cable system connection. Connectors of this type are provided as a legacy holdover for use with older televisions, as the image via an RF connector tends to be of a poorer quality.
- *Component analog video (interlaced scan or progressive scan)*. These three video signals (RGB or $Y'P_bP_r$) can connect to a TV monitor or video projector. Although providing a better quality video connection, these component connections do not provide HDCP security, either.
- *IEEE 1394*, aka *FireWire*. This signal type, although in use on a multitude of consumer electronics devices, is generally used to interface devices with computers.
- *DVI (Digital Visual Interface)*. Present on some BD players and some display devices, DVI provides a secure HDCP-compliant data path but is video only.
- *HDMI (High-Definition Multimedia Interface)*. A secure HDCP-compliant data interface, HDMI carries both full-bandwidth uncompressed digital video and multichannel digital audio. This connection type has become the de facto connection standard for the latest consumer electronics devices.
- *DisplayPort*. DisplayPort is another secure HDCP-compliant data interface, that carries both video and audio data on a single cable. DisplayPort technology is compatible with HDMI and DVI, and offers an optional latching mechanism that alleviates the loose cable syndrome.
- *Ethernet, LAN*. An Ethernet signal can be used to interface a player to a computer, a local area network, and to the Internet.

Connection Types

Player connections come in a variety of types. The different audio and video signals may be present with the following types of connections —

- *RCA phono* (Figure 7.1). This is a very common connector type, used for analog audio, digital audio, composite video, and component video.

Figure 7.1 RCA Phono Connector

Players

- *BNC* (Figure 7.2). This connector carries the same signals as RCA connectors and is found on high-end equipment.

Figure 7.2 BNC Connector

- *Phono* or *miniphono* (Figure 7.3). This connector carries stereo analog audio signals. It also may appear on the front of a player for use with headphones.

Figure 7.3 Phono/Miniphono Connector

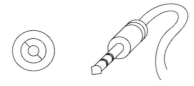

- *S-video DIN-4* (Figure 7.4). This connector, also called Y/C, carries separate chroma and luma video signals on a special four-conductor cable.[4]

Figure 7.4 DIN-4 (S-video) Connector

- *Toslink fiberoptic* (Figure 7.5). This connector, developed by Toshiba, uses a fiber optic cable to carry digital audio. One advantage of the fiber optic interface is that it is not affected by external interference and magnetic fields. The cable should not be more than 30 to 50 feet (10 to 15 meters) in length.

Figure 7.5 Toslink Connector

---

[4] Interestingly, the DIN4 connector is the same cable format as ADB (Apple Desktop Bus), which was Apple's predecessor to USB cables.

- *IEEE 1394* (Figure 7.6). Also known as FireWire or i.Link, these connectors are actually an external bus carrying all types of digital signals using only one cable for all the signals, including audio and video. The larger 6-pin cable also carries power; the smaller 4-pin cable does not.

Figure 7.6 IEEE 1394 Connectors

- *DB-25* (Figure 7.7). This 25-pin connector, adapted from the computer industry, is used by some audio systems for multichannel audio input.

Figure 7.7 DB-25 Connector

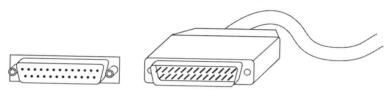

- *SCART* (Figure 7.8). This 21-pin connector, used primarily in Europe, carries many audio and video signals on a single cable: analog audio, composite RGB video, component video, and RF. Also called a Euro or Peritel connector.

Figure 7.8 SCART Connector

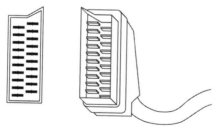

- *Type F* (Figure 7.9). This connector typically carries a combined audio and video RF signal over a 75-ohm cable. A 75- to 300-ohm converter may be required.

Figure 7.9 Type F Connector and Adapters

Players

- *RJ-45* (Figure 7.10). This connector type contains eight very fine leads and was initially developed for the telecommunications industry. It is now ubiquitous in Ethernet and network installations. It is also useful for carrying digital video signals and, in fact, is commonly used in long-distance video cable systems.

Figure 7.10 RJ-45 Connector

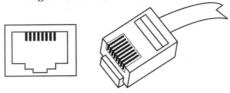

- *DVI* (Figure 7.11). This connector accommodates uncompressed high-definition video connections between DVI-capable devices. The initial application for DVI was to connect a computer with a fixed pixel monitor, such as an LCD display or a projector.

Figure 7.11 DVI Connector

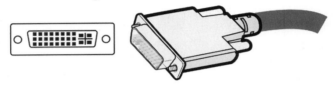

- *HDMI* (Figure 7.12). This connector supports the transmission of standard, enhanced, and high-definition video combined with multichannel digital audio on a single cable.

Figure 7.12 HDMI Connector

- *DisplayPort* (Figure 7.13). A recent single-cable digital A/V connector for computers, graphics cards, and other CE devices.

Figure 7.13 DisplayPort Connector

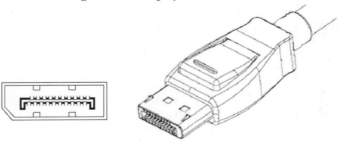

There are two more connection types which do not require physical connectors —
- *Wireless networking (WiFi)*. These connections, based in the IEEE 802.11 local area network standard, transmit data without cables in the 2.4 GHz radio-frequency band.
- *Wireless video* (*Wireless HD [WiHD]*, *Wireless HDMI*, *Wireless Home Digital Interface [WHDI]*, *Ultra-wideband [UWB]*). Various technologies and formats for in-room or multi-room *wireless video area network* (WVAN) transmission of compressed or uncompressed digital video and audio without cables.

Audio Connections

To hear the audio from a BD player, the player must be connected to an audio system — a receiver, a control amp or preamp, a digital-to-analog converter, an audio processor, an audio decoder, an all-in-one stereo, a TV, a boombox, or other equipment designed to process or reproduce audio.

Digital Audio

The digital audio outputs provide the highest-quality audio signal. This is the preferred connection for audio systems that have them.

For multichannel audio output, the encoded digital signal bypasses the player's internal decoder, and the appropriate decoder is required in the audio receiver or separate audio processor. For PCM audio output, the PCM signal is sent directly to the digital audio output. Alternatively, the multichannel decoder in the player may produce the PCM signal, but it is restricted to a maximum of two output channels when using S/PDIF or Toslink. In either case, a digital-to-analog converter (DAC) is required, which may be built in to the receiver or may be a peripheral device connected to the receiver. Some players provide separate outputs for multichannel audio and PCM audio. Other players have either a switch on the back or a section in their onscreen setup menu where you can choose between PCM output and multichannel output (undecoded Dolby Digital or DTS audio). The multichannel output menu option is usually labeled "AC-3" or "Dolby Digital/DTS."

The digital audio output must be connected to a system designed to accept, at least, PCM, Dolby Digital (with optional Dolby Digital Plus and Dolby TrueHD), or DTS Digital Surround (with optional DTS-HD Audio). Most modern digital receivers can automatically sense the type of incoming signal.

Some players include a dynamic range control setting (also called *midnight mode*) that boosts soft audio and reduces loud sound effects. It is recommended that this setting should be turned off to achieve the best effects with a home theater system, but you may wish to activate it for situations where the dialogue cannot be clearly heard, such as when everyone else in the house has gone to bed and the volume is down low.

Additionally, Dolby Digital has a feature called *dialog normalization*, which is designed to match the volume level from various Dolby Digital programs or sources. Each encoded audio source includes information about the relative volume level. Dialog normalization automatically adjusts the playback volume so that the overall level of dialog remains constant. It

Blu-ray Disc Demystified

7-13

Players

makes no other changes to the audio, including dynamic range; it is equivalent to manually turning the volume control up or down when a new program is too soft or too loud. Usually, only one setting exists in a program, which means the volume control does not change in the middle. Dialog normalization is especially useful with a digital television source to handle variations in volume when changing channels.

Connecting Multichannel Digital Audio

BD players can send uncompressed (PCM) multichannel digital audio to a display or to the sound system through an HDMI connection. This is the easiest and the highest quality option, better even than sending Dolby Digital and DTS to the built-in decoders in the receiver, as it includes the complete audio mix from the player. Arguments abound about what kind of cable provides the best sound, but as long as the digital signal makes it through intact, the audio will be faithfully reproduced. (See *A Few Timely Words about Jitter* in *Chapter 2*).

Simply connect the HDMI cable between the player and an HD display or projector, or an A/V receiver that can do HDMI video switching.

Connecting Two-channel Digital Audio

Two different standards govern the standalone digital audio connection interface: coaxial and optical. The arguments are many and varied as to which is superior, but since they are both digital signal transports, high-quality cables and connectors will deliver the exact same data. Some players have only one type of connector, although many players have both.

Coaxial digital audio connections use the IEC-60958-1 for PCM, also known as *S/PDIF* (Sony/Philips Digital Interface Format). Most players use RCA phono connectors, but some use BNC connectors. Use a 75-ohm rated cable to connect the player to the audio system. Multichannel connectors usually are labeled "Dolby Digital" or "AC-3." PCM audio connectors usually are labeled "PCM," "digital audio," "digital coax," "optical digital," etc. Dual-purpose connectors may be labeled "PCM/AC-3," "PCM/Dolby Digital," or something similar. Be sure to use a quality cable; a cheap RCA patch cable may degrade the digital signal to the point that it will not allow the signal to pass.

Optical digital audio connections may use the EIAJ CP-340 standard, known as *Toslink*. Connect an optical cable between the player and the audio processor. The connectors are labeled "Toslink," "PCM/AC-3," "optical," "digital," "digital audio," or the like.

If the connection (either coaxial or optical) is made to a multichannel audio system, select Dolby Digital/AC-3 (or DTS or MPEG multichannel) audio output from the player's setup menu or via a switch on the back of the player. If the connection is to a standard digital audio system (including one with a Dolby Pro Logic processor), select the PCM audio output instead. In cases where a player has an optical (Toslink) connection but the audio system has a coaxial (S/PDIF) connection, or vice versa, a low-cost converter may be purchased.

Multichannel Analog Audio

A component multichannel audio connection can potentially be as good as a digital audio connection. However, such outputs use the digital-to-analog converters that are built into a player, and they may not always be of the highest quality, especially on a low-cost player. The

Players

analog signal must be converted back to digital when connected to a digital receiver. If you have an amplifier with six, seven, or eight multichannel inputs, individual analog audio connections are an appropriate choice, because a Dolby Digital decoder will not be required.

All BD players include a built-in two-channel Dolby Digital audio decoder. Only some players include a full six-channel decoder along with the multichannel digital-to-analog converters and external connectors necessary to make the decoded audio available. Some players support the Dolby EX or DTS ES formats, which add a rear center channel. A few players support eight channels out, decoding DTS Master Audio and Dolby TrueHD, with either two, six, or eight output connections.

Connecting Multichannel Analog Audio

A multichannel capable player typically has six, seven, or eight RCA or BNC jacks, one for each channel. Hence, a receiver/amplifier with six to eight audio inputs — or more than one amplifier — is needed. Hook individual audio cables to the connectors on the player and to the matching connectors on the audio system. The connectors typically are labeled for each speaker position: L, LT, or Left; R, RT, or Right; C or Center; LR, Left Rear, LS, or Left Surround; RR, Right Rear, RS, or Right Surround; Subwoofer or LFE; and sometimes, CR, Center Rear, CS, or Center Surround. Some receivers use a single DB-25 connector instead of separate connectors. An adapter cable would be required to convert from DB-25 on one end to multiple RCA connectors on the other.

Stereo/Surround Analog Audio

A two-channel audio connection is the most widely used option, but it does not have the quality and discrete channel separation of a digital or multichannel audio connection. All players include at least one pair of RCA (or sometimes BNC) connectors for stereo output. Any disc with multichannel audio will be downmixed automatically by the player to Dolby Surround output for connection to a regular stereo system or a Dolby Surround/Pro Logic system.

When making connections for a stereo/surround analog audio environment, connect two audio cables with RCA or BNC connectors to the player. Connect the other ends to a receiver, an amplifier, a TV, or other audio amplification system.

In some cases, the audio input on the stereo system (such as a boom box, if it can be rightly called a stereo system) will be a phono or miniphono jack instead of two RCA jacks, thus an adapter cable becomes necessary. If the player is a portable player with a miniphono connector, a phono-to-RCA adapter cable is usually required to connect the player to the audio system. If the player includes a phono or miniphono connector for headphones, it generally is not recommended that the headphone output be used to connect the player to a stereo system because the line levels are not appropriate.

Bass and Low-Frequency-Effects

The heart-thumping, seat-shaking excitement of action movies relies heavily on deep, powerful low-frequency audio effects. BD provides audio quality that is actually superior to

Players

what comes on film for theaters. It is up to the home theater owner (subject to spousal approval and neighbor tolerance) how close he or she wants to get to a theater sound system.

All the ".1" sound encoding formats provide special channels for low-frequency effects (LFE). Despite becoming a standard feature, the LFE channel is frequently misunderstood by producers and listeners. Part of the problem is the LFE channel's overuse by audio engineers. It is possible, and quite normal, for all of the bass in a movie to be mixed in the five main channels, because all are full frequency channels. The LFE channel should be reserved for extraordinary bass effects, the type that only work well in a full discrete surround system with at least one subwoofer. Again, though, the same bass effects could be mixed in a 5.0 configuration with no loss or compromise, because modern receivers, particularly those with Dolby Digital and DTS decoders and a separate subwoofer output, have integrated bass management. Depending on the speaker configuration, the receiver automatically filters and routes bass below a certain frequency to the speakers that can reproduce it. For example, if an audio system has five small bookshelf speakers and a subwoofer, the receiver should send all the bass below 80 Hz or so to the subwoofer. In an audio system with a few large speakers, some smaller speakers, and a subwoofer or two, the receiver will route low-frequency audio from all channels to the large speakers and the subwoofers. It does not matter what channel the bass comes from; all low frequencies from the main channels and the LFE channel will be sent to every speaker that can handle them.

This is the key to understanding why certain complaints about bass and LFE are groundless. A 5.0 mix does not compromise the audio or cheat owners of high-end audio systems, because all the necessary bass is still in the mix. Omitting the LFE channel when downmixing to two channels is not a terrible thing, because nothing vital rests in the LFE channel — only, extra "oomph" effects that few two-channel systems can do justice to. This does assume that the engineer creating the audio mix understands the purpose of the LFE channel and does not blindly move all low frequencies into it.

Video Connections

To see the video from a BD player, the player must be connected to a video system — a television, a video projector, a flat-panel display, a video processor, an audio/visual (A/V) receiver or other equipment capable of displaying or processing a video signal. For those with a widescreen TV, connection details can become more confusing.

With the exceptions of RF video or HDMI-capable players, audio cables are also required when connecting a BD player to your system, in addition to a video connection because the video connections do not carry audio. And, communication becomes a bit trickier when there are multiple devices fighting over a single display.

For example, a BD player, VCR, cable box, and video game console may all need to be connected to a TV that has only one video input. In this case, the best option is to use an A/V receiver, which will switch the video along with the audio. When buying a new A/V receiver, get as many video inputs as you can afford. You will almost always end up with more video sources than you initially plan for. If an A/V receiver is out of your price range, then you may get either a new TV with more video inputs or a manual video switching box.

Players

Most BD players provide three or four video connection options, which are detailed in the following sections.

Digital Connections

With the emergence of a variety of digital display technologies, and with the adoption of DVI, HDMI, and DisplayPort connection standards, players may now be connected to a digitally equipped display, with no loss of video quality. The digitally encoded data on the disc stays in a digital realm for presentation on the display.

HDMI and DisplayPort are fully backward compatible with DVI, which allows a DVI capable display to be connected with HDMI or DisplayPort, using a format adapter cable.

The primary difference between DVI and HDMI (and DisplayPort) is that DVI is video only. The HDMI and the DisplayPort advantage is that they can contain multichannel digital audio, too.

Component Video

Prior to the introduction of digital connections, component video was the preferred method for connecting a player to a video system. A component video output provides three separate video signals in RGB or $Y'P_bP_r$ (or Y', B-Y', R-Y') format. These are two different formats that are not directly interchangeable.

Unlike composite or s-video connections, component signals are discrete and do not interfere with each other. Thus, a component connection is not subject to the picture degradation that might be caused by video crosstalk.

There are two scan variations for video — interlaced scan and progressive scan. Interlaced-scan is almost exclusive to standard television.[5] Progressive-scan component video produces a picture with significantly more detail than interlaced scan component video and requires a progressive scan display. Depending on the disc content, including content protection constraints, 1080p video may not be allowed on component outputs.

Connecting Component Video

Some players, notably U.S. and Japanese models, have $Y'P_bP_r$ component video output in the form of three RCA phono or three BNC connectors. The connectors may be labeled "Y," "U," and "V," or "Y," "P_b," and "P_r," or "Y," "B-Y," and "R-Y,"[6] they may be colored green, blue, and red, respectively.

Some players, notably European models, have RGB component video output via a SCART connector or via three RCA phono or three BNC connectors. The RGB connectors are generally labeled "R," "G," and "B" and may be colored to match. Hook a SCART cable from the player to the video system, or hook three video cables from the three video outputs of the player to the three video inputs of the video system.

---

[5]Do not be confused by the difference between an interlaced signal (or source) and an interlaced display. Blu-ray Discs can contain 1080-line video in either interlaced (1080i) or progressive (1080p) format. Almost all high-definition display technology — including LCD, DLP, and plasma — is progressive. Even a display that only accepts a 1080i signal must convert it to progressive format — often at a lower resolution — before it can display it.

[6]Many players label the YPbPr connectors as YCbCr. This is incorrect because YCbCr refers only to digital component video signals, not analog component video signals.

Blu-ray Disc Demystified

Players

S-Video

Some BD players have an s-video (Y/C) output, which generally provides a better picture than composite video output, unless the s-video cable is very long. In fact the picture from an s-video connection is only slightly inferior to a component connection. S-video provides more detail, better color, and less color bleeding than composite video. Another advantage of s-video is that the luma (Y) and chroma (C) signals are carried separately. The player downscales HD video when sent over an s-video connection. It also letterboxes widescreen video to fit into a 4:3 picture ratio unless you choose the player setting that tells it a widescreen display is connected.

Connecting S-Video

When making an s-video connection, use an s-video cable from the player to the video system. The round, 4-pin connectors may be labeled Y/C, s-video, or S-VHS.

Composite Video

This is the most common but lowest-quality connection. All players have standard baseband video connectors, the same type of output provided by VCRs, low-cost camcorders, and video game consoles. This signal is also called *composite video baseband signal* (CVBS). The player downscales HD video when sent over a composite video connection. It also letterboxes widescreen video to fit into a 4:3 picture ratio unless you choose the player setting that tells it a widescreen display is connected.

Connecting Composite Video

For composite connections, use a standard video cable from the player to the video system. If the connector is an RCA phono type, it is usually yellow. Connectors may also be BNC type. The connector may be labeled "video," "CVBS," "composite," or "baseband," etc.

RF Audio/Video

This is the worst way to connect a player to a television and is only provided by a few players for compatibility with older televisions that only have an antenna or cable connection. The RF signal carries both audio and video modulated onto a VHF carrier frequency, the type of output frequently provided by VCRs and cable boxes.

Connecting RF Audio/Video

For an RF connection, use a coaxial cable with type F connectors from the player to the antenna input of the TV. The connectors may be labeled "RF," "TV," "VHF," "antenna," "Ch. 3/4," or something similar. If the TV antenna connection has two screws rather than a screw-on terminal, a 75- to 300-ohm adapter is needed.

If you have a TV with only RF antenna inputs, you will need to either get a player with RF output or an RF modulator ($20 to $30). The RF modulator is connected to the player composite video output. (And, we may have mentioned it already, but if you have a TV with only RF connectors it may be time to get a new TV.)

BD Player User Tools

BD players will provide an increased level of user interaction with the player and with the disc content, including support for added devices, such as mice, joysticks, game controllers and keyboards. Further, many of the applications envisioned for players require a far more sophisticated level of control than what the traditional remote control offers. Many players offer standardized Universal Serial Bus (USB) ports to allow connections from a wide range of devices. BD also supports generic controller programming interfaces that make it possible for advanced applications to support new types of controllers yet to be conceived.

One of the most notable additions to the BD players is the ability to access data from locations other than the disc itself. Specifications for BD include persistent storage and network downloads. This makes it possible to deliver applications and presentation data (video, audio, graphics, and subtitles) via home networks and the Internet, as well as to store and later retrieve that data from persistent storage. The amount of persistent storage and storage type varies from player to player. Some players also may provide a slot for flash memory cards and/or a USB socket for an external hard drive.

BD players have a complex demultiplexer and presentation engine that includes both primary and secondary decoders, and complex data pathways that quickly adapt to different content scenarios. For instance, the format supports secondary video that can play simultaneously with the feature video in a picture-in-picture (PIP) configuration. Likewise, a secondary audio decoder can be used to stream network-delivered audio commentaries that are decoded and mixed in realtime with the movie as it plays back. The BD players also support realtime audio mixers that can combine both primary and secondary audio content along with additional sound effects.

Perhaps the most obvious addition to a BD player differentiating it from the standard DVD model is an interactive graphics engine. The BD format supports strong general processing capabilities for creating advanced, media-rich user experiences. This includes a script processor or virtual machine, large amounts of system memory, complex input controllers, and a whole suite of graphics scalers and compositors that can blend high definition video, sub video, interactive graphics, and subtitles, for realtime presentation.

BD Player Remote Control

Blu-ray players are being built by a number of companies. And, the primary interface to a BD player is a wireless remote control. As with DVD, button functions can be blocked or rendered inactive by the programming of the disc.

The button layout and additional functions on a remote control are left to the individual manufacturer. Pioneer, for example, provides television control buttons for displays that can be programmed by an alternative remote control. Figure 7.14 presents a sample remote control layout. A remote control should contain —

- Number keys – zero thru nine
- Arrow keys – left, right, up, down

- Enter or OK key
- Menu keys – top or title menu, popup menu
- Playback control keys – play, stop, fast forward, rewind, pause, skip forward, skip backward, step advance, step reverse
- Function select keys – audio track, subtitle, angle, display
- Color keys – color0, color1, color2, color3

Figure 7.14 Example BD Remote Control

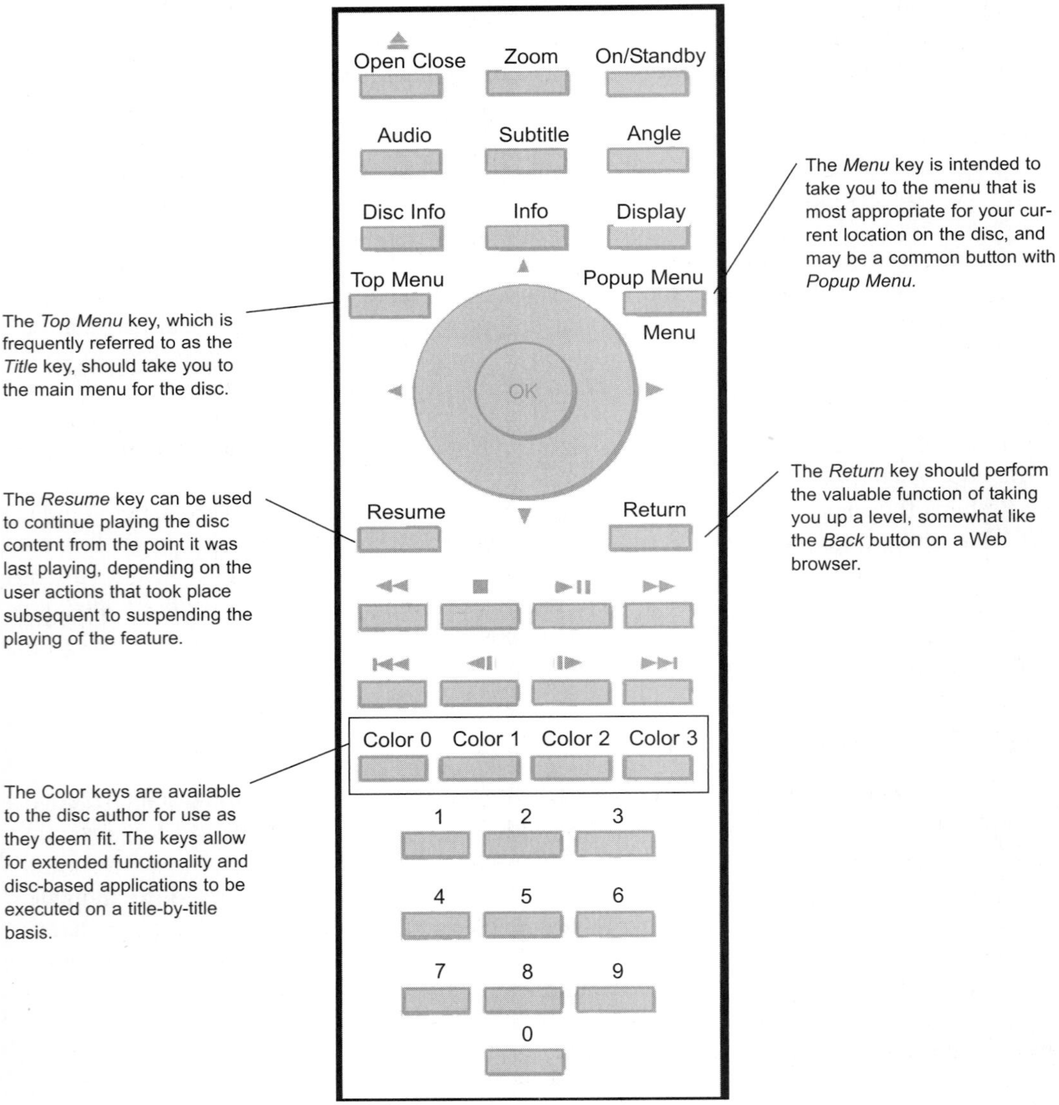
Open Close
Zoom
On/Standby
Audio
Subtitle
Angle
Disc Info
Info
Display
Top Menu
Popup Menu
Menu
OK
Resume
Return
Color 0
Color 1
Color 2
Color 3
1
2
3
4
5
6
7
8
9
0
The *Menu* key is intended to take you to the menu that is most appropriate for your current location on the disc, and may be a common button with *Popup Menu.*
The *Top Menu* key, which is frequently referred to as the *Title* key, should take you to the main menu for the disc.
The *Resume* key can be used to continue playing the disc content from the point it was last playing, depending on the user actions that took place subsequent to suspending the playing of the feature.
The *Return* key should perform the valuable function of taking you up a level, somewhat like the *Back* button on a Web browser.
The Color keys are available to the disc author for use as they deem fit. The keys allow for extended functionality and disc-based applications to be executed on a title-by-title basis.

Players

How to Get the Best Picture and Sound

Now that digital connections are possible between a player and a display, it requires new purchases of digital-capable devices. Digital connections provide a surety of copyright protection that is not possible with analog signals. We certainly recommend that you create an all-digital connection environment.

Beyond the connection scheme that you choose, one of the most important steps to getting better picture and sound is to adjust the television display properly. Reduce the sharpness control on the display. Video from BD is much clearer than from traditional analog sources. The display's sharpness feature adds an artificial high-frequency boost and it exaggerates the high frequencies and causes distortion, just as the treble control set too high for a CD causes it to sound harsh. An overly sharp display can create a shimmering or ringing effect. Reduce the brightness control, as it is usually set too high, as well.

If you get audio hum or noisy video, it is probably caused by interference or a ground loop. Interference can be reduced with an adequately shielded cable. The shorter the cable, the better is the result. You may be able to isolate the source of the interference by turning off all equipment except the pieces you are testing. Try moving things farther apart or plugging them into a different circuit. Wrap your entire house in tinfoil. If those steps do not remove the hum or noise, make sure all equipment is plugged into the same circuit. If nothing works, then it may be time to get a new display, as the culprit may be the set's power supply.

It may be hard to believe, but displays are not necessarily adjusted properly in the factory. Get your display professionally calibrated, or calibrate it yourself. A correctly calibrated display is adjusted to proper color temperature, visual convergence, and so on, resulting in accurate colors and skin tones, straight lines, and a more accurate video reproduction than is generally provided by a television when it comes out of the box. Organizations such as the *Imaging Science Foundation* (ISF) train technicians to calibrate televisions using special equipment. Another option is to use Joe Kane's *Digital Video Essentials* disc.

Connect the player to a good sound system. This may sound like a strange way to improve the picture, but numerous tests have shown that when viewers are presented with identical pictures and two different quality levels of audio, they perceive the picture that is accompanied by high-quality audio to be better than the picture associated with low-quality audio.

Viewing Distance

A great deal of research has gone into human visual acuity and how it relates to viewing distance, information size, and display resolution. Viewing distance is often measured in display screen heights. In general, the best viewing distance is about 3 to 5 times the screen height. Some industry recommendations vary from 2 to 10 times the screen height (the lower the number, the more likely it's the optometry industry). If you sit too close to the screen, you will see video scanline structures and pixels. If you sit too far away, you will lose visual detail and will not engage enough of your field of view to draw you into the experience. The optimal viewing experience is to fill about 35 degrees of your field of vision while staying beyond the limit of picture resolution. The THX recommendation is a 36 degree viewing angle, whereas, SMPTE recommends a 30 degree viewing angle. There is disagreement

Players

about the dimensions of the average human field of view, especially as it relates to watching video, but the vertical range is about 60 to 90 degrees, whereas the horizontal angle varies from 100 to 150 degrees, depending on the individual.[7] Viewing angle fields broader than about 40 degrees, however, can make the viewer uncomfortable after a period of time because human vision tends to focus toward the center.

Often the resolution of the display determines the distance at which viewers naturally sit. Psychophysical studies have shown that viewers tend to position themselves relative to a scene so that the smallest detail of interest subtends an angle of about 1 minute of arc, which is the limit of discrimination for normal vision. This gives an ideal viewing distance of 3,438 pixels, which for a 1920-pixel picture is 1.8 times the screen width, providing a 32-degree field of view ($32 \times 60 = 1920$). For a 1280-pixel image it is 2.7, but that provides only a 21-degree viewing angle. Moving closer in to the SMPTE-recommended distance (1.9 times width for 30 degrees) means you may see the individual pixels at 1.4 times resolution.

Compatibility

Attempting to understand optical disc compatibility could wring tears from a granite plinth. It does not help that compatibility has many facets and meanings that become increasingly complex as formats have been developed for CD, DVD, HD DVD, BD, and beyond. There is backward compatibility, such as Blu-ray players being able to play DVDs and CDs. There is forward compatibility (or usually lack of), such as DVD players not being able to play Blu-ray discs, or Blu-ray players perhaps being able to play future multi-layer discs or 3D movies. There is sideways compatibility — being able to play related formats, such as, DVD players that play DivX files or Blu-ray players that play AVCHD recordings from disc or memory card. And, there is also compatibility within a format or format family, which is always expected but does not always work, such as CD players that cannot play CD-Rs or Blu-ray players that cannot play some BD-RE discs.

The Blu-ray specifications focus more on the discs, file systems, and application formats than on the players. There are profiles, performance minimums, and certain functional requirements but, beyond that, player makers are free to implement the specifications as they see fit. Unfortunately, failures of compatibility occur at various levels across the range of optical disc technology — at the physical level, the logical level (file system and application), and the implementation level (see Table 7.2). Compatibility problems are also discussed in *Chapter 9*, including problems outside of the player, such as HDMI snags.

Table 7.2 Examples of Compatibility Problems

Problem	Incompatibility	Explanation
A BD player cannot play a BD-R or a BD-RE disc	Application	The disc might be in the old BD-RE 1.0 format, which most BD players do not recognize, or it might be in SESF (self-encoded stream format) or RREF (real-time recording and editing format), which not all players can handle.

continues

---

[7]The visual angle is calculated as arctan (height/2/distance) × 2, where size is the width or the height of the display. Dividing by two playce the viewer at the center of the angle.

Table 7.2 Examples of Compatibility Problems (continued)

Problem	Incompatibility	Explanation
A BD player cannot play a BD-R or a BD-RE disc	Implementation/ Physical	There are some compatibility problems, especially with early players, where the disc is not quite in spec and a finicky player cannot read it
A BD player cannot play a BD	Application	The disc may contain some other form of data not recognized by the player, such as, a Sony PlayStation 3 game.
A BD player can only play part of a BD	Application	The disc may use Profile 2 features, which are not fully supported by Profile 1.1 or Profile 1 players.
	Implementation	The BD formats, especially BD-J, are very complex. It's possible the disc has errors on it or the player may have bugs that cause problems with some discs.
A BD player cannot play a disc recorded in a camcorder	Application	Many BD camcorders record in the AVCHD format, which only some BD players can play.
A BD player cannot play a disc recorded in a computer	Application	Computers can record in standard BD formats, but the disc may contain only computer data that the player cannot recognize.
A BD computer cannot play a disc recorded in another computer	Application	The second computer may not have the software needed to play the disc.
	Implementation	Bugs may be present in the drivers or formatting software of either computer.
A BD player cannot play an HD DVD	Physical	Although they are both HD formats, HD DVD is physically very different from BD, and they are not compatible.
A BD player cannot play a DVD or CD	Application	Essentially all BD players can play standard DVDs and CDs, but the disc may have computer data on it or otherwise not be in standard DVD-Video or Audio CD format.
A BD player cannot play a recorded DVD	Application	The disc might be in DVD-VR format or in AVCREC (BDAV recording on recordable DVD media) or AVCHD format, which not all BD players recognize. The disc may contain computer data or other non-standard format.
A DVD player cannot play a BD	Physical	BDs are very different from DVDs and are not compatible with DVD players, except for hybrid discs that combine both formats on a single disc. But even these discs sometimes do not work on DVD players that are confused by the BD layer.
	Application	The AVCREC format stores BD video on a standard, red-laser DVD, but although the player can read data from the disc, it does not understand the format.

Blu-ray Disc Demystified

7-23

Physical Compatibility

Physical compatibility is almost never an issue with the base BD-ROM format. Every player and drive is physically able to read data from BD-ROM discs, the kind that are mass replicated in a manufacturing plant. Players may not know what to do with the data or may not attempt to read certain sections of the disc, but they are capable of reading the bits. Physical compatibility is sometimes a problem with the recordable formats, as illustrated in Table 7.2. It is not as bad as with CD and DVD, where the recordable formats were not developed until after the ROM format was launched but, unfortunately, there are still reports of Blu-ray players not reading recorded discs. Physical compatibility may apply to other formats, such as, DVD and CD, where itis up to the manufacturer to decide whether or not to support a particular physical medium. This is especially true with some BD recordable drives that cannot write to recordable CDs or DVDs.

File System Compatibility

File system compatibility is generally not a problem. The file system determines how the data is organized and accessed. BD-ROM discs use UDF 2.5, and BD-RE (v2 and v3) discs use UDF 2.6. Because recordable Blu-ray discs are, at heart, simply storage media, other file systems such as Microsoft NTFS, Macintosh HFS, and Unix can potentially be used to write data files to the disc, but these OSes normally use UDF when writing to BD. The original 1.0 version of BD-RE, released in 2003, used a custom file system, BD-FS, which is not compatible with much of anything beyond original BD-RE recorders.

Application Compatibility

Application compatibility is the most confusing area. It is not always clear why a disc does not work in a given player, since it may not be obvious which application formats the player supports. For example, a Blu-ray player can read the data on an AVCREC disc recorded by a Blu-ray recorder, but it may not recognize the format, at all, or it may be able to read the format but be incompatible with peculiarities of the video streams on the disc. BD players are not required to support the BDAV format used on recordable discs — some do and some do not. Both the player and the disc carry the Blu-ray Disc logo, but it may not be clear why they do not work together. The proliferation of related application formats such as BDAV, AVCREC, and AVCHD, along with other custom formats for computers and game consoles, places a burden on the consumer to understand which discs use which application formats and which players can play them.

Profile Compatibility

Blu-ray profile compatibility is a form of application compatibility. The Blu-ray specification defines several profiles for players that have different feature sets (see Table 7.1). The good news is that the discs themselves are usually designed to play on a less-featured player by hiding or blocking parts that require advanced features to play. Still, it can be confusing to buy a disc that touts features on the package, only for some of those features to not work on a basic player.

Implementation Compatibility

Implementation compatibility has to do with flaws or omissions in players (including computer software players) or in discs. Some players are poorly designed, whereas, others behave in unexpected ways with unanticipated content. Each player implements the Blu-ray specifications in slightly different ways, complicated by the fact that the specification is sometimes unclear or confusing. The result is that discs may play differently or not at all in different players.

Implementation errors also can occur on the production side. A bug might exist in the encoder or in the system used to author the disc. Content, such as, graphics, fonts, and file packages, may be malformed. The person who authored the disc may have violated the Blu-ray specification or written improper BD-J code. Programming on some discs is so complex or inefficient that it places too heavy a load on underpowered players. Any of this can result in a disc that will not work on some or all players.

Backward Compatibility

Player manufacturers have it rough in the digital age. No one ever expected cassette tape players to play old reel-to-reel tapes or eight-track tapes, but today it would be suicidal for a manufacturer to produce a Blu-ray player that cannot play DVDs and CDs. And the player is also expected to make DVDs look better with progressive video, upscaling, and image processing. Luckily, anyone who owns a large collection of DVDs can count on backwards compatibility of future players. DVDs will only become "obsolete" in the sense that you might want to replace them with new high-definition versions. However, Blu-ray discs are not playable in older DVD players unless they are hybrid discs containing both Blu-ray and DVD layers.

Dealing with Compatibility Problems

The complexity of the Blu-ray format variations, infused with other optical disc formats, gives Mr. Murphy[8] plenty of room to work. The end result is that consumers cannot assume that any disc with "Blu-ray" or "BD" on it will work in every player or computer with a "Blu-ray" or "BD" label.

The BDA, many of whose members cut their teeth with DVD, have tried hard to avoid the playback incompatibility problems that arose with DVD, especially in the early days. There are significant mandatory testing requirements that come along with the Blu-ray Disc logo. In fact, new test discs are released frequently, which continues to raise the quality bar. From

---

[8]That is Mr. Murphy of the infamous Murphy's Law. In some circles, it is Mr. Finagle who is hard at work.

Blu-ray Disc Demystified

Players

the end of 2005 to the end of 2008, the number of test discs used for BD-ROM logo verification increased from just over 10 to over 30, with the total test cases on the discs numbering in the thousands. In some cases, players are required to be updated to work with new test discs. This is a radical concept in the world of consumer electronics — that a device shipped into the marketplace could have mandatory upgrade requirements. Therefore, virtually all Blu-ray players come with an Ethernet port for connecting to the Internet to download firmware upgrades.

BD-J is at least an order of magnitude more complex than DVD. Because DVD was made up of reasonably straightforward streams of video, audio, and subpictures, combined with a simple command set, it was possible to run a disc through a verifier program that would flag errors and avert problems. BD-J, on the other hand, is made up of more complex video and audio streams, plus secondary video and audio streams, plus often thousands of small components for menus, graphics, menu noises, text, and more, all of which can come from the disc or from local storage after being downloaded, and the whole enchilada is held together by a full-fledged programming language with infinitely variable complexity and potential for bugs. It borders on miraculous that Blu-ray players and discs work together as well as they do.

Chapter 8
Myths

Urban legends have been around as long as urbs, and myths have existed even longer. Inevitably, every new technology brings with it a perplexity of misunderstanding, misinformation, and usually a few conspiracy theories. In the case of Blu-ray Disc™, the war with HD DVD made things even worse as people let their facts wander off without a leash, biting bystanders at random. Official and unofficial supporters of each camp lobbed nuggets of half-truths into the fray — "Blu-ray manufacturing will never make it out of the laboratory;" "Blu-ray has better picture;" "HD DVD has better picture;" "Blu-ray has region control and HD DVD doesn't;" "Blu-ray requires so many complex features such as Java™ and picture in picture that players can't be built;" "Blu-ray doesn't have enough mandatory features such as picture-in-picture and an Internet connection," and so on. After Blu-ray hit the market, and then after the format war ended, some myths quickly met their well-deserved deaths, but many others continue to circulate like urban legends of microwaved cats and stolen kidneys.

This chapter tackles myths that will not die, myths that sound true until you think about them for a minute, and myths that even some respectable people bought into whose names we're keeping secret for a whole bunch of money. Some myths are not strictly wrong, but reflect a misunderstanding of a particular fundamental aspect of Blu-ray technology and raise the hackles of your curmudgeonly book authors. In other words, any sort of incorrect or inaccurate statement about Blu-ray is fair game in this chapter.

Myth: "Blu-ray Is Revolutionary"

The printing press was revolutionary. Television was revolutionary. Even CDs can be considered revolutionary because they were a completely new way of storing digital audio and computer data on a compact optical disc. But Blu-ray is fundamentally an evolution from DVD, which was itself an evolution of CD and a refinement of Video CD.

Sure, Blu-ray provides unprecedented ability to interact with video, but this is merely taking DVD to the next level. BD-Live features, such as, Internet-based updates of actor information and dubbing in new downloaded audio tracks are very cool, but most of these were pioneered by InterActual and others using PC playback of DVD. Blu-ray technology is remarkable, and can be used for some impressive things — controlling movie playback from an iPhone or sending e-mail between Blu-ray players — but it is not a radically different, or an even mildly disruptive, technology.

Myth: "Blu-ray Will Fail"

Early in the lifecycle of DVD, herds of pundits predicted that it would be a flop, joining the ranks of other neglected consumer electronic innovations such as quadraphonic sound, the 8-track tape, the Tandy/Microsoft VIS, and the digital compact cassette. It is safe to say, at this point, that those people should not take up careers as fortune tellers.

Blu-ray Disc Demystified

Myths

As the wheel turns for a new technology cycle, even more couch prognosticators have predicted that Blu-ray will fail. They argue that it is too little too late, that people are turning to the Internet and VOD (video on demand) for content, or that most people are happy with DVD and will not bother moving up to Blu-ray. Some even claim that because Sony was responsible for such failures as Betamax, MiniDisc (in some countries), SACD, and UMD, it has branded a large L on the forehead of Blu-ray. Of course, these Sony-smearers neglect to mention the Trinitron TV, Walkman, CD, and DVD.

It is true that the window of opportunity for new technology grows smaller all the time, as evidenced by such not-quite-failures as S-VHS, DVD-Audio, and DualDisc, and it is true that Blu-ray is not as big a step up as DVD was from VHS, but Blu-ray simply has too much behind it to fail. It will probably not achieve the massive success of DVD, which begat over a billion playback devices around the world in less than a decade and transformed the world of home video, but by almost any measure, Blu-ray is and will continue to be a success. In its first year of existence it sold more players and discs than DVD did in its first year. Granted, Blu-ray had the PlayStation 3 to thank for beating DVD player numbers, and it is riding the coattails of DVD, but that is one of the very reasons for its inevitable success. Dozens of the world's biggest consumer electronics companies and computer companies provide Blu-ray products. And the shift from standard definition to high definition will ineluctably pull Blu-ray along with it.

Myth: "We'll Soon Get Everything From the Internet and Discs Will Go the Way of Dinosaurs"

Change "soon" to "eventually, a few decades after the introduction of Blu-ray," and, well, it's still a myth. It is reasonable to expect that more and more video in homes and workplaces will be delivered over the Internet. Short-format video (think YouTube and news clips) will make the transition sooner than long-format video. In the very long run people will watch more movies and other video from the Internet than from disc. But physical media will not become marginalized before 2020 at the earliest and will not disappear for decades after that. See the concluding section of *Chapter 13, Blu-ray and Beyond* for an analysis that makes this clear.

Myth: "Some Discs Won't Play If the Player Doesn't Have an Internet Connection"

This fable stems from the BD-Live feature that provides for additional content over an Internet connection. Not all Blu-ray players have an Internet connection, but discs that incorporate enhanced features, such as picture in picture, are designed to disable those features if the player cannot support them. It is possible to design a disc so it will not work if there is no Internet connection, but only lunatics or sadists would intentionally do this. Most of the content on every disc, such as the main feature, plays fine on every Blu-ray player.

Myth: "BD-Live Discs Don't Work on All Players"

See the myths preceding and following this one.

Myth: "Profile 2 (BD-Live) Players and Discs Make Previous Players Obsolete"

This myth is fueled by the perception that Profile 2 (*BD-Live*) players have extra features that Profile 1 players do not have. However, the only requirements that Profile 2 adds are that players have a network connection and extra local storage to hold downloaded files. Otherwise, player functionality and performance are essentially the same Profile 1 players will continue to work, and will play all but the "connected" parts of BD-Live discs.

Myth: "Profile 3 and Profile 4 or Future Profiles Will Make Previous Players Obsolete"

Profile 3 (audio-only players) and Profile 4 (recording and editing for camcorders, Blu-ray video recorders, and PCs) are actually subsets of Profile 1. Even though they use a bigger number and sound better, they are restricted versions of the format for specialized products.

It is possible that, in the future, the BDA will define a higher performance profile but, by that time, there will be tens of millions of Profile 1 and 2 players in the marketplace, so discs that took advantage of an extended protocol would be designed to still work, at least in part, on older players.

Myth: "The BDA Will Soon Mandate that All BD Players Have an Internet Connection"

The BDA has no plans to make Profile 2 (BD-Live) or Internet connections mandatory for all players. As the cost of manufacturing players continues to fall and as player makers contrive to differentiate their players from all the others, it may very well come to pass that most or all new Blu-ray players implement Profile 2. In any case, although Profile 2 players have the capacity to connect to the Internet, they are not required to be connected. (See the next myth.)

Myth: "If You Unplug Your Profile 2 (BD-Live) Player From the Internet It Will Stop Working"

The BDA does not send unmarked black vans to cruise neighborhoods and harass people who have not plugged their player into the Internet. In fact, some Blu-ray players provide a setting to prohibit Internet access from BD-Live discs. If your player is not connected or if your Internet connection is not working for any of a dozen possible reasons, only the portions

Myths

of discs that depend on content from the Internet will not work or will not be fully featured. For example, some BD-Live discs download updated information about cast and crew. If there is no Internet connection then they simply show information that was put on the disc when it was produced.

Myth: "All Blu-ray Discs Must Use AACS"

It is true that all pre-recorded (replicated) Blu-ray Discs must use AACS content protection. By requiring that the discs that are most likely to be legitimate always have AACS encryption, it makes it easier to identify illegal copies. However, recordable discs (BD-R and BD-RE) are not required to used AACS. BD-ROM Mark is also mandatory on pre-recorded discs, but BD+ is optional. See *Chapter 4* for more on content protection.

Myth: "AACS Is Required for HDMV, BD-J, Network Access, or Local Storage Access"

HDMV and BD-J can be used on BD-R and BD-RE disc, which do not require AACS. BD-J applets must be signed in order to access the Internet or local storage. Application signing keys are not the same as AACS keys, but this seems to have caused some of the confusion that created this myth. BD-J applications must have a set of certificates with a valid chain of keys (from the content owner, authoring facility, replication facility, and so on) in order to used some of the Profile 2 features. However, this does not guarantee that BD-J content on a BD-R or a BD-RE disc will play on all players, since players are not required to play recordable discs. The best way to be sure of playback compatibility is to check for a BD-R/BD-RE logo on the player.

Myth: "AACS, BD+, and BD-Live Allow Studios to Spy on Consumers"

Internet forums and blogs are full of "Big Brother" conspiracy theories and gripes about Blu-ray players enabling invasion of privacy. Some player manuals contain notes, such as, "When discs supporting BD-Live are played back, the player or disc IDs may be sent to the content provider via the Internet,"[1] but this does not mean that your name, address, and credit card are being collected by insidious marketing machines. In order to download content associated with a particular disc, the online service has to know which disc it is. And, to support features, such as *movie e-mail* between players, the online service has to keep track of the player. Viewing and usage patterns may be tracked and reported in aggregate, to help studios figure out what online features on which disc are most popular, for example, but not with user-identifying information, especially if you have not entered such information into the player. If you are still concerned about this issue, check the privacy policy on the content provider's Web site or ask them to give you a copy of their privacy policy.

---

[1] From the Panasonic DMP-BD50 operating instructions manual.

Myths

Myth: "Blu-ray Doesn't Look Any Better Than Upconverted DVD"

There is some (well, really, only a little) truth to this one, but it is a bit like saying that DVD does not look any better than VHS, which is accurate only if you are watching it on a 13-year-old, 13-inch television. Experts agree that the difference between Blu-ray and DVD picture quality becomes imperceptible below a certain screen size, but the experts do not agree on what that size is, especially since it depends on how close the viewing position is. One thing that is certain is that a Blu-ray picture has five to six times more pixels than a DVD picture. That's a huge difference which is clearly visible on a good quality display at typical big-screen viewing distances. As digital displays get bigger and better, the difference becomes more obvious. Tests have shown that the difference also becomes more obvious after viewers become accustomed to HD and then revert to an SD picture that in comparison looks much blurrier than before. And, it does not matter how good the upsampling feature of the DVD player or the display is, no upscaling technology — no matter how ingenious — can accurately extrapolate 500 percent more information if it was never there to begin with.

Myth: "Blu-ray Players Downconvert Analog Video"

Blu-ray players must respect the *image constraint token* (ICT), a copy deterrent feature of AACS, that allows the content author to set a flag on the disc that makes the player downconvert 1920×1080 digital video to 960×540 resolution on analog outputs. It is true that using a digital video output with HDCP is the only guarantee that the full resolution will always be used, but very few discs have ICT turned on. Around the time Blu-ray was introduced, most Hollywood studios stated they had no plans to use ICT. This may change down the road when digital video connections are more pervasive, but unless high-definition analog video recorders become common — which is highly improbable — content owners still have little reason to discomfit the legitimate users who have analog displays.

Myth: "DVD Players Can Be Upgraded to Play Blu-ray Discs"

Believe it or not there was a class-action lawsuit on behalf of DVD player owners who claimed that the promise of disc playback compatibility meant that their DVD players should play Blu-ray discs. This is akin to expecting a horse-drawn wagon to be easily upgraded to an automobile because they both carry people. It is hard to believe that there are people dumb enough to think this, but it is not so hard to believe that there are lawyers smart enough to pretend they are this dumb. In any case, it is simply not true. Even if DVD players had powerful enough processors to handle a software upgrade for HDMV and BD-J processing (which they don't), DVD players do not have the necessary high-definition video decoders, video circuitry, blue-laser optics, copy protection accoutrements, and many other critical functions required to play Blu-ray Discs.

Myths

Myth: "Older Blu-ray Players Can Be Upgraded to New Profiles"

This is not as absurd as upgrading a DVD player, but it is only possible if the player was specifically designed with the necessary features from the beginning. Profile 1.1 added secondary video and audio playback requirements to Profile 1. An extra decoder chip and video overlay hardware cannot just be popped into a player, and there simply is not enough horsepower in standard Profile 1 players to add decoding support with a software update. Likewise, Profile 2 added Internet access and additional local storage requirements, so unless a Profile 1.1 player is already equipped with a connector and necessary networking features, and memory slots or a USB connector for a hard drive, it cannot just be upgraded with a wave of the magic software wand.

The Sony PlayStation 3 is an exception that proves the rule, as Sony provided a downloadable firmware update in December 2007 to add Profile 1.1 features and, in March 2008, released another firmware update that made PlayStation 3 the first Profile 2 Blu-ray player. These upgrades were possible only because the PlayStation 3 is a high-powered system with plenty of memory, video processing features, and full Internet connectivity.

Myth: "Existing Receivers with Dolby Digital and DTS Decoding Work Perfectly With Blu-ray Players"

In some cases, existing receivers connected to a Dolby Digital or DTS bitstream output only reproduce the primary audio track. Blu-ray has the ability to mix in one or two additional audio sources — a secondary audio track, such as, a director commentary on the disc or streamed over the Internet, and interactive audio, such as button sounds in an HDMV menu or sound effects in a BD-J interactive application. Mixing must be done using uncompressed audio, so the player must first decode the primary and the secondary audio tracks.

In certain cases, with players designed certain ways, and with certain audio formats on the disc[2] the player is required to re-encode the mixed audio in Dolby Digital (not Dolby Digital Plus) or DTS (not DTS-HD) before output. Many Blu-ray players re-encode mixed audio but, to reproduce all the possible audio from a Blu-ray player, without any quality loss from re-encoding, the receiver must have an HDMI input or a multichannel audio input. The upshot is that there is little value in having Dolby Digital and DTS decoders outside of the Blu-ray player.

Myth: "Analog Connections from DVD Players and Blu-ray Players Won't Work with US TVs After the February 2009 Analog Cutoff"

This is a headscratcher, since it makes absolutely no sense, but the US analog cutover is

---

[2]The scenarios are rather complex, such as when the player has a S/PDIF output but not a multichannel analog output and the primary audio is not PCM, Dolby Digital Plus, Dolby TrueHD, or DTS-HD Master Audio, and the moon is waning on a Tuesday night.

apparently confusing enough that even people who ought to know better got caught up in a bit of circular reasoning along these lines: when the US switches from analog to digital, old televisions will not work anymore, so the analog connections will be unusable. Of course this is silly when you think about it — the Federal Communications Commission does not have the power to reach out to all the TVs in the US and make them stop accepting analog inputs. The only thing that stops in February 2009 is analog broadcasting.

Analog video from DVD players, Blu-ray players, and other sources such as cable television still works fine. However, do not get confused in the other direction and think that the HDTV digital-to-analog converter boxes will make a high-definition Blu-ray player work with a standard-definition television. The converter is a digital tuner with analog output, not a general purpose downconverting adapter.

Myth: "Blu-ray Manufacturing Is Too Intricate and Too Expensive"

This was favorite ammunition of Blu-ray bashers during the format war, based on the fact that disc manufacturing lines required all-new Blu-ray equipment, whereas only an upgrade to existing DVD equipment was necessary to handle HD DVD manufacturing. But, it never held much water. Some people claimed that dual-layer Blu-ray disc production was so complicated and prone to low yields that it would never be commercially feasible. Blu-ray disc manufacturing is indeed trickier than DVD production but, within a year of launch, most major titles were released on dual-layer discs. The cost issue is also overblown. Shortly after launch, Blu-ray Disc manufacturing costs were only a dollar or two more than DVD (and less than a dollar more than HD DVD), and the retail price difference more than made up for this. Had HD DVD survived longer, the price difference would have steadily narrowed until it was negligible. A price difference of, say, 50 cents, can be made to look like a mountain or a molehill — for a top-ten motion picture that sells ten million discs it is an extra $5 million, which is nothing to sneeze at, but it is less than three percent of the sales from the disc.[3] Regardless of the relative increase in Blu-ray manufacturing costs compared to DVD or HD DVD, it can be worked into the pricing model so that it makes no difference to studio revenue and does not break the bank for consumers.

Myth: "The Blu-ray Disc Association Prohibits Adult Content"

The BDA does not police content on discs, and does not prohibit any particular type of content, including "x-rated." This myth may hav stemmed from a report that Sony was approached early on by adult content creators interested in putting their product on Blu-ray and promoting the format. The consumer electronics giant purportedly stated that as a company, Sony did not support or wish to promote that kind of content. Individual replication

---

[3]The top ten DVDs in the US in 2007 sold an average of 10.6 million copies each with average sales revenue of $195 million. Put another way, 50 cents is less than 3 percent of the average $22 DVD price. Of course Blu-ray discs started out more expensive than DVDs, so 50 cents is an even smaller share.

Myths

facilities may disallow certain types of content, especially if required by their customers[4], but the simple fact that adult titles are plentiful on Blu-ray puts this one to bed, so to speak.

Myth: "Blu-ray Is a Worldwide Standard"

Not quite. Blu-ray does not have some of the incompatible variants that DVD was saddled with from being closely tied to the standard-definition NTSC and PAL television formats. For example, Blu-ray discs primarily use a 1920 × 1080 picture resolution regardless of the country they are intended to play in[5], but video for "50 Hz" areas — such as most of Europe — is often encoded at 25 frames per second, whereas, video for "60 Hz" areas — such as the US and Japan — is often encoded at 29.97 frames per second. Blu-ray players sold outside of 50 Hz countries are not required to decode 25 fps video or 720 × 576 video. Blu-ray also supports a film-friendly rate of 24 frames per second but, in many cases, the alternative 23.976 rate is used in source that was intended for conversion to NTSC video. These peculiar frame rates are a relic of interlaced television that has carried over to HD. Luckily, more and more digital televisions are designed to handle all four frame rates.

As with DVD, Blu-ray includes region codes that can prevent a disc from being played on players sold in other countries. See *Region Playback Control* in *Chapter 4* for further explanation.

Technically, Blu-ray is not a "standard" in the formal sense. Just like DVD, it is a proprietary but open format created by a group of companies motivated by mutual interests and anticipated profits. Related standards from accredited bodies, such as, MPEG, ITU (H.264), and SMPTE (VC-1), were adopted for Blu-ray, but the important parts of the format — such as the application specifications for video, audio and interactivity — are proprietary to the Blu-ray Disc Association. Blu-ray also includes other proprietary formats, such as Dolby Digital, DTS and Java. Most of the standards and formats used in Blu-ray are subject to patent royalties from various companies or licensing entities (see *Chapter 4*).

Myth: "1080p Video Is Twice the Resolution of 1080i"

In a very theoretical sense, in the extreme worst case, this can be true. However, it is like saying, "You're in danger of being killed by the Sun going nova." This is technically true, but the chances are astronomical (sorry, couldn't resist). It is accurate, in general, to state that progressive video usually has more spatial resolution than interlaced video, since objects that move between fields are captured with only half the total number of scan lines, but progressive video often has less temporal resolution at the same frame rate because the picture is updated half as often. (See *Chapter 2, Technology Primer*, for more on interlaced vs. progressive scanning.)

Most attitudes toward interlaced video come from the days when the dominant display

---

[4]After a few embarrassing gaffes where pornographic material ended up in laserdisc and DVD packages that were supposed to contain family movies, some studios implemented a prophylactic solution (ok, pun intended) by demanding that replication facilities must not accept such content if they wanted their business.

[5]DVD has two different primary resolutions — 720×480 for 525/60 (NTSC) television and 720×576 for 625/50 (PAL or SECAM) television. Blu-ray includes both of these resolutions for standard-definition video.

technology was interlaced television. Those days are fading away. Most Blu-ray players are connected to widescreen displays or projectors that are inherently progressive. They are unable to display an interlaced video signal without first deinterlacing it. Therefore, issues of interlaced video resolution, nowadays, are almost all related to deinterlacing, not to interlaced display. There are many approaches to deinterlacing, and some work better than others.[6] On the other hand, progressive source video, such as film, can be stored in 1080i format. A good deinterlacing processor will recognize progressive source and put it back together with no loss of resolution. In this case there is no difference between 1080i and 1080p.

"Progressive is better than interlaced", is a good rule of thumb. It is true that a motion picture encoded at 1080p will look better than a home video shot at 1080i on an HD camcorder. But it is going too far — and ignoring too many complex factors — to say that progressive is two times better.

Myth: "All Blu-ray Titles Are (or Must Be) Encoded in 1080p"

Blu-ray video can be encoded at many resolutions — high-definition 1080p, 1080i, or 720p, and lower definition 576p, 576i, 480p, and 480i. In fact, many 1080p features are accompanied by documentaries, camcorder footage, old TV shows, and other content that is only available in standard definition (which could be upscaled and encoded at 1080p, but it usually is not worth the bother). Blu-ray is actually a great vehicle for standard-definition video, since a BD-50 can hold over 20 hours.

Myth: "Blu-ray Players Only Output 1080i Video"

This was true for a few early players, but almost all modern Blu-ray players output a full 1080p video signal. An HDMI connection is required for protected content that does not allow 1080p output over analog component video connections. Even when the video is encoded on the disc in 1080i, some players deinterlace it and output 1080p.

Myth: "Blu-ray Does Not Support Mandatory Managed Copy"

Managed copy is a feature of AACS. (See *Chapter 3* for basics and *Chapter 4* for details.) Managed copy is mandatory in the sense that every Blu-ray disc protected with AACS must allow at least one authorized copy (with a few exceptions). However, it is not mandatory that Blu-ray playback devices provide the managed copy feature. Very few Blu-ray players will implement it, although it will be popular on computers.

When this book was written, at the end of 2008, Blu-ray actually did not yet support AACS managed copy. AACS was released in 2006 under an interim license. Managed copy was only provided in the final license, which was not expected to be issued until 2009.

---

[6]Intrafield line replication, weighted line averaging, and vertical interpolation are simple approaches that discard temporal resolution unless the output runs at double the frame rate. More sophisticated, and expensive, interfield approaches, such as, vertical-temporal interpolation, motion-adaptive deinterlacing, and motion-compensated deinterlacing, produce better results but are still subject to artifacts and can never fully preserve all the detail of an interlaced capture.

Myths

Myth: "Managed Copy Means Every Blu-ray Disc Can Be Copied for Free"

Managed copy does require that almost all Blu-ray discs be copyable, at least, once. However, the copy must be authorized by the content owner, which is free to charge whatever it wishes or to require any transaction it wishes before the copy can be made. Some titles may be copied for free. Others may require watching an ad, signing up for something, or buying something else.

Myth: "BD+ Interferes With Managed Copy"

Because BD+ scrambles the video and audio streams on the disc, some of those guys who slump in their seats in the back of the classroom and only partly pay attention to the teacher but still think they know everything have claimed that managed copy is impeded by the BD+ *Media Transform* process. What these guys would understand, if they were paying attention, is that the *Managed Copy Machine*, the component that makes the copy after it has been authorized, acts like a player — it carries out AACS authentication and decryption along with BD+ reverse transforms to get proper access to the content before it makes a copy.

Myth: "Region Codes Don't Apply to Computers"

Regional codes apply to Blu-ray video discs played in BD-ROM drives. Every BD-ROM player application is set to a region by the software vendor or is automatically set by the first region-coded disc that it plays. Of course, there are ways around regional restrictions, just as with DVD players and drives. Regional codes do not apply to computer data on Blu-ray Discs, only to the HDMV and BD-J video formats, although not to HDMV or BD-J on BD-R or BD-RE discs.

Myth: "Blu-ray Players Can't Play CDs or DVDs"

Even though the Blu-ray specification makes no mention of CD or DVD compatibility, virtually all Blu-ray players can play audio CDs and DVD-Video discs. Compatibility with other CD formats varies. Some Blu-ray players can play Video CDs, and a few can play Super Video CDs, MP3 CDs, WMA CDs, and DivX DVDs. Of course, Blu-ray players cannot play CD-ROMs or DVD-ROMs (discs containing only computer data).

Some Blu-ray recordable drives are unable to burn recordable CDs or DVDs, but they can read them. Essentially all Blu-ray computers can play DVDs, audio CDs, Video CDs, and MP3 CDs.

Myth: "Blu-ray Is Better Because It Is Digital"

Nothing is inherent to digital formats that magically makes them better than analog formats. The celluloid film used in movie theaters is analog, yet few people would say that Blu-ray video is better than film. The way Blu-ray stores audio and video in digital form has advantages, not the least of which is the ability to use compression to extend playing times by lowering data rates. The quality and flexibility of Blu-ray stands out when compared with similar analog products. It is a mistake, however, to make the generalization that anything digital is automatically superior to anything analog.

Myth: "Blu-ray Video Is Poor Because It Is Compressed"

You will inevitably hear about "digital artifacts" that plague Blu-ray video. While it is true that digital video can appear blocky or fuzzy, a well-encoded Blu-ray disc will exhibit few or no discernible artifacts on a properly calibrated display. The term *artifact* refers to anything that was not in the original picture. Artifacts can come from film damage, film-to-video conversion, analog-to-digital conversion, noise reduction, digital enhancement, digital encoding, digital decoding, digital-to-analog conversion, signal crosstalk, connector problems, impedance mismatch, electrical interference, waveform aliasing, signal filters, television picture controls, video scaling in flat-screen displays, and much more. Many people blame all kinds of image deficiencies on the digital video encoding process. Occasionally, this blame is placed accurately, but usually it is not. Only those with training or experience can tell for certain the origin of a particular artifact. If an artifact cannot be duplicated in repeated playings of the same sequence from more than one copy of a disc, then it is clearly not a result of video encoding. Here are a few of the most common artifacts —

- *Blocks* are small squares in the video. These may be especially noticeable in fast moving, highly detailed sequences or video with high contrast between light and dark. This artifact appears when not enough bits are allocated for storing block detail.
- *Halos* or *ringing* are small areas of distortion or dots around moving objects or high-contrast edges. This is called the *Gibbs effect* and is also known as *mosquitoes*, or *mosquito wings*. This is an artifact of video encoding but is easy to confuse with edge enhancement.
- *Edge enhancement* is a digital picture-sharpening process that is frequently overdone, causing a "chiseled" look or a ringing effect like halos around streetlights at night. Edge enhancement happens before encoding.
- *Posterization* or *banding* manifests as bands of colors or shading in what should be a smooth gradation. This can come from video pre-processing (as when converting from 10-bit to 8-bit video), the video encoding process, or the digital-to-analog conversion process in the player. It also can happen on a computer when the number of video colors set for the display is too low.
- *Aliasing* refers to angled lines that have "stair steps" in them. This artifact is usually caused by detail that is too sharp or too contrasty to be properly represented in video, or by poor scaling in the display.
- *Noise* and *snow* refer to the gray or white spots scattered randomly throughout the picture, or graininess. This may come from film grain or low-quality source video.

Myths

- *Blurriness* refers to low detail and fuzziness of video. This results from low-quality source video or too much filtering of the video before encoding.
- *Worms* or *crawlies* are squirming lines and crawling dots. Usually this results from low-quality source video or bad digitizing. This also may be the result of electrical interference.

All this said, the number one cause of bad video is a poorly adjusted display! The high fidelity of Blu-ray video demands much more from the display. Turning up the sharpness control was the standard operating procedure when viewing fuzzy analog video, but this artifically enhances edges, which makes hidden compression artifacts in digital video stand out and actually makes the picture look worse. Lower the sharpness value and turn the brightness down, and you will be surprised at the improvement these simple adjustments may bring. See **Chapter 7, Players** for more information.

Myth: "Video Compression Does Not Work for Animation"

It is often claimed that animation, especially hand-drawn cell animation such as cartoons and Japanese anime, does not compress well. Others claim that animation is so simple that it compresses better. Neither is generally true.

Supposedly, jitter between frames caused by differences in the drawings or in their alignment causes problems. Modern animation techniques produce very exact alignment, so usually no variation occurs between object positions from frame to frame unless it is an intentional effect. And of course, computer-generated animation has no misalignment between frames. Even when objects change position between frames, the motion estimation feature of the encoder can easily compensate. That said, encoders are typically optimized for live action and may have trouble with animated content.

Because of the way Blu-ray video is compressed, it may have difficulty with the sharp edges common in animation. This loss of high-frequency information can show up as ringing or blurry spots along high-contrast edges. However, at the data rates commonly used for Blu-ray, this problem normally does not occur. The complexity of sharp edges tends to be balanced out by the simplicity of broad areas of single colors. As with DVD, some of the Blu-ray Discs with the most gorgeous pictures are from animated features.

Myth: "Discs Are Too Fragile to Be Rented"

This myth was pretty well demolished by Netflix, which not only rents DVDs and Blu-ray Discs but has the audacity to send them through the US Postal Service, one of the most destructive forces known to modern civilization.

Discs are, of course, vulnerable to scratches, cracks, accumulation of dirt, and fingerprints. But these occur at the surface of the disc where they are out of focus to the laser. Damage and imperfections may cause minor channel data errors that are easily corrected. A common misperception is that a scratch will be worse on a Blu-ray Disc than on a CD or a DVD because of higher areal density and because the audio and video are more compressed. Blu-ray data density is about five times that of DVD, and about 65 times that of CD-ROM, so it is true that a scratch will affect more data. But Blu-ray error correction is more effective than DVD and CD error correction. This improved reliability makes up for the density increase. More importantly, Blu-ray discs include a special hardcoat that makes them much less susceptible to scratches and other surface damage. However, major scratches on a disc may cause uncorrectable errors that will cause a read error on a computer or show up as a momentary glitch in the picture from a player.

DVDs and CDs are likewise subject to scratches, but many rental stores and libraries offer them. Manufacturers are fond of taking a disc, rubbing it vigorously with sandpaper, and then placing it in a player, where it plays perfectly. Disc cleaning/polishing products can repair minor damage. Commercial polishing machines can restore a disc to pristine condition after an amazing amount of abuse.

Myth: "Dolby Digital or DTS Means 5.1 Channels"

Do not assume that the "Dolby Digital" or the "DTS" label is a guarantee of 5.1 channels (or 6.1 channels or 7.1 channels) of digital audio. These audio encoding formats can carry anywhere from one to eight discrete channels, so a Dolby Digital or a DTS soundtrack may be mono, stereo, Dolby Surround stereo, and so on.[7]

Most movies produced before 1980 had a monophonic soundtrack. When these movies are put on Blu-ray, unless a new soundtrack is mixed, the original soundtrack is encoded into a single channel of digital audio, usually the center channel.

In some cases, more than one Dolby Digital version of a soundtrack is available: a 5.1 channel track and a track specially remixed for two-channel Dolby Surround. It's normal for a player to indicate playback of a Dolby Digital audio track while the connected receiver indicates Dolby Surround. This means that the disc contains a two-channel Dolby Surround signal encoded in Dolby Digital format.

Myth: "The Audio Level from Blu-ray Players Is Too Low"

People complain that the audio level from Blu-ray and DVD players is too low. In truth, the audio level on everything else is too high! Movie soundtracks are extremely dynamic, ranging from near silence to intense explosions. In order to support an increased dynamic range and to hit peaks (near the 2V RMS limit) without distortion, average sound volume must be lower. This is why the line level from Blu-ray and DVD players is lower than from many other sources. The volume level among Blu-ray discs varies, but it is more consistent than on CDs or from MP3 files. If the change in volume when switching between Blu-ray and other audio sources is annoying, check the equipment to see if you are able to adjust the output signal level on the player or the input signal level on the receiver.

---

[7]Yes, it is quite possible to encode Dolby Surround audio into a 2-channel DTS track, since it is a matrixing technique that is independent of the digital encoding process.

Myth: "Downmixed Audio Is Not Good Because the LFE Channel Is Omitted"

The low-frequency effects (LFE, or "subwoofer") channel is omitted for a good reason as multi-channel soundtracks are mixed down to two channels in the player. The LFE channel is intended only for extra bass boost, as the other channels carry full-range bass. Audio systems without Dolby Digital or DTS capabilities generally do not have speakers that can properly reproduce very low frequencies, so the designers of the audio coding technologies chose to have the decoders throw out the LFE track to avoid muddying the sound on an average home system. Anyone who truly cares about the LFE channel should invest in a receiver with bass management and a separate subwoofer output.

Myth: "Blu-ray Lets You Watch Movies as They Were Meant to Be Seen"

This may refer to Blu-ray video's 1.78 widescreen aspect ratio, which is close to the most common movie aspect ratio (1.85). However, many movies have a wider aspect than widescreen TVs, such as 2.35. Thus, even though they look much better on a widescreen TV, they still have to be formatted to fit the less oblong shape, usually with black bars at the top and bottom. See *Aspect Ratios* in *Chapter 2* for more information.

Myths: "Java and JavaScript™ Are the Same Thing"

Java is an island in Indonesia, south of Borneo, as well as the name of an object-oriented programming language primarily used for network applications and applets. Programs written in Java do not rely on a specific operating system, so they are platform independent. The BD-J application format for Blu-ray is an implementation of Java. BD-J applications are specific to BD-J players but, if properly written, can be readily adapted to run on other GEM-based platforms.

JavaScript is an unfortunately confusing name given to a dynamic scripting language. JavaScript initially was used to expand the flexibility, look and style of HTML pages on websites and in browser windows. It was formalized as *ECMAScript* and later adopted for use with HD DVD.

The Java programming language was developed by *Sun Microsystems*, while JavaScript was parented by *Netscape*. Reportedly, Netscape was allowed to co-opt Java and create the word JavaScript as part of a deal with Sun in exchange for providing support of Java applets. Sun has regretted the confusion and misbranding ever since.

Chapter 9
What's Wrong with Blu-ray Disc™

"Problems are the price you pay for progress."
— Branch Rickey[1]
"Progress might have been all right once, but it has gone on too long."
— Ogden Nash[2]

    In 2002, following the success of DVD, which proved to be the fastest growing consumer media of all time, the DVD Forum started talks to create a new medium that would meet the needs of consumers for the next 10 to 25 years. Almost at the same time, a second group, composed of many of the same companies as in the DVD Forum, began separate discussions with the same goal. Although the design objectives and the user requirements were similar, the primary differences between the rival groups came down to who controlled the intellectual property rights of the to-be-created technology and could a group member get a better deal by promoting one solution over the other. The efforts by the groups devolved and resulted in the competing high definition optical disc formats — HD DVD and Blu-ray Disc. The leading proponent of HD DVD was Toshiba, while Sony, Philip and, later, Panasonic led the Blu-ray effort.

    In the Spring of 2006, the two next-generation disc formats were introduced to consumers. Playing on the base human nature attributes of greed and ego, neither group was willing to back down or to compromise on a unified format. In fact, the format war was introduced in March, 2006 with such statements as this one from Kazuhiro Tsuga, an executive at Panasonic, during an interview with Reuters, "We are not talking and we will not talk, the market will decide the winner." Giving a hint to the ultimate fate of HD DVD, in June, 2006, Reuters quoted Toshiba's president Atsutoshi Nishida, during the company's annual shareholders' meeting in Tokyo, as saying, "We have not given up on a unified format. We would like to seek ways for unifying the standards if opportunities arise."

    Although the market conflict lasted far longer than was good for any party, the end of the war came far quicker than anyone had expected. In early January 2008, Warner Bros. Company, with about 20% of the DVD market, decided to exclusively support Blu-ray after having straddled the fence by offering titles in both next generation formats. This was followed in February with the leading retailers Best Buy, Walmart, Blockbuster, and Netflix, joined by the Paramount and Universal studios, saying they would only support Blu-ray.

---

[1]Branch Rickey (1881-1965), US baseball player, manager, and executive with the Brooklyn Dodgers and the Pittsburgh Pirates.

[2]Ogden Nash (1902-1971), US humorist, lyricist and poet.

What's Wrong with Blu-ray Disc

Toshiba threw in the towel on February 19, 2008, announcing that it would, "...no longer develop, manufacture and market HD DVD players and recorders." HD DVD "lost" the war and Blu-ray was the victor. Among the costs of the war was the prolonged delay where consumers did not even consider either new HD disc format, while the battle raged.

Even though Blu-ray has won, it is far from perfect (nor, for that matter, was HD DVD). This chapter presents some of the shortcomings of this latest retooling in optical disc technology.

Copy Protection

A 2005 study by the Motion Picture Association of America (MPAA) claims that pirated DVDs were costing the movie industry over $6.1 billion per year. For Blu-ray to succeed, some method of content protection more effective than the thoroughly compromised CSS (Content Scrambling System, used for DVDs) was needed.

Advanced Access Content System

The industry, along with DVD Forum members and Blu-ray Disc Association members, came up with the Advanced Access Content System (AACS). Yet, within six months of the first shipments of HD DVD and Blu-ray titles, a hacker posted software to decrypt the AACS scheme, allowing one to unlock disc content and to make copies or post to the Internet, albeit only for the hacked disc. Technically, this hack took advantage of some poorly implemented naming schemes by the disc producer, but the scheme had been hacked. So much for a secure system.

One of the features of AACS, unlike CSS, is that specific keys used to lock and unlock content can be revoked and new keys can be issued. This has created a cat and mouse game between the studios and the hackers, and it gives the studios some level of confidence that their titles cannot be easily and widely copied.

Movie studios and consumer electronics companies promoted legislation to make it illegal to defeat DVD content protection. Their efforts helped bring about the *World Intellectual Property Organization* (WIPO) *Copyright Treaty* (December, 1996), the *WIPO Performances and Phonograms Treaty* (December, 1996), and the compliant *Digital Millennium Copyright Act* (DMCA), passed into US law October, 1998. Software or devices intended specifically and primarily to circumvent content protection are now illegal in the United States and many other countries. A co-chair of the legal group of the content protection committee stated, "In the video context, the contemplated legislation should also provide some specific assurances that certain reasonable and customary home recording practices will be permitted, in addition to providing penalties for circumvention." Blu-ray with AACS has made this sort of copying, even single copies for personal use, very difficult and, if replicators and player manufactures follow the Blu-ray specification, all but impossible.

Unlike DVD, where all content protection systems are optional for DVD publishers, Blu-ray requires that all replicated discs use AACS. The added production complexity and the licensing costs, currently over a thousand dollars in various fees, have restricted the release of titles to mainly those from the top studios. As with DVD, we may see players that more loosely interpret the specifications and will play unprotected movies from replicated discs. Although, most players will play HDMV and BD-J programmed content on BD-R/RE.

Blu-ray Disc Plus (BD+)

The Blu-ray camp expected that eventually (maybe not so soon, though) such a breach of the AACS security and added another layer of protection called BD+ (BD Plus). Unlike AACS, BD+ allows content producers to change the copy protection scheme from disc to disc. So, if a hacker breaks one disc, they have just broken one disc and not all discs. The evident success of BD+ has been borne out by postings on various forums on the web, that the hackers have only broken discs one at a time. The level of sophistication has made copying discs a serious game. BD+ does what the content owners and format developers wanted it to do and it does not interfere with the user experience, but it does make it very difficult for endusers to make backup copies of their purchases.

Image Constraint Token

Image Constraint Token (ICT) is a feature that lets a content owner restrict the resolution of disc playback on non-secure signal paths. If the High-bandwidth Digital Content Protection (HDCP) path is not used to connect a player to a display, for example, a DVI (Digital Visual Interface) connection without HDCP, then the signal resolution will be down-converted by the player to that of standard definition DVD resolution (720×480). This also means that if you set up your monitor and your Blu-ray player using component cables, which works on some discs and looks great, the discs that use ICT will be downconverted and will look terrible. For the consumer, this can be very confusing. Also, this means that playback on a computer without a secure HDCP path will either be in 720×480 resolution or may not play at all.

Watermarking

Another copy protection gotcha is what can be called the *birthday scenario*. This relates to watermarking (which is a part of AACS). Hypothetically, let's say you are filming your child's birthday party, and a movie is playing on a TV set in the background. If the audio is watermarked (or if there is a video watermark and you get the TV in the frame) then it could be possible, after you have transferred your home video to disc and are playing it in a player with watermark detection, for your kid's birthday party to be mistakenly identified as pirated content. Watermarking systems are supposedly designed to prevent this from happening, but it is an interesting illustration of how complicated the balance is between protecting content owner rights while not impinging on the freedom of consumers.

Region Playback Control

The purpose of region playback control is to allow studios to control the release of their movies in the different markets of the world. The problem this creates for the consumer is that if you buy a disc while traveling in one region and you live in another region, you can-

What's Wrong with Blu-ray Disc

not use the disc if you purchased a player in your home region. That is similar to buying a book in London and finding that the pages are stuck closed and you cannot open it when you return home to San Francisco.

As opposed to the six regions in use for DVD releases, Blu-ray makes this a bit easier by dividing the world into three regions, A, B, and C . The only upside of RPC (Regional Playback Control) is that the studios do not *have* to use this feature and on many of the early discs, RPC was not used. One of the differences between DVD and Blu-ray, is that the RPC is handled by the application layer (with the HDMV or BD-J programming) instead of between the player and the disc. The player can be queried by the Blu-ray title and the programming can be made to display something more appealing than "Wrong Region". For example, the disc could display a pretty graphic announcing — "This disc was sold for the "A" market. If you wish to play this movie title in this player, you can purchase a copy at your favorite retailer." Or, perhaps, an even more advanced announcement — "This disc was purchased for playback in another region, if you wish to play this disc in this player, please click here, enter your credit card number and for a small fee we will let you play this title on this player". The latter example assumes that your Blu-ray player is connected to the Internet.

Hollywood Baggage on Computers

As more computers gain the ability to play movies from Blu-ray, the region playback control and content protection requirements need to be implemented in computers as well as home players. Manufacturers of hardware or software involved in the playback of encrypted movies are required to obtain a license from the AACS Licensing Authority (AACS LA). The manufacturers must ensure that decrypted files cannot be copied, that digital outputs contain proper content protection information, and that analog outputs are also protected.

The upside is that these safeguards assure Hollywood that its property will not be plundered and spread illegally from computer to computer. The downside is that most law-abiding computer owners are inconvenienced and may pay slightly more because of these protection measures. This is especially irritating to those who have no interest in watching commercial movies on their computer screens.

NTSC versus PAL is also 60 Hertz versus 50 Hertz

If you are a content producer you still have to worry about NTSC versus PAL for standard definition video and, now, will also have to worry about 60 Hertz (Hz) versus 50 Hertz for high definition content. In addition, you will have to chose whether to produce your content at 1280×720 resolution (progressive) and risk that some displays will automatically downscale to 720×480 because that's the only resolution they have that supports your frame rate. Or, you'll consider going with 1920×1080 progressive resolution and run the risk that the display either has to interlace the video or scale to a lower resolution, or both, because the display is not capable of showing 1920×1080 progressive video.

In standard definition DVD terms, we have NTSC countries — 60 hertz (Hz) power, 30 frames per second (fps) sized at 720×480 pixels, and PAL countries — 50 hertz power, 25 frames per second at 720×576 pixels. What this means is that you could have discs that would not play because you did not have the right player or the right display. With Blu-ray, one might have hoped that this condition would have been remedied. Well, to a degree it has. With DVD, a player did not have to support both 25 (24.97) fps and 30 fps (29.97). In practice, though, almost all players in PAL standard countries supported NTSC discs.

For Blu-ray, all players must support 23.976/24 hz, 29.97/30 hz and 59.94/60 hz. Optionally, players in 50 hz countries may also support 24.97/25 hz. This means that a disc can be created for all regions unless the video is specifically targeted for the 24.97/25 hz markets.

It remains to be seen if new HD material will be created in a format that is compatible for all regions (of the world and the universe). The bottomline is that endusers should not have to sort out these arcane technical nuances of video.

Connection Incompatibilities

HDMI is the newest interconnect format and it is most promising. With one cable carrying all of the signals it definitely makes the cable clutter behind audio and video components much less of a mess. Alas, even though the standard has evolved from versions 1.1 to 1.2 to 1.3, getting all of the components to talk with one another is not troublefree. The connection *handshake* between a player and a receiver and a monitor, when successful, can take around four seconds. But, there is a lot that can go wrong. The monitor may not support the video resolution that the player is outputting, for example, 720p vs 1080p. In that case the monitor may not display anything. Getting the monitor to display a picture again can become very tricky. For example, on the Sony PlayStation3 (PS3), you would hold the front panel power on button for ten seconds while turning the back panel on/off switch to "on". At one point, eBay listings for Sony PS3s not working because of "no display" were very plentiful.

Some players, with some firmware revisions, will stop playing the disc whenever the HDMI display connection is removed. For example, if you are checking the latest score of a ballgame from your satellite or cable settop box, the disc may have to go through the lengthy loading process to get back to the movie. Getting the player to sync up with the monitor can be very frustrating. Keep a note on your player explaining how to reset the player and monitor to a compatible display setting so that when they go out of sync you can get everything working again. Also, because of audio copy protection concerns, if the disc has a feature which uses audio mixing (e.g., button sounds with high definition DTS or Dolby Digitial audio), the button sounds will not be heard when using HDMI audio. Bizzare, huh? It may leave one longing for the clutter and simplicity of the old fashioned RCA connectors and cables that offer the refreshing option of *plug it in and see it and hear it*.

Playback Incompatibilities

There are two programming modes for Blu-ray — HDMV and BD-J. Generally, there tends to be fewer incompatibility issues with HDMV playback than with BD-J, mainly because the BD-J environment has many more options and is based on a wide array of specifications.

What's Wrong with Blu-ray Disc

With BD-J, the chance for misinterpretation and differences in interpretation, are great. Also, there are many different ways to accomplish the same thing in BD-J. At the core of BD-J is the Java Virtual Machine (JVM). Some manufacturers have licensed the JVM code from Sun Microsystems. Others have decided to implement their own version of JVM. This is reminiscent of the days when you had MS-DOS and DR-DOS and PC-DOS. Often, programs would work as expected on one platform but not on others. Within the BD-J specification, there are several libraries or packages that do virtually the same function. Some of the early players did not implement all libraries properly, but the manufacturers are working diligently with the BDA to improve testing routines to remedy these situations.

To address these differences, the studios and the authoring houses in Hollywood have taken a couple of approaches. One approach is take no prisoners, program to the specifications and if the players don't work, let the manufacturers fix their players. If the manufacturers don't update their players to conform, the consumers will help evolve the survival of the fittest. Or, the other approach is to program to the lowest common denominator. However, this is difficult to achieve because if just one manufacturer's implementation differs then you don't use that feature. Some of the incompatibilities may be frustrating to the user in that they could see poorly drawn graphics, and poor or non-functioning features, such as, network capability.

As if the inherent challenges in making a player conform to the specifications weren't enough, the BDA has specified five different profile levels of players —

Profile 1, version 1.0, players introduced before November 1, 2007,

Profile 1, version 1.1. players introduced after October 31, 2007,

Profile 2 (optional),

Profile 3 (optional audio-only player), and

Profile 4 (realtime recording and editing).

What this means is that depending on when the player you bought was introduced to the market, the latest Blu-ray disc that you have bought may not fully work. For example, *Bonus View*, a marketing name for Profile1, version 1.1 players, includes the features PiP, secondary audio, local storage, and luma value, but these features are not available on any of the early players except the Sony PS3. Also, the network connected feature, called *BD-Live*, requires a player that conforms to the optional Profile 2.

Compatibility is a challenge to all involved whether it is the player manufacturers trying to make their players do all of the cool things in the specifications, or the studios trying to decide which features to put on the discs at the risk of disappointing some of their endusers, or the authoring houses trying to figure out how to implement the feature in such a way that it works on all of, or even just the majority of, the players, or the consumer trying to buy and use the "right" cheapest player.

As was the case for DVD, expect that BD players will continue to have special anomalies for years to come. The difference with Blu-ray players vs DVD players is that all of the Blu-ray players can be updated with firmware updates, either by inserting an update disc or by direct network connection to the player. Note that the ability to upgrade the firmware does not help the limitations of the hardware configuration made by a given manufacturer. For

What's Wrong with Blu-ray Disc

example, a player with no Ethernet port can never be upgraded to Profile 2. Most manufacturers are updating their players on a regular basis so the consumer will have to check often for updates. Fortunately, most discs that push the limits of the specification capabilities come with a reminder note in each disc saying, "check your player manufacturer's website for the latest update."

Poor Performance

The BD specifications, for the most part, do not require any minimum performance for the players and the minimum hardware is described as a 300 Mhz RISC based processor. This level of processor has a speed rating of about 2 Megaflops (two million floating point operations per second). But, Sony introduced the PS3, which plays Blu-ray discs in addition to playing games, and gave it a processor that has an overall speed rating of 2.18 Teraflops[3]. This difference, in the extreme case, means that the PS3 is 6,000 times faster than the minimum player hardware requirement. Put another way, the PS3 can perform a set of operations in one second that would take a minimum hardware system 6,000 seconds (100 minutes). Floating point operations is only one metric, and probably not the best metric, for an overall comparison of performance. Other performance metrics could include integer operations, class-loading time, image decoding, drawing, font rendering, and so on.

In actual cases, the true disparity is less than this, and a PS3 user may see a popup menu of 24 frame animation look very smooth and take one second, while on a minimum hardware compliant player the animation may take more than 7 to 8 seconds. The result, however, is that the animation just doesn't have the same cool look. Also, on a PS3 it might take 7 to 10 seconds to get to the opening credits of a movie, but it might take up to 3 to 5 minutes on some of the settop players. The manufacturers that use the minimum required specifications as their target have been overwhelmingly outclassed by the PS3. The PS3 has, as of 2008, as much as 75% of the market of Blu-ray players. Over time, the performance gap may be closed but, for the early adopters, the PS3 was the way to go.

Feeble Support of Parental Choice Features

Hollywood included a parental management feature for DVD but was very slow to actually use it. And, although present for Blu-ray, it is yet to be seen if Hollywood will use the feature for Blu-ray.

One approach by some clever entrepreneurs from Utah was to create a DVD player that allowed for specially-filtered movies. But it required an act of the US Congress to make it legal. The Family Movie Act is part of the Family Entertainment and Copyright Act of 2005. Legislators have explicitly allowed systems that change DVD playback, as an alternative to

---

[3] The PS 3 overall specs are: vProcessor Core Spec 1 Core, 7 x SPE 3.2GHz (256KB SRAM per SPE (Synergistic Processing Element)), 7 x 128b 128 SIMD GPRs Marketing Performance Measurement 2.18 TFLOPs Processor Clock Speed 3.2GHz L2 Cache 512KB L2 cache, 256KB per SPE). With this system configuration, the conservative speed could be as great as 12 GFLOPs.

What's Wrong with Blu-ray Disc

the absence of multi-ratings feature on the disc itself. This allows special playback devices that filter content for objectional material by either skipping or muting the material. For DVD, this required special players. For Blu-ray, this is, on one hand, easier and, on the other hand, more difficult to implement.

Technically, an enduser could create a custom playlist that skips objectional material, but not all titles support creating custom playlists. This feature may become more popular with movies that take advantage of BD-Live players that allow users to create and share playlists. It is possible to create an application that provides this filtering, but Blu-ray requires any additional application (new software code) to be properly signed, in programming parlance. The intent of this signing process is to prevent unauthorized and, therefore, presumed malicious, applications from taking over a Blu-ray player, just as a virus would with a personal computer. Unless, these new applications are created and/or approved by the content owners, they cannot be signed correctly and, thus, will not work.

Yet, Blu-ray discs can be designed to play a different version of a movie depending on the parental level that has been set in the player. By taking advantage of the seamless branching feature of Blu-ray, objectionable scenes can be skipped automatically or replaced during playback. This requires that the disc be carefully authored with alternate scenes and branch points that do not cause interruptions or discontinuities in the soundtrack.

Hollywood remains unconvinced that there is a large enough demand to justify the extra work involved — shooting extra footage, recording extra audio, editing new sequences, creating branch points, synchronizing the soundtrack across jumps, submitting new versions for MPAA rating, dealing with players that do not properly implement parental branching, having video store chains refuse to carry discs with unrated content, et cetera. Seamless branching is being used quite a lot more these days, but for special editions and unrated versions rather than parental control; this is because it is done with the filmmaker's complicity, to extend their vision.

Not Better Enough

In an age when smaller, lower-quality video imagery, such as, iTunes downloads and YouTube videos, have become more popular and acceptable to consumers because of their convenience and easy access, the BD format faces an uphill battle, even with its often stunning picture quality. And, many consumers cannot tell the difference between upconverted DVD images and high definition video. Further, it requires that they have a 1920 × 1080-capable display to really appreciate the difference, and those monitors did not come down to a price that was acceptable to a large portion of the market until late 2007. Additionally, 1080p displays were not readily available at attractive prices — US$1,000 for a 42 inch display — until mid-2008.

So, if upconverted DVD is good enough and you cannot see the difference anyway, why do you need Blu-ray? One answer may lay in having compelling content with network interactivity that brings out the content on a disc. But, not every Blu-ray player has, or will have, built-in network connectivity.

The challenge is finding and exploiting the new feature capabilities of BD. In time, people will have a playback system that allows them to see the difference between DVD and high definition video. For the present, there is a lot of confusion in consumer's minds about the difference. Most people are looking for "big" screen TV not high definition TV. Until consumers see the difference, the biggest competitor to Blu-ray is DVD.

Blu-ray is designed for HD displays supporting $1920 \times 1080 \times 24$ fps. On these displays, the difference between upconverted DVD and Blu-ray is clearly visible. If you have a less capable display, you probably could stick with your old DVD player. To get the best results from Blu-ray, you really need to invest in a new HD display.

No Reverse Play

Because of the way compressed video builds frames by using the differences from previous frames, it is difficult to smoothly play in reverse. To accomplish that feat, a playback system must have either a very large memory buffer in which to store a set of previous frames or use a very complex high-speed process of jumping back and forth on the disc to build a frame sequence based on the nearest key frame. Although RAM for video buffers has gotten less expensive, there are only DVD computers and some very high-end DVD players that have the capacity to play backwards at normal speed.

Some players can play backwards through a disc by skipping between key frames. Attempting to display these frames at the proper time intervals results in jerky playback with delays of about 1/2 second between each frame. Smooth scan can only be achieved by showing the frames at 12 to 15 times normal speed, thus speeding up the action.

One of the best techie gag gifts is the DVD rewinder. But, one of the things we love about VHS tape is that when you put a tape back in the player it would continue from where you last stopped playing. No copyright notice, no trailers for last season's releases, no cute menu, no having to advance to the video so that you could resume watching the movie. Because the next generation players have memory, it is possible for content authors to create a resume function so that you can have a VHS experience. Progress? Maybe not.

Only Two Aspect Ratios

As with DVD, Blu-ray disc is limited to the two aspect ratios of 1.33 (4:3) and 1.78 (16:9), even though MPEG-2 allows a third aspect ratio of 2.21. Because of the need for a standard physical shape, displays essentially come in two shapes — 4:3 and 16:9. With HDTV, the standard aspect ratio is 16:9 and yet there is all that legacy 4:3 material out there. So, there are black bars on either the top/bottom or the sides of the display depending on the original source aspect.

Better yet would be the ability to support any aspect ratio. If the player or TV were able to unsqueeze an anamorphic source of any ratio, it would provide better resolution because pixels would not be wasted on the letterbox mattes. Letterbox mattes may still need to be generated by the player or the display, but high-resolution displays would be able to make the most of every pixel of an anamorphic signal. The obvious disadvantage to this feature is

What's Wrong with Blu-ray Disc

that variable-geometry picture scaling circuitry is more expensive than the fixed-geometry scaling of DVD players and existing widescreen TVs.

No Barcode Standard

In their day, one of the most powerful features of laserdisc players used in training and in education was barcodes. Printed barcodes could be scanned using a handheld scanner or wand that sent commands to the player, telling the player to search to a specific picture or to play a particular segment. The player became a powerful interactive presentation tool when combined with a barcode reader. In their time, barcodes were added to textbooks, charts, posters, lesson outlines, storybooks, workbooks, and much more, enhancing the presentation material with quick access to pictures and/or movies. So, although Blu-ray players theoretically can support USB keyboards and, therefore, barcode readers, this feature has not, as yet, made it into any of the Blu-ray players.

No External Control Standard

Most consumer laserdisc players included an external control connector, and all industrial laserdisc players included a serial port for connection to a computer. An entire genre of multimedia systems evolved during the 1980s using laserdisc players to add sound and video to computer software. Reaching back more than 20 years may be a real stretch and, admittedly, this is less important today as the multimedia features of computers improve, but many applications of optical discs, such as, video editing, kiosks, and custom installations are limited by the lack of an external control standard. Each player manufacturer uses a different proprietary command protocol.

Poor Computer Compatibility

Early compatibility problems plagued the previous generations of optical media. The CD-ROM industry was long plagued by incompatibility problems with return rates in 1995 as high as 40 percent. Compatibility problems were caused by incorrect hardware or software setups, defects in video and audio hardware, bugs in video and audio driver software, and hardware that was not powerful enough for the envisioned tasks.

And, the problems with DVD-ROM were even worse. DVD-ROMs have to deal with defects in video and audio decoder hardware or software, incompatibilities of proprietary playback implementations, decoder software that could not keep up with full-rate movies, DVD-Video navigation software that did not correctly emulate a DVD-Video player, and so on. In the early days, some computers with a DVD-ROM drive would not play movies from a DVD-Video disc, especially computers that were upgraded from a CD-ROM to a DVD-ROM drive.

Now, getting a Blu-ray disc to play on a computer has been an issue on all except the lat-

What's Wrong with Blu-ray Disc

est, fastest, most powerful computers that were designed with all of the requirements for Blu-ray built-in. Simply adding a Blu-ray drive to an existing, even high-end computer, does not ensure successful results. The video decoding demands often require an upgrade of the video card which may also require a new motherboard. Support for navigating a Blu-ray disc using a computer mouse is often turned off, by default, by some software players. In time, as more computers are built around the Blu-ray system requirements this may be a much more satisfying experience. Upgrading an existing computer system by adding a Blu-ray drive is not recommended. You are better off with a new system that has been designed and built for the task. One must realize that Blu-ray is a great leap forward, so it is unfair to demand that your old 90 MHz Pentium PC be able to play a BD.

Limited Web Standard

One of the most exciting possibilities for Blu-ray is the ability to connect to the Internet with a Profile 2 enabled player. Hollywood studios, corporations, educators, and other Blu-ray creators are exploring the potential applications achievable when combining the contents of a Blu-ray Disc with the best of the Internet.

Unfortunately, the Blu-ray specification only defines the TCP, UDP, HTTP, and HTTPS protocols and does not define any higher transaction level standards. There is no web browser defined in the specification. With a Java program use of TCP and UDP, a Blu-ray Disc is perfectly able to implement higher transaction level standards, e.g., a p2p, download, online sale, chat, or mail protocol. Of course, the burden is on the author/programmer to choose the transaction standard(s) to implement and how. So, Web browsing is something the Blu-ray author has to create. The possibility for player and title incompatibility is high. This lack of standardization may make the incompatabilities of the browser wars — when web producers were seriously constrained by lack of reliable design standards — look tame and simple, in retrospect. On the other hand, the browser in Blu-ray is not burned into the player but, instead, can be included on the disc, or even replaced with a new one with a simple network update. For this reason, compatibility problems may be less, particularly for an application used as a portal to content controlled by the studio.

Too Many Encoding Formats

Beyond the variety of video formats that have to be supported looms a variety of video codecs — MPEG-2, VC-1, or AVC. It is not too bad for those producing discs because they can pick one and stick with it, but for those making players — either hardware consumer devices or software for PCs — have to implement, test, and pay royalties for the different video decoders.

New audio codecs have been added as well, including a new version of Dolby Digital, a new version of DTS, a renamed version of MLP, a lossless variation of DTS, plus a set of optional audio codecs for secondary audio streams — DD+ and DTS-HD LBR (low

Blu-ray Disc Demystified

9-11

What's Wrong with Blu-ray Disc

bitrate), aka DTS Express. While it is good to have choices, and the inclusion of lossless audio is more preferable to an incompatible second format such as DVD-Audio, the end result is that players and authoring tools are more complicated and more expensive.

Too Many Inputs

Even before serious work began on the new disc formats, the seeds were sown for new problems. In the period of time that has transpired since the introduction of HD television sets, a multitude of new video input types have also been created. In addition to the 75-ohm VHF RF-carrier (RF), composite, S-video (Y/C), and component ($Y'P_bP_r$) connectors defined for standard defintion video, HD television sets now may incorporate HD analog component (also $Y'P_bP_r$), DVI (including DVI-A and DVI-D), and HDMI, among others. Due to content protection issues, it is possible that HD displays previously sold as "HD Ready" (no HD tuner but with DVI or HDMI and, maybe, component video inputs at less than 1080p resolution), or "HD Compatible" (HDMI capability but with lower than HD-ready resolution), may display 1080p video at a degraded, less than pixel for pixel resolution. Consumers will have to educate themselves about these issues in order to get what they have paid for from their displays.

Too Many Channels

What about audio? The Blu-ray specification supports high-bandwidth multichannel lossless audio compression, giving the ability to deliver up to eight channels of pristine audio ecstasy. However, it is still not clear how all these channels will get to your speakers. Most current A/V receivers support S/PDIF or Toslink digital connections, which have enough bandwidth to deliver two channels of uncompressed PCM audio data or 5.1 channels of Dolby Digital or DTS-encoded audio data. In order to carry eight channels of uncompressed PCM audio, you would need four S/PDIF or Toslink connections all working together.

All versions of HDMI have the ability to deliver eight channels of uncompressed PCM audio to an A/V receiver. But, if you want to use HDMI for a bitstream output of the other lossless audio codecs (DTS-HD MA and TrueHD), you need to update your receiver to HDMI version 1.3. Versions 1.0 to 1.2 do not have the bandwidth to support the higher bitrates of these advanced codecs. Another approach could be to fall back to using eight analog RCA phono plugs. Alternatively, the audio data can be delivered encoded, though it still requires more bandwidth than most receivers currently support, as well as requiring built-in support for DTS-HD and/or Dolby TrueHD decoding.

Once you have decided on a viable solution (or two) for how to get the audio data to your receiver, there is still the question of where to send it from there. In other words, where are those eight speakers? There is no specific standard for 7.1 speaker placement. In some configurations, the typical 5.1 configuration of Left, Center, Right, Left Surround, Right

What's Wrong with Blu-ray Disc

Surround may be augmented with either additional side or center speakers. In other configurations, the extra channels are used to provide altitude. Integrating eight-channel audio into the home environment means making a great number of decisions.

Please note that the companion disc to this book includes a DTS 7.1 piece that strikingly demonstrates the upside of 7.1 audio. Too bad there are not more uses of 7.1 channels in theatrical productions.

Not Enough Interactivity

The new BD format is leaps and bounds beyond old DVD when it comes to interactivity. BD uses a powerful programming language (Java), multiple planes of video, and sophisticated layout control to provide animated menus that can popup on top of running video, game-like control of onscreen graphics, hyperlinked content, Internet connectivity, and much more. However, many limitations have been set and a huge numbers of compromises have been made during the process of defining these interactive features, most in order to meet the restrictions of insuring that low-cost consumer devices will be produced. Next-generation players will do very cool things, but they will most likely still be a far cry from game consoles and BD-capable computers.

Too Much Interactivity

Notwithstanding the previous paragraph, there is a danger in power. The more complex and flexible a system is, the higher the probability that something will go wrong. The old DVD format was simple, yet there were still many cases where discs — sometimes million-sellers from major studios — would not play properly in some players.

Virtual machines, complex programs, mixed video planes, and other interactive features multiply the potential for playback problems by a few orders of magnitude. Unless player manufacturers and content creators are a lot more careful when designing and testing their wares, a serious backlash could result from consumers perceiving the newfangled format to be unreliable and finicky.

Conclusion

In some ways, the original DVD format suffered from many of the same problems. It, too, had too many video formats (NTSC and PAL). Likewise, it supported too many audio codecs. After all, was it really necessary to require PCM and Dolby Digital and MPEG (for PAL players) and also have optional support for DTS and SDDS audio? But, in the end, the market decided which of these capabilities and features were most important, and which could be ignored. Perhaps we can expect new and creative solutions to the "too many" problems of Blu-ray, as well.

Chapter 10
Interactivity

DVD was largely responsible for introducing consumers to the idea of a menu on their TV screen. In the mid-90's TVs and VCRs had begun to provide on-screen menus for changing device settings and computers and video games have menus, but the widespread adoption of DVDs around the turn of the millennium was what familiarized Joe and Jane Couchpotato to the idea that six little buttons — menu, up, down, left, right, and enter— could provide interactive, nonlinear access to assorted programs and features on a disc.

Most people look back at the early DVD titles and think how simple they were— a movie, a few trailers, perhaps a director commentary, a set of subtitles, and some basic menus to access the disparate elements. Of course, none of this was completely new. Laserdiscs had provided high quality audio and video, along with secondary audio tracks, such as, commentaries, for years before DVD. So what made DVD so successful? More speculation has been offered on that subject than is worth putting in print, but one point stands out — DVD offered a new look to and a new feel for content. As simple as the early menus were, and as basic the features, DVD offered a quantum leap beyond slow, linear videotape machines, and a level of interactivity that jumped far beyond laserdisc. The fact that upon insertion, most DVDs immediately display a set of menus that ask a viewer what they would like to see or hear, coupled with the convenience of no longer needing to rewind the tape, represents a complete transformation in how people interact with content. While this new interaction paradigm was not the only driving force behind DVD's success, it is certainly important when distinguishing DVD from previous mainstream video products.

Recreating DVD's Success

DVD's star is hitched to entertainment content. The unparalleled growth of DVD into a $24 billion industry in the U.S. alone was driven by consumer demand for what's on the discs, not by interest in the technology itself. That growth began to slow in 2005 as saturation became a factor — most people who wanted to replace their VCR and videotape collection had done so but, after its tenth birthday, DVD continued to be a success by almost any measure. By 2007 more than 80 percent of U.S. households had a DVD player, more than half had two or more, and the average number of DVDs per household was about 85.

DVD quickly became responsible for the lion's share of entertainment revenue, accounting for more money than box office, network television, cable, and other sources put together (see Figure 10.1). As a judge of the 2005 DVD Awards put it, "Hollywood is no longer in the business of making movies but is now in the business of making DVDs."

Interactivity

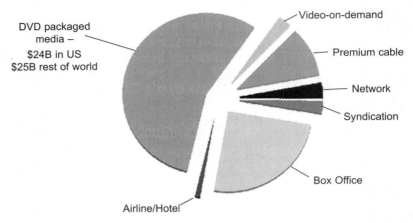

Figure 10.1 Studio Revenue in 2007

Sources: ABN Amro, Veronis Suhler, Adams Media Research

So what happens when customers have bought most of the old movies they are interested in? Will they buy them over again if the movies are "new and improved?" Hopes have been pinned to high-definition formats reinvigorating home video sales, largely based on two elements — high-definition quality and advanced interactivity. Opinions vary on how significantly interactivity will contribute to Blu-ray adoption but it is a fundamental part of the format.

A New Kind of Interactivity

A basic characteristic of DVD interactivity is that everything can be done via menus (making audio and subtitle selections, playing the movie, watching bonus features, and so on). This interaction, however, is generally a modal experience — you are either on a menu or playing a feature, but not both. You must leave the movie to get to a menu to do something else with the disc content. This somewhat jarring activity interferes with the continuity of the experience. Even with stylized, thematic menus the experience can be disruptive. Some attempts have been made with DVD to improve the menu experience. The InfiniFilm™ feature on some New Line Home Video DVDs, for example, uses the subtitle track to provide popup menus over the film. The Follow the White Rabbit feature on the original Warner Bros. *The Matrix* DVD used a similar technique to provide interactive jump points within the movie. Other discs followed suit but, in each case, the extra content caused complications with the format functionality. Having buttons appear over the video required a subpicture stream. More importantly, it meant that no other subpicture stream (such as subtitles) could play while the menu feature was active. As a result, if subtitles were needed in addition to the interactive buttons, much more complex programming and asset preparation was required.

Interactivity

Perhaps the most successful attempts to supersede DVD's inherent modal experience came through the use of the format's multiangle feature. Pioneered by companies like MX Entertainment and Technicolor in the US and the BBC and The Yard in the UK, titles began to appear on the market that allowed the viewer to make interactive selections, activating alternate video and audio material without interrupting playback. Although the feature has been most often used for concert videos, such as, *U2: Live from Boston* (2001), *Rolling Stones: Four Flicks*, and *Rush: Rush in Rio* (2003), it debuted on a special edition title produced by the BBC for their *Walking with Beasts* television series. Unlike the concert videos, which primarily used the technique by providing an on-screen interface for accessing alternative video angles, the BBC used the feature to create fully interactive popup menus over video that provided access to audio stream selection and alternative, synchronized video content. At any time throughout the 30-minute episode the viewer could switch among four related video programs. The primary video, an episode of the television series, served as the starting point. The alternative programs provided specific information related to the main program, for example, details about the flora of the time or about the creatures being shown. But even this groundbreaking title was not able to fully overcome the constraints of the DVD format. Due to the nature of multiangle video, a half-hour program consumes virtually the entire capacity of a disc. In addition, switching between menus created non-seamless pauses in the video playback — not as bad as jumping out to a separate menu, but still rather disruptive to the video program. Nevertheless, these titles paved the way for the next generation of interactivity.

The Seamless User Experience

Dubbing it *the seamless user experience*, Hollywood studios gravitated toward the convenience of being able to watch a movie and, without interrupting the film (without even a glitch in the audio or video), use thematic popup menus that provide access to special features, audio and subtitle selections, and alternate scenes or camera angles, as well as, to launch highly interactive applications that may appear to be part of the film itself, such as, a video game.

Imagine, for example, putting a mystery film into a Blu-ray Disc player. Instead of going directly to a menu, playback launches directly to the feature film (with requisite logos and warning cards). During the opening credits, a stylized menu appears in the lower third of the screen, offering the option of watching just the film or of receiving periodic, interactive prompts about additional bonus features on the disc. You select the latter option and the movie continues to play effortlessly. Soon, a prompt appears to let you know that a director's commentary starts in ten seconds. Would you like to hear it? As the counter ticks down the seconds, you activate the option and the director appears in the lower left corner of the screen superimposed over the movie. When the timer hits zero, the director says hello, introduces himself, and proceeds to tell you about how he made the film, sharing interesting points and highlighting portions of the screen as the movie plays. Later in the film, an option appears for selecting a game in which you have to identify "who dunnit." You launch the feature and find yourself in control of an interactive magnifying glass that you can move around

Blu-ray Disc Demystified

10-3

Interactivity

the screen during the film and select items for closer inspection. As you collect these "clues" you receive additional insight about each character. In time, between the sequence of events in the film and the clues you have found, you decide that you know the guilty party and you make your selection. At the conclusion of the film you find out that your selection was not only correct but made in record time.

The hope is that this new paradigm for interactivity provides a compelling new way in which fans can interact with the content. At the same time, a simple movie-watching experience is preserved for those who just want to see the film. After all, it is the film that ultimately drives the purchase of the content, yet many argue that it is the look and feel of the content that sells the platform.

Target Applications

Beyond the fundamental approach of how content may be accessed, a range of advanced applications were considered for the platform including (but certainly not limited to) —

- The ability to select objects displayed in the feature video, adding them to a wishlist for later purchase.
- Viewer-generated bookmarks that make it easy to jump to any part of the disc.
- Popup trivia questions tied to events taking place in the film.
- The screenplay text displayed in a scrolling window next to or over the video, remaining in sync with the video.
- Multi-player games in which two or more direct viewers can play together simultaneously using multiple controllers or multiple players connected over the Internet.
- Director's commentary in which the director appears over the video and is able to point out key areas in the video with graphical indicators (like a chalk board).
- Scrapbook feature in which the viewer is able to take "snapshots" of the film by pressing a button while the movie plays, making a custom montage of the film.
- Coloring book application in which the viewer can chose a scene from the movie, and bring up the image for coloring with interactive paint brushes and colors.

The list is a very small set of the types of applications that have been envisioned for the new disc formats, all of which are designed around the feature itself.

Creating the Paradigm

A BD player, although vastly more advanced than a DVD player, is still a constrained device, with specific limits on memory capacity, processing speed, and bandwidth. In part, this is due to the realities of trying to build a reasonably priced player, because a $10,000 player would not sell very well. Further, it is also due to the need for interoperability across the different players that will be manufactured.

To ensure that a piece of authored content operates consistently on all players, minimum requirements have to be put on the players to allow files of a certain size to be handled and to be able to handle a certain amount of graphics in real time. These minimum requirements place constraints on content since a producer will generally want the material to work on all players, and staying within the minimum requirements of the players is the only way that can be guaranteed.

This concept of building applications within the constraints of the platform is often quite different than what is actually done in today's software development community. With the ever-growing and ever-changing PC market, there is no single reference specification and, therefore, no guarantee of cross-compatibility. As a result, application developers must define the platform for which they will build their application. These requirements are often listed on the product packaging so that the consumer can make sure that the software will work in their PC. If their system does not meet the minimum requirements, then they know that the software will probably not work, and they may chose to upgrade their system so it will. This could be thought of as a best effort approach, in which the PC is only expected to make its best effort to run the software. If the PC does not meet the minimum requirements, then its best effort is not likely to be good enough. This allows a user to enhance their PC and to grow their capabilities at their own pace, but it also creates an extremely complex market with frequent incompatibility problems and general market confusion.

However, because people have grown up with this model of computer development, expectations for software seem to be much lower than for other types of products. For example, you would not expect your car to crash (literally or figuratively) by itself now and then; in fact, that would be completely unacceptable — and, yet, one usually expects software to crash from time to time, like it or not.

Let's put it this way, if you knew that your car's steering wheel or brakes would lock up once every few hundred miles and could only be fixed by closing all the windows, shutting off the ignition, and then restarting the car, you would probably sue the manufacturer for creating a faulty automobile. However, we don't see software users up in arms about applications that crash every few weeks and require them to close all the windows and restart the machine.

DVD and BD players fall on the automobile side of the equation — they have to work faithfully. So we return to constraints. The trick for content producers is to think imaginatively about how to get interesting new content to look different from other content on the market, while still ensuring their content will work within the minimum system constraints. It is also important that the device be designed from the start to handle the target application and, in the case of BD players, the seamless user experience.

Designed for Interactivity

The BD format has been designed for the seamless user experience by defining an independent interactive graphics plane that overlays the video. This graphics plane can be controlled by authoring and can be filled with popup graphics of all sorts, including those nec-

Interactivity

essary for overlaying menus with alpha-blending so that portions of the graphics can appear translucent or completely transparent. In addition, application resources such as image data can be cached so that they can be accessed seamlessly during disc playback. (Caching, or preloading of content, is necessary to prevent the disc pickup head from having to seek to another part of the disc to read the data, thus interrupting the main feature playback.) There is also a timing mechanism that allows applications to trigger different functions at specific points during feature playback, such as showing the option for a director's commentary or video game that is synchronized to the movie. By building these features into the format, the designers helped to ensure that the interaction model operates smoothly even on constrained devices.

Additional Features

In addition to the seamless user experience, several other key features have a significant effect on how people interact with the disc content. For example, the BD format includes Internet connectivity as a key feature in the player, as well as local, persistent storage and provides for multiple types of input devices.

Internet Connected Players

Although the player itself is not required to be attached to the Internet, many BD producers are creating exciting new forms of content that should make it worth the effort to establish an Internet connection.

In addition to self-updating content, such as cast and crew biographies or upcoming movies, an Internet connection allows downloadable supplemental content, such as audio and subtitle streams that were not available when the disc was released (see *Network Connection* in *Chapter 3* for more information). This use of networked content may not only make a BD title seem boundless, it also introduces new interaction characteristics that DVD never had to address. The fundamental shift is that the BD-J model is much more like Web design than the DVD model, and it requires a very different skill set.

An Internet connection also makes it possible for users to connect to other users and to other devices, computers and Internet-connected mobile phones. In a sense, every connected player is a simple Web device. With the right kind of BD-J programming on a disc or downloaded from a server, users can send e-mail, participate in multi-player chats or games, or send content to mobile phones.

Key differences between networked content and traditional disc-based material are the access time and the quality of service. DVD has established expectations for disc-based content. For instance, each jump from one point to another, though usually not seamless, completes in a few seconds or less. This, however, is rarely the case for access to network content, as it may take quite some time to identify the server, locate the content, and download enough of it into the player's buffers to start playback. Likewise, there is always the possibility that something in the chain fails — the connection to the player could be faulty, the home router could have problems, the external connection to the Internet may go down, or the server itself may be down or overloaded with incoming requests. As a result, applications that use networked content must be designed to handle such situations.

In general, there can be a delay between action and response of about ten seconds before a person thinks a system is broken.[1] Ten seconds may sound like a short period of time but imagine pressing a doorbell and having to count 1, 2, 3, ... up to 10, before you finally hear the doorbell sound. Ten seconds seems like a long time now, doesn't it? With networked content, it can easily take ten seconds or longer to begin playback, so it is imperative that the system provide feedback that an operation is underway. Likewise, because delays associated with network access tend to be somewhat random in duration, it becomes very important that the feedback provided is informative enough that the user will realize what is happening. For example, take the standard computer approach of having an hourglass cursor indicate that the computer is busy doing something. This does provide feedback to the user, but it is not very informative. After a few seconds, the user starts to wonder if the system has broken down. On the other hand, if a progress bar is used to show how much content has actually downloaded, then the user can clearly tell if the content is coming or if the system hung up at the start, before any of the material had actually downloaded. But, on the third hand, a progress bar that pops up for only a second is distracting and confusing, so the feedback system needs to respond to actual network conditions if at all possible, providing feedback (such as a sound or a graphic change) immediately and providing a dynamically calibrated progress bar or a "please wait" notice only if the delay stretches past a few seconds.

Not only does BD content need to provide clear feedback for network operations, but it also must be built with the expectation that the networked content may not be accessible. Imagine a photo gallery with images that are periodically updated online. If the feature is authored on the disc so that you can go to the photo gallery even when there is no network connection, the experience will be pretty bad as the gallery will present an empty display. One way to address this would be to simply disallow the feature, or to hide it altogether, when the server is unreachable. However, that can lead to confusion when the viewer sees the feature available on the same disc in a friend's player who has a network connection. A third alternative which, perhaps, offers the best experience for the non-networked user, is to have a collection of photos for the gallery already stored on disc. Those with a network connection can download additional photos, while those without a connection may still have an acceptable experience.

A helpful approach is to have network access occur in the background while the movie is playing. Since such transfers require relatively low data rates, processing time, and memory, BD players can generally perform background transfers fairly easily. For instance, one might download the latest movie trailer to local storage while the movie plays so that when the

---

[1]Robert B. Miller, in his seminal 1968 paper, *Response Time in Man-Computer Conversational Transactions*, found that a response time of one tenth of a second seems instantaneous and gives the viewer a sense of direct control; a response time of up to one second is fast enough for viewers to feel they are interacting freely without interruption albeit less in control; and a response time of ten seconds or less is needed to keep the user's attention focused on the task. For more information, including the history of research into computer response times, see *Speed up Your Site*, by Andrew B. King.

Interactivity

movie is finished the download will have completed and the trailer may be played immediately following the film (making it a trailer in the true sense of the word).

Another way in which network-connected players may affect the interaction model is through the use of hooks, which can be authored into a disc prior to their scheduled release such that the disc automatically checks the network server for any new applications or content that it should download. Such hooks can be placed at key locations within the playback sequence, such as when the disc is first started, just before the film plays, just after the film plays, and so on. Likewise, the hooks can be designed so that if the server does not have any content for them, they continue playing normally, but if the server does have available content it will download and launch. In this way, one can "future proof" a disc so that new content and applications can be added after release of the product. For instance, let's say a movie is released on disc with one of these hooks. Two years later a sequel for the movie is made. Using the hook on the disc, a notification or a full-blown movie trailer can appear whenever the disc is played in a networked player, alerting viewers who would most likely want to know about a sequel.

Local Storage

Another key Blu-ray feature is local storage, which can range from built-in 64 MB of flash memory up to a multi-gigabyte hard drive. And, removable flash memory and network storage are options that players can support in addition to or instead of the built-in storage. Local storage allows information and content to be saved between playback sessions. In contrast, with DVD, as soon as you eject a disc from the player, all memory associated with that disc is cleared. As a result, there is no way to distinguish the first time a disc is played from the hundredth time. If you were playing a DVD video game and just reached your highest score, it would be forgotten by the player the moment the disc was removed. However, with Blu-ray's local storage feature, the player will be able to save your score so that it is remembered the next time you insert the disc.

Local storage can have other interesting uses, as well, many of which improve the way discs work. For example, multi-disc box sets are very popular. Whenever it is necessary to eject one disc and load another, the local storage can be used to save information about what to do when the new disc is loaded, and the *unbound* BD-J application can continue to run during the disc change. It is possible for each disc to have a complete menu for the set, so that when something is selected that is on another disc, the menu can ask the viewer to insert the appropriate disc, and when the other disc is inserted it can automatically check local storage to see if it should jump immediately to the chosen point, bypassing logo sequences, warning screens, menus, and such.

Local storage can also be used to store disc status information. A producer creates a disc with many different elements and it may be helpful for the viewer to keep track of which elements have already been viewed, even across multiple sessions. For example, imagine an exercise program in which the disc helps you set up an ideal workout session with appropriate time assigned to different exercises, mapped out over a period of months. The disc can present the appropriate exercises for each session and can even be used to help track the performance (weight, strength, etc.) of multiple users throughout the course of the program.

Local storage is not unlimited, so when it fills up the player may delete the oldest entries first or may allow the user. Therefore, use of local storage must be designed to respond gracefully if no data is present.

Specialized User Input

Perhaps one of the greatest challenges that producers of next-generation content must deal with is the increased complexity of user input. Many applications require keyboard input, such as entering an e-mail address or a name and address for transactions. However, most players do not have a physical keyboard. Therefore, the BD-J application may need to provide a virtual keyboard. Likewise, some applications work much better with a mouse or a game controller, but most players do not come with a mouse or trackball so, again, the standard remote control must be accommodated. Designers must pay careful attention to the lack of *direct manipulation* caused by the simplistic and indirect up/down/left/right/enter controls. It is possible for Blu-ray players to support a wide range of input devices, which is a tremendous benefit that allows for more creative applications, but it complicates interoperability because the application developer usually does not know what controllers each user will have. Another challenge that content producers face is inter-application navigation. Blu-ray makes it possible to have multiple applications active and displayed simultaneously, each with its own user interface. Due to the complexities of navigating among different interactive elements (even within a single application), the specifications do not generally provide for automatic inter-application navigation. As a result, an author must remain constantly aware of what applications might be active at any time and determine how to navigate among them (or always use just one application).

One of the more forward-thinking features is support for a generic device framework. This allows new types of controllers to be created and used after the launch of the format. By supporting a method of acquiring raw data input from a controller, applications can be written to support new and innovative types of input devices such as specialized kiosk controllers, new game pads, biometric devices, or other types of input devices that have yet to be conceived.

Chapter 11
Use in Business and Education

The primary focus for marketing DVD and Blu-ray Disc™ is, of course, home entertainment. However, the effectiveness of video has long been recognized for instruction and learning. DVD-Video has been hugely successful in business and education for advertising, marketing and communications, training, self-instruction, museum installations, video signs, and much, much more, especially because of the extremely low cost of DVD players. Digital video on hard disks or flash memory and Internet-delivered video have begun to make a dent in the roles that DVD carved out, but there remains an important role for Blu-ray, especially with its superior interactivity, high-definition video, and relatively low cost.

The Appeal of Blu-ray

Blu-ray has distinct advantages over other media, such as videotape, print, CD-ROM, DVD, and the Internet. For example, the capability of Blu-ray discs to carry large amounts of high definition, full-screen video makes it more compelling, more effective, and more entertaining. Blu-ray players are actually sophisticated computers that a decade ago would have cost thousands of dollars.[1] Much of what can be done with a computer and Web technology can be done more simply, in a more standardized fashion, more accessibly, and less expensively with Blu-ray. Other benefits include —

- **Low cost**. Production and replication costs of Blu-ray discs are surprisingly low compared to videotape, CD and, even, DVD, especially when cost calculations take the larger capacity into account. Corporate and government databases that would fill dozens or hundreds of DVD-ROMs or CD-ROMs can be put on a few Blu-ray discs, with one BD-50 taking the place of 70 CDs.
- **Simple, inexpensive, reliable distribution**. Five-inch discs are easier and cheaper to mail than tapes or books. Optical discs are not susceptible to damage from magnetic fields, x-rays, or even cosmic rays, which can damage tapes and magnetic discs in transit. A single disc is easier to store than bulky videotapes or audiotapes and multiple CDs or DVDs. Production is quicker, logistics are simpler, and inventory is streamlined. Although flash memory cards are becoming ubiquitous, they are not suited for inexpensive mass distribution. At 2008 prices

---

[1]Amazingly, the Silicon Graphics Indy workstations used in the late 90's to run DVD authoring software such as Scenarist, with CPUs at less than 200 MHz and RAM less than 512MB, cost around thirty thousand U.S. dollars. A typical Blu-ray player, at a fraction of that cost, has a CPU running faster than 300 MHz and a few hundred megabytes of RAM, not to mention multistream hardware decoding for video and audio.

Use in Business and Education

of around $5 per gigabyte, flash memory was over 100 times more expensive than Blu-ray discs. The cost to stream video over the Internet in 2008 ranged from around 10 cents per gigabyte (at high volumes above 500 terabytes per month) to over 50 cents per gigabyte. By comparison, BD-50 replication costs ranged from around 2.5 cents per gigabyte (at volumes over 100,000) to over 10 cents per gigabyte, including AACS license, certificate, and per-disc fees. Add the cost to mail a disc via two-day U.S. Priority Mail and the total delivered cost per gigabyte was 12 to 19 cents. Consider that you need an Internet connection faster than 2.5 Mbps to download one BD-50's worth of data in two days.[2] If you used next-day express delivery to cut the time to 24 hours or less (equivalent to a 5.8 Mbps Internet connection) then the cost per delivered gigabyte would increase to around 30 to 38 cents. Ship ten discs in one envelope (equivalent to a 58 Mbps Internet connection) and the cost would be only 5 to 12 cents per gigabyte. These comparisons vary for different countries, of course, and for retail distribution there are other factors such as distribution and inventory management, but the bottom line is that the cost to deliver a movie over the Internet is not significantly less than — and in some cases it is much greater than — the cost to replicate and ship the same movie on a Blu-ray disc.

- **Broad availability**. Blu-ray players are relatively inexpensive, available from many sources and manufacturers, and have a variety of outputs to hook to almost any high-definition display and associated audio equipment. Because virtually all Blu-ray players can also play DVD-Video discs and audio CDs, it is a simple matter to replace an existing player.

- **High capacity**. A dual-layer BD-50 can hold more than six hours of high-definition video, over 25 hours of standard-definition video, and over 75 hours of video compressed at VHS videotape quality (which is still much better than what we have become accustomed to from YouTube and other Internet sources). This much video would take more than a day to transmit over the Internet with a typical 3 to 5 Mbps cable connection, or more than seven days with a 768 kbps DSL connection. As detailed above in the distribution section, dozens of hours of video and hundreds of gigabytes of data can be sent anywhere in the world by slipping a few discs into an overnight express mailer for less than the actual cost of Internet transmission.

- **Sophisticated interactivity**. Other sections of this book go into detail about what can be done with HDMV and BD-J, so suffice it to say that Blu-ray has a major advantage over DVD and all other mainstream packaged media formats — with

---

[2] Actual Internet data transfer speeds are lower than the "rated" speed that your ISP touts. 50 gigabytes of data can be downloaded in 24 hours at 2.3 Mbps, but transmission efficiency taking into account protocol overhead, retries, and concurrent traffic is typically around 70 to 80%, which means you actually need a 2.9 to 3.3 Mbps connection to achieve 2.3 Mbps download speed. According to speedmatters.org, the median broadband download speed in the U.S in 2008 was, coincidentally, 2.3 Mbps, which is 15th place after countries such as Japan (63 Mbps), South Korea (49 Mbps), France (17 Mbps), and Canada (7.6 Mbps). The Speed Matters report pointed out that the increase from 2007 to 2008 in the U.S. was a mere 0.4 Mbps, at which rate it would take the U.S. until after 2108 to catch up to Japan's speed in 2008.

the obvious exception of computer and video game software — by virtue of a rich and powerful interactive model with a very simple remote-control interface. Just about anything that can be done on the Internet with Java and HTML can be done on a Blu-ray disc — minus the download delays and poor quality video. With BD Live it can be updated and expanded ad infinitum.

- **Self-contained ease of use**. Blu-ray programs can include integrated instruction with on-screen text, popup help, and full interaction. Rather than being tied to a linear presentation, the viewer can select appropriate material, instantly repeat any piece, or jump from section to section. Unlike older commercial media such as videotape, audiotape, and laserdisc, Blu-ray Discs need no ancillary printed material for training or user education — everything explaining how to use the disc can be put on the disc itself.
- **Portability**. A portable Blu-ray player, about the size of a hardback book, can be slipped into a briefcase and hooked up to any high-definition television or video monitor (and some players can downscale for standard televisions). One-on-one or small-group presentations can be done using a portable player with an integrated LCD video screen or a laptop computer with a Blu-ray drive. Large-audience presentations can be done with a video projector and a player or laptop computer. There are even high-definition video projectors with built-in DVD or Blu-ray players. Beyond presentations, portable players and laptops can be used for training and learning in any location.
- **Desktop HD video publishing**. Desktop video editing and recordable DVD did to the video industry what desktop publishing and laser printers did to the print industry. The same thing is happening to the film and HD video industries with HD equipment and Blu-ray technology. Video production can be done from beginning to end with inexpensive consumer or prosumer equipment and a computer. A complete setup is remarkably affordable — a consumer-grade high-definition digital video camera (under $2000) can be plugged into a digital video editing computer with a BD-recordable drive (under $2000), and the final product can be assembled and recorded onto a BD-R disc with BD authoring software (less than $200). This constitutes an entire high-definition video publishing system on a desktop for under $4,000, or under $2000 at the low-end.
- **Mixed media**. As with DVD, Blu-ray is a melting pot for many different information sources. A single disc can contain all the information normally provided by such disparate sources as videotapes, newspapers, computer databases, audiotapes, printed directories, and information kiosks. Training videos can be accompanied with printable manuals, product demonstrations can include spec sheets and order forms, databases can include Internet links for updated information, product catalogs can include video demonstrations, and so on. Blu-ray is the perfect medium for this because it provides high-definition video (unlike DVD or the Internet); searchable, dynamic text (unlike paper); hours and hours of random-access audio (unlike tapes); and, a rapidly growing base of devices to read it all.

The Appeal of Blu-ray for Video

The conveniences of Blu-ray video, which in the home are enjoyable but not essential, are translated in the office into efficiency and effectiveness. Blu-ray also brings computers into the picture. Unlike in the home, where some may question the value of being able to play a movie on a computer, the usefulness of computers in the office doing double duty as video players is clearly apparent.

The natural inclination when developing material for computers is to take advantage of the additional features, such as keyboard entry, graphic interactivity, and so on, but in many cases this is counterproductive and inefficient. Simpler may be better.

There are certain advantages to Blu-ray based multimedia compared to traditional computer multimedia, especially the more straightforward HDMV format —

- Easier development at a lower cost. For simple titles there are fewer programming and design requirements. Creating a set of menu screens and related video can be done with low-end, low-cost Blu-ray authoring packages.
- Easier for the customer. The limited interface is simple to learn and is usually accessible using a remote control. Hooking a player to a TV is much simpler than getting a multimedia computer to work.
- Familiar interface. Menus and remote controls are similar.
- Larger audience and no cross-platform complications. Blu-ray-equipped computers, Blu-ray players, and even video game consoles can all play standard Blu-ray Discs.

Certain kinds of programs lend themselves better to Blu-ray video, including programs with large amounts of video, programs intended for users who may not be comfortable with computers, programs with still pictures accompanied by extensive audio, and so on. Here are a few examples of material well-suited for Blu-ray video —

- Employee orientation and sensitivity training
- Press kits, corporate reports, or newsletters
- Emergency response training or information systems
- Product demonstrations and catalogs
- Information kiosks - product/service searches, traveler's aid, and way-finding
- Product training
- Sales and marketing tools
- Educational learning kits
- Video tours or video brochures
- Collections of conferences and lectures for churches and associations
- Video "billboards"
- Video greeting/holiday "cards"
- Testing, including licenses and professional certification
- Video portfolios (ads, promo spots, and demo reels)
- Trade show demo discs

Use in Business and Education

- Point-of-sale displays
- Ambient video and music
- Video "business cards"
- Video "yearbooks"
- Lecture support resources
- Repair and maintenance manuals
- Medical informed consent information
- Video tributes for anniversaries, awards, and funerals
- Patient information systems and home health special needs instructions
- Language translation assistance
- Home videos and photos

Blu-ray video, of course, has its disadvantages when compared with other media. Following are a few examples of material where Blu-ray may not be well suited, especially related to computer applications. (Of course, a Blu-ray Disc can contain computer software in addition to player-based video.)

- **Interface**. Other devices, such as computers, allow keyboard entry and point-and-click graphic interface. Some Blu-ray players can be equipped with mice, keyboards, and even game-console-style hardware such as infrared batons, balance boards, and steering wheels, but these are not typically available on standard players.
- **Text entry**. An on-screen keyboard can be used with the remote control for entering text, but it is cumbersome compared to a computer keyboard.
- **Documentation and databases**. Text may be difficult to read from normal TV viewing distance. Searching requires unwieldy text entry using the remote control.
- **Productivity applications**. Word processing, checkbook balancing, e-mail, et cetera, can actually be implemented in rudimentary fashion using BD-J, but are generally impractical compared to computer implementations.
- **Memory**. Blu-ray players have persistent data storage, but it is limited in size and may be deleted by the user.

The Appeal of Blu-ray for Data

Because a Blu-ray Disc can contain any type of computer data and software, the content possibilities are practically endless. Video and audio, of course, are simply another form of digital data but, in general, when people talk about data on optical discs they are referring to computer data such as software applications, text files, and so on. Often the term *BD-ROM* is used to refer to data-only discs, even when they are recordable, which is a misnomer.

Although CD-ROMs and recordable CDs are still widely used for PC applications and data storage, DVD-ROM and recordable DVDs have become more popular. DVD drives have become even more popular, as they can read and write CDs. Blu-ray will displace DVD

Use in Business and Education

for applications requiring large amounts of data storage and, eventually, when Blu-ray drives approach the price of DVD drives, they will become the dominant hardware format. That said, recordable DVD discs will continue to be widely used because of lower cost and compatibility with older systems.

Sales and Marketing

Blu-ray is an excellent sales and presentation tool. A complete presentation system can be contained in a portable player. DVD proved popular for this, and Blu-ray has the advantage of much more detail and readability on high-definition displays. Home sales presentations, for example, can be enhanced greatly by professional video supplements provided on disc. The presenter simply plugs a portable player into the customer's TV or puts a disc in the customer's player. Unlike videotape, which must be watched in a linear manner, a DVD or Blu-ray Disc can contain different segments for different scenarios, with answers to common questions, and so on, all quickly accessible from menus. The need to train sales and marketing representatives is reduced by having them rely on prepared presentations to be called up, as needed.

A point-of-information kiosk or a trade show video presentation is vastly improved with Blu-ray. The disc can be set to loop forever — customers will not wander off as they did when a black screen appeared after the tape ran out or while it rewound. Full interactivity, either directly for the person using the kiosk or for a representative working an exhibit, can be provided on the disc.

Product catalogs with thousands of photographs and video vignettes can be put on a single disc for a fraction of the cost of printed catalogs. Of course, the video catalog cannot be read at the kitchen table — at least not until thin "videopad" players become available. Environmental resource waste from printed catalogs is becoming a big concern. Environmentally conscious companies can replace tons of paper with polycarbonate discs. By producing discs that connect back to the company Web site via the Internet, the life of the discs can be extended with updated prices, product information, new promotions, and other supplements.

A Blu-ray Disc can contain literally hundreds of hours of audio, any part of which can be accessed in seconds, making it the perfect vehicle for instructional audio programs.

Communications

Companies spend billions of dollars a year producing printed information, much of which requires unwieldy indexes and other reference material merely to make it accessible. In addition, much of it becomes out of date in a very short time. Corporate and product information is moving more and more to the Internet but, for large publications or those that benefit from a graphic or video element, Blu-ray provides a cost-effective means of distribution coupled with improved access to the content.

Blu-ray is also a high-impact business-to-business communications tool. It can be used as a standalone "no instructions needed" communication device or it can back up an in-person

Use in Business and Education

presentation. The message can include high-quality corporate videos, advertising clips, interviews with staff, video introductions, dynamic video press releases, visual instruction manuals, and documentaries of corporate events such as new office openings and seminars.

Companies can send free discs in the mail to targeted audiences. Unlike VHS tapes, which people often ignore because they don't want to have to sit through the whole thing, video on disc can be broken into small, easily digested segments that the viewer can select from on-screen menus. This revolutionizes the way businesses can communicate with their customers and with other businesses. The discs can be played in both settop players and computers, so traveling executives might be tempted to pop a free disc into their Blu-ray-equipped laptop computer while flying or when sitting in a hotel room, for instance.

Highly effective video-supported presentation demands on-the-fly, instant, context-sensitive access to any point in the footage. Linear presentations are woefully inadequate for this task. Even most computer-delivered video is, by comparison, stuck in a previous decade with limited random access, a single audio track, and no subtitle tracks. You can pop a video onto *YouTube*, but the clunky accessibility makes it useless for much more than casual viewing. By supporting multiple language tracks a single Blu-ray Disc eradicates geographic and cultural barriers. Whether distributed to individuals for playback on PCs, shown to groups via set-top players or portable PCs connected to video projectors, or mounted in network servers for remote viewing, Blu-ray helps get the message to every recipient. Obviously, DVD can do much of this, as well, albeit with a less sophisticated interface, but in environments where visual clarity is important, such as, in large-screen presentations, Blu-ray has the advantage.

Training and Business Education

Now that sufficient numbers of DVD players and DVD-equipped computers are established in businesses, there is no question that DVD has become a leading delivery format for business training. Blu-ray also has a role to play wherever there is a need for longer playing times, higher video quality, and more advanced interaction combined with Internet connectivity. The Internet alone has certain advantages such as low cost and timeliness, but the demand for high-quality multimedia far exceeds the capabilities of the typical Internet connection for the near future. Blu-ray and DVD are better suited to deliver quick access to large amounts of high-quality video and audio, and can be integrated easily with the Internet to provide the best of both worlds.

The Blu-ray format, supported by Blu-ray players and by Blu-ray navigation software in computers, is strictly defined in hundreds of pages of technical specifications. Blu-ray content development requires specialized authoring systems. As with all content authoring systems, the ease of development is inversely proportional to the flexibility of the tools (Figure 11.1).

Figure 11.1 The Authoring Environment Spectrum: Utility vs. Ease of Use

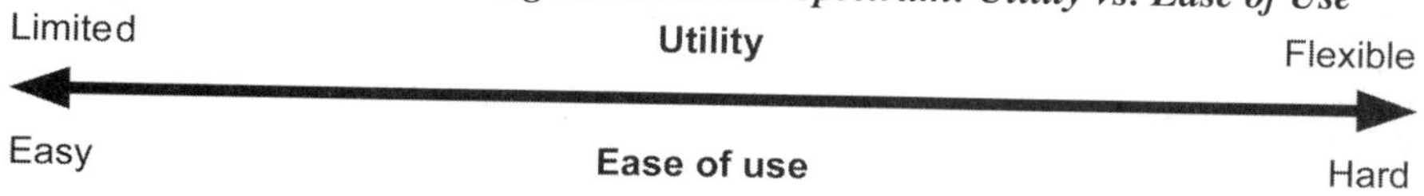

Limited   Utility   Flexible

Easy   **Ease of use**   Hard

Use in Business and Education

As the parameters of the system are constrained, the complexity of the task is reduced. Since HDMV is quite constrained, it is relatively easy to develop for it, given the proper tools. BD-J provides more flexibility and power, but at a cost of a steeper learning curve and more complexity. Anyone considering video training programs should decide if the features of BDMV or BD-J meet their requirements. A simple product containing menus, pictures, and video may be developed using Blu-ray in less time and for less money than with a complex computer authoring system.

DVD-Video worked so well for certain corporate education and training applications that the content sold the hardware. With player prices well under $50, companies spending thousands or millions of dollars on video-based education and training programs do not need to think twice about equipping their employees or learning laboratories with players. Many companies that used expensive or specialized systems, such as study kiosks or computers, have switched to DVD players. As Blu-ray prices drop, it likewise becomes an amazingly inexpensive and standardized device for video-based instruction.

Blu-ray for computer data, on the other hand, covers the entire spectrum from custom-programmed software to fill-in-the blank lesson templates. Practically any authoring or software development system can produce material to be delivered on BD-ROM or BD-recordable. The main advantages of Blu-ray over other media are space, data transfer speed, and cost per byte. Flash memory prices continue to drop as capacities increase, but BD-ROM is literally orders of magnitude cheaper — cents per gigabyte rather than cents per megabyte. For distributing hundreds or thousands of copies, flash memory is simply not cost effective compared to Blu-ray. Byte-per-byte, Blu-ray is even cheaper than DVD. Blu-ray drive transfer rates are similar to hard disks, and even though the access rates are slower, in many cases Blu-ray's higher capacity and lower cost more than compensate.

Industrial Applications

Both Blu-ray and DVD are being used increasingly in specialized applications. Custom material produced for recordable discs can be unique and focused. Installations can use inexpensive, off-the-shelf players. More demanding installations may use professional-grade players, which are more reliable and can be connected to specialized hardware, such as multiplayer controllers, video synchronizers, video walls, touch screens, custom input devices, lighting controllers, robotic controllers and more. Cheap kiosks can be made from an inexpensive player and an input device such as a trackball or touch screen.

A few examples of industrial applications include —
- Kiosks and public learning stations
- Point-of-purchase displays
- Museums
- Video walls and public exhibits
- Vocational skills training
- Government kiosks for voting, licenses, and citizen information
- Corporate presentations and communications

Use in Business and Education

- House video in a store, bar, or dance club
- Theme park and amusement park exhibits
- Patient education programs in waiting rooms at hospitals health practitioner offices
- Closed-circuit television
- Video simulation and video-based training
- Tourism video on buses, trains, and boats
- Hotel video channels

Classroom Education

Laserdiscs were a success in education almost from day one. Teachers saw the advantage of rapid access to thousands of pictures and high-impact motion video sequences. They began investing in laserdisc players and discs after seeing the effectiveness of laserdisc-based instruction in the classroom. In 1998, 20 years after their debut, more than 250,000 laserdisc players were still found in schools in the United States. Computer multimedia has largely replaced laserdisc in the classroom, but the ease of popping in a laserdisc and pressing play, or scanning a few barcodes, may never be matched by computers with their complicated cables and software setups and daunting troubleshooting requirements.

Educational publishers were slow to embrace DVD-Video. They did not see a large market, and many of them were hurt by the mass flocking of teachers to the Internet as the new source of free educational technology. Educational video titles require significant work to develop and must be designed to meet curriculum standards. Purpose-built educational titles will be even slower to appear for Blu-ray than for DVD. However, DVD and Blu-ray titles from many other sources help fill the void— documentaries, historical dramas, newsreel archives, educational television programming, and even popular movies, TV shows, and "edutainment" programs work well for classroom as well as home education. The particular advantage of DVD and Blu-ray for this application is random access. Teachers can quickly jump to any desired video segment and can skip sections during playback. If a computer is used to play the disc, it can be programmed to show particular sections in a specific order.

As Blu-ray eventually becomes a mainstream vehicle for delivering video, a larger base of educational content will build up. It will be able to take advantage of Blu-ray features such as random access to hundreds of video segments, subpicture overlays to enhance video presentation, on-screen quizzes, games, multiple-scenario presentations, adaptation to learner needs, multiple languages, tailored commentary tracks, and much more.

Chapter 12
Production Essentials

This chapter covers the basics of producing BD-ROM video and computer-based titles. It is not a detailed production guide, but rather an overview of the production processes which can serve as a guide and a checklist, as well as an introduction for those who want to understand what is involved. Aspects of BD authoring that are confusing, misunderstood, or overlooked are also covered.

One of the major changes that has happened since the introduction of DVD is that videotape is no longer the predominant format for capturing and working with audio and video. Production workflows are now file-based, not tape-based. Often, content will likely never exist on tape (or film) — from the initial production shoot to the final delivery on Blu-ray Disc™, content will exist as files. Thus, interchange formats are important to be aware of throughout the production process. You will want to insure that the next step in the process can accept the output of the previous step.

There are three modes of programming titles — BDAV, HDMV and BD-J. Both HDMV and BD-J disc elements reside in a directory named BDMV. As a result, there is frequent confusion as to proper terms, with BDMV being used to mean both HDMV and BD-J. Extreme care should be exercised when using these terms, and the term BDMV should not be used for defining a programming mode.

BDAV is used for simple playback, with chapter marks automatically generating thumbnail-based menus. For interactivity that takes advantage of all of the capabilities of BD, programming will be in either HDMV or BD-J mode. The choice of programming mode determines the process steps needed to create a Blu-ray Disc master. Note that HDMV or BD-J mode programming is required for replication.

At this point in time, replication is generally reserved for large volume, high budget titles. If you want to replicate discs you must include the AACS copy protection scheme. The fees to use this required scheme are considerable — as of 2008, the fess consist of an annual content provider fee, a per disc layer fee, and a per replicated disc fee. For small replication runs, for example less than 2,000 discs, the fees will be more than 50% of your total replication costs. Also note, if you have to update the disc, for example for a fix, you have to pay the per layer fee again. Check the AACS site for the latest fees and required agreements at `http://www.aacsla.org`.

The quality of HD video is very obvious when it is shown on a big screen, high-definition monitor. This means that video production must consider everything, from using tripods for image stabilization to giving added attention to unwanted details, from facial warts and nose hairs to stray cables, clutter and dirt. "Realistic" takes on a new meaning with HD. Production capture options include film, HDV, HD SDI, HD-AVC, or HD-MPEG.

Production Essentials

Blu-ray Disc Project Examples

BD projects can take a variety of forms. The list in Table 12.1 is not by any means exhaustive, but it does illustrate the ways that BD can go far beyond a presentation of linear video transferred from videotape.

Table 12.1 BD Project Examples

Project	Example
Video and audio without menus, graphics, or options	Archived video
Video and audio with a single menu to choose selections	Home videos
Video and audio with a main menu and submenus with supplements	Simple movie, educational disc
Audio with a few menus and stills	Music albums (DVD-Video or DVD-Audio)
Multichannel audio and music videos with menus and supplemental video	Music video albums (DVD-Video or DVD-Audio)
Main menu and submenus for video format, audio language, subtitles, angles, chapters, and so on	Special edition movie
Any of the above with computer content or Web connectivity on hybrid disc	BD-Live enabled discs or web connected PC applications
BD-ROM with data or applications	Auto parts database, computer game, and application suite
Multi-disc video library	Episodic television series compiliation
Quizzes, tests	Ability to send scoring data to a grading center

General BD Production

Creating a Blu-ray Disc is a complex process. Blu-ray production demands special knowledge, skills and training in a number of disciplines including HD video production, interactive design, animation, programming and quality control. BD production is surprisingly more difficult than DVD-ROM or CD-ROM production. BD builds upon the earlier disc formats and has more capabilities and more flexibility, and thus more potential for things that can go wrong. Just as in the early days of DVD, tools are still being defined and refined. What is considered "best practice" is rapidly evolving. The workflow and the tools will improve and the process will become easier, but in the meantime do not assume that making a BD is the same as making a DVD.

Simple AV Playback Mode

Many cameras record in AVCHD or BDAV format. Also, many computer-based video editing programs support output to DVD-R or BD-R using either of these formats. This mode is relatively easy to create and gives you full HD video but with limited audio and video codec support and limited interactivity. But, it is high definition, and it can be played back on most BD players. For AVCHD mode, simple menus are possible with some implementations. For BDAV mode, you can get thumbnail menus of each clip. This mode is not bad, but it is not all of what is promised for Blu-ray. For high definition video use at home or simple playback applications, it is a great start.

The rest of this chapter will focus on the process required to create the more complex interactive titles.

HDAV and BD-J Modes

Creating titles in HDMV or BD-J mode requires more extensive tools and processes than for the simple BDAV playback mode.

HDMV and BD-J require interactive design, careful planning and execution of a number of skilled operations including compression, multiplexing, graphics creation, authoring/programming and, finally, testing to realize the finished title.

The HDMV and BD-J modes of interactive programming utilize the compressed and multiplexed audio, video and subtitle streams. These streams are declared as clips which, in turn, are used to create playlists. The playback of a title, whether in HDMV or BD-J mode, calls on these playlists. The interactive components of the user interface are composed of many graphical elements. The programming and the design steps must work closely together to implement the desired experience of a highly interactive, entertaining user interface. HDMV mode programming is more closely related to DVD authoring in that you are assembling and connecting the graphics elements to define the navigational experience via buttons, interactions, animations and button sounds. Even with HDMV mode, though, there is a higher level of programming available than with DVD. For example, instead of the 16 registers (called GRPMs) that are available for DVD, you now have access to 2,048 registers (now called GPRs) with BD.

BD-J Mode programming supports all of the capabilities of the Blu-ray specification including the *Bonus View* (Profile 1.1) and *BD-Live* (Profile 2.0) features. Creating a disc using BD-J code is on a difficulty level that is on a par with creating a video game for a game console or writing software applications. Discs created using BD-J can also be complemented with the use of HDMV code. For example, the opening language selection logic, copyright notices, studio logo splashes, trailers, et cetera, could use HDMV mode to take advantage of the fact that HDMV mode starts playing faster than BD-J because the player does not have to first load the BD-J code and graphics.

The final step of creating a BD is combining the multiplexed streams and the interactive programming into a master image for replication. Creating the BD cutting master format (BD CMF) image requires specialized software systems. At the introduction of BD, two authoring systems were commercially available, Sonic Solutions' *Scenarist BD*, and Sony's *Blu-print*. Both of these systems require a major commitment (fast, powerful workstations with licensing agreements that cost in a range in the tens of thousands of dollars plus annual support contracts as well as a steep learning curve). These systems support the entire

Production Essentials

process of HDMV mode authoring as well as multiplexing and premastering (output of the BD CMF image for replication). Other lower-cost and basic-capability systems are also available from Sonic, Adobe, NetBlender and others that allow multiplexing, HDMV mode authoring and the creation of the BD CMF images. Many of these systems also allow the integration of BD-J mode programming. The differences between the high-end systems and the lower-cost systems is that the high-end systems support all of the features of the specification, such as secondary audio and video, full animated HDMV menus, slideshows, seamless branching, et cetera. Check out the lower-cost systems. They may meet your needs, but for support of all of the BD specification features for advanced titles, the high-end systems will be required.

For compression, many of the companies who had tools for DVD either extended those tools or created new tools for the new codecs supported by BD. Most of the encoding solutions are software-based with a substantial workstation or multi-CPU cluster server platform required. High- end compression solutions include Sonic's *Cinevision* (MPEG2, VC1 and AVC), *CinemaCraft* (AVC), Thomson's *NexCode* (AVC), Inlet Technologies' *Fathom* (MPEG2, VC1 and AVC) and Microsoft's *VC1* encoder (distributed and supported by Sonic Solutions). Many other encoding solutions also support high definition BD-compliant output for MPEG2, VC-1 and AVC (MPEG4, part 10). Look for products from Apple, Main Concept, Canopus Procoder, Sorenson, Avid, Adobe and TMPGEnc for other compression solutions. The quality of the results, control of the encode, the speed of the encode, and the workflow, are all factors that distinguish one solution from another and determine the applicability of any solution for your production. Commercial high-end compression systems range from $50,000 to $250,000. Just as with the early days of DVD, expect that new solutions will be introduced by many of the DVD mainstay companies as well as many newcomers. After a few years, expect the tools to range from barebones, consumer-oriented authoring applications priced at less than $75 to full-featured compression, authoring and multiplexing packages that cost thousands of dollars but do far more than the initial, first-generation systems that cost tens of thousands of dollars.

Producing a Blu-ray Disc is significantly different from producing older formats, such as videotape, laserdisc, audio CD, or CD-ROM. It has many similarities to producing a DVD-Video disc or producing a video game. There are basically six stages —

- interaction design and graphics creation
- encoding/compression
- authoring and/or programming
- multiplexing and premastering
- testing, and
- replication (mastering)

Blu-ray production costs are currently much higher than the costs for DVD. The list of extra features for Blu-ray, such as multiple sound tracks, camera angles, synchronized lyrics, subtitles, secondary audio and video, network connectivity capabilities, interactive programming and so on, can be long and the costs expensive. On a per-byte basis, Blu-ray costs may be about the same as DVD, but Blu-ray can hold 5-10 times as many bytes. The actual pro-

Production Essentials

ject costs have tracked this per-byte metric for production costs through replication on the early titles. Over time, just as with DVD and, before that, CD, Blu-ray production costs will come down as tools are developed and refined.

Authoring/programming costs are proportionately the most expensive part of BD production. Video, audio, and subtitles must be encoded, menus have to be laid out and integrated with audio and video, and control information has to be created. This all has to be multiplexed into a single data stream and finally laid down in a low-level format.

In comparison, videotapes did not have significant authoring or formatting costs, aside from video editing, but they were more expensive to replicate. The same was true for laserdiscs, which also included premastering and mastering costs. CDs require formatting and mastering, but cost less than DVDs.

For the replication factories, BD production requires a major new investment in laser beam recorders, stampers, and replication lines that support the new 405nm laser wavelength of Blu-ray as opposed to the 605nm for DVD. Dual-layer BDs cost slightly more to replicate because they are more difficult to produce. Eventually, mastering and replication costs will likely drop to levels slightly higher than that of DVD. Don't forget that the total cost for a BD or DVD includes all of the packaging, and these packaging costs can easily exceed the actual disc replication costs.

BD-5 or BD-9 or BD-25 or BD-50?

The type of disc to be used is generally mandated by the size of the content. BD-5 and BD-9 are the cousins to DVD-5 and DVD-9 and are single and dual layer discs, respectively, that have been premastered for the Blu-ray file structure. Using BD-5 or BD-9 means that you lose the copy protection scheme of *ROM mark* which is required for BD-25 and BD-50. Smaller projects (25 to 45 minutes of video) may be very appropriate for BD-5 and BD-9 discs and can be duplicated using recordable DVDs.

BD-25 is a single layer disc and BD-50 is a dual layer disc. The added costs for BD-50 replication, which run about 25-50 percent more, mean that you want to pay attention to your bit budget. Exceeding a single layer disc's bit budget by even a few bits means a big difference in the replication costs if you have to move up to a BD-50 disc.

Recordable discs, either DVD+/-R/RW or BD-R/RE, can be used for distribution of the final project as well as for prototypes and testing during development. One advantage of using recordable media is that AACS is not required. Beware that BD-R/RE is still very expensive compared to DVD recordable media. Also, the record times can be as long as 3 hours for a full BD-50 disc.

Tasks and Skills

Many skills and processes are needed for BD production. The larger the facility, the more specialized each position is. No facility is likely to have all facets down to perfection, but many have multiple areas of expertise or relationships with other service providers. Smaller BD production shops may rely on outside facilities to do much of the technical production work.

Production Essentials

The following list covers the essential job positions of BD production. These could be combined into a few positions or spread over a number of people. Rare is the person who excels at all tasks of BD production, although it is a good idea to cross-train personnel so they can assist in other areas, as needed.

- A *project designer* works with clients and is responsible for layout, design, and general budget, and therefore must understand the steps of the production process.

- A *project manager* deals with minimizing bottlenecks and keeping the project moving forward. This person should be familiar with all of the production tasks, works with clients for testing and signoff and is responsible for schedule, coordination, budget details and asset management.

- A *colorist* transfers film to video, is responsible for color correction and proper framing, and often works directly with the director or the director of photography.

- A *video editor* is responsible for assembling, editing, and preparing all the video assets for encoding; uses video editing systems, usually digital video editing software; may need to restore and clean up footage; may do compositing of video and computer graphics; and should be familiar with video encoding and BD authoring. Also, this person must work with the potential for varying video resolutions and framerates of the source content.

- A *video compressionist* must understand the codec(s) and encoders — MPEG-2, VC-1 or AVC compression. Each of these codecs and the compression hardware or software have their own idiosyncracies that must be well understood to get great results. Often DVNR or preprocessing is required to get the quality of encode at the desired/required bitrate; should be familiar with colorist and video editor jobs; have working knowledge of Blu-ray multiplexing and authoring process; and must be familiar with professional video equipment.

- An *audio compressionist* must understand multichannel audio and principles of perceptual audio encoding; must have in-depth knowledge of details of DTS and/or Dolby Digital encoding; must be familiar with professional digital and (possibly) analog audio recording equipment; should be familiar with audio editor's job; and have working knowledge of Blu-ray multiplexing and authoring process to ensure that they can deliver specification-compliant streams that work within the constraints of the disc size and bitrate budgets.

- A *graphic artist* is responsible for graphic design, including menus and stills; may be responsible for user interface design; may also work on disc label and package design; uses graphics packages such as *Adobe Photoshop*; and must understand nuances of integrating various resolutions of HD and SD video. May need to be familiar with BD authoring system(s) to ensure properly prepared assets with layer and file names conforming to delivery specification.

- A *2D or 3D animator* produces animation sequences for menus and transitions; uses graphics packages, such as *Adobe Creative Suite* including products, such as *After Effects* and *Flash*; must understand nuances of video, audio and the capabilities and limitations of the Blu-ray playback devices in rendering animations.

Production Essentials

- A *multiplexor* uses specialized Blu-ray applications to create the streams, playlists, clips that define the structure of the audio, video and subtitles of a disc; must have detailed knowledge of audio, video, subtitles and their Blu-ray specifications. This person must also be knowledgeable in navigation design. Often, this person is also responsible for running and interpreting the results of the verification software that ensures a disc is specification compliant.

- An *author* uses specialized Blu-ray authoring applications; must have detailed knowledge of Blu-ray specification and navigation design. Advanced authoring requires programming knowledge.

- A *Java programmer* knowledgeable on the BD-J specification and the underlying specifications is essential to creating BD-J based titles unless a packaged commercial or proprietary framework is being used. The development of BD-J-based features resembles the workflows, timeframes and costs associated with game development.

- A *tester* is responsible for quality control (QC) of the product; reviews encoded audio, video, and subtitles; tests navigation; and should be familiar with digital video and audio. For larger projects, this person will be experienced in the science of quality testing and be able to develop test strategies and plans, include unit and regression testing.

The BD Production Process

The BD video and audio production process can be divided into basic steps, which are shown in Table 12.2. Unlike DVD production, more of the work can — and in some cases must — be done in parallel. Often the encoding and the multiplexing are happening at the same time as the authoring and programming steps. The QC and testing steps are also spread throughout most of a project. Proper project design and careful asset management are keys to the successful creation of a BD title. Project management in larger, multi-person projects becomes a must. Many of the steps require multiple iterations to ensure that everything will work together. For example, after encoding the audio and video, it may be discovered that the peak bitrates are too high and that either the audio or the video or both needs to be re-encoded.

Table 12.2 The BD Production Process

Task	Description
Project planning	Develop schedule and milestones, storyboard video, lay out disc, design navigation, prototype, budget bits.
Asset preparation	Collect, create, capture, process, edit, and encode video, audio, subtitles, graphics (menus, stills), and data. Check source assets, digitizing, and encoding.
Multiplex	Define title structure and multiplex audio, video, and subtitles to ensure bit budget and bandwidth constraints met. Ensure assets match.
Programming	Program navigation logic and widgets

continues

Production Essentials

Table 12.2 The BD Production Process (continued)

Task	Description
Authoring	Import, synchronize, and link together assets; create additional content, define and describe content; check compliance with BD specification.
Testing	Viewing QC, navigation testing, disc emulation, player compatibility testing and specification compliance testing.
Premastering	Final multiplex, integrate programming and authoring, and create volume image. Check BD spefication compliance, run check discs, and compare checksums. Create master BD CMF Image or BD-R.
Replication	Add AACS to master image, record glass master, mold discs, bond, print, and QC final copies for physical specification compliance and playability on test players.
Packaging and distribution	Insert discs and printed material, insert source tags, shrink wrap, box, and ship.

Authoring refers to the process of designing and creating the interactivity of a BD title. The more complex and interactive the disc, the more authoring is required. If a title is created using BD-J mode, extensive Java programming may be required. Creating an interactive title is similar to other creative processes such as making a movie, writing a book, creating a software game or building an electronic circuit. It may include creating an outline, designing a flowchart, writing a script, sketching storyboards, filming video, recording and mixing audio, taking photographs, creating graphics and animations, designing a user interface, laying out menus, defining and linking menu buttons, writing captions, creating button highlight graphics, programming animation and interactivity, encoding audio and video, and then assembling, organizing, synchronizing, and testing the material. For HDMV mode titles, authoring is done on an authoring system, which also includes the multiplexing and premastering functions.

At the professional level of tools, encoding of video and audio elementary streams is handled by separate specialized encoders that are separate from the authoring system. Some of the prosumer- or consumer-level systems will include the encoding function in an integrated workflow. Using externally encoded streams usually means a higher level of quality.

BD premastering refers to the final multiplexing of the streams, formatting a *UDF 2.5* image, padding for AACS data and creating the CMF (*Cutting Master Format*) image. The CMF image, which is actually a collection of files, is sent to the replication factory via a very fast network connection or via shipping as a hard disk drive.

Production Decisions

During the production process, many decisions between different options need to be made. The choices that are made depend on the particular project's technical and creative requirements, the budget, the timeframe, and, occasionally, personal preference.

Service Bureau or Home Brew?

You must decide whether to do some, most or all of the work yourself, or to outsource the project to a BD production house or to partners who can take on some of the specialized

Production Essentials

work needed to create a BD. Even the larger facilities have a network of partners to do some of the specialized work whether it be interactive menu design and graphics creation or BD-J programming for custom widgets or special encoding with a particular codec. All but the very largest facilities use a partner for their subtitle needs.

The advantages of producing a title yourself are that you have full control of the project and can make changes, as needed. If you intend to produce a number of BDs, you can build a base for long-term production. Service bureaus have the advantage of knowledge, experience, process and the hardware and software tools needed to produce a BD. They have trained personnel and can usually produce a disc quickly, if necessary. They have tools for testing and should have a collection of players and BD-drive-equipped PCs for compatibility testing and quality control. The specification verifier software, for example, is very expensive, has sometimes difficult to interpret test results, and only needs to be run before replication. This is a good case for having a third-party partner perform this step.

Service facilities also usually have connections with replicators. They can also guide you through the AACS licensing process. To find an appropriate service facility for your BD project, check with other media producers for referrals, comments and recommendations, or look in trade magazines and features for knowledgeable BD personnel and service bureaus. The better companies are well known and are often featured in articles. Advertisements display the companies with marketing budgets but often the best service providers remain so busy that they don't need to advertise for more business.

With a list of several companies in hand, do a comparative analysis between them. Request company information, including background, personnel, equipment, BD experience (whether with HDMV or BD-J mode authoring/programming), video and audio encoders, production capacity, and other available services that can help you with your project. Ask for a demonstration or sample discs. Once you choose a facility, develop a good relationship and communicate regularly during projects.

There are many more variables available for the VC-1 and AVC codecs than with MPEG-2. Depending on the encoder, you may have control of bitrate (CBR, VBR, or constant quality), GOP structure as well as motion estimation. Variable bitrate (VBR) compression always enables better quality, especially for video with varying scene complexity. VBR allows using more bits when needed. Most encoders will only encode with VBR. Since source material is now often in the form of digital files and not videotape, one- or two-pass VBR does not take that much longer than a CBR encode. Good VBR encoding is somewhat of a black art and requires a compressionist experienced with the encoder being used. Constant bitrate (CBR) compression is cheaper and often faster because it does not require multiple passes, although many encoders do not offer CBR due to the complexities of the VC-1 and AVC codecs. If the encoder offers CBR encoding, for projects without a lot of video and with enough bandwidth for the high data rate, CBR can achieve the same quality as VBR. Beware, though, that when using lossless compressed audio, you will not have the bandwidth for high video bitrates, so you will have to drop the average bitrate to allow the combined audio and video rates to be within the maximum allowable bitrate. A high quality encoder with a highly skilled compressionist will be needed to get streams that can be multiplexed and still be within BDA specifications.

Production Essentials

One point to remember is that video source files formatted in 12- to 16-bit 4:2:0 colorspace consume about one terabyte of disk space for 60 minutes of material. Compressing a file this size takes a lot of time and finesse. Just encoding without reviewing and then re-encoding segments can take five to twenty times realtime (1 minute could take 20 minutes to encode). With viewings and re-encodes, the elapsed time for encoding can easily take a week or more for a high end Hollywood title. As encoders are further optimized and workstation hardware gets faster and cheaper, the time to get a great-looking encode will become shorter and the cost will become less.

720 or 1080?

High definition video comes in a number of frame sizes, including 1920×1080, 1440×1080 or 1280×720. Many broadcasters are using 720p, asserting that it looks better than 1080i, which it may. But, in our opinion, true HD is defined as 1080p, meaning 1920×1080 pixels, progressive usually at 24 fps or 23.976 fps. Review all of your content and ensure that it is of the highest resolution and, hopefully, all the same resolution. Mixing frame sizes and frame rates, although possible using separate playlists (you cannot mix resolutions within a playlist), will result in some very unpleasant user experiences. When the player is set to pass through the video (with no upscaling), the BD player and the monitor will re-sync every time the resolution changes. The re-sync can take anywhere from half a second to more than five seconds.

50 Hz or 60 Hz?

The BD specification allows for 25 and 50 Hz frame rates for players sold in countries with 50 Hz power. If you develop content for 60 Hz (e.g., 25 or 50 fps) then the material will most likely not play back in players that are sold in 60 Hz power system countries. 60 Hz content (29.97, 30, or 59.94 fps) will play back on all displays. So to make things simple, use either 24 fps (film frame rate) or 30/60 fps when creating or preparing content.

PCM or Dolby Digital or DTS?

With Blu-ray, DTS is the preferred audio codec due to its optimal balance of file size and sound quality. If you have the bandwidth or disc space, multi-channel PCM is feasible and appropriate for titles that will be appreciated by audiophiles. The Dolby Digital audio format is still quite a good format as well and streams that were prepared for DVD can be used with BD. Every BD player, including computer based software players, can play PCM, DTS, and Dolby Digital audio tracks. The lossless audio codecs and PCM do not leave much room for video, so remember to do both disc space budgeting (bit budgeting) as well as bandwidth budgeting. Special encoding software may be required for the advanced forms of the DTS and Dolby codecs. PCM is pretty straightforward and is a widely-compatible format that can be output from most audio editing tools.

Multiregion and Multilanguage Issues

Unlike with DVD, where you had a region declared for the entire disc, with BD, you can program the disc to check the region on a title by title basis (e.g., each playlist can have different region settings). Also, note that the regions for BD are different than for DVD. There are only three BD regions — A, B, and C.

If you decide to use regional management, then you need to decide which regions. Again, broadening the scope of the disc early may be a lifesaver later, when marketing plans become more ambitious. Table A.1 lists what regions apply to which countries. You must also decide on menu languages, and choose which audio languages and which subtitle languages to put on the disc. Carefully check the audio and subtitle tracks to make sure they synchronize to the video you are using. In many cases, an existing audio dub or subtitle set may be from a different cut of the movie, and will have to be modified or redone to match the cut you are using. Audio from film usually has to be sped up (approximately 4 percent) for 50 hz video, so make sure the audio properly matches the video.

If you cover multiple regions, or even if you only cover a single region such as Europe/Japan, you may need to consider certification. Each country has its own requirements for classifications and certification. A disc with a "children's" rating in one country may end up with a "teen" rating in another country. If you have to cut scenes to get a particular classification, you will have to deal with conforming the audio and any subtitles to the revised edit. The parental management features usually will not work. The classification boards rate a work according to all the versions on the disc, regardless of how you have restricted access to them.

You must also decide if you want a single package (often called a SKU — stock-keeping unit) or multiple packages for multiple regions. In some cases, creating multiple SKUs is easier, since you can concentrate on making each version appropriate for its region rather than trying to make a single version work everywhere. For example, you might want to make three versions of a disc for release in Europe — one for the UK and Ireland, one for Germany and Scandinavia, and one for central and southern Europe (France, Spain, Italy, and so on). Multiple SKUs can also help with legal restrictions concerning how soon the home video version can be sold or rented after theatrical release.

Regardless of the number of regions, if you expect to use more than one language on the disc, especially for menus, keep all the graphic and video elements as language-neutral as possible. Keep text separate from graphics so it can be easily changed for other languages. Use subpictures, including forced subpictures, to put text on video, rather than permanently setting the text into the video. Keep in mind that American English is different from British English regarding usage and spelling. Likewise there are differences between American Spanish and Castillian Spanish, between Canadian French and continental French, between Brazilian Portuguese and Portugal Portuguese, and so on. In all cases, check to ensure that you have proper rights for the content. A particular work may have only been licensed for distribution in one country, or a foreign language audio dub may be restricted for use in certain countries.

Persistent Storage

In BD-J mode, you have access to *Persistent Storage* on the player. This area is not available with HDMV mode programming.

Persistent storage, which can be either the *application data area* (APDA) or the *binding unit data area* (BUDA), is used for storing information or content for the disc that does not get lost when the disc is ejected or the player stopped or shut off, and that is available to the

Production Essentials

user from one time to the next when using a disc. In your production process, you want to address a couple of things. First, you need to ensure that you have assigned an *Org_ID* and a *Disc_ID* for each disc. These values define where the information is stored in the player. You will also need to address how to delete old information or data as the space is limited.

The BUDA is the area of persistent storage where content, including program code, is stored and invoked via the *virtual file system* (VFS). Plan for how this area is used and how it will work with your disc-unbound or title-unbound application(s).

All profile players, 1.x and 2.0, support the APDA storage. Be aware, though, that the APDA is only specified to be a minimum of 64 Kbytes which means that that is the maximum size you can count on being available. This area is shared by all discs that are played on the player, so you want to use and share the space in a friendly and efficient way.

The BUDA is supported only on Profile 1.1 and 2.0 players. The space is a minimum of 256 Mbytes for Profile 1.1 players and 1 Gbyte for Profile 2.0 players. Many players allow the user to expand this space with external devices such as USB flash or hard disk drives. You need to address in your design and implementation that this space varies in size and can be removable. If there is no BUDA storage at all then you may want ot prompt the user to insert a storage device (such as a USB stick). If there is storage but it is filled up you may want to display a message, e.g., "Please clear storage and try again...". Again, you need to use this area in a friendly way with other discs. Data and content from the same content provider can be shared but cannot be accessed by other discs unless authorized. You may want to include a utility in your disc to allow the viewing and deletion of the content that you have stored here.

Network-Connected Features — BD-Live

Including network-connected features on a disc means that you have several items that need to be addressed in the design and implementation of these titles. When a disc is accessing the network, be sure to check —

1. Is the player Profile 2.0 capable?
2. Is the player connected to the Internet?
3. Is there updated content on the server?
4. Does the connection have the bandwidth to complete the download (or upload) in a reasonable time? A one gigabyte file (about 5 minutes of HD video and audio), could take about 45 minutes at 3 Mbps.
5. Can the download be restarted if the connection is interrupted?
6. Is there enough space available on local storage for any files that you will download into APDA or BUDA?
7. Can the server and the infrastructure handle the volume of concurrent users and the required bandwidth to support their interactivity and downloads?

For development and testing prior to shipment, you may need to set up a test environment that mimics what endusers will experience in terms of connection speed, interruptions and player capabilities. This is to test not only what will ship on the disc, but the backend infrastructure needed to support the endusers.

BD-Live opens up the possibility for discs that are never "finished". They can evolve over time for both content and functionality. Not only can the content be updated, but the programming code can also be updated. The updates can not only add new functionality/features but update/fix code anomalies. It is conceivable that at some point all discs will ship with a BD-Live capability if for no other reason than to update the programming code on the disc. With the right planning and backend infrastructure, all discs can be fixed after they have shipped. (Don't you wish you had that capability on DVD???)

Supplemental Material

One of the great appeals of BD is that a disc can hold an incredible variety of added content to enhance the featured program. Being able to watch a movie a second time and listen to the comments of the director, writer or actors gives viewers a deepened appreciation for the movie and for the filmmaking process. Adding more video, audio, graphics, and text information can greatly enrich a disc. Anyone producing a BD should consider how the product can be enhanced with supplements; not just because it is neat or expected, but to truly improve the viewing experience. The following list provides some examples of added content for BD. Some types of added content, especially text and information that can become out of date, are best delivered using network connected capabilities of BD-Live. Any of this content could be accessed in a BD-Live mode. The more likely BD-Live content is indicated with an asterisk (*):

- Multiple languages and subtitles, menus in other languages
- Biographies: information about actors, directors, writers, and crew, describing careers, influences, and other interesting data*
- Filmographies: information about actors, directors, writers, and crew describing previous films, acting experience, awards, and current projects*
- Commentaries: audio or subtitle commentaries from people who worked on the film, people who have studied the film, people with historical information, and even fans
- Song list: a chapter index to all the songs on the soundtrack
- Backgrounders: historical context, timeline, important details, related works*
- "Making of" documentary
- Production notes and photos, behind-the-scenes footage, lobby cards, memorabilia
- Live production details, such as synchronized storyboards, screenplay, or production notes*
- Trivia quiz or game
- "Easter eggs:" hidden features that are found by selecting disguised buttons on menus
- Related titles: "If you liked this disc, you should see...," lists or previews of discs with similar characteristics or with the same actors. For BD-Live discs, this could include discs that have been released after the initial release of this disc.*
- Trailers or ads for other titles

Production Essentials

- Bibliographies: where to find more information on subject matter*
- Web content that can be either show the links or can link directly to special versions of the Websites for the movie, actors, studio, producer, fans, and the like that can be presented on the network connected BD players*
- Excerpts: screenplay, documents, scanned photographs, sound bites, and other text or graphic information*
- Related computer applications: screen savers, games, quizzes, digitized comic books, novelizations, and even applications for texting and chatting*
- Additional content more easily programmed or accessed through ROM on a computer
- Product catalog and ordering system to sell related goods such as other DVD titles, action figures, et cetera*
- Newspaper and magazine articles or reviews (with appropriate permissions, of course!)*

Scheduling and Asset Management

The secret to a smooth and successful project is a carefully-monitored production schedule. The schedule enables you to allocate resources and determine bottlenecks. Project planning software such as *Microsoft Project* is very helpful. In many cases you will start with a rough high-level schedule, then create an updated schedule based on the detailed project tasks. More so than with DVD production, BD projects can have many activities that are concurrent. For example, after the title has been designed and architected, compression and multiplexing can be happening in parallel with the graphics creation and the BD-J programming. This concurrency is especially important where there are complex interactive elements that require extensive programming to implement. Some of the programming of the interactive elements may be created before other assets of a title are even conceived. For example, features such as a progress bar with bookmarks and user-defined playlists may be created and completed even before the rest of the disc content is complete.

Throughout the production process, keep track of the assets. Make a list of each asset, when it is due, when it arrived, what format the source is in, when it was digitized and encoded and by whom, what the filename is, where the files are stored, when the client signed off on each stage of asset conversion or production, and so on. Clarify who owns the copyright for new content that may be created during production.

Project Design

A project design document is the blueprint from which everyone will work; it is the important first step for creating a disc. Decide upfront the scope and character of your title. Make a plan and try to stick to it. Authoring is a creative process, but if you continually change the project as it goes along, you will waste time and money. Determine the level of interactivity you want. The more interactive the project, the more complex the layout and, thus, the more time you need to spend at an early stage.

Laying out the disc generally involves the following —

Production Essentials

■ **Flowchart** Diagram each part of the title. Decide what appears when the disc is first inserted. Does the disc go straight to a video element or to a menu? Diagram each menu and how it links to other menus and video segments. For a complex title, unless you create a clean design with clear relationships between the various hierarchical levels, your viewers will become lost when trying to navigate the disc. Each box on the flowchart should indicate how the viewer will move to other boxes and what keys or menu buttons they will use to do so (Figure 12.1).

Figure 12.1 Sample Disc Flowchart

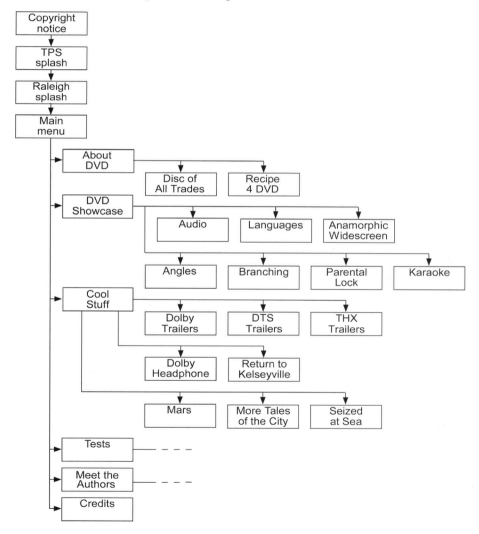

Blu-ray Disc Demystified

Production Essentials

- **Storyboard** Draw pictures of the menus and the video segments. You may want to draw a storyboard picture for each chapter point. Make sure that transitions between menus and from menus to video are smooth.

- **Prototype** Create rough versions of the menus in *PhotoShop*, *PowerPoint* or another graphics program. Simulate jumping from one menu to another by using layers or by opening new files. Bring in other people to test the simulation. Make sure the experience is smooth, clear, and easy. *Adobe Flash* is another tool that can be used to create the menu interactions and the animations.

- **UOP controls** For every menu and video segment, determine what the user will be able to do. Start with all user operations enabled, and only disable them for a good reason. Do not needlessly frustrate your viewers by restricting their freedom. Specify user operation results.

- **If you have a dual-layer disc** You will need to plan the layer break point with your compressionist so that the layer change is seamless.

The layout process serves as a preflight check to make sure you have accounted for all the pieces and that they work together. Good layout documents provide a roadmap to be used by the entire team throughout the remainder of the project — a checklist for coordinating assets and schedules, a framework for menus and transitions, a guide for bit budgeting, an outline for the authoring program, a checklist for testing, and so on.

Many authoring systems attempt to provide layout and storyboarding tools in the application. Some succeed better than others. In general, at least a basic layout should be done on paper or in a separate program, such as *Adobe Flash,* before jumping into the authoring process, especially since the authoring stations are often key pieces of expensive equipment that are in constant demand.

Menu Design

Menus make the disc; they set the tone and define the user experience. They are also fun to design. The possibilities are endless, so you must make many decisions up front. Do you want still menus or animated menus? If you have an animated menu, consider whether to use animated graphics or a pre-composited video or a combination of both. If you are using video, carefully consider how to design the video so that it loops smoothly. If a great deal of movement occurs, consider starting with a still and ending with the same still. Fade the audio down just before the end, and fade it back up at the beginning.

Do you use only popup menus or a combination of full screen and popup menus? Consider the best way to allow the user to have logical interactions with the disc. Sometimes full screen, DVD-like menus are more appropriate for the content than a popup menu.

What kind of animated transitions make sense? For full screen menus, can you use a video or animated segment that appears at the beginning of the main menu, or between menus and video segments? Make sure the transitions are smooth and appropriate. Even if you do not use motion transitions, jumping from one design and color scheme to a completely different one can be very jarring. HDMV and BD-J mode programming both support animations for

Production Essentials

both popup and for full screen menus that are much more flexible and dynamic than pre-rendered, video-based animations. When using full screen menus, jumping from one type of audio codec in the menu to a completely different type of audio codec in the video program can be grating.

Do you want background or button audio in the menu? Be very careful with this. One of the most obnoxious features of BD discs are menus with audio that is too loud and too intrusive. Drop the audio to a pleasingly low level; then cut the level in half. Provide a way for the user to turn off any background or button sounds. Play your menus sounds at least 20 times in a row to discover just how annoying your creation can be.

Do you want multilanguage menus? If so, keep the text separate from the graphics and video so that it can be easily changed for each language version of the menu. If you are creating the original backgrounds in English, leave 30 percent more space for translations.

Will you use animated buttons? Consider that each state of a button needs to be either a separate PNG or programmed/authored. The number of graphic assets can be huge.

Do you want to idle out? That is, do you want the menu to automatically jump somewhere else after a particular amount of time? If so, choose a long timeout period. Make sure the menu does not time out while the viewer is still reading it. You may want the menu to time-out to a screen saver to save from any burn-in on monitors.

Menu Creation

Menus are much more dynamic with BD than with DVD. Instead of a graphic layer that defines a background and a subpicture layer for the highlight states of buttons, each button and its associated highlight states are separate graphics. These graphics are often stored as a set of graphic PNG files. These PNGs are then composited realtime onto the graphics plane by the player (see Figure 12.7, presented at the end of this chapter).

A large part of the BD menuing experience is the use of animation. There are a number of ways to create the animations, either with a set of PNG files to represent each state of the animation or via movement of graphical areas or programatic calculations that either expose areas (such as a progress bar) or a programmatic scaling of a graphic or other creative rendering. The challenge in BD-J mode programming of the menus is that the player speeds vary significantly so that an animation sequence on one player may take seven seconds while on another less than a second. Animations need to be tested to ensure that the user has a desirable, pleasant experience on the player they are using. With BD, a highlight state may be represented with a series of PNGs that animate a button to indicate that it has the focus.

Menus can be full screen for the top level menu as well as popup menus that appear over the video. Creating the menu experience means that you need to address what is happening on all the planes, e.g., the background plane, the video plane and the interactive graphics plane. Creating the best user experience means taking advantage of the best capabilities of each of these planes. An animation can be either programmed or it could be also composited into the video plane. There are many more creative choices than with DVD.

Remember that you also have sounds that can be associated with button presses or the animations or randomly.

Blu-ray Disc Demystified

Production Essentials

One other consideration is the space that the graphics will use in the graphics buffer. HDMV has a 16 Mbyte limit and is restricted to 8-bit indexed graphics. Careful planning and creative authoring make this a workable framework but many large graphical elements will quickly consume the graphic buffer and cause a menu design to be very difficult or impossible to implement.

With BD-J mode programming, the graphics buffer is 45.5 Mybtes for Profile 1.x players and 65.5 Mbytes for Profile 2.0 players. But these are 24-bit graphics with an (optional) 8-bit alpha channel. Good planning and clever programming will allow the creation of a great interactive experience. BD-J has the capability to dynamically load/unload graphic assets so this can be less of a restriction than with HDMV, although without proper implementation, the load/unload activity can cause the video playback to stutter.

When you lay out a menu, consider interbutton navigation. The user has four directional arrow keys to use in jumping from button to button. Lay out the menu so that it is obvious which button will be selected when a particular arrow key is pressed. At the same time, do not forget that your disc inevitably will be played in a computer with a mouse interface, so do not rely on the user moving sequentially from button to button. In general, it is best to have button links wraparound, so when the user presses the down key on the bottom button, the top button should be selected. When the user presses the up arrow key on the top button, the bottom button should be selected, likewise for left and right movement. The reasoning is that it is better to have a keypress do something than nothing. The brief confusion experienced by novice users when the arrow key functions wraparound is preferable to the angry frustration of users who expect arrows to wrap when they don't.

Tips and Tricks

The following will help in creating your menu and animation graphics —

- If you use button sounds, provide a way to turn them off. It is easy to go from being entertaining to annoying.

- A user should be able to tell within less than two seconds which button is highlighted. The number one mistake that BD and DVD menu designers make is failing to differentiate between a highlighted and an unhighlighted button. For example, the viewer sees a screen with two buttons — one is yellow, one is green. Which one is highlighted? If the user presses an arrow key, the yellow button turns green and the green button turns yellow. The user still cannot tell which one is highlighted. This can even be confusing with more than two buttons if some of the buttons are different colors in their unhighlighted state. In many cases, it is better to put a box or a circle around a button, or have an animated icon or pointer appear next to it, rather than only changing the color or the contrast.

- Make sure the highlighted color is different enough from the unhighlighted color so that button selections are visible. Do not rely on menu simulation in the computer because the highlights will appear different on different resolution HD monitors.

Production Essentials

- Each menu should pre-highlight the button that the user is most likely to select; this is called the "default" selection for any given menu. For example, if the main menu is shown before the feature, the *play* button on the menu should be preselected so the user can simply press *Enter* or *Play* to start. This is especially important since the Play key on most remote controls does the same thing as the Enter key when on a menu. Beginning DVD users will intuitively press the Play key when they reach the first menu.

- Author menus so that the button that takes the viewer out of the menu is highlighted upon return. If a video segment or another menu jumps to a menu, the button that returns the viewer to where they just came from should be highlighted. Note that this is a user interface design decision. There may be cases where you want the highlighted button to advance from the button for segment A to segment B as the user watches each video segment.

- Design your menus for numerical access. Each button on a menu is internally assigned a button number in programming, and many remote controls enable the user to press the 1 through 9 keys on a menu to directly choose a button. Arrange the buttons in proper numerical order. Be careful with invisible buttons. On a disc with several menus and complex navigation, consider putting a number next to each button.

- An introductory video sequence for the main menu will become a flaming irritant if the user has to sit through it every time they return to the menu. Use the intro for initial play only, then stifle it for later menu accesses. With persistent storage, it is possible to have this behavior carryover from one play of the disc to another.

- Animated menus can be enjoyable, but keep them short and sweet; never make animations longer than two seconds. When creating the animations, play them over and over and imagine yourself pressing a menu button each time the video loops. If it seems like a long time between button presses, then it is too long a time.

- HD video is 16:9, but when using 4:3 SD mode video, you may need to create another set of menus for when the monitor is in SD mode. Downscaled HD menus, often do not look pleasant and are difficult to read in standard definition.

- Make sure the button highlight/selection rectangle is over the top of the button art in the background. If you put the rectangle somewhere else on the screen, the disc will still work with the arrow keys but the button will not work when a computer user clicks it.

- Doublecheck that button graphics are properly aligned on a 1920 × 1080 monitor. It is easy to misalign a graphic if you use a less than full HD monitor in your production workflow.

- Temper creativity with good user interface design. Weigh functionality over aesthetics and experimentation. Frustrated users will head for the eject button and will never see your avant-garde designs.

Blu-ray Disc Demystified

Navigation Design

Navigation is one of the hardest aspects of Blu-ray Disc production to get right. In addition to the confusion of having a TOP Menu key, a Pop-UP Menu key, and the inconsistency of navigation on different discs, it is difficult to balance creativity and playfulness against ease of use. And, please note, selecting *Top Menu* will often require a reload of the BD-J code and resources, depending on the player. The time to reload the Top Menu functionality can vary from less than ten seconds to well over 2 minutes depending on the player and the size of your graphics.

When you create the flowchart, identify the effect of important menu keys for each menu. Clearly indicate where the viewer will go when they press *Top* (Top Menu), *Pop-Up Menu*, or *Return* (GoUp). If you have several menus, use a hierarchical design that the user will be able to intuitively understand. If you use more than three levels in your hierarchy, try combining menus to flatten the structure. However, try not to have more than seven to ten buttons on each menu in order to avoid overwhelming the user with options. Keep all selectable features and options as close to the main menu page as possible, with a minimal number of button selections. Make it as easy to navigate back up through menus as it is to navigate in the down direction.

Understand the difference between the *Top* (Title) key and the *Menu* key. "Top" means "take me to the very top, to the table of contents." "Menu" means "take me to the most appropriate menu for where I am." For multilevel menu structures, the Menu key can be used to go up one level, or the Return key can accomplish this feature. A well-designed disc will program the *Return* (GoUp) function to act somewhat like the back button on a Web browser. Each time the viewer presses *Return*, they move up one level until they reach the top (see Figure 12.2). Every remote should include the three basic navigation keys — *Top* (Title), *Menu*, and *Retu*rn. Take advantage of the intended function of each button (Figure 12.3).[1]

Always enable navigation with on-screen buttons as an alternative to relying on remote control keys. Assume that the viewer has only the directional arrow keys and the Enter (OK) key, or assume that the viewer has only a mouse. Make sure that buttons labeled *back* or *main menu* or something similar are included to enable the viewer to navigate through all the menus without relying on any other keys.

Figure out where each chapter point will be. Anywhere the viewer might want to jump to should be a chapter point. Identify the chapter points by timecode on the storyboard or flowchart. Do this before the video is encoded, since each chapter entry point in the video must be encoded as an *I*-frame at a GOP header.

---

[1] Unfortunately, some player interfaces, especially on computers, leave out the Top (Title) key or the Return key. However, the incompetence of some player designers should not cause you to castrate your disc. The more discs there are that properly use the navigation keys, the sooner players will be fixed to provide them all. In the meantime, you can overcome the problem by providing on-screen buttons for navigation. See the following paragraph.

Production Essentials

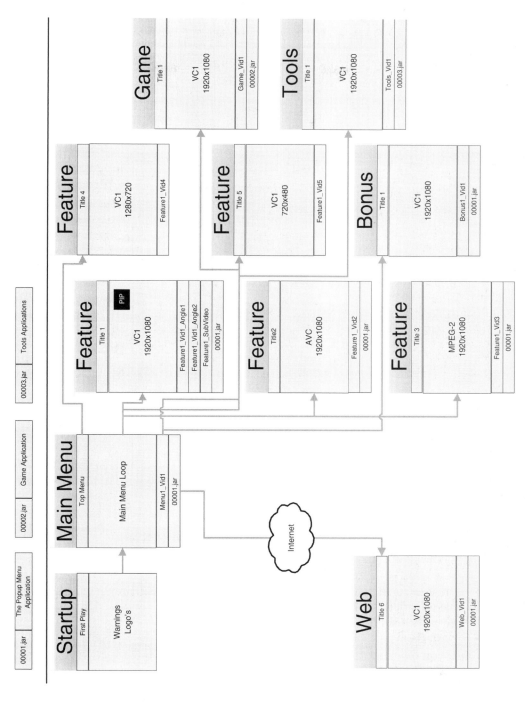

Figure 12.2 Navigation Flowchart Example

Production Essentials

Figure 12.3 Blu-ray Player Remote Control Example

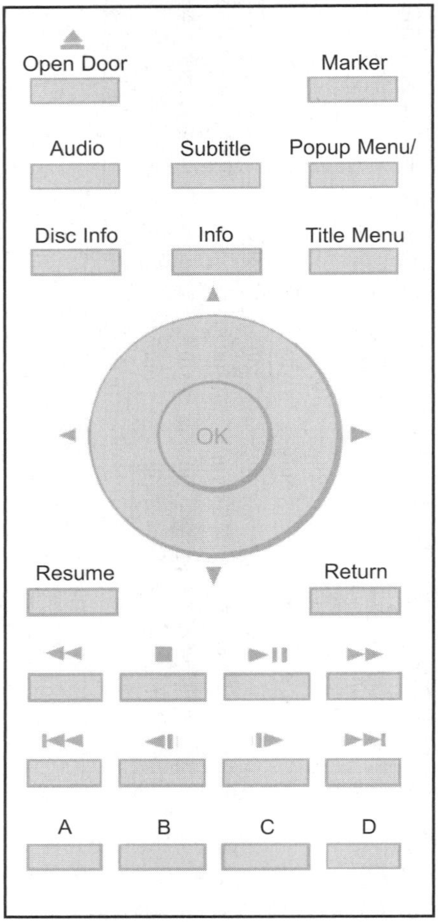

Tips and Tricks

The following tips may assist your efforts to make user navigation easier and clearer —

- If you are going to use multiple BD-J or HDMV titles, you might want to give the user some indication that there will be some delay before the next title/module is loaded. A black screen that is displayed for more than a couple of seconds gives the user the impression that something is broken, especially if you did not warn them.
- If you only have one menu, you may want to disable the Top Menu key as this will fully reload the disc which, as previously noted, can take minutes on some players.

Production Essentials

- You should not disable the Top Menu or the Popup Menu keys if you have both menu sets available.
- Do not author the disc so that pressing the Top Menu (Title) key starts over and forces the viewer to watch logos and FBI warnings again. You can store the information in either the application data area or in the player registers (GPRMs).
- Make sure the *Next* and the *Previous* keys always work. At minimum, they should move through chapters.
- Consider placing a "How to use this disc" element on every disc you make. It can explain the basics of your navigation when using the arrow keys, *Enter*, *Top Menu*, *Popup Menu*, and *Return*.
- Keep in mind that not all player remotes present all of the keys, such as *audio*, *subtitle*, or others on the remote. It is better to build these functions into the user interface of the title. With BD-J, any key that might be on the remote can be controlled programmatically and you can, therefore, create a more coherent way to use the disc than trying to find a particular key on the remote.

Balancing the Bit Budget

Bit budgeting is a critical step before you encode the audio and video. You must determine the data rate for each segment of the disc. If you underestimate the bit budget, your assets will not fit and you will have to re-encode and re-multiplex. If you overestimate the bit budget, you will waste space on the disc that could have been used for higher quality encoding. However, be conservative — using less than the entire disc works, but exceeding its capacity, even by a small amount, forces you to rework the project.

The idea is to list all the audio and video assets and determine how many of the total bits available on the disc can be allocated to each. Part of the early bit budgeting process involves balancing program length and video quality. Two axes of control exist — data rate and capacity. Within the resulting two-dimensional space, you must balance title length, picture quality, number of audio tracks, quality of audio tracks, number of camera angles, amount of additional footage for seamless branching, and other details. The data rate (in megabits per second) multiplied by the playing time (in seconds) gives the size (in megabits). At a desired level of quality, a range of data rates will determine the size of the program. If the size is too large, reduce the amount of video, reduce the data rate (and thus the quality), or move up to a bigger capacity disc.

Once you have determined the disc size and the total playing time, maximize the data rate to fill the disc, which will provide the best quality within the other constraints, such as the maximum bitrate for video of 40Mbps. Note that increasing the bitrate above a certain threshold will technically give better video or audio quality, but you may not see or hear the difference. The added bits will only result in longer production times to copy and process the larger filesizes.

Production Essentials

See Figure 12.4 to get an idea of needed disc sizes. A DVD-5 holds 15 minutes of the highest bitrate video to over 6 hours of video at the lowest practical rate. A BD-25 can hold 1.5 hours of high bitrate video to over 25 hours at the lowest bitrate. Of course, the video quality at the longer playing times would likely be for SD resolution sub-DVD quality video.

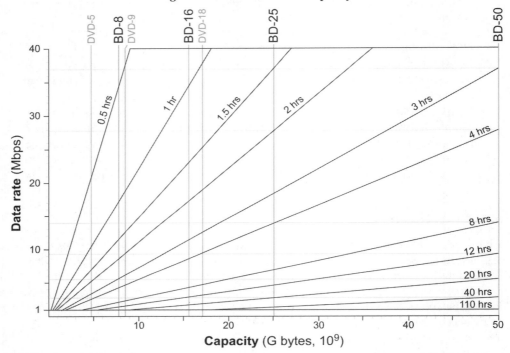

Figure 12.4 Data Rates vs Capacity

The easiest way to make a bit budget is to use a spreadsheet. One is included on the sample disc that comes with this book. For simple projects with only one video segment, use the *DVDCalc* spreadsheet on the sample disc. Some authoring programs will calculate bit budgets for you when you use their layout features.

To simplify calculations, keep track of sizes in megabits, rather than megabytes (see Table 12.3). Allow an overhead of 5-7 percent for control data and backup files which are added during formatting and multiplexing. This also allows a bit of breathing room to make sure everything fits on the disc.

Table 12.3 Disc Capacities for Bit Budgeting

	BD-5	BD-9	BD-25	BD-50
Billions of bytes	4.7	8.54	25.0	50.0
Megabits	37,600	68,320	75,200	105,920
7% overhead	2,632	4,782	5,264	7,414
Adjusted capacity	34,968	63,538	69,936	98,506

Production Essentials

In addition to fitting the assets into the available space on the disc, also make sure that the combined data rate of the streams in a video program do not exceed the maximum data rate (instantaneous bitrate) of 40.08 Mbps. Subtract the audio data rates from the maximum to get the remaining data rate for video. Lowering the maximum data rate reduces the ability of a variable bitrate encoder to allocate extra bits when needed to maintain quality.

A typical bit budgeting process for a disc with a single main video program, plus a few ancillary segments, is depicted in Table 12.4.

Table 12.4 Sample Bit Budget

Element	Minutes	Data Rate	Number of Streams	Size (Mbits)	Formula – minutes × 60 seconds × data rate × number of streams
English audio track (7.1) DTS-HD 96kHz/24-bit	110	6.000	1	39,600	$110 \times 60 \times 6.0 \times 1 = 39{,}600 \text{Mbits}$
English audio track (2.0)	110	0.192	1	1,267	$110 \times 60 \times 0.192 \times 1 = 1{,}267 \text{Mbits}$
Spanish, French, and German audio tracks (5.1 at 448 kbps)	110	0.448	3	8,870	$110 \times 60 \times 0.448 \times 1 = 1{,}109 \text{Mbits}$
4 subtitle tracks	110	0.040	4	1,056	$110 \times 60 \times 0.040 \times 1 = 1{,}056 \text{Mbits}$
Subtotal (audio and subtitles)				50,794	
Top level menu with video and audio	3	18.192	1	3,275	$3 \times 60 \times (18 + .192) \times 1 = 3{,}275 \text{Mbits}$
Intro logos (HD video at 18Mbps/audio 5.1 at 448kbps)	0.5	18.448	3	1,660	$0.5 \times 60 \times (18 + .448) \times 3 = 1{,}660 \text{Mbits}$
Movie Previews (HD video at 18Mbps/audio 5.1 at 448kbps)	2.5	18.448	3	8,302	$2.5 \times 60 \times (18 + .448) \times 1 = 8{,}302 \text{Mbits}$
Interview (SD video at 6Mbps/audio 2.0 at 192 kbps)	30	6.192	1	11,146	$30 \times 60 \times (6 + .192) \times 1 = 11{,}146 \text{Mbits}$
Subtotal (non-movie elements)				24,382	
Total for disc without movie				75,176	
BD-25 capacity (25 Gbytes leaving 5% open)				196,000	$25{,}000 / 8 \times .95 = 196{,}000 \text{Mbits}$
Disc space remaining for movie				120,824	$196{,}000 - 75{,}176 = 120{,}824$
Calculated bitrate for movie on a BD-25 disc	110	18.307			$120{,}824 / (110 \times 60) = 18.307 \text{Mbps}$

Calculate the total bitrate and size of the main program's audio tracks and subtitle tracks. If there are additional video segments — animated logos, top menus with video backgrounds, previews — calculate their sizes (including audio), and add them to the total size of the main program's audio and subtitle tracks. Still logos and copyright warnings should be authored as timed stills, which have no impact on disc space. Subtract the total size of the ancillary material from the disc capacity minus overhead. This gives the remaining capacity for the main program.

Production Essentials

Calculate the average video bitrate for the main program by dividing the remaining capacity by the running time. Also, calculate the maximum video bitrate by subtracting the combined audio and subtitle bitrate from 40.08 Mbps. These numbers will be set in the encoder when compressing the video for the main program. Note that minor elements, such as, subtitle tracks and motion menus, are usually so small that they often can be omitted from calculations. If they were left out of the example in Table 12.4, the average data rate would come out to 24.17 Mbps, which is close enough for budgeting purposes.

An alternative bit budgeting approach is to divide the disc capacity (in bits) by the total running time of all segments (in seconds) to get the average data rate (in Mbps). Adjust up or down for quality of different program elements, as needed. For each program, subtract the combined data rate of its audio and subtitle streams to get a video rate. It is a good idea to follow up by summing the sizes of each segment (including audio) to check that the total does not exceed the disc capacity.

Tips and Tricks

The following list of tips will help you in balancing the bit budget —

- Don't forget top menu videos and video based menu transitions, as well as logo sequences.
- Don't forget to leave room for any added computer-based content or for managed copy files. A few programs or some HTML files are not worth worrying about, but large multimedia applications and installer packages need to be accounted for.
- Each camera angle increases the data; two camera angles double it, three angles triple it, and so on (including the corresponding audio and subtitles). Reduce the maximum data rate according to the number of angles (see Table 6.11).
- If you have two or more video segments that contain much of the same data, consider using the seamless branching feature to combine the duplicated segments (but only if you have an authoring tool that supports the multistory feature).
- Still images take up so little space that you don't need to include them in your calculations unless you have dozens of them. Unless you are including more than four subtitle streams, you may wish to omit subtitles from the calculations, since the space they take up is negligible.

Asset Preparation

A typical project requires dozens, if not hundreds, of assets from a variety of sources. A complicated project may require thousands (see Table 12.5). Each asset must be properly prepared before it can be fed into the BD authoring process. The following sections cover the details for preparing each type of asset.

Production Essentials

Table 12.5 Typical Project Assets

Type	Assets	Tasks
Video	Source video files (movie, trailer, supplements)	Clean up, encode
	Already encoded video (logos, computer graphics)	View, verify codec, bitrate, framerate, frame size
Audio	Original language source file	Conform, confirm length with video length
	Foreign language source files	Conform, confirm length with video length
	Commentary or other supplemental audio sources	Conform, confirm length with video length
	Already encoded audio	Check synchronization
Graphics	Menu graphics	Create/integrate graphics, slice graphics, optimize palette index and create button highlight states
	Supplemental graphics	Create/integrate graphics, slice graphics, optimize palette index and create button highlight states
	Production stills and other photos	Digitize, correct color, create button highlight states
Subtitles	Text files	Add timecodes at proper framerate
	Graphic files	Create timecode-filename mapping file

Preparing Video Assets

The final quality of BD video is primarily dependent on four things — the condition and quality of the source material, the visual character of the material, the codec used, and the data rate allocated to the video during encoding. The better the quality of the source, the better job the encoder can do. Always use the best quality source and insist on an uncompressed file-based format, whenever possible. Tape-based material can be on HDCAM SR, D5-HD, DVCPro-HD, or HDV, while some video material may be in AVCHD format. Keeping the video in a digital format without conversion will help to ensure the best quality. Verify that you have correctly identified all of the sources as 720p, 1080i, 1080p, or one of the supported standard-definition resolutions.

If the project video is not all high definition, you may want to consider upconverting the SD video to HD (1080p) to ensure that the viewer will have the highest consistent quality video to view. Although the BD player and/or the monitor can upconvert the video, you no longer have control over the quality of the conversion. The enduser's playback setup may do a great job or a poor job. And, rest assured, the poor job will be perceived to be your fault, not the player's.

Production Essentials

If the video is of poor quality or contains noise, use some form of digital video noise reduction (DVNR) to improve it. Noise is random, high-entropy information, which is antithetical to compression with AVC, VC-1 or MPEG-2 encoding. Efficient encoding depends on reducing redundant information, which is obscured by noise. Noise can come from grainy film, dust, scratches, video snow, tape dropouts, cross-color from video decoders, satellite impulse noise, and other sources. It may also be intentional as a creative style. Complex video, including high detail and high noise levels, can be dealt with in two ways — increase the data rate or decrease the complexity. The DVNR process compares the video across multiple frames and removes random noise. After DVNR, the video compresses better, resulting in better quality at the same data rate. With HD video, many flaws in the image are exposed that did not show up in even the best DVD productions.

If possible, keep the production path short and digital. A frequent irony of the process is that video is edited on a computer, then output to an analog or digitally lossy tape format in order to be encoded back into a computer file. If you use a digital nonlinear editing system (NLE), look into options for directly outputting BD-compliant AVC, VC-1 or MPEG-2 streams. Most HD video is 16:9. If you are using other sourced material, such as, standard definition video or film, anticipate the need for aspect ratio conversion. Make sure the various sources are in the proper aspect ratio (4:3 or 16:9), especially for assets that need to be edited together.

You may need to choose between variable bitrate encoding (VBR) or constant bitrate encoding (CBR). VBR does not directly improve the quality, it only means that quality can be maintained at lower average bitrates. For best quality of long-playing video, VBR is the only choice. For short video segments, CBR set at or near the same data rate of VBR will produce similar results with less work and lower cost. For VBR encoding, set the average and the maximum bitrate as determined in the bit budgeting step. These will be largely determined by your choice of encoding, whether AVC, VC-1, or MPEG-2 and the encoder.

Tips and Tricks

The following tips will help you in preparing video assets —

- The display frame rate for high definition video can be 23.796, 24, 25, 29.97, 50, and 59.94 fps. 25 and 50 fps video are only required in 50 hz video countries (mainly the countries where PAL was the SD standard). True 30 and 60 fps are not allowed.
- If utilizing secondary video (for PIP), the frame rate of the PIP video needs to match that of the primary video — you cannot have 29.97 video, which may have been previously used on DVD, as the PIP video on top of 24p primary video!
- BD supports different frame rates and frame sizes. Mixing and matching at the end of a project will create results that are noticeable on HD monitors. Plan ahead in production and in post-video production so that you have a good looking master for the BD compression.
- Ensure that the encoder can produce BD-compliant streams, with proper GOP size, display rates, peak bitrates, and so on. Also look for an encoder that can do segment-based re-encoding, which enables you to re-encode small segments, as

Production Essentials

needed, rather than reprocess hours of video to correct a small isolated problem. Not all MPEG-2, AVC (MPEG-4, part 10) or VC-1 streams are BD compliant. For high profile projects that will be replicated, you will need to run the multiplexed streams through a specification verifier to ensure that the streams will be compliant to the specification, thereby ensuring the best playback compability.

- If you are working with film transferred to video, make sure the workflow maintains the 24p framerate and progressive frame format.
- Chapters and programs must start at a sequence header or GOP header, and the GOP must be closed, not open. Each cell must also start at a sequence header. If you have very precise points for chapter marks, make sure they are encoded properly, otherwise the authoring software may move the chapter point to the nearest I-frame.
- Stills are encoded as I-frames using one of the video codecs (MPEG-2, AVC, or VC-1) for either timebased or browsable slideshows. You can also use PNGs that you can load/unload and control playback with BD-J code.
- Static video, where the picture does not change over time, such as FBI warnings, logos, and so on, should be authored as a timed still. Not only does this look better, with no movement or variation when played back, it can save space on the disc.
- Make sure the video does not have burned-in subtitles. Use the BD subtitles instead, particularly for multilingual discs. Even narrative subtitles can be better displayed with BD subtitles. This will ease the localization of your disc.
- Presentation Graphics (PG)-based subtitles will look better than player-rendered subtitles (termed TextST, for text subtitle). PG subtitles are 8-bits/pixel with an indexed color palette which are multiplexed in the A/V stream. Each player may render text fonts differently, so you will have more control of the quality and the timing of the subtitles if you use the PG pre-rendered subtitle approach. Subtitling software that creates the PG-based subtitles vary widely in quality. Work with your software or service bureau to ensure that you have great-looking subtitles. With HD resolution, subtitles can look great.
- If you want to pack a large amount of HD video on a disc, do not try to encode to MPEG-2 under 12 Mbps. Use either SD video at 16:9 or use an AVC or a VC-1 codec. The AVC or VC-1 codec will allow you to use a bitrate that is 20-40% lower for the equivalent quality of a MPEG-2 encoded video stream. Before encoding the video, use heavy digital video noise reduction (DVNR) and liberal application of blurring filters in video processing programs. These steps reduce the high-frequency detail in the video, which makes MPEG encoding more efficient. Also try reducing the color depth from 24 bits to something between 16 bits and 23 bits. Do not use a dither process, which increases high-frequency detail. This can help with encoding efficiency and can reduce posterization artifacts.

Production Essentials

- If video quality is a top consideration, create or request your content in progressive format if at all possible — shoot on film or use a progressive digital video camera. If your video assets come from an interlaced source, consider converting to progressive format before encoding. Depending on the nature of the video, progressive conversion may cause artifacts that unacceptably degrade the video. For some types of interlaced content, however, a high-quality progressive conversion will result in a BD that looks significantly better when played on progressive BD players and monitors See Chapter 2 for more on progressive video.

Preparing Animation and Composited Video

Video produced on a computer or camera-source video that is composited with computer graphics have their own sets of issues. Preview all of your video at the full resolution of 1920x1080. If you are working at lower resolutions than this, there will be artifacts that your enduser may see. Conversely, preview your material at lower resolutions also. You may need to decide which wreaks havoc on interlaced displays, even at 1080i. 720p displays will lose some of the fine detail that you have at 1080p. Avoid thin horizontal lines; use the anti-aliasing and motion blur features of your animation or editing software, whenever possible. Use a low-pass or a blur filter, but don't overdo it because too much blurring reduces resolution on progressive scan players and on computer displays.

For SD content, consider upconverting to 1080p. Although this will result in more disc space being used, this ensures that the enduser will have the best consistently-displayed video. If you need to use SD video — for 525-line (60hz) video, you can render animations at 720x540 and scale to 720x480 before encoding; For 625-line (50hz) video, you can render animations at 768x576 and scale to 720x576 before encoding, or if your software supports it, render with D1 or DV pixel geometry. Render at 24 fps progressive as this will yield the best looking video on high definition 1080p displays.

Player Rendered Animations and Compositions

With DVD, any animations had to be pre-rendered as video. You would then use these pre-rendered clips as transitions between either still or motion menus. This technique still has applications with BD, but most of the animations can be rendered by the player as the user interacts with the disc. Some things to remember, though, unless the playback is on a very fast player (e.g., Sony PS3), you need to pay attention to the number of pixels per second that you are moving on the display at any one time. Keep the pixel rendering under about a fifth of the screen size per second and you will not have tearing or stuttering on any of the players. Test your animations to ensure that they work on the players that you have targeted.

Preparing Audio Assets

Contrary to most expectations, audio is usually just as complex as video and often more complex. You must deal with multiple tracks, synchronization, audio levels across the disc,

Production Essentials

and so on. As with video, the better quality the source, the better the results after encoding. Audio assets usually come on modular digital multitrack (MDM) formats, or you may receive PCM (WAV) files on a CD-R, DVD-R, BD-R or HDD, or already-encoded Dolby Digital or DTS files. With a multitrack tape source, make sure the track-to-channel assignments are correct. Dolby has specified that the sequence is L (left channel), R (right channel), C (center channel), LFE (low frequency effects), Ls (left surround), Rs (right surround, L_t (left total), R_t(right total), but not all production houses follow this standard.

Capture your audio mixes at the highest sample rate and word size possible. If someone else is doing the audio mixing, request that they do the same. As recording and mastering studios acquire the new equipment needed to work with high-resolution audio, 96 kHz 24-bit audio will become commonplace. Unlike with DVD-Audio or for audio-oriented DVD-Video, a 96 kHz 24-bit PCM audio track can have plenty of remaining headroom on BD.

When converting from NTSC or PAL video, or transferring from film, the audio may need to be conformed to match the transfer imagery, preferably via a pitch shift to restore proper pitch. If the audio is from a CD, it must be upsampled from 44.1 kHz to 48 kHz for BD. Many audio files produced on PCs are also at 44.1 kHz.

Do all of the necessary processing before encoding — clean up and sweeten the audio, remix and equalize, adjust phase, adjust speed and pitch, upsample or downsample, and so on. Audio levels can be adjusted before encoding or, for Dolby Digital, they can be adjusted during encoding by using the dialog normalization feature. If the mix was done for the theater, re-equalize for the home environment. Establish consistent audio levels and equalization throughout all of the audio on the disc, including menus and supplements. This is especially important with multiple audio tracks from various sources, since the viewer will be able to jump at will between them and will be bothered by inconsistencies in level or harmonic content. Resist the temptation of locking out the audio key on the remote control, which will only annoy your viewers. Instead, get all of the sources to PCM format so they can be matched and harmonized before being encoded.

Tips and Tricks

The following tips will assist you in preparing the audio assets —

- Make sure timecode is included with the audio so that it can be synchronized with the video (and make sure the timecode for the audio actually matches that of the video).
- It is almost always a good idea to mute the audio for a second or so at the beginning of each segment. This ensures that the player (or receiver) does not clip the audio as it begins the decoding process.
- If you are doing 7.1 or 5.1 channel, Dolby or DTS encoding, or using multi-channel PCM tracks, test the downmix.
- If you do your own encoding, request a copy of the *Dolby Digital Professional Encoding Manual* from Dolby Labs.

Production Essentials

The Zen of Subwoofers

If you understand the proper use of the LFE channel, you belong to an elite minority. As Dolby Labs has pointed out over and over, apparently with little success, LFE does not equal subwoofer. All five channels of Dolby Digital are full range, which means they are just as capable of carrying bass as the LFE channel. Most theaters have full-range speakers, and do not need separate subwoofers since most of the speakers have a built-in subwoofer. In the home, it is the responsibility of the receiver or the subwoofer crossover to allocate bass frequencies to the subwoofer speaker. This is not the responsibility of the BD player (unless it has a built-in 5.1 channel decoder) or of the engineer mixing the audio in the studio.

The LFE channel is heavily overused in most audio mixes. A case in point is a certain famous movie released on laserdisc with Dolby Digital 5.1 tracks in which the booming bass footsteps of large dinosaurs were placed exclusively in the LFE channel. When downmixed for two channel audio systems, the bass was discarded by the decoder, leaving mincing, tiptoeing dinosaurs; this was not the fault of downmixing or the decoder. It could have been fixed by a separate two-channel mix, but it also could have been fixed with a proper 5.1 mix or a proper 5.0 mix. Had the footsteps been left in the main five channels, which are all full range, the bass effects would have come through properly on all home systems.

Dolby Digital decoders ignore the LFE channel when downmixing. This is because the typical stereo or four channel surround system does not have a subwoofer. Again, this does not mean that the LFE channel represents the subwoofer, only that the extra "oomph" intended to be supplied in the LFE channel is inappropriate and could muddy the audio. Dolby Digital decoders all have built-in bass management[2], which directs low-frequency signals from all six channels to the subwoofer, if one exists. Therefore, the LFE channel should be reserved only for added emphasis to very low frequency effects such as explosions and jets.[3]

As illustrated by the tiptoeing dinosaurs, moving low-frequency audio to the LFE channel does not make the soundtrack better in the home environment. This mistaken approach is what has caused complaints about the LFE channel being omitted in downmixing. Nothing important in a film soundtrack, including low-frequency audio, should ever be moved out of the main five channels. Isolating low-frequency audio in the LFE channel does not remove any sort of burden from the center and main speakers since the built-in bass management already does this.

It's possible to mix all "subwoofer sounds" without an LFE channel at all. Some theatrical releases are mixed this way. Such a mix will play the same in 5.1 channel home theaters as in cinemas, since the decoder and receiver automatically route the bass to the speakers that can best reproduce it. The .1 channel is present only for an extra kick that the audio engineer might decide should only appear in full 5.1 audio systems.

---

[2]Most A/V receivers support the ability to disable all post processing, including bass management. It sometimes has names like, "Pure audio mode" or "Direct mode". If a consumer has all their speakers configured as "full range", this will usually bypass all bass management.

[3]Roger Dressler, Technical Director for Dolby, has remarked that the LFE channel should be renamed as the "explosions and special effects" channel in order to avoid confusion.

Slipping Synchronization

An occasional problem with BD is lack of proper synchronization between the audio and the video). Players are, sometimes, partly to blame for the problem but steps can be taken during production to keep synchronization tight.

Check the timecodes on the audio to make sure they are correct. A handy trick for testing audio sync, especially when no timecodes exist and it has to be aligned by hand, is to bump the audio track forward several frames, check playback, bump it back several frames from the original position, and check playback again. If moving it one direction had little or no effect, while moving in the other direction had a very big effect, chances are that the original position was not optimal.[4] Further, make sure the master print was not incorrectly dubbed before being transferred to video.

If sync steadily worsens, check to see if non-drop timecode is being used on the audio master. The difference between 29.97 and 30 fps will slowly move the audio out of sync with the video. Also make sure the film chain in the telecine machine is run at the proper video speed. If sync suddenly changes, suspect a slip during a reel change in the telecine process or an edit in the NLE system, independent of the audio.

Preparing Subtitles

Subtitles can be prepared graphics, like with DVD, or they can be text that is rendered by the player. Graphics or text subtitles are presented on top of the video using the Presentation Graphics plane, and behind any Java graphics (which are in the Interactive Graphics plane).

Subtitles are basically the timed display of a text file or a graphic. The possible uses for subtitles are endless. They can expand and alter existing video, emphasize regions on the screen, cover or censor, tint the video a different color, add a logo bug, and so on.

A handy feature of subtitles is that they can be forced to appear under program control. This is especially useful for multilanguage discs because the video can be clean and the subtitle can be made to appear in the appropriate language during playback of the disc.

Subtitles can be used to create limited animation overlaying the video. A graphic subtitle rate of 15 frames per second is the practical limit for many players, although anything faster than 2 per second may cause problems on some players. Lowering the maximum data rate for the other streams that are present can help avoid problems. Likewise, using simple pictures that encode more efficiently than large, complex ones can make all the difference.

Effects such as fade, wipe, and crawl can be performed by the player. Good subtitling software will be able to add the display commands needed to create these effects.

Tips and Tricks

The following tips will assist you in preparing subtitles —
- Subtitles cannot cross chapter points or mark IDs.

---

[4] When using this trick, keep in mind that synchronization problems in audio delayed behind the video are less perceptible than in audio that precedes the video.

Production Essentials

- Pre-rendered subtitles (graphic subtitles), need to be rendered for HD as Presentation Graphics (8-bit indexed palette graphics) and setup for a 1920 × 1080 display. SD subtitles that were made for DVD will look low resolution and aliased when attempting to reuse for HD.
- Subtitles that will be rendered by the player (TextST) are limited to 25-50 characters per second (cps). You may have to adjust/edit the dialog and/or translations to meet this rate. Graphic based subtitles do not have this cps limitation.

Subtitling, particularly creating foreign language subtitles, is a tricky job best given to a subtitling house. Provide the subtitling service with a proxy video copy that has clear audio and *burned-in timecode (BITC)* that exactly matches the BD video master. This is critical for frame-accurate subtitling work; timing is very important.

You may request that the subtitle source be provided in text form rather than graphic form, as text is much easier to change, but text subtitles must then be turned into graphics unless you are using player-rendered text. If you are using player-rendered text, the subtitling house must use the same font as you will distribute on the disc. Please note that when you distribute a font on a disc, you must either license the font (and pay the appropriate fee, unless it is a free font) or create your own font. In most cases, you are better off using subtitles that have been pre-rendered to graphics.

If you are creating your own subtitles, consider the following —

- Don't use serif fonts. Sans serif fonts are easier to read, as the serifs tend to create single-pixel horizontal lines that produce interlaced twittering.
- Use a drop shadow or an outline, either black or translucent (using the alpha channel of PNGs).
- Each new subtitle should appear at the scene change, not at the dialog start. This provides more on-screen time, and studies show that it reduces reading fatigue.
- Center the subtitles. If there is more than one line, left-justify the lines within the centered position (or right-justify the lines for right-to-left writing systems). Justifying the lines makes them easier to scan.
- Try to limit subtitles to two lines. Keep the first line shorter, which covers less video and clues in the eye that there is another line.
- If there is more than one speaker, indicate speaker change with a hyphen at the beginning of the line.
- Display subtitles for no less than one second and no longer than seven seconds.
- Keep in mind that jumping to the middle of a scene, or using fast forward or rewind, will not refresh a subtitle. Subtitles with long periods will take a long time to appear. To avoid this problem, you may wish to repeat the same subtitle at intervals of five to ten seconds, if necessary.
- Proofread! With thousands of subtitles, especially in foreign languages, it is easy for misspellings to occur.

Production Essentials

Preparing Graphics

Graphics are much more granular for BD than with DVD. For BD, graphics can be created for a full-screen menu, but graphics are also created for buttons, animation sequences (such as load icons) and highlight/selector indicators. Graphics are also used for slideshows, such as photo galleries and information pages. Graphics can be tricky and the traditional tools for creating computer graphics are not always well suited for video-based graphics.

Colors

Graphics should be created in 24-bit mode. Choose RGB colorspace (or YUV colorspace, if available) rather than CMYK or other print-oriented colorspaces. The file format should be either PSD, PNG or JPEG. PNG is the basic format and it has an advantage over JPEG in that it has an alpha channel. Whichever format(s) that you use is ultimately determined by your authoring system or programming framework. Please be aware that JPEG compression can cause artifacts that may be compounded by MPEG compression. If you must use JPEG, choose the highest quality level that is compatible for BD. With *Photoshop*, for some odd reason, this means choosing a quality level of 6 to ensure compabitility.

HD video uses the ITU-R BT.709 colorspace (RGB values in the 0-255 range). While any SD mode video on a disc is ITU-R BT.601 colorspace (RGB values in the 15-235 range) and saturated colors, especially bright reds and yellows, will "bleed" on the display. With a 1920 × 1080 display many, if not all, of the color issues that existed with DVD are gone. The only thing to keep in mind is that there are multiple playback resolutions — 1920 × 1080, 1440 × 1080, 1280 × 720 and 720 × 480. Choose and then test the resolutions that you want to support with your design. It is possible to use some very small fonts with the HD resolutions. These fonts may not be readable on lower resolution monitors or at the viewing distance of a user's display.

Image Dimensions

The pixel density of the image does not matter. Whether you use 72 dpi or 96 dpi or 300 dpi, what matters is the final pixel dimensions of the output. However, pixel geometry does matter when using SD material. Most computer graphics programs use square pixels. SD material on BD is based on the ITU-R BT.601 format that does not use square pixels. In *Photoshop*, choose rectangular pixels for your SD graphics. For HD graphics, use square pixels (the default) and you're good. The screen size for HD is 1920 × 1080 pixels, so it is best to work on a computer monitor that has at least this resolution so that you see an exact pixel representation with no scaling.

Safe Areas

Most SD televisions employ overscan, which hides the edges of the picture. Overscan can hide as much as ten percent of the picture. Two common *safe areas* are used to account for this. The *action-safe area* defines a five percent picture perimeter as the guideline for the outer boundary for action and important video content. The *title-safe area* defines a ten percent picture perimeter as the guideline for the outer boundary for text and other vital information (see Figure 12.5).

Production Essentials

Figure 12.5 Video Safe Areas

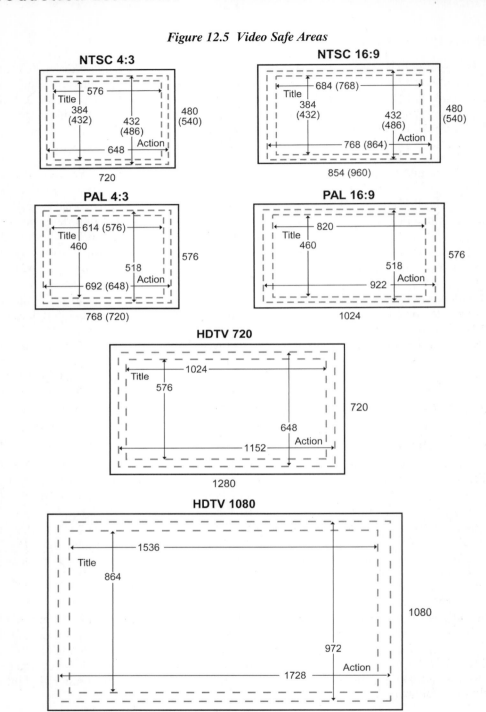

Production Essentials

Some programs have built-in video safe templates. In *After Effects*, for example, press the apostrophe key to turn the safe area markers on and off. Most modern TVs do not have more than 5 percent overscan, but older models may approach 10 percent. Computers, LCD or plasma screens, and many video projectors have little or no overscan. Although, unless the display is 1920 × 1080, you still need to account for action and title safe areas. Most often, 7.5% is a good allowance for action and title safe on the newer high definition displays.

Video Artifacts

Interlaced SD televisions are being replaced in large numbers around the world with the move to HDTV, but as we have said before, not all HDTV monitors are created equal. You may still need to address that your high definition Blu-ray Disc video might be displayed on a non-progressive display, and therefore you must consider interlace artifacts. The fundamental problem is that as only every other horizontal scan line is displayed at a time, thin horizontal lines will disappear and reappear 30 times a second. This interlaced *twitter* effect (also called *flicker*) can be especially bad with computer-generated graphics. In addition to interlace artifacts, *chroma crawl* and *color crosstalk* from composite video signals make thin lines or sharp color transitions problematic, regardless of their orientation.

To reduce these effects, use the antialiasing and feathering features of your graphics software, whenever possible. Make sure that all horizontal lines are at least two pixels wide. To be even more careful, keep the thickness of horizontal lines at multiples of two pixels.

Use gradual color transitions instead of sharp contrasts between dark and light colors. To adapt existing graphics, or as a final pass before encoding, apply a blur filter to the entire image or to offending areas. For example, a *Gaussian blur* of 0.5 to 1 has little visual effect but helps to reduce video artifacts. Also, because blurring reduces high-frequency detail, compression will be more efficient, possibly resulting in higher quality.

Text is also affected by interlacing and other video artifacts. Small point sizes produce single-pixel lines, which are bad, especially the horizontal lines created by serifs. Small text sizes are also difficult to read from standard viewing distances. The smallest object that normal human vision can discern subtends one minute of arc on the retina. Studies have shown that for legibility, the height of a lower case character must subtend at least nine or ten minutes of arc; more as the viewer moves off axis. ANSI standards recommend 20 to 22 minutes of arc, with a minimum of 16 minutes. Angular measurements are independent of screen size. This means, for example, that at a viewing distance of five feet, 21 minutes of arc equates to 0.37 inches, while at a viewing distance of 15 feet, 21 minutes of arc results in 1.10 inches. As a rule, stick with 24 point or larger sizes. 14 point, bold text should be the absolute minimum. Use antialiased text whenever possible.

Tips and Tricks

Here are a few tips that will help you in preparing graphics. Table 12.6 also provides some guidelines to follow —

Production Essentials

- Proofread all graphics before importing them into the authoring system. Don't rely on the graphic artist.
- If you work with clients, have them sign off on every graphic and menu before the start of authoring.

Table 12.6 Video Graphics Checklist

	Standard Definition Video	High Definition Video
Pixel geometry	For 60hz video (NTSC): Create at 720×540, resize to 720×480. (For 16:9 anamorphic, create at 854×480 or 960×540, resize to 720×480.) Or create at 720×480 if vertical distortion is not a problem. For 50hz video (PAL): Create at 768×576, resize to 720×576. (For 16:9 anamorphic, create at 1024×576, resize to 720×576.) Or create at 720×576 if horizontal distortion is not a problem.	Use square at 1920×1080. Anamorphic is only a concern for video that is 1440×1080 video.
Colors	No bright reds or yellows. Keep saturation below 90 percent or RGB values below 230. Not an issue for 50 hz video.	Full 0-255 range is available because the video is Rec709.
Signal Range	Check for below black (L 16 in Rec 601 Video) or above white (L 235 in Rec 601 Video).	Full 0-255 range is available because the video is Rec709.
Safe areas	Keep text and important detail about 70 pixels from the sides, and about 50-56 pixels from the top and bottom.	Keep text about 140 pixels from the sides and 80 pixels from the top and bottom.
Small detail	Avoid single-pixel lines, especially horizontal. Use antialising and blurring. Use feathering and color gradations instead of sharp transitions.	Avoid single-pixel lines, especially horizontal. Use antialising and blurring. Use feathering and color gradations instead of sharp transitions.
Text	14 points minimum. Avoid serifs. Always antialias your text.	Varies with the viewing distance. 6 points minimum.

Putting It All Together (Authoring)

Once all the assets are prepared and proofed, bring them into an authoring system and knit them together. A variety of authoring paradigms are available. Some systems stick close to the BD specification, while others try to hide terminology and details "under the hood" of a graphical user interface (GUI). Most provide an onscreen flowchart, timeline, or a combination of the two. Large production environments may benefit from networked authoring workstations, where each station focuses on a specific task (encoding, menu creation, asset integration, simulation, formatting, and so on) and the project data flows from station to station over a high-speed network.

Production Essentials

At the authoring stage, specify basic parameters of the volume such as video format and disc size. Import audio and video, synchronize them, and add chapter points. Then, apportion titles and title sets. Create menus by importing graphics and subpictures, link buttons together for directional highlighting, and link buttons to video segments or to other menus. These are the essential authoring tasks. More complex projects require much more detailed work.

Once everything is imported and connected, simulate the final product. Good authoring systems can show menus, video, and audio, on-the-fly, giving a reasonable facsimile of how the disc will look in a player. Take advantage of the simulation feature to test navigation and overall layout before going further.

The authoring system is responsible for verifying compliance with the BD specification. The better the authoring system, the better a job it will do of warning you when you have violated the spec. Better authoring systems will also provide information or warnings from the "reality spec" — they will warn you about things that are allowed by the spec but may cause problems on certain players. Some authoring systems also implement behind-the-scenes workarounds to avoid known player deficiencies.

Region coding and CSS or CPPM flags may be added in the authoring system. In general, the actual encryption is done by the replicator. This is the best option since the replicator will have already executed a license and will have been given sets of disc keys and media key blocks.

Tips and Tricks

The following tips will help you with the authoring process —

- Make sure you have more hard disk storage space than you think you could possibly need; BD assets are even larger than DVD and take up huge amounts of space. The first time you have to suspend a project because of a bottleneck and start work on a different project, you will be driven mad by the amount of time it takes to backup a project from your RAID storage to near-line or off-line storage, such as tape or hard disk. You may want to consider allocating a 1TB or larger hard disk drive for each project.

- If you must do a layer change in the middle of the video, look for slow-moving pictures, a still, or a fade to black. The secret to a smooth transition is low data rate, which gives more time for the BD player to refocus on the second layer before the buffer underflows. At the encoding stage, make the data rate as low as possible for a few seconds going into the layer change, as well as a few seconds coming out of the layer change. The players have buffers to make the layer change undetectable even when there is motion in the video. This is not as critical for BD as it was with DVD.

Formatting and Output

Once the authoring, debugging and testing process is complete and everything works on the Sony PS3 and your other targeted hardware players, it is time to create the final disc

Production Essentials

image. The authoring system will add the assets together with the programming information. The resultant fileset is stored in the BDMV directory. The multiplex is stored in the files under these directories of the BDMV directory —

    PLAYLIST
    CLIPINF
    STREAM

The following directories include the programming (including for BD-J) —

    BDMV\AUXDATA
    BDMV\BACKUP
    BDMV\BDJO
    BDMV\JAR
    BDMV\META
    CERTIFICATE

The AACS directory will be used when you create the final image for replication.

Your debugging and testing process will have you re-creating these directories and files, repeatedly. Unlike with DVD, you can simply replace the files that are updated and then test again. If you are just updating the programming, you can use a multi-session BD-R/RE disc to record the changed files. Recording just the changed files will saved a lot of time. Whereas, a record of a full disc will take 45-90 minutes for a BD-25, a multi-session record with just the changed BD-J code will take only 1-2 minutes. Plan on going through this step many, many times as you debug and test your project on the various hardware players.

The data files can be used from a hard drive for use with a computer software player to test for problems that do not show up when simulating the disc in the authoring system.

Once everything is working perfectly, you will create a BD CMF image. This image file set includes one file for each layer. The file set can be delivered to the replication factory either via a fast, secure network connection or on hard disk drive.

Testing and Quality Control

Testing is the most important factor in the success of a project. If you work with clients, quality control is vital to your reputation. Testing should never be left to the end; it must be an ongoing process from the beginning. The BD players all have idiosyncracies that necessitate testing on many players to ensure that the disc will work the way you intend.

The first testing stage is source asset checking. Never assume that your supplier will get it right. QC all incoming assets, at least until you have verified which suppliers you can rely on to always get it right. Although it seems redundant and a drain on time and resources, it will save you money and time in the long run. Screen all incoming assets for quality and accuracy. View every video source and encoded video file. Audition all audio files. View every graphic file, checking for spelling mistakes and missing elements. If you cannot check foreign language subtitles, make sure the supplier guarantees accuracy and takes responsibility for costs associated with reauthoring the disc in case of subtitle errors. Check timecodes and synchronization between audio and video and between subtitles and video.

Production Essentials

Review video and audio during and after compression. Seamelss branching, if it exists in the title, should be checked for smooth audio switching, as well. Check encoded audio against the master to spot sync, level, and equalization problems. Check all text before it goes to the graphic artist. Provide text to the graphic artist as text files that they copy and paste (or otherwise import) into the graphics application. Do not let them type text — they always misspell. If you work with clients, have the client proof and sign off on all the prepared assets (video, audio, and graphics) before authoring. Check navigation, layout, and general functionality as much as is supported by the simulation features of your authoring system.

After outputting the BDMV file set, do navigation testing with computer software players. At this point, check whether the computer has reliable playback. Also be sure to check menu functionality, since the mouse-based user interface changes the way things work. Software player testing can catch errors before you spend the time and expense of writing a BD-R/RE.

QC of video and audio, especially synchronization, is best done on hardware settop players. You will want to test on any target display types (CRT, LCD, DLP, et cetera), as well as at 1920×1080 resolution and any other resolutions that you expect endusers to be using.

Although testing on a software player will help find and fix navigation issues, include testing on hardware players early in the process. There is a vast difference in performance and compatibility of the players (whether software- or hardware-based). You can choose a few players to streamline your debugging process and then expand to a more comprehensive list of players before finalizing the project. Reserve some of this *matrix testing* time for replicated check discs. Consider sending a BD-R to a verification service.

Check with your replicator, but many replicators will require that your final disc image has passed a specification verification test. If you do not have this software in-house, consider sending this to a BDA-approved testing center.

When you are sure everything works and complies with the specification, send the master image to the replicator and request check discs. The check discs will require AACS encryption and the fees that go along with that process. Each master will require a new set of fees, so do everything you can to ensure that the check discs have no surprises. Check discs are a short replication run from the stamping master. A check disc is essentially the same as a final mass-produced disc. If you approve the check disc, the replicator can use the existing master for the full replication run. Most replicators include check discs as part of their replication service charges. However, they will charge additional fees if you find errors and have to send in a new master for another set of check discs. Unless you have thoroughly tested with BD-R/REs and are comfortable with the potential of having thousands of shiny coasters at your disposal, do not skip the check disc step.

Tips and Tricks

Table 12.7 and the following tips may help your testing and quality control to be more effective and efficient —

- Test on a cheap TV, but also test on each of the display technologies — LCD, DLP and Plasma (and maybe CRT). Most of the production process is done with high-end, high-quality video equipment, but your customers could watch your work on a cheap, old television or the latest 60-inch display. Make sure that details of video and graphics, menu highlights, and so on are not lost on your

Production Essentials

target user. Each of the display technologies has artifacts that may make your work look terrible. Check your work beforehand so that there are no surprises.

- It is almost impossible to test all the permutations of a highly-interactive disc; try to break things down into testable subsections. If you are using random functions in your programming, you may want to do your testing folks a huge favor and design in a way to bypass/configure them to allow easier QC of the disc.
- If you have computer software that can mount a formatted BD image file from your CMF image on a hard drive, mount the volume for more thorough emulation. This will find file system errors that normal emulation will not find.
- If you put computer applications or large amounts of computer data on the disc, use a checksum program to verify the accuracy of the original data against BD-Rs, CMF images, check discs, and final discs. A file-compare program such as *windiff* is also useful for verifying filenames and data integrity. Be careful when testing BD-Rs, as some BD-ROM drives have problems reading BD-Rs and will report errors that will not be present on a check disc or the replicated copies.
- Although putting Macintosh files on the disc is not as much of an issue as it was before Mac OS X, make sure that the icons display properly and that the application bundle directory is still intact. The application bundle directory name will show when viewing the disc under MS Windows or other operating systems.
- Most of the BD players can play BD-R/RE media. Use multi-session recording mode to reduce the time it takes to record your test discs. Most often you are only changing the programming which are small files (only a few megabytes) vs the multiple gigabytes for the audio and video multiplex files which do not change as much. This will cut your recording time from 45-90+ minutes to 30-60 seconds.

Table 12.7 Testing Checklist

Category	Criteria
Intro sequence	Does the disc do what's intended when inserted? Are the intros annoyingly longer than you realized?
Menus	Check arrow movement in all four directions from each button. Test that Top Menu, and Pop-up Menu functions work as intended. Check motion loops and idle outs. Make sure menu animations are not annoyingly long and menu audio and button sounds are not too loud or annoying.
User operations	Which functions work? Which are locked out? Are they supposed to be locked out?
Video	Verify quality. Check chapter search and Previous/Next functions. Check any angles, branching, and layer switch. Test on interlaced and progressive displays at 480, 720 and 1080 resolutions.
Audio	Make sure languages match the language name displayed by the player. Test multichannel playback and stereo downmix. Carefully look for audio sync problems. Test the audio in the analog, SPDIF and HDMI output paths.

continues

Table 12.7 Testing Checklist (continued)

Category	Criteria
Subtitles	Check that subtitles appear at proper times. Make sure languages match the language name displayed by the player.
Closed Captions	Verify that Closed Captions can be decoded by a TV.
Parental control	Verify that disc or section lockout works. Try the disc at each parental level, and also with no parental level set.
ROM Data filenames	Make sure long filenames were not truncated and that special characters are intact.
ROM data integrity	Verify readability of all files. (If there are errors with a BD-R, try a different drive.) Verify file and directory counts as well as byte size comparison with original file set. MD5 checksum comparisons can be used for key files.

Replication, Duplication, and Distribution

At this point, things are mostly out of your hands. Your BD CMF image delivery, representing weeks or months or years of your life, is in the hands of someone else. Most replicators have excellent processes in place and will take good care of your project.

The replicator will use the BD CMF image to cut a glass master and make stamping masters. Some replicators first verify the BD CMF image formatting and check for specification compliance. The replicator will send you check discs made from the master. Once you approve the check disc, then thousands or millions of copies of your disc will be churned out.

If you want to serialize the individual discs with ID, tracking or serial numbers, or put other custom information in the BCA, a small number of replicators have the necessary equipment. If you need to make less than a hundred discs, you may want to use DVD-R or BD-R duplication instead. You can burn DVD/BDs by hand, buy a DVD-R or BD-R bulk-duplication system, or have a service bureau duplicate the discs.

Most replicators can do *source tagging* — inserting *electronic article surveillance* (EAS) labels — into packaging. The two most popular source tagging technologies are *acousto-magnetic* tags from *Sensormatic* and *RF* tags from *Checkpoint Systems*.

If you need a rush job, schedule replication early, since large orders (especially things such as *Star Wars* discs or Microsoft operating systems) can sometimes swallow all available replication capacity. Establishing a relationship with one or two replicators will help when you have special demands.

Disc Labeling

Many options for disc labeling are available, such as silkscreen printing, offset printing, pit art, and holograms. The area of a disc available for label printing normally extends from the 46 mm to the 116 mm diameter points (the data area runs from 48 mm to 116 mm). For special applications, the print area can start at 34 mm, just past the clamping area. The BCA area is at 44.6 mm to 47 mm, so discs that use the BCA must start the label at about 48 mm.

Production Essentials

Silkscreening applies layers of colored ink to the label side of the disc. If more than two colors are used, or if the disc has large ink coverage areas, care must be taken that the disc does not warp as the ink dries. Offset web printing causes fewer warping problems from ink shrinkage than screen printing, but it is more expensive.

Pit art is produced by creating a stamper that embosses a graphic design onto the back substrate. The black print of the source graphic will appear as a mirrored surface, while the white areas of the graphic will appear as a frosted surface. The pit art area extends from 40 mm diameter to 118 mm. The source graphic is a one-bit monochrome image, usually 2048 × 2048 pixels. Check with your replicator, as this option may not be available from your replicator.

Package Design

Packaging is the spokesperson for your disc. Make it reflect the contents of the disc and grab the attention of the customer. Even with Internet shopping, the disc jacket influences the buyer's impression of the work.

Design with the collector in mind. Part of the reason people buy BDs rather than renting them or taping them (or waiting for the network broadcast) is because they enjoy collecting. Use quality artwork on the package. Produce an informative and enjoyable insert. Consider including little goodies in the package that relate to the disc.

Explain what is on the disc. Do not assume that the customer is familiar with the contents. Include a list of titles, chapters, extras, and so on. Use standard text and icons to indicate the format of the contents. Clearly identify number and type of audio tracks, aspect ratio, disc region, video format, running time, and so on (see Figure 12.6). Sometimes it is a good idea to delay package design until the end because details of aspect ratio, soundtracks, extras, and so on may change during production.

Figure 12.6 Package Icons and Identifiers

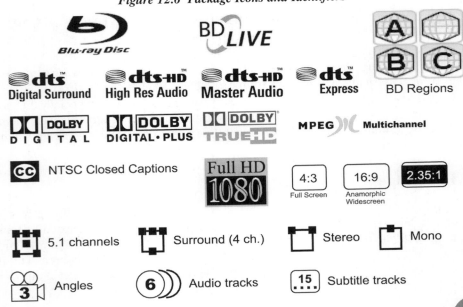

Production Maxims

For your edification and enjoyment, here is a small collection of maxims, reminders, and rules of thumb — many of them learned the hard way — collected from various BD and DVD authors —

- Entry points (chapters and mark items) must start at a GOP header and must be at least 0.4 seconds apart.
- Corollary: An I frame is not necessarily a GOP header. To create an entry point, you must force a GOP header, not just an I frame.
- Subtitles cannot cross an entry point (mark item).
- GPRMs are reset to zero on disc change (with title bound applications) or when the player is stopped.
- Any amount of QC on a replicated check disc is worth more than the cost of paying for a disc run that failed for the simplest problem. If there will not be enough time or budget to go through check disc testing, then you, above all others, *need* to test the check disc.
- Last, but not least, no disc is ever *easy*, *simple* or *trivial*, and saying that it will be usually guarantees that it won't be.

NOTE

Anyone who is serious about producing Blu-ray Discs should invest in most of the following equipment —
- a 1080p 1920 × 1080 LCD monitor,
- a 1920 × 1080 Plasma display,
- a 720p LCD or Plasma display,
- an expensive 1920x1080p CRT monitor,
- a Blu-ray Recorder compatible PC, and
- a full audio playback setup with either 5.1 or 7.1 channels to be able to properly preview the audio.

BD-ROM Production

The following subsections go over a few of the basics of producing a BD for computers. Also, see the testing section earlier in this chapter.

File Systems and Filenames

BD-Video discs use the UDF (*Universal Disk Format*) version 2.5 file system. This file system does not include ISO 9660 and Micro UDF as with DVD-Video. A BD for computer data may be formatted with any file system — UDF 2.0, ISO 9660, Windows NTFS, Apple HFS, EXT2/3 and so on. However, many of the older file systems do not support volumes greater than 4 GBytes or files larger than 2 GBytes or 4 Gbytes. UDF 2.5 solves many of

Production Essentials

these problems and has been optimized for optical disc. UDF 2.5 can handle files up to 16 EiBytes and volumes in excess of 8 TiB which will handle even potential 20 layer 500 Gbyte discs.

Not all operating systems can read the UDF 2.5 file system. For Windows you need Vista. For WinXP, you need a special UDF 2.5 filesystem driver. For MacOS, you need 10.5 or later. For Linux, depending on the distro, you need 2.6.26 or later. UDF 2.5 supports filenames of up to 255 characters using the Unicode character set, which supports characters from almost every language. However, an operating system has no guarantee that it will be able to correctly display a given Unicode character. The pathnames can be up to 1,023 characters.

If you need compatibility with older operating systems such as Windows 95, DOS, and Unix that are unable to read the UDF 2.5 file system on BD volumes, you are probably better off staying with DVD-ROM and using the older techniques such as implementing the Joliet extensions to allow long filenames within the ISO 9660 file system.

In the final premastering step for replication, if this will be a data-only disc, you can deliver a UDF 2.5 image file for a BD25 on a hard disk drive. For a BD50, you will need to produce a CMF image where each layer has its own image file. This is called a *BDCMF type P* image. For a data-only discs (no BD-ROM video), AACS is not needed.

Bit Budgeting

Bit budgeting on BD data discs is a breeze compared to BD-ROM Video. You just add up all the file sizes and see if they fit on a disc; however, a few potential pitfalls might arise. Don't forget the difference between billions of bytes and gigabytes. The operating system will report file sizes in gigabytes, but BD capacities are usually given in billions of bytes. Don't forget to leave some room for the file system directory structure. When there are hundreds or thousands of files, the file system overhead can be significant. Leave at least 4-7% for this overhead. As a final check, for a BD25 disc, create a UDF 2.5 image file and make sure it is less than 25×10^9 bytes minus the 4-7%. Note, not all formating software is created equal, so the same file set will have a different UDF 2.5 image size depending on the formating software. Some software is much more efficient than others. If you are just over the limit with one formating software, try another.

When adding up file sizes, use the actual file size, not the "size on disk" value, since the space taken on the hard disk is dependent on the block size. Likewise, space taken up by each file on a BD can be more than the actual file size. For a very tight bit budget, adjust the size of each file by the BD sector size — divide the file size in bytes by 2048, round up, then multiply by 2048. This will give the sector-adjusted file size, which represents the true amount of space the file occupies on the BD.

Creating a two layer disc, e.g. a BD-9 or a BD-50, is not as simple as it might seem. You need an image file for each layer. Simply dividing your project in half and creating two disc images will not work. A dual layer disc has a single UDF bridge directory section located on the first layer that references files on both layers. If you simply use formatting software to create two images, you will get two volumes, each with its own directory section. The first layer will work fine, but the second layer will be inaccessible. Use a BD authoring/premastering application or a BD-ROM formatting utility that knows how to create BD-9 or BD-50 CMF images, preferably one that lets you specify where the layer break occurs.

A/V File Formats

When creating a multimedia BD for playback on PCs, you have essentially two choices for the A/V format — BD-ROM Video files (.m2ts), or computer-oriented media files (.flv,.avi, .mov, .mpg, et cetera). The advantages of using BD-Video content are that the disc will also play in standard BD players, the features of DVD-ROM Video such as angles, seamless branching, fast chapter access, and subtitles are available, and the video will be more compatible with PC decoders.

The advantages of using other formats are that the development tools are simple, widely available and, often, inexpensive. Since the data rate of most BD drives is at least equivalent to an 80x CD-ROM drive, high data rates can be used with a variety of codecs beyond those that are included in the BD specification — *Windows Media Video*, *Flash*, *2K* or *4K*. Rich Internet applications that have interactivity beyond what is possible within BD-Video may also be much easier to create that work on both Windows, Linux and Macintosh computers by using tools such as *Adobe AIR* or *Microsoft's Silverlight*. Older technologies, such as *Adobe/Macromedia Director*, *Click2Learn Toolbook*, or even HTML authoring applications, can be used to create multimedia experiences that surpass even the vast array of capabilities included in BD Profile 2.0.

Hybrid Discs

If you are adding computer data/files to a BD-ROM Video (or adding BD-ROM Video content to a data disc), be mindful of a few things. If you will be replicating the disc, you will need to use a formatting or premastering software tool that allows for both data files and BD-ROM Video files. Multiplexing and file topology on the disc are critical to ensure proper playback of BD-ROM Video. Be sure to run a verifier on the final UDF 2.5 image to ensure that the added computer data has been properly interleaved and layed out with the BD-ROM Video files and that the appropriate padding has been reserved for the AACS data. These discs will play fine in software players on PCs and work on all of the BD settop players. If you add more than a few data/computer files to a BD-ROM Video disc, put them in a subdirectory. Too many extra files in the root directory may cause some BD players to fail. Many authoring/premastering software tools assume that the BD-ROM Video will take up most of the space on the disc. But, if you have a lot of space consumed by the computer files, make sure that the premastering software lets you have control over where the files will be placed on the disc.

Disc topology or where the files go on the disc is critical especially for multilayer discs. For database applications, you want to ensure that the indices are placed close to the data. You do not necessarily want the indices to be on one layer and the data on another. You want to be able to fine tune the file topology so that your BD-ROM Video content and the computer application files have the best performance when played on the final replicated optical disc.

Production Essentials

Figure 12.7 Sample PNG Mosaic Graphic

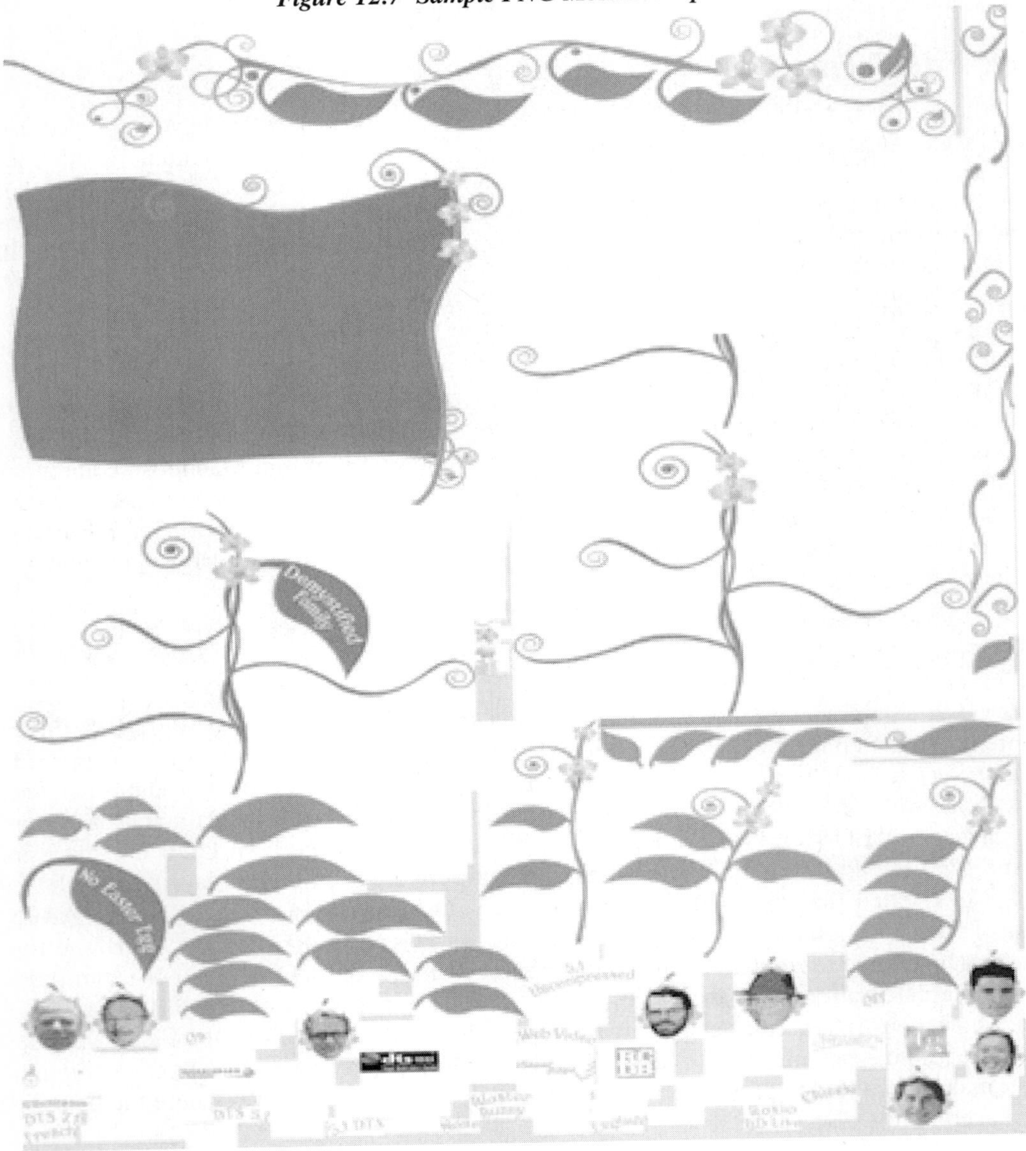

Chapter 13
Blu-ray and Beyond

This chapter examines past and present forecasts[1] and looks ahead to possibilities for Blu-ray over the next decade or so. Questions, such as, "When will the format war end and which format will emerge victorious?" have been answered, but many other questions remain — "When will Blu-ray discs or players begin to outsell DVD discs or players?" "Will we ever need another physical video format?" "What comes after blue laser?" "Will the Internet obviate the need for physical media?" This chapter touches on these and other questions.

In the fall of 2008, as this book was being finished, Blu-ray Disc™ had just turned two years old (counting birthday candles from first disc and player shipments in June 2006). By many measures, it is doing better than DVD did at the same stage, but there are obviously several apples-to-orange factors at play. One major difference is that the Blu-ray format has the *Sony PlayStation 3* (PS3) to give it a major boost.

Looking Back at DVD

In 1997, the first year of DVD, 349,000 players were sold to dealers in the US. This was more than double the first three years of VCR sales from 1975 to 1977, and represented more than 12 times the first-year sales of CD players in 1983 (see Table 13.1). In comparison, in 2006 — Blu-ray's first year — only 25,000 standalone players were sold in the US, although adding PS3 players to the mix brings the total to 425,000. Admittedly, DVD was released in March of its first year, compared to Blu-ray's June launch, but the DVD launch was limited primarily to Warner Bros.' seven-city trial until August.

Table 17.1 Technology Penetration Rates – Years to Reach 50% of US Homes

Technology	Years	Technology	Years
Telephone	71	Cellular phone	14
Radio	10	VCR	11
B&W TV	9	CD	11
Color TV	17	Personal Computer	19
Cable TV	23	DVD	7

Sources: US Census, CEA, IDC, JPR, Technology Futures

[1]Research and projections in this chapter come from many sources including, *Adams Media Research*; *Entertainment Merchants Association* (formerly *Video Software Dealers Association*); *Futuresource Consulting* (formerly *Understanding & Solutions*); *Interpret*; *Kaplan, Swicker & Simha*; *Nielsen Media Research*; *NPD Group*; and *Strategy Analytics*.

Blu-ray and Beyond

Just over one million DVD players were sold in 1998, its second year. Roughly 350,000 Blu-ray players sold in 2007, its second year, although PS3 sales in the US kicked the number up to about 2.6 million. More than four million DVD players were sold in 1999 for a US total of about 5.5 million at the end of 1999. About 2.5 million standalone Blu-ray players sold in 2008, with 5.7 million PS3s lifting the combined sales to around 8.2 million.

Comparing Blu-ray to DVD during the first three years of each format, Blu-ray straddled the line with standalone player sales below DVD player sales while combined BD and PS3 players were significantly higher (see Figure 13.1).

Figure 13.1 US DVD and BD Penetration in the First Three Years (Percentage)

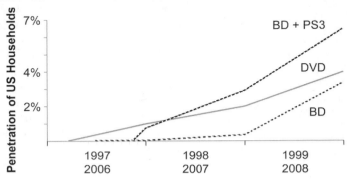

Research results indicate over 80 percent of PS3 owners watch Blu-ray titles and the combined player sales are meaningful, but comparing DVD and Blu-ray player sales head to head does not paint a very precise picture. Disc sales, perhaps, tell a more consequential story.

In the first two years, somewhere between 16.3 and 30.6 million DVDs were sold in the US, depending on who is counting. It seems only fair to combine Blu-ray and HD DVD sales for the period since the two formats were competing for the same customers and were spread fairly equally across the studios. High-definition discs of both flavors sold somewhere between 6.1 and 11.25 million units in the US between June of 2006 and the end of 2007. That is more or less half of DVD sales during the equivalent period.

In October 2000, the ten millionth US DVD player shipped. Three-and-a-half years after their US introduction, DVD players achieved the mark that VCRs and CD players each took eight years to reach. By the end of 2007, over 98 million US households (84 percent) owned a DVD player. This compares to 112 million US households (97 percent) in 2007 that owned a TV. The total number of DVD playback devices — including players, DVD computers, and DVD game consoles — passed the one billion mark in 2007. DVD disc sales reached their peak from 2004 to 2006, generating over $24 billion in US sales and rentals each year. Worldwide DVD sales and rentals in 2006 were over $50 billion.

In June 2008, *Futuresource Consulting* released a report that projected combined sales of Blu-ray players and PS3 consoles in Western Europe would reach 10.48 million units by the end of the year. The report pointed out that only 1.58 million DVD players sold during the similar time period (1997 to 1999). However, DVD was not officially launched in Europe until the middle of 1998, so it was not a fair comparison. And, of course, there is the PS3 effect to consider.

Blu-ray and Beyond

Many dubious and even anecdotal reports floated around, claiming success or disaster for Blu-ray. One wag, upon seeing 21 discs in a store, reported that the format was selling "much better." Even worse than the anecdotal reports were the non-sequiter reports, such as the much bandied claim from the replication equipment maker, Singulus Technologies, that Blu-ray adoption "far exceeded" adoption of DVD. This statement was based on orders for 21 Blu-ray dual-layer machines in the first quarter of 2008 compared to 17 DVD machines in the same period 11 years before. Apparently a four-unit increase in a statistically invalid sample size of different products compared to a dubiously correlated time period under considerably different conditions was taken seriously as an indicator of success. Large numbers of critical thinkers were seen hurling themselves in despair from their ivory towers upon reading the 350 exuberant reports of this assertion that appeared on the Internet.

Nevertheless, Blu-ray was doing fine. It was expected that by the end of 2008 there would be approximately 30 million Blu-ray playback devices worldwide, about two thirds of them PS3s. Over 15 million Blu-ray discs had been shipped by September 2008, with about 1000 different titles available (see Figures 13.2 and 13.3).

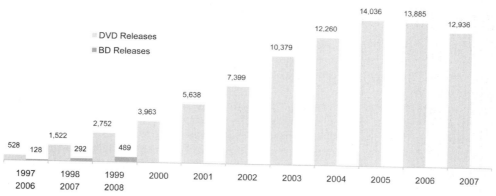

Figure 13.2 DVD and BD Title Releases

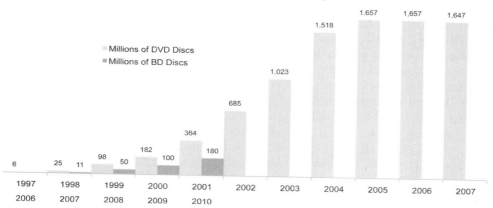

Figure 13.3 DVD and BD Disc Shipments

Blu-ray Disc Demystified

Blu-ray and Beyond

According to the Consumer Electronics Association (CEA), 40 million US households (36 percent) were on track to have HDTV sets by the end of 2008, for a total installed based of 52.5 million sets. Less than half of the sets were set up to receive HDTV programming. HDTV penetration at the end of 2008 was around 35 percent in the UK, 30 percent in Japan, 18 percent in Germany, and 21 percent in France.

Peering Forward...Into the Digital Fog

The variables in predicting the path of Blu-ray are extremely complex, but this has not deterred many people from conjectures ranging from sensible to outrageous. Looking back on how DVD forecasts fared in the face of actual events, it seems that some prognosticators do a reasonable job with ballpark numbers while others should consider becoming weather forecasters. At least they would increase their accuracy to 50 percent.

In 2008, every analyst's favorite setting on the prognostication machine was 2012, four years out. Adams Media Research predicted that by 2012 there would be 100 million Blu-ray players, including 30 million PS3s, worldwide. The Entertainment Merchant Association predicted 115 million. Strategy Analytics predicted 132 million — representing 55 percent household penetration in Japan, 32 percent in Europe, and 44 percent in the US. These were all impressive numbers yet, at the same time, reports appeared about how Blu-ray sales were "stumbling" and Blu-ray would soon be eclipsed by digital delivery and video on demand.

Despite the nattering of naysayers, Blu-ray is guaranteed to succeed. It will be around for decades, especially given its heritage from CD and DVD. Consider the thinking that prevailed in the 1980's and 1990's about technology lifecycles. At the time, almost every new generation of reproduction technology had to displace preceding technologies, requiring a complete changeover of devices and media. This was especially true of analog technologies, but it also applied to digital. Witness portable data storage transitions from 8-inch floppy disks to 5-inch floppies to 3.5-inch diskettes and then to USB drives. Or, video formats from 70 mm film to 35 mm to 16 mm to 8 mm to super 8, along with U-matic tape, VHS, laserdisc, VHS-C, 8 mm videotape, Hi8, Digital 8, Mini DV, and so on, not to mention the litany of professional tape formats. Or, portable music format transitions from 33-, 45-, and 78-rpm vinyl records and reel-to-reel tape to cassette tape, 8-track tape, CD, DAT, DCC, MiniDisc, and eventually MP3 files on portable players. Since conversion between formats was difficult, and formats tended to last ten to twenty years, it was not surprising that CD was expected to peter out after twenty years. But on its 25th birthday, in 2008, CD was far from obsolete. The overlooked difference was that succeeding formats were backwards compatible with CDs. DVD players and DVD-ROM drives, as well as, BD players and BD-ROM drives, all read CDs. The effect carries forward as well. The ability of BD players and drives to read tens of billions of CDs and DVDs makes them significantly more compelling and valuable than so-called disruptive — but incompatible — technologies.

The question is not whether Blu-ray will succeed, it is how successful will Blu-ray be? In other words, how compelling is the format that comes after DVD? DVD is a tough act to follow because it was such a vast improvement over VHS tape. DVD heralded the switch from analog to digital video, added multichannel audio, quick access to programs with chapters, menus and interactivity, multiple audio tracks and subtitles, and much more. Blu-ray offers essentially the same thing. Blu-ray picture quality is better, but DVDs already look pretty good on HDTV sets. Interactivity is significantly improved, but consumers are not exactly rioting in the streets demanding more interactive titles. The promise of fresh new content delivered to Internet-connected players is probably the most interesting new feature. Is it enough to get consumers to make the switch to new players and, more importantly, start changing their collections over to HD format? It depends, in part, on the draw of HDTV. Anyone who has paid a lot of money for a high-definition display will want high quality, high-definition content to feel like they are getting their money's worth.

Some people expect that Blu-ray will be adopted even faster than DVD, which is actually the case if PS3s are counted. However, the format war clearly delayed adoption and allowed consumers to gravitate toward alternative sources, such as, pay-per-view, video on demand, and the Internet. The outlook is brighter on the computer side, where demand for increased storage never abates. Major computer manufacturers such as Apple, Dell, and HP support the Blu-ray format. Notebook computers, now selling more than desktop models, are getting smaller and less likely to have an internal disc drive, but it is still inevitable that prices of BD drives will drop to the point that they replace DVD drives, which largely replaced CD drives after 2005.

Beyond Blu-ray

Blu-ray Disc is almost certainly the last mainstream packaged media format. There were good reasons to raise the bar above DVD, the most important one being high-definition video. Few consumers need anything beyond 1080p high-definition. Digital cinema resolutions, such as, 2k (2048×1152) and 4k (4096×2304), are unlikely to make their way into average homes. Stereovision (3D TV) is gaining momentum, but can be accommodated easily in Blu-ray with minor format updates. In August 2008, Taka Miyama, Sony's product strategy manager for home video marketing in Europe, said, "Blu-ray is the final format for the optical disc. We don't have a shorter laser."

The most promising successor technology to blue-laser optical media is holographic storage, which came on the market in 2007 with commercial products.[2] Holographic systems read and write millions of bits simultaneously in a few milliseconds by arranging them in "pages" stored on 12 cm (5-inch) discs in holographic interference patterns. At introduction, holographic disc capacity was 300 Gbytes, but plans were in place for capacities up to 1.6 trillion bytes (about 32 times the capacity of a dual-layer BD), with data transfer rates over one gigabit per second (more than twice as fast as a 12x BD drive). However, there simply is not a demand for that kind of capacity and data rate in a consumer pre-recorded media format. Computer applications can always use more storage capacity, but without an entertainment format to drive adoption, future computer storage solutions will fragment, as in the days of Syquest, Bernoulli, Zip, Jaz, and MO drives. Holographic storage may flourish, but only in niche markets.

The primary competitor to holographic storage is actually an enhanced version of Blu-ray.

---

[2]This refers to using holographic laser technology to store data, not to three-dimensional holographic video.

Blu-ray and Beyond

In 2005, before blue-laser formats reached the marketplace, variations such as four layers were being demonstrated. In July 2008, Pioneer presented research results for a 20-layer Blu-ray disc holding 500 Gbytes, plenty to store hours of 4k video. Existing Blu-ray readers probably cannot be software-upgraded to read multilayer BDs, but new Blu-ray products could be easily modified to handle multiple layers. Such increased capacity for Blu-ray Discs is especially appealing for smaller 8 cm (3-inch) discs for use in camcorders and the like.

Other future possibilities include superlenses (which focus light to a spot much smaller than its own wavelength, getting past the "we don't have a shorter laser" problem), diamond-based x-ray lasers, charged particle beam recorders, near-field scanning microscopy, atomic force microscope arrays (millipede), memristors, electron spin manipulation (spintronics), nanodots, nanomechanics, carbon nanotube grids, self-assembling polymer arrays, and any number of new technologies for manipulating matter in ways to store vast amounts of data. But, it is likely that any such products will find only niche applications for portable media. Blu-ray will remain the mainstream portable media format until it is eventually displaced by wireless connections and solid-state memory cards, once they become cheap enough.

The Death of DVD?

Just as CD has lived much longer than most people expected, DVD will enjoy a long life in the nurturing shadow of Blu-ray. The total number of DVD devices will eventually pass two billion, which makes for an accumulation that will not dwindle quickly. DVD drives will disappear first, followed by players, but the vast collections of video and data on DVDs, still usable in BD players and drives, will endure long after manufacturers abandon the devices.

Seeing Double

Probably the most significant new development that will affect Blu-ray is *stereovision*, or *3D video*. Beginning in 2005, Hollywood studios discovered that revenue from theaters showing 3D films was two to four times higher than from standard films, partly because the ticket prices were higher and partly because the 3D versions attracted more viewers. (Or maybe it's just that wearing 3D glasses makes you eat more popcorn.) Many studios, especially Disney, Dreamworks, and Paramount, began seriously studying 3D technologies for both theater and home use.

There are, essentially, six technologies that may be used for 3D viewing —

- *Anaglyph*, using the good old red/green or red/cyan glasses;
- *Color shift*, an improvement on anaglyph that filters bands of color into each eye (*Dolby 3D*);
- *Polarization*, usually circular, with polarized glasses (*Real D* and *MasterImage*);
- *Shutter*, using active LCD shutters on glasses synchronized to the display to alternately feed the image to each eye (*XpandD*);
- *Head-mounted displays*, using helmets or glasses with two built-in screens, one for each eye (Sony, Sensics, Nintendo, and many others); and
- *Autostereoscopic*, using image separators, such as, lenticular filters or parallax

barriers in front of the display to produce two slightly different images (Philips, Dimension, SeeReal, Sharp, and others).

Digital cameras, digital processing, and digital projection made the renaissance of 3D movies possible. It is expected that 3D theater screens will grow to 7,000 in the US by the end of 2010, and to about 2500 in the rest of the world.

For theatrical projection, each frame image is flashed six times — instead of the usual two times — to produce a refresh rate of 144 frames per second. For home use, 60- or 120-Hz frame-sequential (left eye/right eye) progressive-scan video at 1080p gives very good results, without the flicker and eyestrain often associated with 3D video. Given the importance of high-definition for good 3D picture quality and the potential need for higher frame rates, Blu-ray is the perfect candidate to carry 3D movies into the living room.

At the CEATEC show in Japan in October, 2008, Panasonic demonstrated a prototype system using a modified Blu-ray player, a 120 Hz 100-inch plasma screen, and polarized glasses. Groups, such as, SMPTE's *3D Home Display Task Force* and the *3D@Home Consortium* are working to facilitate standardization, which could find its way to Blu-ray. The simple approach for Blu-ray is to pass 3D video (using anaglyph or split-resolution approaches) out the HDMI connector to devices that know how to handle it. This has the advantage of potentially working in every Blu-ray player on the market, but the disadvantages of not supporting active 3D technologies, not using higher frame rates, and not integrating onscreen menus and interactivity into the experience by controlling the depth of the Blu-ray graphics planes. It may be that a new generation of BD players (*Profile 5?*) is the best way to add 3D.

The Changing Face of Home Entertainment

The final few years of the 20th century were a watershed period for digital formats. Direct broadcast digital satellite (DBS) systems, such as, DirecTV/USSB and DISH Network, were introduced in the US beginning in 1994. DirectTV signed up more than three million customers in three years, making DBS the most successful consumer electronics format to date. The first DVB-S (satellite) broadcasts began in Europe in 1995 and in Japan in 1996, followed by DVB-T (terrestrial) broadcasts in Europe in 1998. DVD was unveiled in Japan in 1996, launched in the US in 1997, and officially launched in Europe in 1998. It went on to surpass DBS and become one of the most successful consumer electronics formats of all time. The first HDTV sets went on sale in the United States in 1998.

DBS and DVD each had one foot in the past and one foot in the future. They used digital signal recording and compression methods to squeeze the most quality into their limited transmission and storage capacities, then converted the signal to analog format for display on conventional televisions. As digital televisions slowly pushed their creaky predecessors out of living rooms and offices, DVD players sprouted digital video outputs to complete the shift from analog to digital. Blu-ray then took the baton to advance the movement to high-definition, progressive-scan, digital video, leaving behind the flicker and flutter of interlaced video and enabling television personalities everywhere to wear pinstriped fabrics and sit in front of miniblinds.

Consumer electronics devices of the 21st century, such as, televisions, DVD players, Blu-ray players, and mobile phones, now contain computer processors more powerful than super-

Blu-ray and Beyond

computers of the 1980s. As consumer electronics companies and computer makers endeavor to make their products more suitable to the basic entertainment, communication, and productivity needs of home users, features from computers, such as, onscreen menus, touch screens, multiple windows, and even Internet connections, have become commonplace. DVD and Blu-ray are perfect formats for this environment because they are able to carry content for the television, the computer, and all the variations between.

Even movie theaters are being changed by digital technology. The success of digital cinema has shown that film can, and should, be replaced by digital files. Movies are being released simultaneously worldwide, or very shortly after US release dates, as the old model of movie distribution has been changed by digital distribution and by Internet advertising that hits all audiences simultaneously.

Convergence is happening with systems other than desktop computers and TV. The prime candidates are cable settop boxes, satellite receivers, video game consoles, telephones, mobile phones, laptop computers, portable music and video players, PVRs, DVD players, and, of course, Blu-ray players (see Figure 13.4 at the end of this chapter).

In short, because of the flexibility and interconnectability of inexpensive digital electronics, all the independent pieces we were once accustomed to have been tossed in a big hat, shaken around, and pulled back out as new, mixed-and-matched systems with more features at lower prices. Customers are benefiting from personalization, independence from location, and independence from time. Our electronic devices are getting closer to the goal of enabling us to access anything, anytime, in any format we choose.

The Far Horizon

In 2004, two very rich men predicted the imminent death of optical disc technology. In July, 2004, speaking to Germany's Bild magazine, Bill Gates said, "The entertainment of the future will definitely not be on a DVD player, that technology will be completely gone within 10 years at the most." In an August, 2004, Internet blog, HDNet's Mark Cuban said, "Which is the better way to deliver a movie or movies? On a DVD with a boring, lifeless future, or hard drives?" Apparently the amount of money in one's bank account can be inversely proportional to one's connection with the real world, because they are both wrong.[3] Bill Gates is not wrong about the eventual disappearance of packaged media, but he is wrong about the timeframe. His deadline comes only two years after 2012, the year many people expect Blu-ray disc sales to finally catch up with DVD sales.

In September 2008, Andy Griffiths, director of consumer electronics at Samsung UK, ruffled a lot of blue feathers when he told the Pocket-lint Website, "I think [Blu-ray] has five years left, I certainly wouldn't give it ten." He was neither the first nor the last person to predict an early demise for Blu-ray, DVD, and CD before it.

---

[3]Cuban seems to ignore the fact that millions of copies of DVDs or Blu-ray discs can be mass produced for less than a dollar or two, and even the cheapest hard disks cost hundreds of times more. Even putting hundreds of movies on a single hard drive would never make it economical to ship the drives to thousands or millions of customers. As pointed out in **Chapter 11**, the cost of Internet delivery and Blu-ray delivery are not far apart, at all. Lastly, compare the interesting interactive features of Blu-ray with a typical Internet video download where you are lucky if you can even reliably fast forward. Which is the boring one?

If you look back at predictions of the death of CD from Internet-delivered music, it is easy to see how people get carried away by the promise of new technology and new paradigms, seriously underestimating the inertia of established technologies. Consider that MP3 files became popular around 1996, developed into the Internet's "killer app" in 1999 with the release of *Napster*, were legitimized and commercialized in 2001 by Apple's *iTunes*, yet when Blu-ray was released a decade after MP3 came to life, CDs and DVDs still accounted for more than 77 percent of US music sales compared to downloaded songs and mobile-phone ringtones.[4] At the beginning of 2008, Forrester Research projected that US digital music download sales would finally surpass CD sales in 2012. And that is music downloads, which had a roughly ten-year head start on video downloads, which do not lend themselves as well to downloading and playing on portable devices. Speaking of portable devices, consider the number of DVD and Blu-ray entertainment systems in vehicles, which will take much longer to become Internet connected. Add in the fact that most people don't throw away their old DVD or Blu-ray player when they buy a new one — instead they put the old one in another room or give it to the kids. Anyone who thinks that online delivery will soon displace the incredible inertia represented by over a billion DVD playback devices, hundreds of millions of Blu-ray playback devices, and tens of billions of discs is not even dipping a toe into the waters of reality.

Eventually the demand for physical media will decline, but it will require an arduous increase in bandwidth from the completely insufficient typical rate of 2 to 5 Mbps (see Table B.6), plus a new generation of consumers who are comfortable with ephemeral delivery of content rather than ownership of tangible property. This might happen as soon as 2020, but certainly not before 2010 or even 2015, as some people predict.

The Internet has begun to merge with cable TV, broadcast TV, radio, telephones, satellites, and, even, newspapers and magazines. The Internet will eventually take over the communications world. News, movies, music, advertising, education, games, financial transactions, e-mail, and most other forms of information will be delivered via a world-spanning network. Internet bandwidth, currently lagging far behind the load being demanded of it, will eventually catch up, as did other systems, such as, intercontinental telephone networks and communication satellites. Discrete media, such as, Blu-ray will then be relegated to niches, such as, data backups, photo archiving, and collector's editions of movies with fancy packaging. Why go to a store to buy software on a disc or to a movie rental store to rent a DVD or BD when you can have it delivered right to your computer or your TV or your digital video recorder? In the intervening years, however, optical media in all its permutations and generations — including Blu-ray — promises to be the definitive medium for both computers and home entertainment.

---

[4]CDs represented 72 percent of sales, while music DVDs accounted for about 4.5 percent and DVD-Audio and SACD discs each contributed a barely noticeable 0.03 percent. The US leads the world in digital music sales, so the numbers for non-physical sales in other countries are even smaller. Over 30 percent of digital music revenues in 2007 in US — and an astounding 91 percent in Japan — came from the mobile market, which clearly won't have the same effect with movie downloads. US figures are from the Recording Industry Association of America (RIAA). Japan figures are from IFPI.

Blu-ray and Beyond

Figure 13.4 Media Convergence in the Digital Age

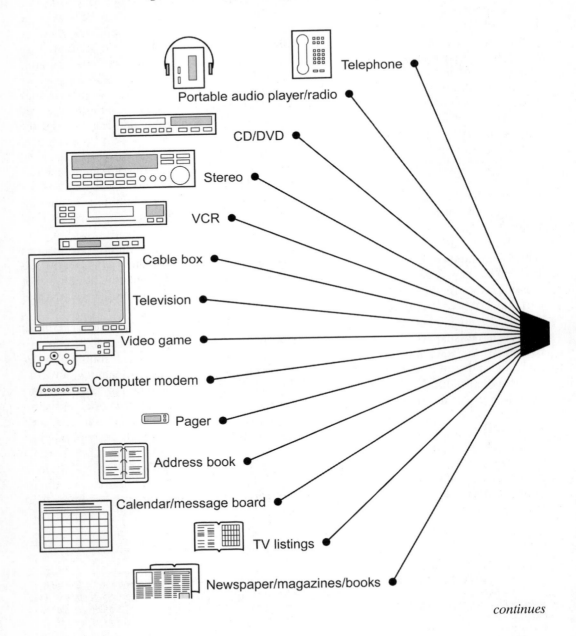

continues

Figure 13.4 Media Convergence in the Digital Age (continued)

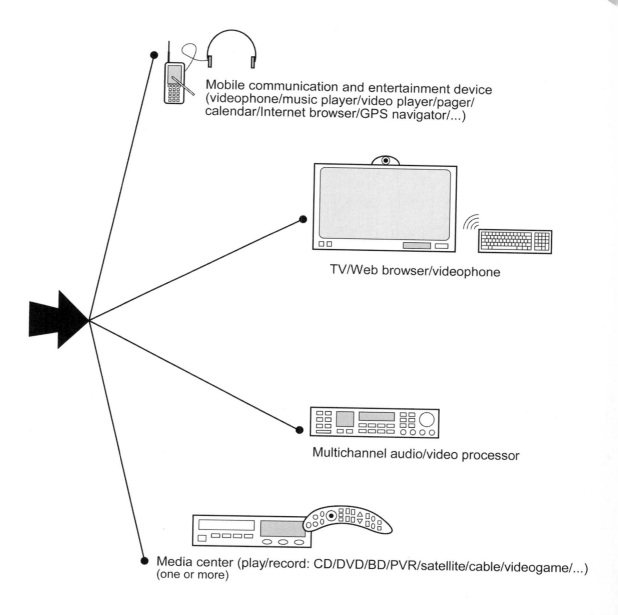

Appendix A
About the Disc

The Blu-ray Disc™ provided with this book includes material that demonstrates many of the features available with the Blu-ray Disc format. As described in **Chapter 7, Players**, the Profile level of a player determines the feature playability of a Blu-ray Disc player. Please read the Profile sections below to understand how some features might work on a Blu-ray player. Figure A.1, Disc Flowchart, presented at the end of this appendix, depicts the contents of the accompanying disc.

Most Blu-ray player users will already be familiar with some of the format features, such as, pop-up menus and button sounds. To demonstrate the picture in picture (PIP) feature, this disc includes Bonus View material that can be displayed when the VC1 1920 × 1080 video element is playing. And, this disc includes BD-Live (Profile 2.0) elements which allow a user to download material via the Internet, located at http://www.bddemystified.com.

The main menu is displayed following the playing of the First Play Logos. The main menu is also available as a popup menu using the popup key on the player remote control. The menu is divided into four primary tiers, or branches, that connect to the disc content —

- Features
- Setup
- Bonus
- Credits

Features

Blu-ray technology accommodates video that may be encoded using any one of three encoding varieties — MPEG2, H.264 AVC, or VC1. The accompanying disc presents the same video element that has been encoded in each of the three encoding flavors, at a 1920 × 1080 video resolution.

PIP video is provided on the disc that has been created to only play with the VC1 feature video. Although acquired in a high definition format, the resolution of this picture in picture video has been decreased to demonstrate video as a 1280 × 720 aspect ratio and as a 720 × 480 aspect ratio, when using the VC1 video feature.

All of the feature video pieces include the audio and subtitle streams that are outlined in the Setup section below. Information about each video piece may appear as a programmatic on-screen display while the video plays.

About the Disc

Setup

Two streams of audio are provided for each version of the feature video — 5.1 uncompressed PCM and DTS 5.1.

Three subtitle streams have been provided for each version of the feature video — English, French and Chinese.

Bonus

The bonus section of the disc offers three bonus components — downloadable material, the Bunny Blaster game, and a bonus piece that presents audio streams in either 5.1 or 7.1 DTS HD.

Downloadable Material

Three high definition web video pieces are available for download, when an Internet connection exists for a Blu-ray Disc player. These elements are accessible via an Internet connection that is designed by RCDb (Related Content Database). The videos are each ten second in length, and demonstrate the ability to gain access to new video content that can accompany the content already present on a disc.

Roxio BD Live is a BD-J application that showcases the interactivity and extensibility of the Blu-ray platform. Roxio BD Live can download an updated BD-J application from a Roxio website, demonstrating that applications that can take advantage of the rich interactivity of the BD format platform.

Bunny Blaster Game

The Bunny Blaster game demonstrates the use of BD-J as a programming language to enable collision detection. Games like this were never possible on DVD!

Bonus Audio Format Selection

A bonus video element provides demonstration audio streams in either 5.1 DTS HD or the new 7.1 DTS HD surround.

Credits

The credits page is a programmed list of companies and people involved in the creation of this disc.

Easter Egg

An Easter Egg is also present on this disc, and it can be revealed using the following key combination on the Blu-ray player remote control — Red, Green, Yellow, Blue. If this button combination is executed incorrectly, a message will reveal that you did not execute the combination correctly. You may try again.

Player Profiles Functionality

This disc demonstrates functionality for the three Blu-ray Disc player profiles. The functionality for Profile 1.1 will not appear on a Profile 1.0 player. The functionality programmed in Profile 2.0 will not function in either a Profile 1.1 or a Profile 1.0 player.

Profile 1.0 Functionality

All players exhibit the Profile 1.0 functions. This disc contains BD-Java popup menus, button sounds, 1920 × 1080 video in VC1, AVC, and MPEG2 video formats, LPCM 5.1 audio, DTS HD Master Audio, DTS HD 7.1 audio, and font rendered text. The Bunny Blaster game will also function in all players.

Profile 1.1 Functionality

Players with Profile 1.1 will exhibit the picture in picture (PIP) video as an example of Profile 1.1 functionality. On the main menu, or on the popup menu, a leaf will appear on the 1920 × 1080 VC1 video button that can be used to enable or disable the PIP video streams for the video feature. The PIP is only available with the 1920 × 1080 VC1 video stream and these buttons only appear with VC1 button. If the player is Profile level 1.0, the graphic leaf will not display.

Profile 2.0 Functionality

Only players with Profile 2.0 allow access to the Internet via an ethernet cable. This disc demonstrates this function allowing the user to download a choice of three video pieces from the website, *http://www.bddemystified.com/*. All three video pieces are ten seconds in length but vary in their compression bitrate. Download speed will vary depending on the throughput of the Internet connection. A player that is not Profile 2.0 enabled will not allow download and will display the message, "Your player is not Profile 2.0 network compatible".

The high bitrate version of the video is the best video quality at approximately 27Mbps but is the largest in size at approximately 38MB. This high bitrate video will look the best but will take the longest to download.

The medium video is encoded at a bitrate of approximately 10Mbps and its size is approximately 9MB.

The low bitrate video is encoded at approximately 6Mbps and is the smallest in size at approximately 6MB.

Each of these files may be downloaded onto a player that has an extended storage capability referred to as BUDA (Binding Unit Data Area). The files will remain in BUDA or they can be deleted using the trashcan icon that may be displayed on-screen. Some players may have an onboard system to delete BUDA material manually, as well.

About the Disc

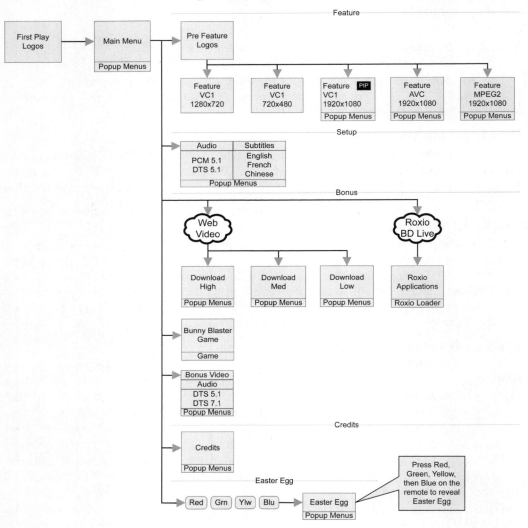

Figure A.1, Disc Flowchart

Appendix B
Reference Data

Figure B.1 Conversion Formulas for Playing Time, Data Rate, and Size

Formula	Factor	Example
Time (hours) = $\dfrac{\text{size (G bytes)} \times 8 \times 1000}{\text{rate (Mbps)} \times 60 \times 60}$	2.22$\overline{2}$	$\dfrac{25}{18} \times 2.22 = 3.1$ hours
Data rate (Mbps) = $\dfrac{\text{size (G bytes)} \times 8 \times 1000}{\text{time (hours)} \times 60 \times 60}$		$\dfrac{50}{4} \times 2.22 = 27.8$ Mbps
Size (G bytes) = $\dfrac{\text{rate (Mbps)} \times \text{time (hours)} \times 60 \times 60}{8 \times 1000}$	0.45	$22 \times 2 \times 0.45 = 19.8$ G bytes

Factors: 8 = bits to bytes, 1000 = G (billions) to M (millions), 60 × 60 = hours to seconds.
For gigabytes (2^{30} instead of G bytes, 10^9) replace 2.22 with 2.39 and 0.45 with 0.42.

Figure B.2 Relationships of BD Formats

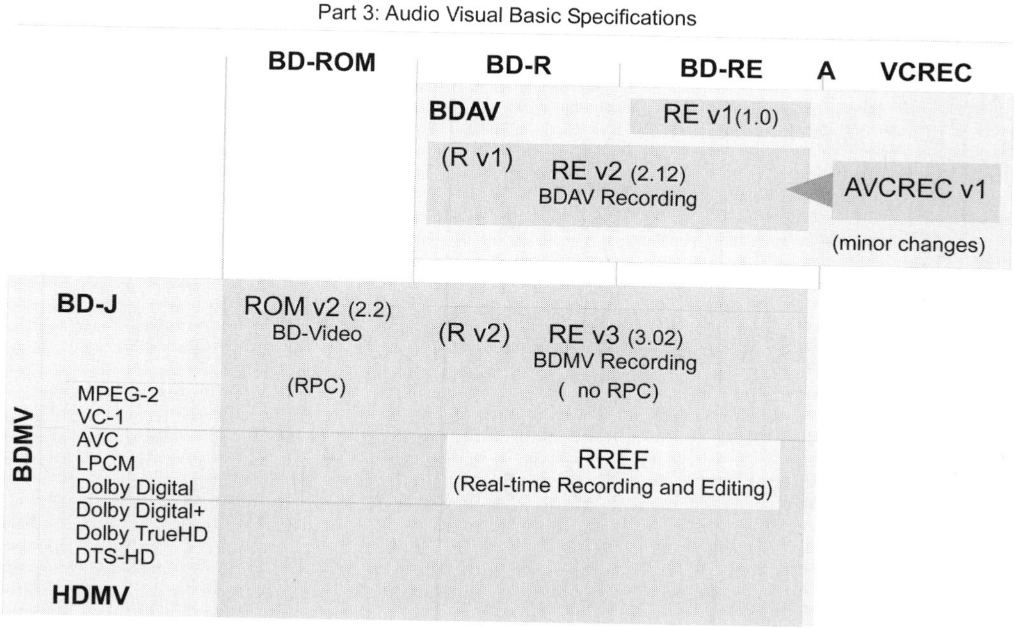

Part 3: Audio Visual Basic Specifications

	BD-ROM	BD-R	BD-RE	A	VCREC
		BDAV	RE v1(1.0)		
		(R v1)	RE v2 (2.12) BDAV Recording		AVCREC v1
					(minor changes)
BD-J	ROM v2 (2.2) BD-Video (RPC)	(R v2)	RE v3 (3.02) BDMV Recording (no RPC)		
MPEG-2 VC-1 AVC LPCM Dolby Digital Dolby Digital+ Dolby TrueHD DTS-HD			RREF (Real-time Recording and Editing)		
HDMV					

(BDMV bracket on left side)

Blu-ray Disc Demystified

Reference Data

Table B.1 BD, DVD, and CD Capacities and Playing Times

Disc type	Sides/layers	Billions of bytes[a]	Gigabytes[a]	CD	DVD	Video playback hours (h) Typical[b]	Video playback hours (h) Min.to.max.[c]	Audio playback hours (h) or days (d) 2-ch PCM[d]	Audio playback hours (h) or days (d) 5.1-ch[e]	Audio playback hours (h) or days (d) Min. to max.[f]
12 cm										
BD-25	SS/SL	25.025	23.306	36.7	5.3	2.3h	1.2h to 37h	12.1h	86.9h	1.9h to 36.2d
BD-27	SS/SL	27.020	25.164	39.6	5.7	2.5h	1.3h to 40h	13.0h	93.8h	2h to 39d
BD-50	SS/DL	50.050	46.613	73.4	10.6	4.6h	2.3h to 74.1h	24.1h	173.8h	3.7h to 72.4d
BD-54	SS/DL	54.040	50.329	79.2	11.5	5.0h	2.5h to 80h	26.1h	187.6h	4h to 78.1d
BD-100[g]	SS/QL	100.100	93.225	146.8	21.3	9.3h	4.6h to 148.2h	48.3h	347.6h	7.4h to 144.8d
DVD-5	SS/SL	4.700	4.377	6.9	1.0	1.9h	1h to 9h	6.8h	27.2h	1.7h to 6.7d
DVD-9	SS/DL	8.540	7.950	12.5	1.8	3.5h	1.9h to 16.5h	12.4h	49.4h	3.1h to 12.3d
DVD-18	DS/DL	17.080	15.910	25.0	3.6	6.9h	3.8h to 33h	24.7h	98.8h	6.2h to 24.7d
CD-ROM[h]	SS/SL	0.682	0.635	1.0	0.1	0.3h	0.2h to 1.3h	1.0h	3.9h	0.2h to 0.9d
8 cm										
BD-8	SS/SL	7.791	7.256	11.4	1.7	0.7h	0.4h to 11.5h	3.8h	27.1h	0.6h to 11.2d
BD-16	SS/DL	15.582	14.512	22.8	3.3	1.4h	0.7h to 23h	7.5h	54.1h	1.2h to 22.5d
DVD-2	SS/SL	1.460	1.360	2.1	0.3	0.6h	0.3h to 2.8h	2.1h	8.4h	0.1h to 2.1d
DVD-3	SS/DL	2.650	2.470	3.9	0.6	1.1h	0.6h to 5.1h	3.8h	15.3h	0.2h to 3.8d
DVD-6	DS/DL	5.310	4.950	7.8	1.1	2.1h	1.2h to 10.2h	7.7h	30.7h	0.4h to 7.6d
Mini CD[h]	SS/SL	0.194	0.180	0.3	0.04	0.1h	0.04h to 0.3h	0.3h	1.1h	0.01h to 6.7h

[a] Reference capacities in billions of bytes (10^9) and gigabytes (2^{30}). Actual capacities can be slightly larger if the track pitch is reduced.

[b] Approximate video playback time with a few audio tracks, given an average data rate of 5.5 Mbps for DVD, 24 Mbps for BD. Actual playing times can be much longer or shorter (see next column).

[c] Minimum video playback time at the highest data rate of 10.08 Mbps for DVD, 48 Mbps for BD. Maximum playback time at the MPEG-1 data rate of 1.15 Mbps for DVD, 1.5 Mbps (standard definition AVC/VC-1/MPEG-2) for BD.

[d] Example audio-only playback time at the two-channel PCM rate of 48 kHz and 16 bits (1.536 Mbps) for DVD, 96 kHz and 24 bits for BD (4.608 Mbps).

notes continued below

Reference Data

Table B.1 notes continued

[e] Example audio-only playback time at the Dolby Digital 5.1-channel rate of 0.384 Mbps for DVD, 0.640 Mbps for BD.

[f] Minimum audio-only playback time at the highest PCM rate of 6.144 Mbps for DVD, 30 Mbps for BD. Maximum audio-only playback time at the lowest Dolby Digital data rate of 64 kbps.

[g] The legendary quad-layer Blu-ray disc.

[h] Mode 1, 74 minutes (333,000 sectors) or 21 minutes (94,500 sectors). Audio/video times are for comparison only, assuming that the data from the CD is transferred at a typical DVD video data rate, about four times faster than a single-speed CD-ROM drive.

Table B.2 Data Rates at Various Playing Times

	Mbps											
	0.5 hrs	1 hr	1.5 hrs	2 hrs	2.5 hrs	3 hrs	4 hrs	5 hrs	8 hrs	10 hrs	20 hrs	30 hrs	50 hrs
BD-8 (7.7G)	31.2	15.6	10.4	7.8	6.2	5.2	3.9	3.1	1.9	1.6	0.8	0.5	0.3
BD-16 (15.5G)	40.0	31.2	20.8	15.6	12.5	10.4	7.8	6.2	3.9	3.1	1.6	1.0	0.6
BD-25 (25G)	40.0	40.0	33.4	25.0	20.0	16.7	12.5	10.0	6.3	5.0	2.5	1.7	1.0
BD-50 (50G)	40.0	40.0	40.0	40.0	40.0	33.4	25.0	20.0	12.5	10.0	5.0	3.3	2.0

Blu-ray Disc Demystified

Reference Data

Table B.3 BD Stream Data Rates

	Mbps		
	Minimum	Typical	Maximum
Video (high-definition)			
MPEG-4 AVC video	4[a]	16	40
SMPTE VC-1 video	4[a]	18	40
MPEG-2 video	4[a]	24	40
Secondary Video (std. def. 525/60 or 625/50)			
MPEG-4 AVC video	1.5[a]	2	8
SMPTE VC-1 video	1[a]	2	8
MPEG-2 video	1[a]	3.5	8
Audio			
Dolby Digital 2.0 audio	0.064	0.192	0.64
Dolby Digital 5.1 audio	0.384	0.448	0.64
Dolby Digital Plus 7.1 audio	0.640	1.024	4.736
Dolby TrueHD 5.1 (lossless) audio	0.800[a]	3.900[e]	18.000
DTS 5.1 audio[b] (CBR, core)	0.192	1.509	1.509
DTS 6.1 audio[b] (CBR, core+XC, DTS ES)	0.640	1.509	1.509
DTS-HD 7.1 audio[b] (CBR, core+XXCH)	0.768	1.509	3.000
DTS-HD 5.1 96 kHz audio[b] (CBR, core+X96)	0.300	1.509	3.000
DTS-HD 5.1 Master (lossless) audio[b] (VBR, core+XLL))	0.800[a]	4.000[c]	24.500
DTS-HD 7.1 Master (lossless) audio[b] (VBR, core+XLL)	0.800[a]	4.000[c]	24.500
Secondary Audio (streaming)			
Dolby Digital Plus 1.0 audio (VBR)	0.032	0.064	0.256
Dolby Digital Plus 2.0 audio (VBR)	0.032	0.128	0.256
Dolby Digital Plus 5.1 audio (VBR)	0.032	0.128	0.256
DTS-HD Secondary (LBR) 2.0 audio	0.048	0.064	0.256
DTS-HD Secondary (LBR) 5.1 audio	0.192	0.256	0.256

[a] Not an absolute limit but a practical limit below which presentation quality is too poor

[b] DTS core rates include 768, 960, 1152, 1344 and 1509 kbps.

[c] The compressed rate of DTS-HD and Dolby TrueHD is heavily dependent on the source material. Values shown here represent movie source material, which can typically be more highly compressed than music.

Reference Data

Table B.4 Limits on BD Elements

Video and audio stream data rates	Limit
Transport stream (from BD)	48 Mbps
Transport stream (from BD format on DVD)	28 Mbps
Video stream (primary or secondary; 1920 × 1080, 1280 × 720)	40 Mbps
Secondary video stream (525/60, 625/50)	8 Mbps
Primary and secondary video stream total	80 Mbps
MPEG-2 MP@ML (525/60, 625/50) video stream	15 Mbps
AVC level 3.2 video stream	8 Mbps
AVC Max GOP length	1.001 secs. (>15 Mbps), 2.002 secs. (≤15 Mbps)
Presentation Graphics stream	48 Mbps
Interactive Graphics stream	48 Mbps
Text subtitle stream	48 Mbps
Dolby Digital audio stream	0.640 Mbps
Dolby Digital Plus audio stream	4.736 Mbps
Dolby TrueHD (MLP) audio stream	18 Mbps
DTS (core) audio stream	1.509 Mbps
DTS-HD audio stream (core + extension)	24.5 Mbps
Secondary audio stream from disc or local storage (DD+ or DTS-HD LBR)	2 Mbps
Secondary audio stream from network (DD+ or DTS-HD LBR)	0.256 Mbps
48kHz/96kHz LPCM audio stream (HDMV)	20 Mbps
192kHz LPCM audio stream (HDMV)	30 Mbps
Number of BDAV/BDMV elements	
Angles	9
Secondary video streams	32
Audio streams	32
Secondary audio streams	32
Text subtitle stream	255
Pop-up menus (Interactive Graphics streams)	32
Chapter (PlayList Marks)	999
Playlists (filenames 00000.mpls to 01999.mpls)	2000
PlayItems in a PlayList	999
SubPaths in a PlayList	255
ClipInfo (.clpi files)	4000
Movie Objects (.bdjo files)	1001 (999 + 2)
JAR files	Limited only by disc space
Fonts (.otf files)	255
Menu sounds (in sound.bdmv)	128

Reference Data

Table B.5 Download Time for Various Payloads and Bandwidths

		Download time in seconds (s), minutes (m), hours (h), or days (d)							
Service Rated Mbps		56K	DSL	DSL	T1, DS-1	Cable	HS Cable	Blu-ray	OC-3
Actual Mbps[a]		0.056	0.384	0.768	1.54	3	10	54	155
		0.045	0.307	0.614	1.23	2.4	8	54	124
Content	Size[b]								
DVD-5	4.700	9.7d	1.4d	17.0h	8.5h	4.4h	1.3h	11.6m	5.1m
DVD-9	8.540	17.7d	2.6d	1.3d	15.4h	7.9h	2.4h	21.1m	9.2m
DVD-18	17.080	35.3d	5.1d	2.6d	1.3d	15.8h	4.7h	42.2m	18.4m
BD-25	25.025	51.7d	7.5d	3.8d	1.9d	23.2h	7.0h	1.0h	26.9m
BD-50	50.050	103.4d	15.1d	7.5d	3.8d	1.9d	13.9h	2.1h	53.8m
10 mins. audio at 64 kbps	0.005	14.3m	2.1m	1.0m	31.2s	16.0s	4.8s	0.7s	0.3s
1 hr. audio at 64 kbps	0.029	1.4h	12.5m	6.3m	3.1m	1.6m	28.8s	4.3s	1.9s
2 hrs. audio at 64 kbps	0.058	2.9h	25.0m	12.5m	6.2m	3.2m	57.6s	8.5s	3.7s
1 hr. audio at 256 kbps	0.115	5.7h	50.0m	25.0m	12.5m	6.4m	1.9m	17.1s	7.4s
2 hrs. audio at 256 kbps	0.230	11.4h	1.7h	50.0m	24.9m	12.8m	3.8m	34.1s	14.9s
2 mins. SD video at 1.5 Mbps	0.023	1.1h	9.8m	4.9m	2.4m	1.3m	22.5s	3.3s	1.5s
2 mins. HD video at 12 Mbps	0.180	8.9h	1.3h	39.1m	19.5m	10.0m	3.0m	26.7s	11.6s
10 mins. SD video at 1.5 Mbps	0.113	5.6h	48.8m	24.4m	12.2m	6.3m	1.9m	16.7s	7.3s
10 mins. HD video at 12 Mbps	0.900	1.9d	6.5h	3.3h	1.6h	50.0m	15.0m	2.2m	58.1s
30 mins. SD video at 1.5 Mbps	0.338	16.7h	2.4h	1.2h	36.5m	18.8m	5.6m	50.0s	21.8s
30 mins. HD video at 12 Mbps	2.700	5.6d	19.5h	9.8h	4.9h	2.5h	45.0m	6.7m	2.9m
30 mins. HD video at 22 Mbps	4.950	10.2d	1.5d	17.9h	8.9h	4.6h	1.4h	12.2m	5.3m

[a] Typical real-world efficiency of networks is roughly 80 percent of rated bandwidth.
[b] Payload size in billions of bytes (10^9).

Table B.6 Bandwidth Requirements for Desired Download Times

Content	Size[a]	Desired download time						
		5 mins	15 mins	30 mins	1 hr	2 hrs	24 hrs	48 hrs
		Bandwidth needed (Mbps)						
DVD-5	4.700	156.7	52.2	26.1	13.1	6.5	0.5	0.3
DVD-9	8.540	284.7	94.9	47.4	23.7	11.9	1	0.5
DVD-18	17.080	569.3	189.8	94.9	47.4	23.7	2	1
BD-25	25.025	834.2	278.1	139	69.5	34.8	2.9	1.4
BD-50	50.050	1668.3	556.1	278.1	139	69.5	5.8	2.9
10 mins. audio at 64 kbps	0.005	0.2	0.1	0	0	0	0	0
1 hr. audio at 64 kbps	0.029	1	0.3	0.2	0.1	0	0	0
2 hrs. audio at 64 kbps	0.058	1.9	0.6	0.3	0.2	0.1	0	0
1 hr. audio at 256 kbps	0.115	3.8	1.3	0.6	0.3	0.2	0	0
2 hrs. audio at 256 kbps	0.230	7.7	2.6	1.3	0.6	0.3	0	0
2 mins. SD video at 1.5 Mbps	0.023	0.8	0.3	0.1	0.1	0	0	0
2 mins. HD video at 12 Mbps	0.180	6	2	1	0.5	0.3	0	0
10 mins. SD video at 1.5 Mbps	0.113	3.8	1.3	0.6	0.3	0.2	0	0
10 mins. HD video at 12 Mbps	0.900	30	10	5	2.5	1.3	0.1	0.1
30 mins. SD video at 1.5 Mbps	0.338	11.3	3.8	1.9	0.9	0.5	0	0
30 mins. HD video at 12 Mbps	2.700	90	30	15	7.5	3.8	0.3	0.2
30 mins. HD video at 22 Mbps	4.950	165	55	27.5	13.8	6.9	0.6	0.3

[a] Payload size in billions of bytes (10^9).

Reference Data

Table B.7 DVD, HD DVD, BD, and CD Characteristics Comparison

	DVD	HD DVD	BD	CD
Thickness	1.2 mm (2 × 0.6)	1.2 mm (2 × 0.6)	1.2 mm (0.1 + 1.1)	1.2 mm
Mass (12 cm)	13 to 20 g	14 to 20 g	12 to 17 g	14 to 33 g
Diameter	120 or 80 mm	120 mm	120 mm	120 or 80 mm
Spindle hole diameter	15 mm	15 mm	15 mm	15 mm
Lead-in diameter	45.2 to 48 mm	46.6 mm	44 to 44.4 mm	46 to 50 mm
Data diameter (12 cm)	48 to 116 mm	48.2 mm	39.8 to 48 mm	50 to 116 mm
Data diameter (8 cm)	48 to 76 mm	-	-	50 to 76 mm
Lead-out diameter	70 to 117 mm	115.78 mm	116 mm	76 to 117 mm
Outer guardband diameter (12 cm)	117 to 120 mm	115.78 mm to 120 mm	116 to 120 mm	117 to 120 mm
Outer guardband diameter (8 cm)	77 to 80 mm	-	-	77 to 80 mm
Reflectivity (full)	45% to 85%	-	35 to 70% (DL),12 to 28% (DL)	70% min.
Readout wavelength	650 or 635 nm	405 nm	405 nm	780 nm
Numerical aperture	0.60	0.65	0.85	0.38 to 0.45
Focus depth	0.47 μm	-	0.1 mm	1 (± 2 μm)
Track pitch	0.74 μm	0.4 μm	0.32 μm	1.6 mm (1.1 μm[a])

continues

Blu-ray Disc Demystified

Reference Data

Table B.7 DVD, HD DVD, BD and CD Characteristics Comparison (continued)

	DVD	HD DVD	BD	CD
Pit length	0.400 to 1.866 μm (SL), 0.440 to 2.054 μm (DL)[b]	0.204 to 1.020μm	0.149 to 0.695μm	0.833 to 3.054 μm (1.2 m/s), 0.972 to 3.560 μm (1.4 m/s); [0.623 to 2.284 μm[a](0.90m/s)]
Pit width	0.3 μm	-	-	0.6μm
Pit depth	0.16 μm	0.10 μm	0.10 μm	0.11 μm
Data bit length	0.2667 μm(SL), 0.2934 μm(DL)	0.154 μm	0.11175 μm	0.6 μm(1.2 m/s), 0.7 μm(1.4 m/s)
Channel bit length	0.1333 μm(SL), 0.1467 μm(DL)	-	0.745 μm	0.3 μm
Modulation	8/16	ETM, RLL(1,10) (8:12)	1-7 PP[e], RLL(1,7) (2:3)	8/14 (8/17 w/merge bits)
Error correction	RS-PC	RS-PC	LDC and BIS picket	CIRC (CIRC7[a])
Error correction overhead	13%	-	17%	23%/34%[c]
Bit error rate	10^{15}	-	2×10^4	10^{14}
Correctable error (1 layer)	6 mm (SL), 6.5 mm (DL)	7.1 mm	7.0 mm	2.5 mm
Speed (rotational)[d]	570 to 1600 rpm	-	-	200 to 500 rpm
Speed (scanning)[d]	3.49 m/s (SL), 3.84 m/s (DL)	-	4.917 m/s	1.2 to 1.4 m/s (0.90 m/s[a])
Channel data rate[d]	26.15625 Mbps	64.8	66.0 Mbps	4.3218 Mbps (8.6436 Mbps[a])
User data rate[d]	11.08 Mbps	36.55	35.965 Mbps	1.41 Mbps/1.23 Mbps[c]
User data: channel data	2048:4836 bytes	2048:3631	2048:3758	2352:7203/2048:7203[c]
Format overhead	136 percent	77%	83%	206 percent/252 percent[c]
Capacity	1.4 to 8.0 GB per side	15 to 30 GB per side	25 to 50 GB per side	0.783/0.635 GB[c]

[a]Double-density CD
[b]SL 5 single layer, DL 5 dual layer
[c]CD-DA / CD-ROM Mode 1.
[d]Reference value for a single-speed drive.
[e]PP=Parity preserve/Prohibit RMTR. RMTR=Repeated minimum transition run length.

Reference Data

Table B.8 Video Resolutions

Format		VHS (1.33)	VHS (1.78)	VHS (2.35)	LD (1.33)	LD (1.78)	LD (1.85)	LD (2.35)	VCD (1.33)	VCD (1.78)	VCD (2.35)
NTSC	TVL	250	250	250	425	425	425	425	264	264	264
	H pixels	333	333	333	567	567	567	567	352	352	352
	V pixels	480	360	272	480	360	346	272	240	180	136
	Total pixels	159,840	119,880	90,576	272,160	204,120	196,182	154,224	84,480	63,360	47,872
PAL	TVL	240	240	240	450	450	450	450	264	264	264
	H pixels	320	320	320	600	600	600	600	352	352	352
	V pixels	576	432	327	576	432	415	327	288	216	163
	Total pixels	184,320	138,240	104,640	345,600	259,200	249,000	196,200	101,376	76,032	57,376

Format		DVD (1.33/1.78)	DVD (1.85)	DVD (2.35)	BD 720 (1.33)	BD 720 (1.78)	BD 720 (2.35)	BD 1080 (1.78)	BD 1080 (2.35)
NTSC	TVL	540/405	540/405	540/405	720	720	720	1,080	1,080
	H pixels	720	720	720	1,280	1,280	1,280	1,920	1,920
	V pixels	480	461	363	960	720	545	1080	817
	Total pixels	345,600	331,920	261,360	1,228,800	921,600	697,600	2,073,600	1,568,640
PAL	TVL	540	540	720					
	H pixels	720	720	720					
	V pixels	576	554	436					
	Total pixels	414,720	398,880	313,920					

continues

Reference Data

Table B.8 Video Resolution (continued)

Notes:

1. BD is neither PAL nor NTSC. The values are placed in the NTSC rows for convenience.
2. Wide aspect ratios (1.78 and 2.35) for VHS, LD, and VCD assume a letterboxed picture. For comparison, letterboxed 1.66 aspect ratio resolution is about 7 percent higher than 1.78. Letterbox is also assumed for DVD and BD at a 2.35 aspect ratio. DVD's native aspect ratio is 1.33; it uses anamorphic mode for 1.78. BD's native aspect ratio is 1.78.
3. The very rare 1.78 anamorphic LD has the same pixel count as 1.33 LD. Anamorphic LD letterboxed to 2.35 has almost the same pixel count as 1.78 LD (567 × 363). The almost non-existent 1.78 anamorphic VHS has the same pixel count as 1.33 VHS. Anamorphic VHS letterboxed to 2.35 has almost the same pixel count as 1.78 VHS (333 × 363). No commercial 2.35 anamorphic format exists and no corresponding stretch mode exists on widescreen TVs.
4. TVL is lines of horizontal resolution per picture height. For analog formats, the customary value is used; for digital formats, the value is derived from the actual horizontal pixel count adjusted for the aspect ratio. DVD's horizontal resolution is lower for 1.78 because the pixels are wider. Pixels for VHS and LD are approximations based on TVL and scan lines.
5. Resolutions refer to the medium, not the display. If a BD or DVD player performs automatic letterboxing on a 1.85 movie (stored in 1.78), the displayed vertical resolution on a standard 1.33 TV is the same as from a letterboxed LD (360 lines).

Reference Data

Table B.9 ISO 3166 Country Codes, BD and DVD Regions

ISO 3166 Codes			Region (BD, DVD)		Country Name
AF	AFG	4	C	5	Afghanistan
AX	ALA	248	B	2	Åland Islands
AL	ALB	8	B	2	Albania
DZ	DZA	12	B	5	Algeria
AS	ASM	16	A	1	American Samoa
AD	AND	20	B	2	Andorra
AO	AGO	24	B	5	Angola
AI	AIA	660	A	4	Anguilla
AQ	ATA	10	-	-	Antarctica
AG	ATG	28	A	4	Antigua And Barbuda
AR	ARG	32	A	4	Argentina
AM	ARM	51	C	5	Armenia
AW	ABW	533	A	4	Aruba
AU	AUS	36	B	4	Australia
AT	AUT	40	B	2	Austria
AZ	AZE	31	C	5	Azerbaijan
BS	BHS	44	A	4	Bahamas
BH	BHR	48	B	2	Bahrain
BD	BGD	50	C	5	Bangladesh
BB	BRB	52	A	4	Barbados
BY	BLR	112	C	5	Belarus
BE	BEL	56	B	2	Belgium
BZ	BLZ	84	A	4	Belize
BJ	BEN	204	B	5	Benin
BM	BMU	60	A	1	Bermuda
BT	BTN	64	C	5	Bhutan
BO	BOL	68	A	4	Bolivia
BA	BIH	70	B	2	Bosnia and Herzegovina
BW	BWA	72	B	5	Botswana
BV	BVT	74	-	-	Bouvet Island
BR	BRA	76	A	4	Brazil
IO	IOT	86	B	5	British Indian Ocean Territory
BN	BRN	96	A	3	Brunei Darussalam
BG	BGR	100	B	2	Bulgaria
BF	BFA	854	B	5	Burkina Faso
BI	BDI	108	B	5	Burundi
KH	KHM	116	A	3	Cambodia
CM	CMR	120	B	5	Cameroon
CA	CAN	124	A	1	Canada
CV	CPV	132	B	5	Cape Verde
KY	CYM	136	A	4	Cayman Islands
CF	CAF	140	B	5	Central African Republic
TD	TCD	148	B	5	Chad
CL	CHL	152	A	4	Chile

Reference Data

Table B.9 ISO 3166 Country Codes, BD and DVD Regions (continued)

ISO 3166 Codes			Region (BD, DVD)		Country Name
CN	CHN	156	C	6	China
CX	CXR	162	A	4	Christmas Island
CC	CCK	166	A	4	Cocos (Keeling) Islands
CO	COL	170	A	4	Colombia
KM	COM	174	B	5	Comoros
CG	COG	178	B	5	Congo
CD	COD	180	B	5	Congo, the Democratic Republic of the
CK	COK	184	A	4	Cook Islands
CR	CRI	188	A	4	Costa Rica
CI	CIV	384	B	5	Côte d'Ivoire
HR	HRV	191	B	2	Croatia
CU	CUB	192	A	4	Cuba
CY	CYP	196	B	2	Cyprus
CZ	CZE	203	B	2	Czech Republic
DK	DNK	208	B	2	Denmark
DJ	DJI	262	B	5	Djibouti
DM	DMA	212	A	4	Dominica
DO	DOM	214	A	4	Dominican Republic
EC	ECU	218	A	4	Ecuador
EG	EGY	818	B	2	Egypt
SV	SLV	222	A	4	El Salvador
GQ	GNQ	226	B	5	Equatorial Guinea
ER	ERI	232	B	5	Eritrea
EE	EST	233	B	5	Estonia
ET	ETH	231	B	5	Ethiopia
FK	FLK	238	A	4	Falkland Islands (Malvinas)
FO	FRO	234	B	2	Faroe Islands
FJ	FJI	242	A	4	Fiji
FI	FIN	246	B	2	Finland
FR	FRA	250	B	2	France
GF	GUF	254	A	4	French Guiana
PF	PYF	258	A	4	French Polynesia
TF	ATF	260	B[a]	2[a]	French Southern Territories
GA	GAB	266	B	5	Gabon
GM	GMB	270	B	5	Gambia
GE	GEO	268	C	5	Georgia
DE	DEU	276	B	2	Germany
GH	GHA	288	B	5	Ghana
GI	GIB	292	B	2	Gibraltar
GR	GRC	300	B	2	Greece
GL	GRL	304	B	2	Greenland
GD	GRD	308	A	4	Grenada

Reference Data

Table B.9 ISO 3166 Country Codes, BD and DVD Regions (continued)

ISO 3166 Codes			Region (BD, DVD)		Country Name
GP	GLP	312	A	4	Guadeloupe
GU	GUM	316	A	4	Guam
GT	GTM	320	A	4	Guatemala
GG	GGY	831	B[b]	2	Guernsey
GN	GIN	324	B	5	Guinea
GW	GNB	624	B	5	Guinea-Bissau
GY	GUY	328	A	4	Guyana
HT	HTI	332	A	4	Haiti
HM	HMD	334	A	4	Heard Island and McDonald Islands
HN	HND	340	A	4	Honduras
HK	HKG	344	A	3	Hong Kong
HU	HUN	348	B	2	Hungary
IS	ISL	352	B	2	Iceland
IN	IND	356	C	5	India
ID	IDN	360	A	3	Indonesia
IR	IRN	364	B	2	Iran, Islamic Republic of
IQ	IRQ	368	B	2	Iraq
IE	IRL	372	B	2	Ireland
IM	IMN	833	B	2	Isle of Man
IL	ISR	376	B	2	Israel
IT	ITA	380	B	2	Italy
JM	JAM	388	A	4	Jamaica
JP	JPN	392	A	2	Japan
JE	JEY	832	B[b]	2	Jersey
JO	JOR	400	B	2	Jordan
KZ	KAZ	398	C	5	Kazakhstan
KE	KEN	404	B	5	Kenya
KI	KIR	296	A	4	Kiribati
KP	PRK	408	A	5	Korea, Democratic People's Republic of (North)
KR	KOR	410	A	3	Korea, Republic of (South)
KW	KWT	414	B	2	Kuwait
KG	KGZ	417	C	5	Kyrgyzstan
LA	LAO	418	A	3	Lao People's Democratic Republic (Laos)
LV	LVA	428	B	5	Latvia
LB	LBN	422	B	2	Lebanon
LS	LSO	426	B	2	Lesotho
LR	LBR	430	B	5	Liberia
LY	LBY	434	B	5	Libyan Arab Jamahiriya (Libya)
LI	LIE	438	B	2	Liechtenstein
LT	LTU	440	B	5	Lithuania
LU	LUX	442	B	2	Luxembourg

Table B.9 ISO 3166 Country Codes, BD and DVD Regions (continued)

ISO 3166 Codes			Region (BD, DVD)		Country Name
MO	MAC	446	A	3	Macao
MK	MKD	807	B	2	Macedonia, the Former Yugoslav Republic of
MG	MDG	450	B	5	Madagascar
MW	MWI	454	B	5	Malawi
MY	MYS	458	A	3	Malaysia
MV	MDV	462	A	5	Maldives
ML	MLI	466	B	5	Mali
MT	MLT	470	B	2	Malta
MH	MHL	584	A	4	Marshall Islands
MQ	MTQ	474	A	4	Martinique
MR	MRT	478	B	5	Mauritania
MU	MUS	480	B	5	Mauritius
YT	MYT	175	B	5	Mayotte
MX	MEX	484	A	4	Mexico
FM	FSM	583	A	4	Micronesia, Federated States Of
MD	MDA	498	B	5	Moldova
MC	MCO	492	B	2	Monaco
MN	MNG	496	C	5	Mongolia
ME	MNE	499	B	2	Montenegro
MS	MSR	500	A	4	Montserrat
MA	MAR	504	B	5	Morocco
MZ	MOZ	508	B	5	Mozambique
MM	MMR	104	A	3	Myanmar
NA	NAM	516	B	5	Namibia
NR	NRU	520	A	4	Nauru
NP	NPL	524	C	5	Nepal
NL	NLD	528	B	2	Netherlands
AN	ANT	530	A	4	Netherlands Antilles
NC	NCL	540	A	4	New Caledonia
NZ	NZL	554	B	4	New Zealand
NI	NIC	558	A	4	Nicaragua
NE	NER	562	B	5	Niger
NG	NGA	566	B	5	Nigeria
NU	NIU	570	A	4	Niue
NF	NFK	574	A	4	Norfolk Island
MP	MNP	580	A	4	Northern Mariana Islands
NO	NOR	578	B	2	Norway
OM	OMN	512	B	2	Oman
PK	PAK	586	C	5	Pakistan
PW	PLW	585	A	4	Palau
PS	PSE	275	B[a]	2[a]	Palestinian Territory, Occupied
PA	PAN	591	A	4	Panama

Reference Data

Table B.9 ISO 3166 Country Codes, BD and DVD Regions (continued)

ISO 3166 Codes			Region (BD, DVD)		Country Name
PG	PNG	598	A	4	Papua New Guinea
PY	PRY	600	A	4	Paraguay
PE	PER	604	A	4	Peru
PH	PHL	608	A	3	Philippines
PN	PCN	612	A	4	Pitcairn
PL	POL	616	B	2	Poland
PT	PRT	620	B	2	Portugal
PR	PRI	630	A	1	Puerto Rico
QA	QAT	634	B	2	Qatar
RE	REU	638	B	5	Réunion
RO	ROU	642	B	2	Romania
RU	RUS	643	C	5	Russian Federation
RW	RWA	646	B	5	Rwanda
BL	BLM	652	A[a]	4[a]	Saint Barthélemy
SH	SHN	654	B	5	Saint Helena
KN	KNA	659	A	4	Saint Kitts and Nevis
LC	LCA	662	A	4	Saint Lucia
MF	MAF	663	A[a]	4[a]	Saint Martin
PM	SPM	666	A	1	Saint Pierre and Miquelon
VC	VCT	670	A	4	Saint Vincent and the Grenadines
WS	WSM	882	A	4	Samoa
SM	SMR	674	B	2	San Marino
ST	STP	678	B	5	Sao Tome And Principe
SA	SAU	682	B	2	Saudi Arabia
SN	SEN	686	B	5	Senegal
RS	SRB	688	B	2	Serbia
SC	SYC	690	B	5	Seychelles
SL	SLE	694	B	5	Sierra Leone
SG	SGP	702	A	3	Singapore
SK	SVK	703	B	2	Slovakia
SI	SVN	705	B	2	Slovenia
SB	SLB	90	A	4	Solomon Islands
SO	SOM	706	B	5	Somalia
ZA	ZAF	710	B	2	South Africa
GS	SGS	239	B[a]	4	South Georgia and the South Sandwich Islands
ES	ESP	724	B	2	Spain
LK	LKA	144	C	5	Sri Lanka
SD	SDN	736	B	5	Sudan
SR	SUR	740	A	4	Suriname
SJ	SJM	744	B	2	Svalbard and Jan Mayen
SZ	SWZ	748	B	2	Swaziland

Reference Data

Table B.9 ISO 3166 Country Codes, BD and DVD Regions (continued)

ISO 3166 Codes			Region (BD, DVD)		Country Name
SE	SWE	752	B	2	Sweden
CH	CHE	756	B	2	Switzerland
SY	SYR	760	B	2	Syrian Arab Republic
TW	TWN	158	A	3	Taiwan
TJ	TJK	762	C	5	Tajikistan
TZ	TZA	834	B	5	Tanzania, United Republic Of
TH	THA	764	A	3	Thailand
TL	TLS	626	A	3	Timor-Leste
TG	TGO	768	B	5	Togo
TK	TKL	772	A	4	Tokelau
TO	TON	776	A	4	Tonga
TT	TTO	780	A	4	Trinidad and Tobago
TN	TUN	788	B	5	Tunisia
TR	TUR	792	B	2	Turkey
TM	TKM	795	C	5	Turkmenistan
TC	TCA	796	A	4	Turks and Caicos Islands
TV	TUV	798	A	4	Tuvalu
UG	UGA	800	B	5	Uganda
UA	UKR	804	C	5	Ukraine
AE	ARE	784	B	2	United Arab Emirates
GB	GBR	826	B	2	United Kingdom
US	USA	840	A	1	United States
UM	UMI	581	A	1	United States Minor Outlying Islands
UY	URY	858	A	4	Uruguay
UZ	UZB	860	C	5	Uzbekistan
VU	VUT	548	A	4	Vanuatu
VA	VAT	336	B	2	Vatican City (Holy See)
VE	VEN	862	A	4	Venezuela
VN	VNM	704	A	3	Viet Nam
VG	VGB	92	A	4	Virgin Islands, British
VI	VIR	850	A	1	Virgin Islands, U.S.
WF	WLF	876	A	4	Wallis and Futuna
EH	ESH	732	B	5	Western Sahara
YE	YEM	887	B	2	Yemen
ZM	ZMB	894	B	5	Zambia
ZW	ZWE	716	B	5	Zimbabwe

a Not documented for Blu-ray. Assumed from geographic location.

b Guernsey and Jersey are listed in Blu-ray documentation as Channel Islands.

Reference Data

Table B.10 ISO 639 Language Codes

639-2/T	639-1	Hex	Dec	Language Name
aar	aa	4141	6565	Afar
abk	ab	4142	6566	Abkhazian
ace				Achinese
ach				Acoli
ada				Adangme
ady*				Adyghe; Adygei
afa				Afro-Asiatic
afh				Afrihili
afr	af	4146	6570	Afrikaans
ain*				Ainu
aka	ak	414B	6575	Akan
akk				Akkadian
ale				Aleut
alg				Algonquian languages
alt*				Southern Altai
amh	am	414D	6577	Amharic
ang				English, Old (ca.450-1100)
anp*				Angika
apa				Apache languages
ara	ar	4152	6582	Arabic
arc				Official Aramaic (700-300 BCE); Imperial Aramaic (700-300 BCE)
arg*	an	414E	6578	Aragonese
arn				Mapudungun; Mapuche
arp				Arapaho
art				Artificial
arw				Arawak
asm	as	4153	6583	Assamese
ast*				Asturian; Bable; Leonese; Asturleonese
ath				Athapascan languages
aus				Australian languages
ava	av	4156	6586	Avaric
ave	ae	4145	6569	Avestan
awa				Awadhi
aym	ay	4159	6589	Aymara
aze	az	415A	6590	Azerbaijani
bad				Banda languages
bai				Bamileke languages
bak	ba	4241	6665	Bashkir
bal				Baluchi
bam	bm	424D	6677	Bambara
ban				Balinese
bas				Basa
bat				Baltic

Reference Data

Table B.10 ISO 639 Language Codes (continued)

639-2/T	639-1	Hex	Dec	Language Name
bej				Beja; Bedawiyet
bel	be	4245	6669	Belarusian
bem				Bemba
ben	bn	424E	6678	Bengali
ber				Berber
bho				Bhojpuri
bih	bh	4248	6672	Bihari
bik				Bikol
bin				Bini; Edo
bis	bi	4249	6673	Bislama
bla				Siksika
bnt				Bantu
bod	bo	424F	6679	Tibetan
bos*	bs	4253	6683	Bosnian
bra				Braj
bre	br	4252	6682	Breton
btk				Batak languages
bua				Buriat
bug				Buginese
bul	bg	4247	6671	Bulgarian
byn*				Blin; Bilin
cad				Caddo
cai				Central American Indian
car				Galibi Carib
cat	ca	4341	6765	Catalan; Valencian
cau				Caucasian
ceb				Cebuano
cel				Celtic
ces	cs	4353	6783	Czech
cha	ch	4348	6772	Chamorro
chb				Chibcha
che	ce	4345	6769	Chechen
chg				Chagatai
chk				Chuukese
chm				Mari
chn				Chinook jargon
cho				Choctaw
chp				Chipewyan; Dene Suline
chr				Cherokee
chu	cu	4355	6785	Church Slavic; Old Slavonic
chv	cv	4356	6786	Chuvash
chy				Cheyenne
cmc				Chamic languages
cop				Coptic
cor	kw	4B57	7587	Cornish

Reference Data

Table B.10 ISO 639 Language Codes (continued)

639-2/T	639-1	Hex	Dec	Language Name
cos	co	434F	6779	Corsican
cpe				Creoles and pidgins, English based
cpf				Creoles and pidgins, French-based
cpp				Creoles and pidgins, Portuguese-based
cre	cr	4352	6782	Cree
crh*				Crimean Tatar; Crimean Turkish
crp				Creoles and pidgins
csb*				Kashubian
cus				Cushitic
cym	cy	4359	6789	Welsh
dak				Dakota
dan	da	4441	6865	Danish
dar*				Dargwa
day				Land Dayak languages
del				Delaware
den				Slave (Athapascan)
deu	de	4445	6869	German
dgr				Dogrib
din				Dinka
div	dv	4456	6886	Divehi; Dhivehi; Maldivian
doi				Dogri
dra				Dravidian
dsb*				Lower Sorbian
dua				Duala
dum				Dutch, Middle (ca.1050-1350)
dyu				Dyula
dzo	dz	445A	6890	Dzongkha
efi				Efik
egy				Egyptian (Ancient)
eka				Ekajuk
ell	el	454C	6976	Greek, Modern (1453-)
elx				Elamite
eng	en	454E	6978	English
enm				English, Middle (1100-1500)
epo	eo	454F	6979	Esperanto
est	et	4554	6984	Estonian
eus	eu	4555	6985	Basque
ewe	ee	4545	6969	Ewe
ewo				Ewondo
fan				Fang
fao	fo	464F	7079	Faroese
fas	fa	4641	7065	Persian
fat				Fanti
fij	fj	464A	7074	Fijian

Table B.10 ISO 639 Language Codes (continued)

639-2/T	639-1	Hex	Dec	Language Name
fil*				Filipino; Pilipino
fin	fi	4649	7073	Finnish
fiu				Finno-Ugrian
fon				Fon
fra	fr	4652	7082	French
frm				French, Middle (ca.1400-1600)
fro				French, Old (842-ca.1400)
frr*				Northern Frisian
frs*				Eastern Frisian
fry	fy	4659	7089	Western Frisian
ful	ff	4646	7070	Fulah
fur				Friulian
gaa				Ga
gay				Gayo
gba				Gbaya
gem				Germanic
gez				Geez
gil				Gilbertese
gla	gd	4744	7168	Gaelic; Scottish Gaelic
gle	ga	4741	7165	Irish
glg	gl	474C	7176	Galician
glv	gv	4756	7186	Manx
gmh				German, Middle High (ca.1050-1500)
goh				German, Old High (ca.750-1050)
gon				Gondi
gor				Gorontalo
got				Gothic
grb				Grebo
grc				Greek, Ancient (to 1453)
grn	gn	474E	7178	Guarani
gsw*				Swiss German; Alemannic; Alsatian
guj	gu	4755	7185	Gujarati
gwi				Gwich'in
hai				Haida
hat*	ht	4854	7284	Haitian; Haitian Creole
hau	ha	4841	7265	Hausa
haw				Hawaiian
heb	he	4845	7269	Hebrew
her	hz	485A	7290	Herero
hil				Hiligaynon
him				Himachali
hin	hi	4849	7273	Hindi
hit				Hittite

Reference Data

Table B.10 ISO 639 Language Codes (continued)

639-2/T	639-1	Hex	Dec	Language Name
hmn				Hmong
hmo	ho	484F	7279	Hiri Motu
hrv	hr	4852	7282	Croatian
hsb*				Upper Sorbian
hun	hu	4855	7285	Hungarian
hup				Hupa
hye	hy	4859	7289	Armenian
iba				Iban
ibo	ig	4947	7371	Igbo
ido*	io	494F	7379	Ido
iii*	ii	4949	7373	Sichuan Yi; Nuosu
ijo				Ijo languages
iku	iu	4955	7385	Inuktitut
ile	ie	4945	7369	Interlingue; Occidental
ilo				Iloko
ina	ia	4941	7365	Interlingua (IALA)
inc				Indic
ind	id	4944	7368	Indonesian
ine				Indo-European
inh*				Ingush
ipk	ik	494B	7375	Inupiaq
ira				Iranian
iro				Iroquoian languages
isl	is	4953	7383	Icelandic
ita	it	4954	7384	Italian
jav	jv	4A56	7486	Javanese
jbo*				Lojban
jpn	ja	4A41	7465	Japanese
jpr				Judeo-Persian
jrb				Judeo-Arabic
kaa				Kara-Kalpak
kab				Kabyle
kac				Kachin; Jingpho
kal	kl	4B4C	7576	Kalaallisut; Greenlandic
kam				Kamba
kan	kn	4B4E	7578	Kannada
kar				Karen languages
kas	ks	4B53	7583	Kashmiri
kat	ka	4B41	7565	Georgian
kau	kr	4B52	7582	Kanuri
kaw				Kawi
kaz	kk	4B4B	7575	Kazakh
kbd*				Kabardian
kha				Khasi
khi				Khoisan

Table B.10 ISO 639 Language Codes (continued)

639-2/T	639-1	Hex	Dec	Language Name
khm	km	4B4D	7577	Central Khmer
kho				Khotanese
kik	ki	4B49	7573	Kikuyu; Gikuyu
kin	rw	5257	8287	Kinyarwanda
kir	ky	4B59	7589	Kirghiz; Kyrgyz
kmb				Kimbundu
kok				Konkani
kom	kv	4B56	7586	Komi
kon	kg	4B47	7571	Kongo
kor	ko	4B4F	7579	Korean
kos				Kosraean
kpe				Kpelle
krc*				Karachay-Balkar
krl*				Karelian
kro				Kru languages
kru				Kurukh
kua	kj	4B4A	7574	Kuanyama; Kwanyama
kum				Kumyk
kur	ku	4B55	7585	Kurdish
kut				Kutenai
lad				Ladino
lah				Lahnda
lam				Lamba
lao	lo	4C4F	7679	Lao
lat	la	4C41	7665	Latin
lav	lv	4C56	7686	Latvian
lez				Lezghian
lim*	li	4C49	7673	Limburgan; Limburger; Limburgish
lin	ln	4C4E	7678	Lingala
lit	lt	4C54	7684	Lithuanian
lol				Mongo
loz				Lozi
ltz	lb	4C42	7666	Luxembourgish; Letzeburgesch
lua				Luba-Lulua
lub	lu	4C55	7685	Luba-Katanga
lug	lg	4C47	7671	Ganda
lui				Luiseno
lun				Lunda
luo				Luo (Kenya and Tanzania)
lus				Lushai
mad				Madurese
mag				Magahi
mah	mh	4D48	7772	Marshallese
mai				Maithili

Reference Data

Table B.10 ISO 639 Language Codes (continued)

639-2/T	639-1	Hex	Dec	Language Name
mak				Makasar
mal	ml	4D4C	7776	Malayalam
man				Mandingo
map				Austronesian
mar	mr	4D52	7782	Marathi
mas				Masai
mdf*				Moksha
mdr				Mandar
men				Mende
mga				Irish, Middle (900-1200)
mic				Mi'kmaq; Micmac
min				Minangkabau
mis				Uncoded languages
mkd	mk	4D4B	7775	Macedonian
mkh				Mon-Khmer
mlg	mg	4D47	7771	Malagasy
mlt	mt	4D54	7784	Maltese
mnc*				Manchu
mni				Manipuri
mno				Manobo languages
moh				Mohawk
mol	mo	4D4F	7779	Moldavian; Moldovan
mon	mn	4D4E	7778	Mongolian
mos				Mossi
mri	mi	4D49	7773	Maori
msa	ms	4D53	7783	Malay
mul				Multiple languages
mun				Munda languages
mus				Creek
mwl*				Mirandese
mwr				Marwari
mya	my	4D59	7789	Burmese
myn				Mayan languages
myv*				Erzya
nah				Nahuatl languages
nai				North American Indian
nap*				Neapolitan
nau	na	4E41	7865	Nauru
nav	nv	4E56	7886	Navajo; Navaho
nbl	nr	4E52	7882	Ndebele, South; South Ndebele
nde	nd	4E44	7868	Ndebele, North; North Ndebele
ndo	ng	4E47	7871	Ndonga
nds*				Low German; Low Saxon; German, Low; Saxon, Low
nep	ne	4E45	7869	Nepali

Reference Data

Table B.10 ISO 639 Language Codes (continued)

639-2/T	639-1	Hex	Dec	Language Name
new				Nepal Bhasa; Newari
nia				Nias
nic				Niger-Kordofanian
niu				Niuean
nld	nl	4E4C	7876	Dutch; Flemish
nno	nn	4E4E	7878	Norwegian Nynorsk
nob*	nb	4E42	7866	Bokmål, Norwegian
nog*				Nogai
non*				Norse, Old
nor	no	4E4F	7879	Norwegian
nqo*				N'Ko
nso				Pedi; Sepedi; Northern Sotho
nub				Nubian languages
nwc*				Classical Newari; Old Newari; Classical Nepal Bhasa
nya	ny	4E59	7889	Chichewa; Chewa; Nyanja
nym				Nyamwezi
nyn				Nyankole
nyo				Nyoro
nzi				Nzima
oci	oc	4F43	7967	Occitan (post 1500)
oji	oj	4F4A	7974	Ojibwa
ori	or	4F52	7982	Oriya
orm	om	4F4D	7977	Oromo
osa				Osage
oss	os	4F53	7983	Ossetian; Ossetic
ota				Turkish, Ottoman (1500-1928)
oto				Otomian languages
paa				Papuan
pag				Pangasinan
pal				Pahlavi
pam				Pampanga; Kapampangan
pan	pa	5041	8065	Panjabi; Punjabi
pap				Papiamento
pau				Palauan
peo				Persian, Old (ca.600-400 B.C.)
phi				Philippine
phn				Phoenician
pli	pi	5049	8073	Pali
pol	pl	504C	8076	Polish
pon				Pohnpeian
por	pt	5054	8084	Portuguese
pra				Prakrit languages
pro				Provençal, Old (to 1500)
pus	ps	5053	8083	Pushto; Pashto

Reference Data

Table B.10 ISO 639 Language Codes (continued)

639-2/T	639-1	Hex	Dec	Language Name
que	qu	5155	8185	Quechua
raj				Rajasthani
rap				Rapanui
rar				Rarotongan; Cook Islands Maori
roa				Romance
roh	rm	524D	8277	Romansh
rom				Romany
ron	ro	524F	8279	Romanian
run	rn	524E	8278	Rundi
rup*				Aromanian; Arumanian; Macedo-Romanian
rus	ru	5255	8285	Russian
sad				Sandawe
sag	sg	5347	8371	Sango
sah				Yakut
sai				South American Indian
sal				Salishan languages
sam				Samaritan Aramaic
san	sa	5341	8365	Sanskrit
sas				Sasak
sat				Santali
scn*				Sicilian
sco				Scots
sel				Selkup
sem				Semitic
sga				Irish, Old (to 900)
sgn*				Sign Languages
shn				Shan
sid				Sidamo
sin	si	5349	8373	Sinhala; Sinhalese
sio				Siouan languages
sit				Sino-Tibetan
sla				Slavic
slk	sk	534B	8375	Slovak
slv	sl	534C	8376	Slovenian
sma*				Southern Sami
sme*	se	5345	8369	Northern Sami
smi				Sami languages
smj*				Lule Sami
smn*				Inari Sami
smo	sm	534D	8377	Samoan
sms*				Skolt Sami
sna	sn	534E	8378	Shona
snd	sd	5344	8368	Sindhi
snk				Soninke

Table B.10 ISO 639 Language Codes (continued)

639-2/T	639-1	Hex	Dec	Language Name
sog				Sogdian
som	so	534F	8379	Somali
son				Songhai languages
sot	st	5354	8384	Sotho, Southern
spa	es	4553	6983	Spanish; Castilian
sqi	sq	5351	8381	Albanian
srd	sc	5343	8367	Sardinian
srn*				Sranan Tongo
srp	sr	5352	8382	Serbian
srr				Serer
ssa				Nilo-Saharan
ssw	ss	5353	8383	Swati
suk				Sukuma
sun	su	5355	8385	Sundanese
sus				Susu
sux				Sumerian
swa	sw	5357	8387	Swahili
swe	sv	5356	8386	Swedish
syc*				Classical Syriac
syr				Syriac
tah	ty	5459	8489	Tahitian
tai				Tai
tam	ta	5441	8465	Tamil
tat	tt	5454	8484	Tatar
tel	te	5445	8469	Telugu
tem				Timne
ter				Tereno
tet				Tetum
tgk	tg	5447	8471	Tajik
tgl	tl	544C	8476	Tagalog
tha	th	5448	8472	Thai
tig				Tigre
tir	ti	5449	8473	Tigrinya
tiv				Tiv
tkl				Tokelau
tlh*				Klingon; tlhIngan-Hol
tli				Tlingit
tmh				Tamashek
tog				Tonga (Nyasa)
ton	to	544F	8479	Tonga (Tonga Islands)
tpi				Tok Pisin
tsi				Tsimshian
tsn	tn	544E	8478	Tswana
tso	ts	5453	8483	Tsonga
tuk	tk	544B	8475	Turkmen

Table B.10 ISO 639 Language Codes (continued)

639-2/T	639-1	Hex	Dec	Language Name
tum				Tumbuka
tup*				Tupi languages
tur	tr	5452	8482	Turkish
tut				Altaic
tvl				Tuvalu
twi	tw	5457	8487	Twi
tyv				Tuvinian
udm*				Udmurt
uga				Ugaritic
uig	ug	5547	8571	Uighur; Uyghur
ukr	uk	554B	8575	Ukrainian
umb				Umbundu
und				Undetermined
urd	ur	5552	8582	Urdu
uzb	uz	555A	8590	Uzbek
vai				Vai
ven	ve	5645	8669	Venda
vie	vi	5649	8673	Vietnamese
vol	vo	564F	8679	Volapük
vot				Votic
wak				Wakashan languages
wal				Walamo
war				Waray
was				Washo
wen				Sorbian languages
wln*	wa	5741	8765	Walloon
wol	wo	574F	8779	Wolof
xal*				Kalmyk; Oirat
xho	xh	5848	8872	Xhosa
yao				Yao
yap				Yapese
yid	yi	5949	8973	Yiddish
yor	yo	594F	8979	Yoruba
ypk				Yupik languages
zap				Zapotec
zbl*				Blissymbols; Blissymbolics; Bliss
zen				Zenaga
zha	za	5A41	9065	Zhuang; Chuang
zho	zh	5A48	9072	Chinese
znd				Zande languages
zul	zu	5A55	9085	Zulu
zun				Zuni
zxx*				No linguistic content; Not applicable
zza*				Zaza; Dimili; Dimli; Kirdki; Kirmanjki; Zazaki

Reference Data

* Language code is not supported by the player text subtitle capability bitmap stored in PSRs 48-61. Sadly, Klingon is one of these.

Note 1: The Blu-ray specifications refer to ISO 639-2:1988, which has been updated regularly since 1998. Because players have been in production since 2006, it is recommended that codes as of 2006 (as shown in this table) be used in disc production to be compatible with older players. It is recommended that players recognize old codes and map them to any corresponding new codes.

Note 2: Blu-ray uses ISO 639-2/T, which lists roughly 480 three-letter codes for languages and language groups, many of them obscure or no longer used. The "T" refers to the codes for terminology purposes, which sometimes differ from the "B" codes for bibliographic purposes. ISO 639-1 lists 185 two-letter codes for more general terminology purposes. These are included for reference.

Note 3: DVD uses ISO 639-1 codes, represented as pairs of two-digit decimal or hexadecimal numbers representing each letter. These are not formally used in Blu-ray formats but may be used by Blu-ray players to set the preferred language for both BD and DVD discs.

Appendix C
Related Standards and Specifications

The Blu-ray Disc™ format is based on or has borrowed from dozens of standards developed over the years by many organizations. Most of the standards in this appendix are a normative part of the format. Some, such as CD and Video CD, are for players that choose to implement the format. Many standards and specifications have been updated since most of the Blu-ray format was finalized in 2005. The best practice is for players to implement the latest version (as long as it does not change things in a way that would cause compatibility problems) and for discs to implement the older version specified here. Blu-ray format updates after 2008 may refer to newer versions of the standards and specifications listed here.

Physical and Device Interface Specifications

DVD

The following DVD specifications apply to BD format on DVD 4.7GB and 8.5GB discs (including AVCREC) and BD hybrid discs.

- *DVD Specifications for Recordable Disc, Part 1: Physical Specifications Ver. 1.0*
    - Also *ECMA 267* or *ISO/IEC 16448*
    - *3x-Speed DVD-ROM Rev. 1.1*
- *DVD Specifications for Recordable Disc for General, Part1: Physical Specifications Ver.2.1* [DVD-R]
- *DVD Specifications for Recordable Disc for Dual Layer, Part1: Physical Specifications Ver.3.0* [DVD-R]
- *DVD Specifications for Re-recordable Disc, Part1: Physical Specifications Ver.1.2* [DVD-RW]
    - Also *ECMA-338* or *ISO/IEC 17342*
- *DVD Specifications for Re-recordable Disc for Dual Layer, Part1: Physical Specifications Ver.2.0* [DVD-RW]
- *DVD Specifications for Rewritable Disc, Part1: Physical Specifications Ver.2.2* [DVD-RAM]
    - Also *ECMA-330* or *ISO/IEC 17592*
- *DVD+R Part 1 Single Layer: DVD+R 4.7 Gbytes, Basic Format Specifications Version 1.3*
    - Also *ECMA-349* or *ISO/IEC 17344*
- *DVD+R Part 2 Dual Layer: DVD+R 8.5 Gbytes, 8x Basic Format Specifications Version 1.1*

Related Standards and Specifications

- *DVD+RW Basic Format Specifications Part 1: Single layer, Volume 1 Version 1.3, Volume 2 Version 1.0*
    - Also *ECMA-337* or *ISO/IEC 17341*
- *DVD+RW Basic Format Specifications Part 2: Dual layer, Volume 1: 2.4x Version 1.0*

Compact Disc (CD Audio and CD-ROM)

- *IEC 60908 (1987-09) Compact disc digital audio system* [Red Book]
- *ISO/IEC 10149:1995 Information technology-Data interchange on read-only 120 mm optical data discs (CD-ROM)* [Yellow Book]
  (Note: Equivalent to *ECMA 130, 2nd Edition, June 1996*)
- *Philips/Sony Orange Book part-II Recordable Compact Disc System - Philips/Sony Orange Book part-III Recordable Compact Disc System*
- *IEC 61104: Compact Disc Video System, 12 cm* [CDV Single]

Laserdisc

- *IEC 60857: Pre-Recorded Optical Reflective Videodisk System "Laser Vision" 60 Hz/525 Lines - M/NTSC*
- *IEC 60856: Pre-Recorded Optical Reflective Videodisk System "Laser Vision" 50 Hz/625 Lines - PAL*

Drive Interface

- *ANSI INCITS 430-2007 MultiMedia Command Set - 5 (MMC-5)* [Mt. Fuji 5]
- *ANSI INCITS (Draft) MultiMedia Command Set - 6 (MMC-6)* [Mt. Fuji 6, Revision 0.7 or later]
- *ANSI X3.131-1994: Information Systems-Small Computer Systems Interface-2* [SCSI-2]
- *ANSI X3.277-1996: Information Technology-SCSI-3 Fast-20*
- *ANSI X3.221-1994: Information Systems-AT Attachment Interface for Disk Drives* [ATA/IDE]
- *ANSI X3.279-1996: Information Technology-AT Attachment Interface with Extensions (ATA-2)* [EIDE]
- *ANSI X3.298-1997: Information Technology - AT Attachment-3 Interface* [ATA-3]
- *ANSI NCITS 317-1998: AT Attachment with Packet Interface Extension (ATA/ATAPI-4)* [EIDE; Ultra DMA 33]
- *ANSI NCITS 340-2000: Information Technology - AT Attachment with Packet Interface-5 (ATA/ATAPI-5)* [EIDE; Ultra DMA 66]
- *ANSI NCITS 361-2002: Information Technology - AT Attachment - 6 with Packet Interface (ATA/ATAPI-6)* [EIDE; Ultra DMA 100]
- *ANSI NCITS Serial ATA: High Speed Serializing AT Attachment Revision 1.0,* Serial ATA Working Group
- *ANSI NCITS Information Technology: AT Attachment with Packet Interface - 6 Rev. 3b,* T13 Committee

Related Standards and Specifications

System Specifications

File System

- *OSTA Universal Disc Format Specification Revision 1.02: 1996* (OSTA UDF Compliant Domain of ISO/IEC 13346:1995 *Volume and file structure of write-once and rewritable media using non-sequential recording for information interchange*)
    - *ISO/IEC 13346* is equivalent to *ECMA 167*
- *OSTA Universal Disc Format Specification Revision 2.50: 2003* (OSTA UDF Compliant Domain of ECMA 167 *3rd edition*)
- *OSTA Universal Disc Format Specification Revision 2.60: 2005* (OSTA UDF Compliant Domain of ECMA 167 *3rd edition*)
- *ISO 9660:1988 Information processing - Volume and file structure of CD-ROM for information interchange* [ISO image format]
    - Equivalent to *ECMA 119, 2d edition, 1987*
- *ECMA TR/71 DVD Read-Only Disk File System Specifications* [UDF Bridge for DVD]
- *Joliet CD-ROM Recording Specification, ISO 9660:1988 Extensions for Unicode* (Microsoft)

MPEG-2 System

- *ISO/IEC 13818-1:2000 Information technology - Generic coding of moving pictures and associated audio information: Part 1: Systems* [program streams for DVD, transport streams for BD]
    - *ISO/IEC 13818-1:2000/Amd 2:2004: Support of IPMP on MPEG-2 systems*
    - *ISO/IEC 13818-1:2000/Amd 3:2004: Transport of AVC video data over ITU-T Rec H.222.0 | ISO/IEC 13818-1 streams*

Video Specifications

MPEG-1 Video

- *ISO/IEC 11172-2:1993 Information technology - Coding of moving pictures and associated audio for digital storage media at up to about 1.5 Mbit/s - Part 2: Video*

MPEG-2 Video

- *ISO/IEC 13818-2:2000 Information technology - Generic coding of moving pictures and associated audio information - Part 2: Video*
    - Equivalent to *ITU-T H.262 Information technology - Generic coding of moving pictures and associated audio information: Video*
- *SMPTE RP 202-2000: Video Alignment for MPEG-2 Coding*
- *ISO/IEC 13818-9:1996 Information technology - Generic coding of moving pictures and associated audio information - Part 9: Extension for real time interface for systems decoders*

Related Standards and Specifications

MPEG-4 AVC
- *ISO/IEC 14496-10:2005 Information technology - Coding of audio-visual objects - Part 10: Advanced video coding*
    - Equivalent to *ITU-T H.264: Advanced video coding for generic audio-visual services (2005-03)*

VC-1
- *SMPTE 421M Television - VC-1 Compressed Video Bitstream Format and Decoding Process*
- *SMPTE RP227 VC-1 Bitstream Transport Encodings*
- *SMPTE RP 228 VC-1 Decoder and Bitstream Conformance*

NTSC Video
- *SMPTE 170M-1994 Television - Composite Analog Video Signal - NTSC for Studio Applications*
- *ITU-R BT.470-4 Conventional Television Systems*

PAL Video
- *ITU-R BT.601-5 Studio encoding parameters of digital television for standard 4:3 and widescreen 16:9 aspect ratios (10/1995)*
- *ITU-R BT.709-5 Parameter Values for the HDTV Standards for Production and International Programme Exchange (2002-04)* [1080-line video]

ATSC
- *A/53: ATSC Digital Television Standard, Parts 1-6, 2007*
- *A/65: ATSC Program and System Information Protocol for Terrestrial Broadcast and Cable (23 Dec 1997)*
- *A/90: ATSC Data Broadcast Standard with Amendment 1 and Corrigenda 1 and 2 (26 July 2000)*

DVB
- *ETSI EN 300 468 V1.3.1: Digital Video Broadcasting (DVB); Specification for Service Information (SI) in DVB systems (1998-02)*
- *ETSI EN 300 472 V1.2.2: Digital Video Broadcasting (DVB); Specification for conveying ITU-R System B Teletext in DVB bitstreams (1997-08)*
- *ETSI ETR 211: Digital Video Broadcasting (DVB); Guidelines on implementation and usage of Service Information (SI) (August 1997, Second Edition)*
- *ETSI TR 101 154 V1.4.1: Digital Video Broadcasting (DVB); Implementation guidelines for the use of Video and Audio Coding in Broadcasting Applications based on the MPEG-2 Transport Stream (July 2000)*
- *ETSI TR 101 162: Digital Video Broadcasting (DVB); Allocation of Service Information (SI) and Data Broadcasting Codes for Digital Video Broadcasting (DVB) systems (October 1995)*

Related Standards and Specifications

ARIB

- *ARIB STD-B10: Service information for digital broadcasting system (Version 1.2)*
- *ARIB STD-B20: Digital broadcasting system and related operational guidelines for broadcasting satellites (Version 1.1)*
- *ARIB STD-B21: Digital receiver for digital satellite broadcasting services using broadcasting satellites (Version 1.1)*
- *ARIB TR-B15: Operational guidelines for digital satellite broadcasting services using broadcasting satellites (Version 1.1)*

Additional Video Signals

- *SMPTE 274M-1998 Television - 1920 x 1080 Scanning and Analog and Parallel Digital Interfaces for Multiple Picture Rates*
- *SMPTE 293M-1996 Television - 720 x 483 Active Line at 59.94-Hz Progressive Scan Production - Digital Representation*
- *SMPTE 296M-2001 Television - 1280 x 720 Progressive Image Sample Structure - Analog and Digital Representation and Analog Interface*
- *CEA-708-B Digital Television (DTV) Closed Captioning (1999)*
- *ETSI ETS 300 294: Television Systems; 625-line television: Wide Screen Signalling (WSS) (Edition 2:1995-12)*
- *ITU-R BT.1119-1 Widescreen signaling for broadcasting. Signaling for widescreen and other enhanced television parameters*
- *ITU-R BT.1358 Studio Parameters of 625 and 525 Line Progressive Scan Television Systems (1998)*
- *IEC 61880: Video systems (525/60) - Video and accompanied data using the vertical blanking interval - Analogue interface (1998-01)* [CGMS-A; NTSC line 20; PAL/SECAM/YUV line 21]
- *EIA-608 (September 1994): Recommended Practice for Line 21 Data Service*
- *EIA/CEA-608-B: Line 21 Data Services (October 2000)* [Closed Captions, XDS, CGMS-A; NTSC line 21; YUV line 21]
- *EIA-708-B: Digital Television (DTV) Closed Captioning (December 1999)*
- *EIA-744-A: Transport of Content Advisory Information using Extended Data Service (XDS) (1998)*
- *EIA-770.1-A: Analog 525 Line Component Video Interface - Three Channels*
- *EIA-770.2-A: Standard Definition TV Analog Component Video Interface*
- *EIA-770.3-A: High Definition TV Analog Component Video Interface*
- *EIA-775-A: DTV 1394 Interface Specification*
- *EIA/CEA-775.2: Service selection information for digital storage media interoperability (September 2000)*
- *EIA/CEA-805: Data Services on the Component Video Interfaces* [CGMS-A for progressive-scan component video, including YPbPr, RGB, and VGA]
- *ETS 300294* [PAL/SECAM/YUV CGMS-A]

Related Standards and Specifications

- *EIA-608: Recommended Practice For Line 21 Data Service* [NTSC Closed Captions]
- *EIA-746: Transport Of Internet Uniform Resource Locator (URL) Information Using Text-2 (T-2) Service* [TV links; ATVEF triggers]
- *ETS 300 294 Edition 2:1995-12* [Film/camera mode]
- *ITU-R BT.1119-1: Widescreen signaling for broadcasting. Signaling for widescreen and other enhanced television parameters* [PAL CGMS-A]
- *IEC 61880: Video systems (525/60) - Video and accompanied data using the vertical blanking interval - Analogue interface (1998-01)* [NTSC VBI line 20 CGMS-A]
- *IEC 61880-2: Video systems (525/60) - Video and accompanied data using the vertical blanking interval - Analogue interface - Part 2: 525 progressive scan system (2002-09)* [520p line 41 CGMS-A]
- *JEITA CPR-1204* [NTSC widescreen signaling and CGMS-A, progressive-scan; formerly EIA- CPX-1204]
- *SMPTE 259M-1997: Television - 10-Bit 4:2:2 Component and 4fsc Composite Digital Signals - Serial Digital Interface* [SDI]
- *SMPTE 292M-1998: Television - Bit-Serial Digital Interface for High-Definition Television Systems* [HD-SDI]

Audio Specifications

Dolby Digital Audio (AC-3)

- *ATSC A/52 1995* [Dolby Digital]
- *ATSC A/52B: Digital Audio Compression (AC-3) (E-AC-3) Standard, Rev. B* [Dolby Digital Plus]
- *ETSI TS 102 366 V1.1.1 (2005-02) Digital Audio Compression (AC-3, Enhanced AC-3) Standard* [Dolby Digital Plus]

Dolby TrueHD Audio

- *Meridian Lossless Packing - Technical Reference for FBA and FBB streams Version 1.0*

DTS Audio

- *DTS Coherent Acoustics Core, Version 3.0, December 20, 2005, Document #F335; DTS Coherent Acoustics Extensions, Amendment to DTS Coherent Acoustics Decoder Development Manual, Release Version 3.0, January 4, 2006, Document #F413*
- *DTS-HD Sub-stream and Decoder Interface Description Version 1.0 (12/6/2005)*

MPEG

- *ISO/IEC 11172-3:1993 Information technology-Coding of moving pictures and associated audio for digital storage media at up to about 1,5 Mbit/s-Part 3: Audio* [MPEG-1 Audio]
- *ISO/IEC 13818-3:1995 Information technology-Generic coding of moving pictures and associated audio information-Part 3: Audio* [MPEG-2 Audio Extensions]

Related Standards and Specifications

Interface Specifications

Digital Interface

- *IEEE 1394-1995 IEEE Standard for a High Performance Serial Bus* [FireWire]
- *IEC 61883 Standard for Digital Interface for Consumer Electronic Audio/Video Equipment (transport protocol for IEEE 1394) - 1394 Trade Association Audio/Video Control Digital Interface Command Set (AV/C)* [control protocol for *IEEE 1394*]
- *1394TA 1999029: AV/C Disc Subunit Enhancements for Hard Disk Drive Specification (July 10, 2000)*
- *DVD Forum: Guideline of Transmission and Control for DVD-Video/Audio through IEEE1394 Bus*
  - Originally *1394TA 2000002, DVD-Video Stream Specification using MPEG-TS and IEEE1394 Transmission*

Digital Audio Interface

- *IEC 60958-1: Digital audio interface - Part 1: 1999 - General*
  - Also *EIAJ CO-1201*
- *IEC 60958-2: Digital audio interface-Part 2: 1997 - Software information delivery mode*
- *IEC 60958-3: Digital audio interface - Part 3: 2000 - Consumer applications [S/PDIF,* "type II" consumer-use version of *AES3]*
- *IEC 60958-4: Digital audio interface - Part 4: 2000 - Professional applications*
- *AES3-1992: AES Recommended practice for digital audio engineering - Serial trans-mission format for two-channel linearly represented digital audio data* [formerly *AES/EBU;* complement of *IEC 60958*]
- *IEC 61937-1 Interfaces For Non-Linear PCM Encoded Audio Bitstreams Applying IEC 60958 - Part 1: Non-Linear PCM Encoded Audio Bitstreams For Consumer Applications* [supersedes *ATSC A/52 Annex B: AC-3 Data Stream in IEC 958 Interface*]
- *EIAJ- CP-340 A Digital Audio Interface (1987)* [optical digital audio; "Toslink"]
- *IEC61937-2: Digital audio - Interface for nonlinear PCM encoded audio bitstreams applying IEC 60958 - Part 2: Burst-info*
- *IEC61937-3: Digital audio - Interface for nonlinear PCM encoded audio bitstreams applying IEC 60958 - Part 3: Nonlinear PCM bitstreams according to the AC-3 and enhanced AC-3 formats*
- *IEC 61937-5: Digital audio - Interface for Nonlinear PCM encoded audio bitstreams applying IEC 60958 - Part 5: Nonlinear PCM Bitstreams According to the DTS (Digital Theater Systems) Format(s)*
- *IEC 61937-9: Digital audio - Interface for nonlinear PCM encoded audio bitstreams applying IEC 60958 - Part 9: Nonlinear PCM bitstreams according to the MAT format*

Digital Video Interface

- *VESA-2008-1: DisplayPort Standard - Version 1.1a*

Related Standards and Specifications

- CEA-861-E A DTV Profile for Uncompressed High Speed Digital Interfaces
- High-Definition Multimedia Interface Specification Version 1.3
- HDCP Specification Rev. 1.3

Other Specifications

Graphics

- ITU-T.81 Information technology - Digital compression and coding of continuous-tone still images - Requirements and guidelines (1992-09) [JPEG]
    - Equivalent to ISO/IEC 10918-1
- JPEG File Interchange Format Version 1.02

Markup and Metadata

- Extensible Markup Language (XML) 1.0 (Third Edition): W3C Recommendation 04 February 2004
- ETSI TS 102 822-3-1 v1.1.1 Broadcast and On-line Services: Search, select, and rightful use of content on personal storage systems ("TV-Anytime Phase 1"); Part 3: Metadata; Sub-part 1: Metadata schemas (2003-10)
- ISO 8601:2004 Data elements and interchange formats - Information interchange - Representation of dates and times

Network

- RFC 2396: Uniform Resource Identifiers (URI): Generic Syntax
- RFC 2246: The TLS Protocol. Version 1.0

Interactive Applications

- ETSI TS 102 819 V1.3.1: Digital Video Broadcasting (DVB); Globally Executable MHP (GEM1.02), including GEM 1.0.2 Errata 1 - TM3443r3
- ETSI TS 101 812 V1.3.1: Digital Video Broadcasting (DVB); Multimedia Home Platform (MHP) Specification 1.0.3, including MHP Specification Version 1.0.3 errata #2
- HAVi 1.1 (HAVi v1.1 Chapter 8, 15-May-2001; HAVi v1.1 Java L2 APIs, 15-May-2001; HAVi v1.1 Chapter 7, 15-May-2001)
- JSR 36: Connected Device Configuration 1.0b [included in JSR 129]
- JSR 46: Foundation Profile 1.0b
- JSR 129: Personal Basis Profile 1.0b
- JSR 217: Personal Basis Profile 1.1
- JSR 218: Connected Device Configuration (CDC) 1.1
- JSR 927: Java TV(tm) API 1.1
- JAR File Specification: Sun Microsystems, version 1.5

Related Standards and Specifications

Recording Codes
- *ISO 3901:1986 Documentation - International Standard Recording Code (ISRC)*
- *IFPI SID CODE Implementation Guide*

Language Codes
- *ISO 639-2 Codes for the Representation of Names of Languages - Part 2: Alpha-3 Code* (see Table B.10)

Country Codes
- *ISO 3166-1:1997 Codes for the representation of names of countries and their subdivisions - Part 1: Country codes* (see Table B.9)

Fonts and Text
- *OpenType™ Specification version 1.4* (Adobe & Microsoft)
- *ISO/IEC 646:1991 Information technology-ISO 7-bit coded character set for information interchange*
- *ISO 8859-1:1987 Information processing-8-bit single-byte coded graphic character sets-Part 1: Latin alphabet No. 1*
- *ISO 8859-2:1987 Information processing-8-bit single-byte coded graphic character sets-Part 2: Latin alphabet No. 2*
- *ISO/IEC 10646-1:1993, Information technology - Universal Multiple Octet Coded Character Set (UCS) - Part 1: Architecture and Basic Multilingual Plane with Amendment 1, 2, 3, 4, 5, 6, and 7*
- *RFC2279 - UTF-8, a transformation format of ISO 10646*
- *RFC2781 - UTF-16, an encoding of ISO 10646*
- *ISO/IEC 2022:1994 Information technology-Character code structure and extension techniques*
- *BIG5 - Institute for Information Industry, "Chinese Coded Character Set in Computer" (March 1984)*
- *GB18030-2000: Information technology - Chinese ideograms coded character set for information interchange - Extension for the basic set*
- *GB2312-80: Coding of Chinese Ideogram Set for Information Interchange Basic Set*
- *RFC 1922: Chinese Character Encoding for Internet Messages (CN-GB)*
- *JIS X0208:1997 Appendix 1 (Shift JIS)*
- *KS C 5601-1987 - Korea Industrial Standards Association, "Code for Information Interchange (Hangul and Hanja)," Korean Industrial Standard, 1987*
- *KS C 5861-1992: Korea Industrial Standards Association, "Hangul Unix Environment," Korean Industrial Standard, 1992 [EUC-KR]*

Cryptography and Content Protection
- *Advanced Access Content System (AACS) - Introduction and Common Cryptographic Elements*

Related Standards and Specifications

- *Advanced Access Content System (AACS) - Pre-recorded Video Book*
- *Advanced Access Content System (AACS) - Managed Copy Book*
- *Advanced Access Content System (AACS) - Recordable Video Book*
- *Advanced Access Content System (AACS) - Prepared Video Book*
- *Advanced Access Content System (AACS) - Blu-ray Disc Pre-recorded Book*
- *Advanced Access Content System (AACS) - Blu-ray Disc Recordable Book*
- *NIST, FIPS PUB 180-1: Secure Hash Standard*
- *RFC 2315 - PKCS#7: Cryptographic Message Syntax, version 1.5*
- *IETF RFC 2313 - PKCS #1: RSA Encryption Version1.5*
- *DTLA, Digital Transmission Content Protection Specification Revision-1.1, Appendix B*
- *Digital Transmission Licensing Administrator, Digital Transmission Content Protection Specification Volume 1 Revision 1.4*
- *A/70A: ATSC Standard: Conditional Access System for Terrestrial Broadcast (22 July 2004)*
- *ITU-T X.509: Information technology - Open Systems Interconnection - The Directory: Public-key and attribute certificate frameworks (2005-08)*
- *ANSI X9.31-1998, Digital Signatures Using Reversible Public Key Cryptography for the Financial Services Industry (rDSA) (September 9, 1998)*
- *National Institute of Standards and Technology (NIST), Digital Signature Standard (DSS), FIPS Publication 186-2 (+Change Notice) (January 27, 2000)*
- *National Institute of Standards and Technology (NIST), A Statistical Test Suite for Random and Pseudorandom Number Generators for Cryptographic Applications, NIST Special Publication 800-22, with revisions dated May 15, 2001*
- *National Institute of Standards and Technology (NIST), Recommendation for Block Cipher Modes of Operation - Methods and Techniques, NIST Special Publication 800-38A, 2001 Edition*
- *National Institute of Standards and Technology (NIST), Secure Hash Standard, FIPS Publication 180-2, August 1, 2002*
- *National Institute of Standards and Technology (NIST), Advanced Encryption Standard (AES), FIPS Publication 197, November 26, 2001*
- *National Institute of Standards and Technology (NIST), Cipher-based Message Authentication Code (CMAC), NIST Special Publication 800-38B, May, 2005*
- *RSA Laboratories, PKCS #1 (v2.1): RSA Cryptography Standard, June 14, 2002*

Appendix D
References and Information Sources

For an up-to-date list of references and information sources visit the Blu-ray Disc Demystified Website at www.bddemystified.com.

Recommended References

Dunn, Julian. *Sample Clock Jitter and Real-Time Audio over the IEEE1394 High-Performance Serial Bus*. Preprint 4920, 106th AES Convention, Munich, May 1999.

Dunn, Julian, and Ian Dennis. *The Diagnosis and Solution of Jitter-Related Problems in Digital Audio*. Preprint 3868, 96th AES Convention, Amsterdam. February 1994.

Jack, Keith. *Video Demystified: A Handbook for the Digital Engineer* (5th ed.). Newnes, 2007. ISBN: 0750683953.

Johnson, Mark R., Charles G. Crawford, and Christen M. Armbrust. *High-Definition DVD Handbook: Producing for HD DVD and Blu-ray Disc*. McGraw-Hill, 2007. ISBN: 0071485856.

Morris, Steven, and Anthony Smith-Chaigneau. *Interactive TV Standards: A Guide to MHP, OCAP, and JavaTV*. Focal Press, 2005. ISBN: 0240806662.

Negroponte, Nicholas and Marty Asher. *Being Digital*. Vintage Books, 1996. ISBN: 0679762906.

Pohlmann, Ken C. *Principles of Digital Audio* (5th ed.). McGraw-Hill, 2005. ISBN: 071441565.

Poynton, Charles A. *Digital Video and HDTV Algorithms and Interfaces*. Morgan Kaufmann, 2003. ISBN: 1558607927.

- *A Technical Introduction to Digital Video*. John Wiley & Sons, 1996. ISBN: 047112253X.

Solari, Stephen J. *Digital Video and Audio Compression*. McGraw-Hill, 1997. ISBN: 0070595380.

Symes, Peter. *Digital Video Compression*. McGraw-Hill, 2003. ISBN: 0071424873.

Taylor, Jim, Mark R. Johnson, and Charles G. Crawford. *DVD Demystified* (3rd ed.). McGraw-Hill, 2006. ISBN: 0071423966.

Watkinson, John. *The MPEG Handbook* (2nd ed.). Focal Press, 2004. ISBN: 024080578X.

- *The Art of Digital Video* (4th ed.). Focal Press, 2008. ISBN: 024052005X.
- *The Art of Digital Audio* (3d ed.) Focal Press. ASIN: B001CD663E.
- *An Introduction to Digital Audio* (2nd ed.). Focal Press, 2002. ASIN: B001B0MQ50.

Zink, Michael, Philip C. Starner, and Bill Foote. *Programming HD DVD and Blu-ray Disc: The HD Cookbook*. McGraw-Hill, 2007. ISBN: 007149670X.

Glossary

1080i 1080 lines of interlaced video (540 lines per field). Usually refers to 1920×1080 resolution in 1.78 aspect ratio. 1080i30 refers to 30 interlaced frames (60 fields) per second, which is often confusingly written as 1080i60.

1080p 1080 lines of progressive video (1080 lines per frame). Usually refers to 1920×1080 resolution in 1.78 aspect ratio. 1080p30 refers to 30 frames per second. 1080p24 refers to 24 frames per second (film source).

17PP 1-7 parity preserve, the RLL(1,7) modulation technique used by BD, where each set of 2 bits is replaced by a 3-bit code before being written onto the disc.

2-2 pulldown The process of transferring 24-frame-per-second film to video by repeating each film frame as two video fields. When 24-fps film is converted via 2-2 pulldown to 25-fps 625/50 (PAL) video, the film runs 4 percent faster than normal.

2-3 pulldown The process of converting 24-frame-per-second film to video by repeating one film frame as two fields, then the next film frame as three fields.

3-2 pulldown An uncommon variation of 2-3 pulldown, where the first film frame is repeated for three fields instead of two. Most people mean 2-3 pulldown when they say 3-2 pulldown.

4:1:1 The component digital video format with one Cb sample and one Cr sample for every four Y samples. 4:1 horizontal downsampling with no vertical downsampling. Chroma is sampled on every line, but only for every four luma pixels (i.e., 1 pixel in a 1×4 grid). This amounts to a subsampling of chroma by a factor of two compared to luma (and by a factor of four for a single Cb or Cr component). BD and DVD use 4:2:0 sampling, not 4:1:1 sampling.

4:2:0 The component digital video format used by BD and DVD, where there is one Cb sample and one Cr sample for every four Y samples (i.e., 1 pixel in a 2×2 grid). 2:1 horizontal downsampling and 2:1 vertical downsampling. Cb and Cr are sampled on every other line, in between the scan lines, with one set of chroma samples for each two luma samples on a line. This amounts to a subsampling of chroma by a factor of two compared to luma (and by a factor of four for a single Cb or Cr component).

4:2:2 The component digital video format commonly used for studio recordings, where there is one Cb sample and one Cr sample for every two Y samples (i.e., 1 pixel in a 1×2 grid). 2:1 horizontal downsampling with no vertical downsampling. This allocates the same number of samples to the chroma signal as to the luma signal. The input to MPEG-2 encoders is typically in 4:2:2 format, but the video is subsampled to 4:2:0 before being encoded and stored.

4:4:4 A component digital video format for high-end studio recordings, where Y, Cb, and Cr are sampled equally.

480i 480 lines of interlaced video (240 lines per field). Usually refers to 720×480 (or 704×480) resolution.

480p 480 lines of progressive video (480 lines per frame). Usually refers to 720×480 (or 704×480) resolution. 480p60 refers to 60 frames per second; 480p30 refers to 30 frames per second; and 480p24 refers to 24 frames per second (film source).

4C The four-company entity that produces copy-protection schemes: IBM, Intel, Panasonic, Toshiba. See CPRM, CPPM, and www.4centity.com.

525/60 The scanning system of 525 lines per frame and 60 interlaced fields (30 frames) per second. Used by the NTSC television standard.

5C Nickname for the Digital Transmission Licensing Administrator (DTLA), the five-company entity that produces copy-protection schemes: Hitachi, Intel, Panasonic, Sony, and Toshiba. See DTCP and www.dtcp.com.

Glossary

625/50 The scanning system of 625 lines per frame and 50 interlaced fields (25 frames) per second. Used by PAL and SECAM television standards.

720p 720 lines of progressive video (720 lines per frame). Usually refers to 1280×720 resolution in 1.78 aspect ratio. 720p60 refers to 60 frames per second; 720p30 refers to 30 frames per second; and 720p24 refers to 24 frames per second (film source).

7C The group of seven founding companies responsible for defining the HDMI specification: Hitachi, Panasonic, Philips, Silicon Image, Sony, Thomson, and Toshiba. See HDMI and www.hdmi.org.

8/16 modulation The form of modulation block code used by DVD to store channel data on the disc. See modulation.

9C Nickname for the original Blu-ray Disc Founders (BDF).

AAC Advanced audio coder. An audio-encoding standard for MPEG-2 that is not backward-compatible with MPEG-1 audio.

AACS Advanced Access Content System. (Sometimes pronounced "access.") A specification for encryption of content on prerecorded and recordable optical media to restrict copying. The AACS Licensing Administrator, LLC (AACS LA) was founded by Disney, Intel, Microsoft, Panasonic, Warner Bros., IBM, Toshiba, and Sony.

AC Alternating current. An electric current that regularly reverses direction. Adopted as a video term for a signal of non-zero frequency. Compare to DC.

AC-3 The former name of the Dolby Digital audio-coding system, which is still technically referred to as AC-3 in standards documents. AC-3 is the successor to Dolby's AC-1 and AC-2 audio coding techniques.

access time The time it takes for a drive to access a data track and begin transferring data. In an optical jukebox, the time it takes to locate a specific disk, insert it in an optical drive, and begin transferring data to the host system.

ActiveMovie The former name for Microsoft's DirectShow technology.

ADA See Application Data Area.

ADPCM Adaptive differential pulse code modulation. A compression technique that encodes the difference between one sample and the next. Variations are lossy and lossless.

AES Audio Engineering Society. See www.aes.org.

AES/EBU The old name for AES3.

AES3 A digital audio signal transmission standard for professional use, defined by the Audio Engineering Society and the European Broadcasting Union. Uses time-division multiplexing to create a two-channel digital audio signal. Also published as ANSI S4.40-1992. S/PDIF is the consumer adaptation of this standard.

AGC Automatic gain control. A circuit designed to boost the amplitude of a signal to provide adequate levels for recording. Also see Macrovision.

aliasing A distortion (artifact) in the reproduction of digital audio or video that results when the signal frequency is more than half the sampling frequency. The sampling resolution is insufficient to distinguish between alternate reconstructions of the waveform (high-frequency components of the signal resemble low-frequency components), thus admitting noise that was not present in the original signal.

Always-On Menu A variation of the HDMV Pop-Up Menu that is activated by the content on the disc and cannot be removed by the user.

analog A signal of (theoretically) infinitely variable levels. Compare to digital.

anamorphic A widescreen video format where the image is distorted to fit into a thinner aspect ratio.

angle An angle is a scene recorded from different viewpoints. Each angle is equal in time length and an Angle Block may contain up to nine (9) angles.

ANSI American National Standards Institute (see http://www.ansi.org).

Glossary

AOD Advanced optical disc. The original name for HD DVD. A pun on the Japanese word "ao" for blue.

API Application programming interface. A set of publicly defined functions that an operating system or a code library provides to be controlled by other computer programs.

Application Data Area See **Persistent Storage**.

application format A specification for storing information in a particular way to enable a particular use. Examples are HDMV, BDJ, DVD-Video, and CD audio.

APS Analog protection system, designed to prevent copying in analog form, such as to tape. Macrovision is the prime example.

artifact An unnatural effect not present in the original video or audio, produced by an external agent or action. Artifacts can be caused by many factors, including digital compression, film-to-video transfer, transmission errors, data readout errors, electrical interference, analog signal noise, and analog signal crosstalk. Most artifacts attributed to the digital compression of DVD are in fact from other sources. Digital compression artifacts will always occur in the same place and in the same way. Possible MPEG artifacts are mosquitoes, blocking, and video noise.

aspect ratio The width-to-height ratio of an image. A 4:3 aspect ratio means the horizontal size is a third again wider than the vertical size. Standard television ratio is 4:3 (or 1.33:1). Widescreen DVD and HTDV aspect ratio is 16:9 (or 1.78:1). Common film aspect ratios are 1.85:1 and 2.35:1. Aspect ratios normalized to a height of 1 are often abbreviated by leaving off the :1.

assistant vocal A leading guide function for singing a song, as in the vocal part of Karaoke songs in Karaoke equipped DVD Video players. It is also called the "guide vocal."

ATAPI Advanced Technology Attachment (ATA) Packet Interface. An interface between a computer and its internal peripherals such as DVD-ROM drives. ATAPI provides the command set for controlling devices connected via an ATA (IDE) interface. ATA-2 is also known as Enhanced IDE (E-IDE). ATAPI was extended for use in DVD-ROM drives by the SFF 8090 specification, developed by the Mt. Fuji group.

ATSC Advanced Television Systems Committee. In 1978, the Federal Communications Commission (FCC) empaneled the Advisory Committee on Advanced Television Service (ACATS) as an investigatory and advisory committee to develop information that would assist the FCC in establishing an advanced broadcast television (ATV) standard for the United States. This committee created a subcommittee, the ATSC, to explore the need for and to coordinate development of the documentation of Advanced Television Systems. In 1993, the ATSC recommended that efforts be limited to a digital television system (DTV), and in September 1995 issued its recommendation for a Digital Television System standard, which was approved with the exclusion of compression format constraints (picture resolution, frame rate, and frame sequence).

ATV Advanced television. TV with significantly better video and audio than standard TV. Sometimes used interchangeably with HDTV, but more accurately encompasses any improved television system, including those beyond HDTV. Also sometimes used interchangeably with the final recommended standard of the ATSC, which is more correctly called DTV.

audio coding mode The method by which audio is digitally encoded, such as PCM, AC3, DTS, MPEG Audio or SDDS.

authoring For BDMV, BD-J, and DVD-Video, authoring refers to the process of designing, creating, collecting, formatting, and encoding material. For BD-ROM and DVD-ROM, authoring usually refers to using a specialized program to produce multimedia computer software.

autostart If Title on a Blu-ray disc contains associated BD-J applications, one or more of the applications must be marked as autostart. These applications will be executed when the Title is played.

AV stream file File (with .M2TS extension) containing a clip AV stream (MPEG-2 transport stream) used by the BDAV system. Stored in the STREAMS directory on the disc.

Glossary

AVC Advanced Video Coding. The informal name for ITU-T H.264 or ISO/IEC MPEG-4 Part 10. AVC is a digital video codec used by BD and other formats.

AVCHD An AVC/H.264 HD camcorder recording format using a subset BDAV. Originally designed for recording onto DVD, then extended to hard disk drives and memory cards. Released by Panasonic and Sony in 2006. Similar to AVCREC, but without MPEG-2 for video, without Dolby Digital and PCM for audio, data rate maximum of 18 Mbps, only 8.3 filenames, and without AACS for recording protected content.

AVCREC An adaptation of BDAV for playback and recording on recordable DVDs (-R, +R, -RW, +RW, -RAM), released in 2007. Essentially the same as BDAV with the requirement to support AVC/H.264 streams in transcode mode.

B picture (or **B frame**) One of three picture types used in MPEG video. B pictures are bidirectionally predicted, based on both previous and following pictures. B pictures usually use the least number of bits. B pictures do not propagate coding errors since they are not used as a reference by other pictures.

bandwidth Strictly speaking, the range of frequencies (or the difference between the highest and the lowest frequency) carried by a circuit or signal. Loosely speaking, the amount of information that can be carried in a signal. Technically, bandwidth does not apply to digital information; the term data rate is more accurate.

BCA Burst cutting area. A circular section near the center of a DVD disc where ID codes and manufacturing information can be inscribed in bar-code format.

BD See Blu-ray Disc.

BD+ An added layer of content protection that uses cryptographic algorithms provided on the disc and executed in a virtual machine in the BD player to descramble content before it can be played.

BD-16 A dual-layer, 8-cm (3-inch) Blu-ray disc that holds 15.582 billion bytes of data.

BD-25 A single-layer, 12-cm (5-inch) Blu-ray disc that holds 25.025 billion bytes of data.

BD-50 A dual-layer, 12-cm (5-inch) Blu-ray disc that holds 50.050 billion bytes of data.

BD-8 A single-layer, 8-cm (3-inch) Blu-ray disc that holds 7.791 billion bytes of data.

BDA Blu-ray Disc Association. The association formed in October 2004, when the Blu-ray Disc Founders (BDF) group opened up to general membership. The BDA is responsible for defining and promoting the BD format.

BDAV Blu-ray Disc audio/video. The BD format for recording video streams. Also refers to the underlying clip format used by both the BDAV and BDMV (HDMV + BD-J) formats.

BDCMF Blu-ray Disc cutting master format. The specification for a complete BD image, as sent to a replication plant for mastering.

BDF Blu-ray Disc Founders. The group of companies that originally developed the Blu-ray disc format. Began as 9 companies (Hitachi, LG, Panasonic, Mitsubishi, Pioneer, Philips, Samsung, Sharp, Sony, and Thomson). Later joined (in chronological order) by Hitachi, Dell, HP, TDK, 20th Century Fox, Disney, and Apple.

BD-J Blu-ray Disc Java. The programming platform for advanced interactive applications on Blu-ray Discs. Developed by the BDA as an adaptation of GEM, which is itself a profile of MHP. BD-J is usually implemented using Java Platform, Micro Edition specification (JME, formerly known as J2ME), Personal Basis Profile (PBP).

BD-J Object A list of BD-J applications and associated playback, application management, and user event information corresponding to a BD-J Title.

BD-Live The marketing name for BD Profile 2.

BDMV Technically, the directory on a Blu-ray disc that holds all the playback content: Movie Objects, BD-J Objects, JAR files, PlayLists, Clips, AV Streams, and related data. Colloquially, BDMV refers to

Glossary

the complete BD audio/video playback format, both HDMV and BD-J, sometimes called BD-Video or BD-ROM.

BD-R The record-once format of BD that is analogous to DVD-R. The BD-R application formats are defined in the BD-RE specifications. BD-RE v2 defines BD-R v1, and BD-RE v3 defines BD-R v2.

BD-RE The re-recordable format of BD that is analogous to DVD-RW. BD-RE v1 was the original version of Blu-ray. BD-RE v2 defines the BDAV recording application format, and BD-RE v3 defines the BDMV recording application format.

BD-ROM Blu-ray Disc read-only memory. The pre-recorded (replicated) form of BD. Sometimes used in a generic sense to refer to the BDMV (HDMV + BD-J) format, even when recorded onto BD-R or BD-RE media.

bidirectional prediction A form of compression in which the codec uses information not only from frames that have already been decompressed, but also from frames yet to come. The codec looks in two directions: ahead as well as back. This helps avoid large spikes in data rate caused by scene changes or fast movement, improving image quality.

Binding Unit Data Area (BUDA) The large part of the BD player Local Storage that holds audio, video, BD-J applications, and other data download from the Internet. HDMV and BD-J applications can read from the Binding Unit Data Area, but only signed BD-J applications can write to the Binding Unit Data Area.

birefringence An optical phenomenon where light is transmitted at slightly different speeds depending on the angle of incidence. Also light scattering due to different refractions created by impurities, defects, or stresses within the media substrate.

bit A binary digit. The smallest representation of digital data: zero/one, off/on, no/yes. Eight bits make one byte.

bitrate The volume of data measured in bits over time. Equivalent to data rate.

bitmap An image made of a two-dimensional grid of pixels. Each frame of digital video can be considered a bitmap, although some color information is usually shared by more than one pixel.

bits per pixel The number of bits used to represent the color or intensity of each pixel in a bitmap. One bit allows only two values (black and white), two bits allows four values, and so on. Also called color depth or bit depth.

bitstream Digital data, usually encoded, that is designed to be processed sequentially and continuously.

bitstream recorder A device capable of recording a stream of digital data but not necessarily able to process the data.

BLER Block error rate. A measure of the average number of raw channel errors when reading or writing a disc.

block In video encoding, an 8×8 matrix of pixels or DCT values representing a small chunk of luma or chroma. In DVD MPEG-2 video, a macroblock is made up of 6 blocks: 4 luma and 2 chroma. In AES3 or S/PDIF digital audio, a group of 192 consecutive frames.

block-based video compression A video compression technique that identifies blocks of pixels that are similar between frames and stores them only once to save space. Common video artifacts seen from this technique are mosquito wings and blocking. MPEG, AVC, and VC-1 are block-based codecs.

blocking or blockiness A term referring to the occasional blocky appearance of compressed video (an artifact). Caused when the compression ratio is high enough that the averaging of blocks of pixels becomes visible.

Blue Book The document that specifies the CD Extra interactive music CD format (see also Enhanced CD). The original CDV specification was also in a blue book.

Glossary

Blu-ray Disc   The second-generation DVD format created by the BDF/BDA. Uses a 0.1-mm cover layer (compared to the 0.6-mm cover layer of DVD and the 1.2-mm cover layer of CD) and 405-nm blue laser to store 4 to 5 times as much data as DVD.

Book A   Archaic name for DVD Specification Part 1, the document specifying the DVD physical format (DVD-ROM). Finalized in August 1996.

Book B   Archaic name for DVD Specification Part 3, the document specifying the DVD-Video format. Mostly finalized in August 1996.

Book C   Archaic name for DVD Specification Part 4, the document specifying the DVD-Audio format.

Book D   Archaic name for DVD Recordable Specification Parts 1 and 2, the documents specifying the DVD record-once format (DVD-R). Finalized in August 1997.

Book E   Archaic name for DVD Rewritable Specification Parts 1 and 2, the documents specifying the rewritable DVD format (DVD-RAM). Finalized in August 1997.

bound copy   A copy of content, such as from a Blu-ray disc, to a hard drive or other storage media that is bound to the device or to the media using encryption derived from a value that is unique to the device or media.

BPDG   Broadcast Protection Discussion Group. A technical discussion group chartered by the CPTWG in November 2001 to explore issues of protecting digital broadcasts, with a particular eye toward preventing Internet retransmission of broadcast content. Dissolved in June 2002 after making recommendations for, among other things, a broadcast flag to be embedded in digital transmissions and checked by authorized receiving equipment.

bps   Bits per second. A unit of data rate.

brightness   Defined by the CIE as the attribute of a visual sensation according to which area appears to emit more or less light. Loosely, the intensity of an image or pixel, independent of color; that is, its value along the axis from black to white.

BUDA   See **Binding Unit Data Area**.

buffer   Temporary storage space in the memory of a device. Helps smooth data flow.

burst   A short segment of the color subcarrier in a composite signal, inserted to help the composite video decoder regenerate the color subcarrier.

button   An active rectangular area in a menu associated with a specific action.

Button Sound   A feature of HDMV that can play a sound when a button is selection and another sound when the button is activated.

B-Y, R-Y   The general term for color-difference video signals carrying blue and red color information, where the brightness (Y) has been subtracted from the blue and red RGB signals to create B-Y and R-Y color-difference signals.

byte   A unit of data or data storage space consisting of eight bits, commonly representing a single character. Digital data storage is usually measured in bytes, kilobytes, megabytes, and so on.

CABAC   Context-adaptive binary arithmetic coding. An entropy coding mechanism used after quantization in the AVC (H.264) video compression standard. This mechanism requires more processing than other mechanisms like CAVLC but gives a better compression ratio.

caption   A textual representation of the audio information in a video program. Captions are usually intended for the hearing impaired, and therefore include additional text to identify the person speaking, offscreen sounds, and so on.

capture   In video, this is the process of changing a video analog signal into a digital file, usually with video capture hardware and software.

CAV   Constant angular velocity. Refers to rotating disc systems in which the rotation speed is kept constant, where the pickup head travels over a longer surface as it moves away from the center of the disc.

Glossary

The advantage of CAV is that the same amount of information is provided in one rotation of the disc. Contrast with CLV and ZCLV.

CAVLC Context-adaptive variable-length coding. An entropy coding method used after coefficient quantization in the AVC (H.264) video compression standard. CAVLC uses different VLC tables based on the coefficient sequences.

C_b, C_r The components of digital color-difference video signals carrying blue and red color information, where the brightness (Y) has been subtracted from the blue and red RGB signals to create B-Y and R-Y color-difference signals.

CBEMA Computer and Business Equipment Manufacturers Association.

CBR Constant bit rate. Data compressed into a stream with a fixed data rate. The amount of compression (such as quantization) is varied to match the allocated data rate, but as a result quality may suffer during high compression periods. In other words, data rate is held constant while quality is allowed to vary. Compare to VBR.

CCI Copy control information. Information specifying if content is allowed to be copied.

CCIR Rec. 601 A standard for digital video. The CCIR changed its name to ITU-R, and the standard is now properly called ITU-R BT.601.

CD Short for compact disc, an optical disc storage format developed by Philips and Sony.

CD+G Compact disc plus graphics. A variation of CD that embeds graphical data in with the audio data, allowing video pictures to be displayed periodically as music is played. Primarily used for karaoke.

CD-DA Compact disc digital audio. The original music CD format, storing audio information as digital PCM data. Defined by the Red Book standard.

CD-i Compact disc interactive. An extension of the CD format designed around a set-top computer that connects to a TV to provide interactive home entertainment, including digital audio and video, video games, and software applications. Defined by the Green Book standard.

CD-Plus A type of Enhanced CD format using stamped multisession technology.

CD-R An extension of the CD format allowing data to be recorded once on a disc by using dye-sublimation technology. Defined by the Orange Book standard.

CD-ROM Compact disc read-only memory. An extension of the Compact disc digital audio (CD-DA) format that allows computer data to be stored in digital format. Defined by the Yellow Book standard.

CD-ROM XA CD-ROM extended architecture. A hybrid version of CD allowing interleaved audio and video.

CDV A combination of laserdisc and CD that places a section of CD-format audio on the beginning of the disc and a section of laserdisc-format video on the remainder of the disc.

cDVD A CD containing DVD content (such as .VOB and .IFO files in a VIDEO_TS directory). A cDVD will play in most PCs, and in fact often includes a DVD player application. cDVDs will play in only very few consumer DVD players, since the players don't recognize the DVD content on the CD.

CE A common abbreviation for consumer electronics, the popular electricity-based products that are a means of entertainment for humans. Popular CE devices include televisions, DVD players, BD players, video game systems, and digital cameras. Items that do not fit into this category include toasters, light bulbs, and electric toothbrushes.

Cell In DVD-Video, a unit of video anywhere from a fraction of a second to hours long. Cells allow the video to be grouped for sharing content among titles, interleaving for multiple angles, and so on.

CEMA Consumer Electronics Manufacturers Association. A subsidiary of the Electronics Industry Association (EIA).

CGMS Content Generation Management System. A method of preventing copies or controlling the number of sequential copies allowed. CGMS/A is added to an analog signal (such as the line 21 vertical blanking portion of NTSC). CGMS/D is added to a digital signal, such as IEEE 1394.

Glossary

channel   A part of an audio track. Typically there is one channel allocated for each loudspeaker.

channel bit   The bits stored on the disc, after being modulated.

channel data   The bits physically recorded on an optical disc after error-correction encoding and modulation. Because of the extra information and processing, channel data is larger than the user data contained within it.

chapter   In BD and DVD-Video, a division of a title. In BD, chapters are defined by Entry-Marks in a PlayList. In DVD, a chapter is technically called a part of title (PTT).

chroma (C')   The nonlinear color component of a video signal, independent of the luma. Identified by the symbol C' (where prime ['] indicates nonlinearity) but usually written as C because it's never linear in practice.

chroma subsampling   Reducing color resolution by taking fewer color samples than luminance samples.

chrominance (C)   The color component (hue and saturation) of light or a video signal, independent of luminance. Technically, chrominance refers to the linear component of video, as opposed to the transformed nonlinear chroma component.

CIE   Commission Internationale de l'Éclairage/International Commission on Illumination.

CIF   Common intermediate format. Video resolution of 352×288.

Cinavia   Trade name of Verance audio watermark technology used in AACS.

CIRC   Cross-interleaved Reed Solomon code. An error-correction coding method that overlaps small frames of data.

clamping area   The area near the inner hole of a disc where the drive grips the disc in order to spin it.

Clip   A Clip AV Stream file (an MPEG-2 transport stream segment) together with its Clip Information file (access point time stamps). Clips are linked together by one or more PlayLists. A Blu-ray disc may contain up to 4000 Clips.

Clip AV Stream   MPEG-2 transport stream carrying audio and video used by the BDAV system.

Clip Information   A time-ordered list of time stamps defining access points into the associated Clip AV Stream.

closed caption   Textual video overlays that are not normally visible, as opposed to open captions, which are a permanent part of the picture. Captions are usually a textual representation of the spoken audio intended for hearing impaired viewers. In the United States, the official NTSC Closed Caption standard requires that all TVs larger than 13 inches include circuitry to decode and display caption information stored on line 21 of the video signal. BD closed captions are carried in MPEG-2 cc_data() fields (as defined in A/53C, EIA-708-B, and EIA/CEA-608-B), AVC SEI user-data messages, or VC-1 SMPTE 421M user data. DVD-Video can provide closed caption data in MPEG-2 user-data fields.

closed GOP   See **GOP**.

clpi   File extension for BD Clip Information files (in the BDMV/CLIPINF directory on the disc).

CLUT   Color lookup table. An index that maps a limited range color values to a full range of values such as RGB or YUV.

CLV   Constant linear velocity. Refers to a rotating disc system in which the head moves over the disc surface at a constant velocity, requiring that the motor vary the rotation speed as the head travels in and out. The further the head is from the center of the disc, the slower the rotation. The advantage of CLV is that data density remains constant, optimizing use of the surface area. Contrast with CAV and ZCLV.

CMI   Content management information. General information about copy protection and allowed use of protected content. Includes CCI.

codec   Coder/decoder. The circuitry or computer software that encodes and decodes a signal.

Coherent Acoustics   The full name for the multi-channel audio format called DTS.

Glossary

color banding An artifact related to digital graphics with insufficient color depth to represent subtle differences in color. Banding occurs most visibly in slow gradients such as the fade of the blue sky. It appears as hard lines (bands) as the shade of the color changes.

color depth The number of levels of color (usually including luma and chroma) that can be represented by a pixel. Generally expressed as a number of bits or a number of colors. The color depth of MPEG video in DVD is 24 bits, although the chroma component is shared across 4 pixels (averaging 12 actual bits per pixel).

color difference A pair of video signals that contain the color components minus the brightness component, usually B-Y and R-Y (G-Y is not used, since it generally carries less information). The color-difference signals for a black-and-white picture are zero. The advantage of color-difference signals is that the color component can be reduced more than the brightness (luma) component without being visually perceptible.

color model Any of several means of specifying colors according to their individual components. See RGB, YUV.

colorburst See burst.

colorist The title used for someone who operates a telecine machine to transfer film to video. Part of the process involves correcting the video color to match the film.

combo drive A DVD-ROM drive capable of reading and writing CD-R and CD-RW media. May also refer to a DVD-R or DVD-RW or DVD+RW drive with the same capability.

companded In digital data, first compressed and then expanded, resulting in nonlinear code representation.

component video A video system containing three separate color component signals, either red/green/blue (RGB) or chroma/color difference (YCbCr, YPbPr, YUV), in analog or digital form. The MPEG-2 encoding system used by DVD is based on color-difference component digital video. Very few televisions have component video inputs.

composite video An analog video signal in which the luma and chroma components are combined (by frequency multiplexing), along with sync and burst. Also called CVBS. Most televisions and VCRs have composite video connectors, which are usually colored yellow.

compression The process of removing redundancies in digital data to reduce the amount that must be stored or transmitted. Lossless compression removes only enough redundancy so that the original data can be recreated exactly as it was. Lossy compression sacrifices additional data to achieve greater compression.

compression rate A measurement as a percent of how much something has been compressed. For pictures, for example, the higher the percentage is, the better the quality of the image and the bigger the file size.

conform Adjust the beginning, ending, and playing time of an audio track to properly synchronize with the video.

constant data rate or **constant bitrate** See **CBR**.

Content Code Software programs distributed on Blu-ray discs that run on the BD+ VM to examine the security of the player before undoing Media Transforms to make the content playable.

Content Scrambling System (CSS) In DVD-Video, the encryption scheme designed to protect copyrighted material that resides on a disc by scrambling portions of the data using encryption keys.

contrast The range of brightness between the darkest and lightest elements of an image.

core stream In DTS-HD audio, the core stream carries the signal that is backwards compatibility with legacy DTS decoding equipment. Extension substreams add additional audio information for more channels, lossless encoding, and so on.

CPAC Content Protection Advisory Council. The advisory body for DVD CCA, responsible for any

Glossary

changes to CSS. Formed of three groups: motion picture studios, consumer electronics companies, and computer hardware and software companies

CPPM Content Protection for Prerecorded Media. Copy protection developed by the 4C for DVD-Audio. Essentially the second generation of CSS.

CPRM Content Protection for Recordable Media. Copy protection developed by the 4C for writable DVD formats.

CPS Content protection system. A particular technology designed to control access to a work, such as AACS, BD+, CSS, and CPRM.

CPTWG Copy Protection Technical Working Group. The industry body responsible for discussing and occasionally developing or approving content protection systems.

CPU Central processing unit. The integrated circuit chip that forms the brain of a computer or other electronic device. DVD-Video players contain rudimentary CPUs to provide general control and interactive features.

CRC Cyclic redundancy check. A method for computing a signature of a sequence of bytes used to detect the presence of errors in the data.

crop To trim and remove a section of the video picture in order to make it conform to a different shape. Cropping is used in the pan & scan process, but not in the letterbox process.

cross field A frame displaying two different overlapped images (fields). For video edited directly from film sources converted on a telecine via the 3-2 pulldown method, one frame (two fields) may have two fields containing entirely different picture data.

CRT Cathode-ray tube. Since televisions were invented, the moving images were viewed on CRT monitors. CRTs work by shooting electrons from a gun in the back of the TV in horizontal lines down the back of the screen, which is coated with phosphors that glow when excited by the electron stream. CRTs are being inexorably replaced by various flat-screen display technologies.

CSS See **Content Scrambling System**.

cursor A mouse pointer or a text insertion point. Sometimes called a caret.

CVBS Composite video baseband signal. Standard single-wire video, mixing luma and chroma signals together.

D-1 A professional 3/4-inch digital videotape format for uncompressed component video using the ITU-R BT.601 4:2:2 standard. Developed by Sony.

D-2 A professional 3/4-inch digital videotape format for uncompressed composite (Y/C) video with 8-bit sampling. Developed by Sony and Ampex.

D-3 A professional 1/2-inch digital videotape format for uncompressed composite (Y/C) video with 8-bit sampling. Developed by Panasonic.

D-4 There is no D-4 videotape format, since 4 is considered an unlucky number in Japan.

D-5 A professional 1/2-inch digital videotape format for uncompressed high-definition or standard-definition component video with 10-bit sampling. Developed by Panasonic.

D-6 A professional 3/4-inch videotape standard (SMPTE 277/278M) for uncompressed, high-definition video (1920×1080, SMPTE 274M).

D-7 Same as DVCPRO.

DAC Digital-to-analog converter. Circuitry that converts digital data (such as audio or video) to analog data.

DAE Digital audio extraction. Reading digital audio data directly from a CD audio disc.

DAT Digital audio tape. A magnetic audio tape format that uses PCM to store digitized audio or digital data.

data area The physical area of a DVD disc between the lead in and the lead out (or middle area) that contains the stored data content of the disc.

Glossary

data rate The volume of data measured over time; the rate at which digital information can be conveyed. Usually expressed as bits per second with notations of kbps (thousand/sec), Mbps (million/sec), and Gbps (billion/sec). Digital audio date rate is generally computed as the number of samples per second times the bit size of the sample. For example, the data rate of uncompressed 16-bit, 48-kHz, two-channel audio is 1536 kbps. Digital video bit rate is generally computed as the number of bits per pixel times the number of pixels per line times the number of lines per frame times the number of frames per second. For example, the data rate of a DVD movie before compression is usually $12 \times 720 \times 480 \times 24 = 99.5$ Mbps. Compression reduces the data rate. Digital data rate is sometimes inaccurately equated with bandwidth.

dB See **decibel**.

DBS Digital broadcast satellite. The general term for 18-inch digital satellite systems.

DC Direct current. Electrical current flowing in one direction only. Adopted in the video world to refer to a signal with zero frequency. Compare to AC.

DCC Digital compact cassette. A digital audio tape format based on the popular compact cassette. Abandoned by Philips in 1996.

DCT Discrete cosine transform. An invertible, discrete, orthogonal transformation. Got that? A mathematical process used in MPEG video encoding to transform blocks of pixel values into blocks of spatial frequency values with lower-frequency components organized into the upper-left corner, allowing the high-frequency components in the lower-right corner to be discounted or discarded. Also digital component technology, a videotape format.

DD Dolby Digital.

DDWG Digital Display Working Group. See DVI.

deblocking filter A tool in video compression to reduce blocking artifacts. Post filters are optionally used in some decoder implementations for MPEG-2 and other formats. In-loop deblocking filters are a mandatory part of AVC and VC-1 in both the encode and decode process.

decibel (**dB**) A unit of measurement expressing ratios using logarithmic scales related to human aural or visual perception. Many different measurements are based on a reference point of 0 dB; for example a standard level of sound or power.

decimation A form of subsampling that discards existing samples (pixels, in the case of spatial decimation, or pictures, in the case of temporal decimation). The resulting information is reduced in size but may suffer from aliasing.

decode To reverse the transformation process of an encoding method. Decoding processes are usually deterministic.

decoder 1) A circuit that decodes compressed audio or video, taking an encoded input stream and producing output such as audio or video. DVD players use the decoders to recreate information that was compressed by systems such as MPEG-2 and Dolby Digital; 2) a circuit that converts composite video to component video or matrixed audio to multiple channels.

delta picture (or **delta frame**) A video picture based on the changes from the picture before (or after) it. MPEG P pictures and B pictures are examples. Contrast with key picture.

deterministic A process or model whose outcome does not depend upon chance, and where a given input will always produce the same output. Audio and video decoding processes are mostly deterministic.

DigiRise DRA Multichannel audio coding technology developed in China, used as an alternative to Dolby Digital and DTS audio for BD players and discs in China. Claimed to be independent of existing intellectual property rights. One to eight channels of 24-bit, 8 to 192 kHz audio at constant or variable data rates from 32 to 9216 kbps. Chinese Electronic Industry Standard, Specification for Multichannel Digital Audio Coding Technology SJ/T11368-2006.

Glossary

digital Expressed in digits. A set of discrete numeric values, as used by a computer. Analog information can be digitized by sampling.

digital signal processor (DSP) A digital circuit that can be programmed to perform digital data manipulation tasks such as decoding or audio effects.

digital video noise reduction (DVNR) Digitally removing noise from video by comparing frames in sequence to spot temporal aberrations.

Digital Visual Interface (DVI) The digital video interface standard developed by the Digital Display Working Group (DDWG). A replacement for analog VGA monitor interface.

digital zero The case where all digitally encoded audio amplitude data is equal to zero.

digitize To convert analog information to digital information by sampling.

DIN Deutsches Institut für Normung. The German Institute for Standardization.

directory The part of a disc that indicates what files are stored on the disc and where they are located. Often called a folder.

directory structure The organization of folders (directories) and files defined by the BD (or DVD) format.

DirectShow A software standard developed by Microsoft for playback of digital video and audio in the Windows operating system. Replaces the older MCI and Video for Windows software.

discrete cosine transform See **DCT**.

discrete surround sound Audio in which each channel is stored and transmitted separate from and independent of other channels. Multiple independent channels directed to loudspeakers in front of and behind the listener allow precise control of the soundfield in order to generate localized sounds and simulate moving sound sources.

display rate The number of times per second the image in a video system is refreshed. Progressive scan systems such as film or HDTV change the image once per frame. Interlace scan systems such as standard television change the image twice per frame, with two fields in each frame. Film has a frame rate of 24 fps, but each frame is shown twice by the projector for a display rate of 48 fps. 525/60 (NTSC) television has a rate of 29.97 frames per second (59.94 fields per second). 625/50 (PAL/SECAM) television has a rate of 25 frames per second (50 fields per second).

Divx Digital Video Express. A short-lived pay-per-viewing-period variation of DVD, later resurrected as DivX for a series of codecs and the resulting company.

DL Dual layer. To get more physical data on an optical disc, a second layer of data can be added. To read the second layer, the laser refocuses through the first layer to read the second layer.

DLP Digital Light Processing. The trademark name for Texas Instrument's digital micromirror device (DMD) for video image generation and projection.

DLT Digital linear tape. A digital archive standard using half-inch tapes, commonly used for submitting a premastered DVD disc image to a replication service.

Dolby Digital (AC-3) A perceptual coding system for audio, developed by Dolby Laboratories and used for film, BD, DVD, and DTV. Dolby Digital reduces the amount of data needed to represent an audio signal by removing information that usually can't be heard by human ears.

Dolby Digital EX Dolby Digital EX matrix encodes the left and right surround channels to add a rear center channel for 6.1-channel surround audio (7.1 if two rear speakers are used for the added channel).

Dolby Digital Plus An extension to Dolby Digital that supports 8 or more audio channels (limited to 7.1 in Blu-ray) and data rates up to 6 Mbps (limited to 1.7 Mbps in Blu-ray). An optional audio format for Blu-ray discs.

Dolby Lossless An early name for MLP format licensed by Dolby and later extended and named Dolby TrueHD.

Glossary

Dolby Pro Logic   The technique (or the circuit that applies the technique) of extracting surround audio channels from a matrix-encoded audio signal. Dolby Pro Logic is a decoding technique only, but is often mistakenly used to refer to Dolby Surround audio encoding.

Dolby Surround   The standard for matrix encoding surround-sound channels in a stereo signal by applying a set of defined mathematical functions when combining center and surround channels with left and right channels. The center and surround channels can then be extracted by a decoder such as a Dolby Pro Logic circuit that applies the inverse of the mathematical functions. A Dolby Surround decoder extracts surround channels, while a Dolby Pro Logic decoder uses additional processing to create a center channel. The process is essentially independent of the recording or transmission format. Any stereo encoding format, including Dolby Digital, MP3, DTS, and others, is compatible with Dolby Surround audio.

Dolby TrueHD   Lossless audio encoding used in the Blu-ray format. A maximum bit rate of 18 Mbps provides up to eight channels of 24-bit, 96 kHz audio with features such as dialog normalization and dynamic range control. Based on MLP combined with a Dolby Digital stream for backward compatibility.

DOT   Digital only token. Defined by AACS, a flag in the content that restricts video output to protected digital connections only.

double-sided disc   A type of disc on which data is recorded on both sides.

downmix   To convert a multichannel audio track into a two-channel stereo track by combining the channels with the Dolby Surround process. All DVD players are required to provide downmixed audio output from Dolby Digital audio tracks.

downsampling   See **subsampling**.

DRA   See **DigiRise DRA**.

DRC   See **dynamic range compression**.

driver   A software component that enables an application to communicate with a hardware device.

DRM   Digital rights management. A technical system for controlling access to content based on a set of usage rights determined by the content owner. DRMs typically encrypt content and use certificates (public/private keys) delivered over the Internet to control access.

drop-frame timecode   The method of time code computation that accounts for the reality of there being only 29.97 frames of video per second. The 0.03 frame is visually insignificant, but mathematically very significant. A one hour video program will have 107,892 frames of video (29.97 frames per second × 60 seconds × 60 minutes). The drop-frame time code method of accommodating reality was developed where 2 frames are dropped from the numerical count for every minute in an hour except for every 10th minute when no frames are dropped. See also **non-drop-frame timecode** and **timecode**.

DSD   Direct Stream Digital. An uncompressed audio bitstream coding method developed by Sony. An alternative to PCM.

DSP   See digital signal processor.

DSVCD   Double Super Video Compact Disc. Long-playing (100-minute) variation of SVCD.

DTCP   Digital Transmission Content Protection. Technology developed by the 5C to protect video and audio transmission against unauthorized interception or retransmission in the digital home environment. DTCP can be implemented over any high-speed bidirectional digital bus, including IEEE 1394, USB, and the MOST network used in automobiles. DTCP is designed for networks and distributed busses, as compared with HDCP, which is designed for point-to-point connections.

DTHD   Dolby True HD.

DTLA   The Digital Transmission Licensing Administrator (DTLA, also known as 5C), the five-company entity that produces copy-protection schemes: Hitachi, Intel, Panasonic, Sony, and Toshiba. See DTCP and www.dtcp.com.

Glossary

DTS Decode time stamp, an extension to MPEG-2 PTS (presentation time stamp) that indicates when a frame should be decoded. PTS is used for frames that are not in presentation order. DTS and PTS values reside in the packet header.

DTS Digital Surround A multichannel perceptual audio-coding system. Mandatory audio track format for BD. Optional audio track format for DVD-Video and DVD-Audio.

DTS Digital Surround 96/24 A DTS audio format (core + 96/24 extension) where 96 represents the sample rate (96 kHz) and 24 represents the bit depth (24 bits).

DTS Digital Surround ES A version of DTS audio that adds a rear center channel extension stream for 6.1-channel surround.

DTS-HD Audio A set of extensions to the DTS core stream that are backward compatible with all DTS decoders. DTS HD Audio comprises DTS-HD Master Audio (lossless), DTS-HD High Resolution Audio (lossy), and DTS-HD Secondary Audio (low bit rate).

DTS-HD High Resolution Audio Multichannel audio codec that provides lossy compression at up to 6 Mbps constant bit rate for up to 7.1 channels of 24-bit, 96 kHz audio.

DTS-HD LBR Technical name for low-bit-rate DTS-HD Secondary Audio.

DTS-HD Lossless The original name for DTS-HD Master Audio.

DTS-HD Master Audio Multichannel audio codec that provides lossless compression at up to 24.5 Mbps variable data rate for up to 7.1 discrete channels of 24-bit, 96 kHz audio or 2 channels of 24-bit, 192 kHz audio.

DTS-HD Secondary Audio Optional Secondary Audio codec for Blu-ray that provides low bit rate audio at data rates as low as 24 kbps per channel, for up to 5 channels. Also called DTS-HD LBR (low bitrate).

DTV Digital television. In general, any system that encodes video and audio in digital form. In specific, the Digital Television System proposed by the ATSC or the digital TV standard proposed by the Digital TV Team founded by Microsoft, Intel, and Compaq.

dual mono Two independent audio channels in place of a two-channel stereo signal. For example, the left channel might contain English dialog while the right channel contains Spanish. Often identified as 1+1. Dual mono also refers to amplifiers with separate monophonic amplifier blocks.

duplication The reproduction of media. Generally refers to producing discs in small quantities, as opposed to large-scale replication.

DV Digital Video. Usually refers to the digital videotape format developed by Sony, JVC, Panasonic, and others to store DCT-intraframe compressed standard-definition digital video with 4:1:1/4:2:0 sampling, standardized as IEC 61834.

DVB Digital video broadcast. A European standard for broadcast, cable, and digital satellite video transmission.

DVC Digital video cassette. Early name for DV.

DVCAM Sony's proprietary version of DV, aimed at the professional market.

DVCD Double Video Compact Disc. Long-playing (100-minute) variation of VCD.

DVCPRO Panasonic's proprietary version of DV, aimed at the professional market.

DVD Generic name for a family of related disc formats encompassing Video, Audio, and computer file storage on an optical disc format. They share common or similar physical formats, logical data structures, and file formats. DVD is sometimes spelled out as digital video disc or digital versatile disc, but has come to stand on its own.

DVD CCA DVD Copy Control Association. The company that maintains and licenses the CSS content protection system for DVD.

DVD Forum An international association of hardware and media manufacturers, software firms and other users of DVD, created for the purpose of exchanging and disseminating ideas and information about the DVD format.

Glossary

DVD+R An alternative to DVD-R developed by the DVD+RW Alliance (not by the DVD Forum).

DVD+RW An alternative to DVD-RW developed by the DVD+RW Alliance (not by the DVD Forum).

DVDA DVD Association. A non-profit industry trade association representing DVD authors, producers, and vendors throughout the world.

DVD-Audio (DVD-A) The high-fidelity audio format of DVD. Primarily uses PCM audio with MLP encoding, along with an optional subset of DVD-Video features.

DVD-Music A colloquial name (not a defined format) for a DVD-Video disc containing primarily music, usually with accompanying still video images. Plays in all DVD players, rather than requiring specialized a DVD-Audio player.

DVD-R (DVD Recordable) The write-once, read-many DVD format akin to CD-R, used for DVD-Video as well as for data archival and storage applications. The authoring-use version, which uses a 635-nm laser, was introduced in 1998 by Pioneer. The general-use format, which uses a 650-nm laser, was authorized by DVD Forum in 2000.

DVD-RAM (DVD Rewritable) A phase-change rewritable DVD disc format from the DVD Forum. RAM stands for random-access memory. DVD-RAM uses a cartridge for data recording, optionally removable for playback.

DVD-ROM The base format of DVD. ROM stands for read-only memory, referring to the fact that standard DVD-ROM and DVD-Video discs can't be recorded on. A DVD-ROM can store essentially any form of digital data.

DVD-RW (DVD Re-recordable) A phase-change rewritable DVD format from the DVD Forum. Similar to CD-RW.

DVD-Video (DVD-V) The application format for storing and reproducing audio and video on DVD-ROM discs, based on MPEG video, Dolby Digital audio, and other proprietary data formats.

DVD-VR DVD Video Recording format. The application specification designed for real-time video recording and editing. Much simpler than DVD-Video, but unfortunately supported by only a small percentage of DVD players.

DVI See **Digital Visual Interface**.

DVNR See **digital video noise reduction**.

DVR Digital video recorder. Video device that uses a hard drive to record broadcast or cable television shows, usually integrating an electronic program guide (EPG) for programming information. Displacing the use of VCRs to record live programming.

DVS Descriptive video services. Descriptive narration of video for blind or sight-impaired viewers.

dye polymer The chemical used in record-once optical media such as CD-R, DVD-R, and BD-R that darkens when heated by a high-power laser.

dye sublimation Optical disc recording technology that uses a high-power laser to burn readable marks into a layer of organic dye. Other recording formats include magneto-optical and phase-change.

dynamic range The difference between the loudest and softest sound in an audio signal. The dynamic range of digital audio is determined by the sample size. Increasing the sample size does not allow louder sounds; it increases the resolution of the signal, thus allowing softer sounds to be separated from the noise floor (and allowing more amplification with less distortion). Dynamic range refers to the difference between the maximum level of distortion-free signal and the minimum limit reproducible by the equipment.

dynamic range compression A technique of reducing the range between loud and soft sounds in order to make dialogue more audible, especially when listening at low volume levels. Used in the downmix process of multichannel Dolby Digital sound tracks.

Glossary

Easter egg   A hidden feature on a disc, such as a bonus video clip, that requires the user to take certain actions (e.g., choosing an invisible buttons, moving through buttons or menus in a certain sequence, or even finish playing a game) before it can be accessed.

EBU   European Broadcasting Union.

ECC   See Error correction code.

ECC constraint length   The number of sectors that are interleaved to combat bursty error characteristics of discs. 16 sectors are interleaved in DVD. Interleaving combats typical disc defects such as scratch marks by spreading the error over a larger data area, thereby increasing the chance that the error correction codes can conceal the error.

ECD   Error detection and correction code. See error-correction code.

ECMA   European Computer Manufacturers Association (see http://www.ecma.org).

ECMA-262   An ECMA standard that specifies the core JavaScript language (roughly equivalent to JavaScript 1.1). Equivalent to ISO/IEC 16262.

ECMAScript   The standardized version of the JavaScript language. Used as the scripting language of HD-DVD.

edge enhancement   When films are transferred to video in preparation for DVD encoding, they are commonly run through digital processes that attempt to clean up the picture. These processes include noise reduction (DVNR) and image enhancement. Enhancement increases contrast (similar to the effect of the "sharpen" or "unsharp mask" filters in Photoshop), but can tend to overdo areas of transition between light and dark or different colors, causing a "chiseled" look or a ringing effect like the haloes you see around street lights when driving in the rain. Video noise reduction is a good thing, when done well, since it can remove scratches, spots, and other defects from the original film. Enhancement, which is rarely done well, is a bad thing. The video may look sharper and clearer to the casual observer, but fine tonal details of the original picture are altered and lost.

EDS   Enhanced data services. Additional information in NTSC line such as a time signal.

EDTV   Enhanced-definition television. A system that uses existing transmission equipment to send an enhanced signal that looks the same on existing receivers but carries additional information to improve the picture quality on new enhanced receivers. PALPlus is an example of EDTV. (Contrast with HDTV and IDTV.)

EFM   Eight-to-fourteen modulation. This low-level and very critical channel coding technique maximizes pit sizes on the disc by reducing frequent transitions from 0 to 1 or 1 to 0. CD employs pulse width modulation, representing 1's as Land-pit transitions along the track. The 8/14 code maps 8 user data bits into 14 channel bits. In the 1982 compact disc standard (IEC 908 standard), 3 merge bits are added to the 14 bit block to further eliminate 1-0 or 0-1 transitions between adjacent 8/14 blocks.

EFM Plus   DVD's eight-to-sixteen modulation is sometimes called EFM+, since it is a derivative of EFM. It folds the merge bits into the main 8/16 table.

EIA   Electronics Industry Association.

EICTA   EICTA was formed in 1999 as the European Information and Communications Technology Industry Association by the consolidation of the two former European federations of the information and telecommunications industries. In 2001, it merged with EACEM (European Association of Consumer Electronics Manufacturers) and changed its name to European Information, Communications and Consumer Electronics Technology Industry Association.

E-IDE   Enhanced Integrated Drive Electronics. Extensions to the IDE standard providing faster data transfer and allowing access to larger drives, including CD-ROM and tape drives, using ATAPI. E-IDE was adopted as a standard by ANSI in 1994. ANSI calls it Advanced Technology Attachment-2 (ATA-2) or Fast ATA.

elementary stream   A general term for a coded bitstream such as audio or video. Elementary streams are made up of packs of packets.

Glossary

emphasis Amplifying a given range of recorded frequency signals in order to reproduce the original signal during playback.

emulate For DVD and BD, to test the function of a disc on a computer after formatting a complete disc image.

encode To transform data for storage or transmission, usually in such a way that redundancies are eliminated or complexity is reduced. Most compression is based on one or more encoding methods. Data such as audio or video is encoded for efficient storage or transmission and is decoded for access or display.

encoder 1) A circuit or program that encodes (and thereby compresses) audio or video; 2) a circuit that converts component digital video to composite analog video. DVD players include TV encoders to generate standard television signals from decoded video and audio; 3) a circuit that converts multi-channel audio to two-channel matrixed audio.

Enhanced CD A general term for various techniques that add computer software to a music CD, producing a disc that can be played in a music player or read by a computer. Also called CD Extra, CD Plus, hybrid CD, interactive music CD, mixed-mode CD, pre-gap CD, or track-zero CD.

entropy coding Variable-length, lossless coding of a digital signal to reduce redundancy. RLE, Huffman, CABAC, and CAVLC are examples of entropy coding. MPEG-2, AVC, VC-1, DTS, and Dolby Digital (AC-3) apply entropy coding after the quantization step. MLP also uses entropy coding.

EQ Equalization of audio.

error-correction code Additional information added to data to allow errors to be detected and possibly corrected.

Ethernet A common system for connecting computers and devices together in a local area network (LAN) to share data. Many BD players have Ethernet ports to access the Internet.

ETSI European Telecommunications Standards Institute.

EVD Enhanced versatile disc. A high-definition optical disc format in China, officially endorsed by the Chinese government. Originally planned to move from red laser to blue laser and use multilevel recording, but did not achieve the success hoped for by its backers. Competed with HDV and HVD.

export To convert data to another form of data or media.

extension stream For DTS audio, a substream that extends the core stream with additional channels or additional audio resolution.

extent 1) For the volume structure and the ISO 9660 file structure, an extent is defined as a set of logical sectors, the logical sector numbers of which form a continuous ascending sequence. The address, or location, of an extent is the number of the first logical sector in the sequence. 2) For the UDF file structure an extent is defined as a set of logical blocks, the logical block numbers of which form a continuous ascending sequence. The address, or location, of an extent is the number of the first logical block in the sequence.

father The metal master disc formed by electroplating the glass master. The father disc is used to make mother discs, from which multiple stampers (sons) can be made.

field A set of alternating scan lines in an interlaced video picture. A frame is made of a top (odd) field and a bottom (even) field.

file A collection of data stored on a disc, usually in groups of sectors.

file system A defined way of storing files, directories, and information about them on a data storage device. DVD and BD use versions of the UDF file system.

filter (verb) To reduce the amount of information in a signal. (noun) A circuit or process that reduces the amount of information in a signal. Analog filtering usually removes certain frequencies. Digital filtering (when not emulating analog filtering) usually averages together multiple adjacent pixels, lines, or frames to create a single new pixel, line, or frame. This generally causes a loss of detail, especially with complex images or rapid motion. See letterbox filter. Compare to interpolate.

Glossary

FireWire A standard for transmission of digital data between external peripherals, including consumer audio and video devices. The official name is IEEE 1394, based on the original FireWire design by Apple Computer.

fixed rate Information flow at a constant volume over time. See CBR.

FLLA Format and Logo License Agreement from the BDA, required to manufacture BD products and use the BD logo.

forensic watermark A watermark containing information about the origin of video or audio content, such as a particular theater or particular player. It is not intended to protect the content, but is used in tracking down the source of unauthorized copies.

formatting 1) Creating a disc image. 2) Preparing storage media for recording.

fps Frames per second. A measure of the rate at which pictures are shown for a motion video image. In NTSC and PAL video, each frame is made up of two interlaced fields.

fragile watermark A watermark designed to be destroyed by any form of copying or encoding other than a bit-for-bit digital copy. Absence of the watermark indicates that a copy has been made.

frame In video, the piece of a video signal containing the spatial detail of one complete image; the entire set of scan lines. In an interlaced system, a frame contains two fields. In audio, a frame is one or more samples.

frame doubler A video processor that increases the frame rate (display rate) in order to create a smoother-looking video display. Compare to line doubler.

frame rate The frequency of discrete images. Usually measured in frames per second (fps). Film has a rate of 24 frames per second, but usually must be adjusted to match the display rate of a video system.

frequency The number of repetitions of a phenomenon in a given amount of time. The number of complete cycles of a periodic process occurring per unit time.

FRExt Fidelity range extensions. Optional encoding features of AVC (H.264), including an 8×8 transform and quantization matrix, that increase video quality and improve coding efficiency.

FTP File transfer protocol. One of the several Internet protocols used to connect devices over the Internet using TCP/IP so that files can be transferred between them.

full-motion video Video that plays at a rate of at least 24 frames per second.

fullscreen A video image that fills a standard 4:3 TV shape. Contrast with widescreen and letterbox.

FVD Forward versatile disc. A high-definition, red-laser optical disc format developed in Taiwan. Uses tighter track pitch than standard DVD to achieve slightly higher disc capacities. Uses Microsoft Windows Media Video and Windows Media Audio encoding.

G byte One billion (10^9) bytes. Not to be confused with GB or gigabyte (2^{30} bytes).

G or **giga-** An SI prefix for denominations of 1 billion (10^9). Note the uppercase "G."

GB See **gigabyte**.

Gbps Gigabits/second. Billions (10^9) of bits per second.

GEM Globally executable MHP. A core set of MHP APIs (with transmission-related elements removed to create a transport-neutral system) for Java applications on various platforms. BD-J is based on GEM 1.0.2 which is standardized as ETSI TS 102 819 V1.3.1 (2005-10).

General Purpose Register (GPR) Temporary storage area in a BD player used to keep track of user actions and application status. (Similar to GPRMs in DVD-Video.) BD players have 4096 GPRs, each holding a 32-bits (4-byte) unsigned integer.

gigabyte 1,073,741,824 (2^{30}) bytes. The progression of byte measurement is byte, kilobyte, megabyte, gigabyte, terabyte, petabyte, and so on.

GOF Group of audio frames. The data area of 1/30 second that is composed of 20 audio frames of linear PCM audio.

Glossary

GOP Group of pictures. A sequence of frames in an MPEG video stream, usually 15 frames for NTSC DVDs or 12 frames for PAL DVDs, beginning with one or more I pictures followed by I, P, and B pictures. A GOP is the atomic unit of MPEG video access. There are two types of GOPs, closed and open. A closed GOP keeps all references (from I and P frames) within the GOP. This simplifies jumping into the middle of a video stream, since the decoder doesn't have to reference a frames from the previous GOP. In an open GOP, I or P frames near the beginning or the end of a GOP may refer to frames in the adjacent GOP. Open-GOP encoding can be slightly more efficient than closed-GOP encoding.

gray market Dealers and distributors who sell equipment without proper authorization from the manufacturer.

grayscale recording See multilevel recording.

Green Book The document developed in 1987 by Philips and Sony as an extension to CD-ROM XA for the CD-i system.

GUID Globally unique identifier. A number intended to be unique across space and time. Used in various applications to uniquely identify information.

guide melody The melody for the vocal part recorded to assist singers in Karaoke.

guide vocal A leading guide function for singing a song, and the vocal part of Karaoke songs in Karaoke equipped players. Also called the assistant vocal.

H/DTV High-definition/digital television. A combination of acronyms that refers to both HDTV and DTV systems.

Half D1 An MPEG-2 video encoding mode in which half the horizontal resolution is sampled (352×480 for NTSC, 352×576 for PAL).

hard disk drive A storage system using internal fixed magnetic disks. Almost all computers have at least one hard disk. Many Profile 2.0 BD contain an internal hard disk or provide a USB port to connect an external hard disk for Local Storage.

HAVi A consumer electronics industry standard for interoperability between digital audio and video devices connected via a network in the consumer's home. BD-J uses a device model similar to HAVi.

HD 1) High definition. 2) High density.

HD DVD High-definition DVD or high-density DVD. Developed initially by Toshiba and NEC, then standardized by the DVD Forum and introduced in 2006. Largely abandoned (except in China) in 2008.

HD SDI High-definition serial digital interface, a connection standard for transmitting high-definition video with a data rate of 1.485 Gbps.

HDCD High-definition Compatible Digital. A proprietary method of enhancing audio on CDs.

HDCP High-bandwidth Digital Content Protection. Technology developed by Intel to restrict copying of digital video and audio by encrypting the transmission over digital interconnect formats such as DVI and HDMI. HDCP is designed for point-to-point connections, as compared to DTCP, which is designed for networks and distributed busses.

HDD See hard disk drive.

HDi Microsoft's trademark name for its specific implementation of the HD DVD advanced interactivity specification, originally called iHD.

HDMI A digital video and audio connection standard based on DVI but using a smaller connector and adding support for multichannel digital audio, improved color depth, and device control.

HDMV The high-definition movie mode of BD. A simplified system, similar to DVD-Video, for displaying menus with animated buttons and sounds, pop-up menus, video (including multiple angles and seamless branching), audio, subtitles, slideshows, and more.

HDREC An extension of the HD DVD format for use on standard red-laser DVDs. Comparable to AVCREC for BD.

HDTV High-definition television. A video format with a resolution approximately twice that of con-

Glossary

ventional television in both the horizontal and vertical dimensions, and a picture aspect ratio of 16:9. Used loosely to refer to the U.S. DTV System. Contrast with EDTV and IDTV.

HDV High-definition video. A high definition, red-laser optical disc format in China that uses MPEG-4. Promoted by Beijing Kaicheng High-Clarity Electronics Technology Co. Competed with EVD and HVD.

Hertz See **Hz**.

hexadecimal Representation of numbers using base 16.

HFS Hierarchical file system. A file system used by Apple Computer's Mac OS operating system.

High Sierra The original file system standard developed for CD-ROM, later modified and adopted as ISO 9660.

highlight A method of display that emphasizes a selected item on a menu by increasing the brightness level to show which function is executed.

horizontal resolution See **lines of horizontal resolution**.

HQ-VCD High-quality Video Compact Disc. Developed by the Video CD Consortium (Philips, Sony, Panasonic, and JVC) as a successor to VCD. Evolved into SVCD.

HRRA Home Recording Rights Association.

HSF See **High Sierra**.

HTML Hypertext markup language. A tagging specification, based on SGML (standard generalized markup language), for formatting text to be transmitted over the Internet and displayed by client software.

hue The color of light or of a pixel. The property of color determined by the dominant wavelength of light.

Huffman coding A lossless compression technique of assigning variable-length codes to a known set of values. Values occurring most frequently are assigned the shortest codes. MPEG uses a variation of Huffman coding with fixed code tables, often called variable-length coding (VLC).

HVD 1) Holographic versatile disc, as defined by the HVD Alliance. An optical disc format using laser holography to store hundreds of gigabytes on a single disc. 2) High-definition versatile disc. A high definition, red-laser optical disc format in China. Competed with EVD and HDV.

hybrid disc 1) In the general sense, a combination of two different physical discs formats on a single disc, such as the DualDisc combination of CD audio and DVD-Audio. 2) For BD, Hybrid (usually capitalized) means a disc that combines one layer of DVD and one layer of BD.

Hz Hertz. A unit of frequency measurement. The number of cycles (repetitions) per second.

I picture (or **I frame**) In MPEG video, an intra picture that is encoded independent from other pictures (see **intraframe**). Transform coding (DCT, quantization, and VLC) is used with no motion compensation, resulting in only moderate compression. I pictures provide a reference point for dependent P pictures and B pictures and allow random access into the compressed video stream.

i.Link Trademarked Sony name for IEEE 1394.

ICT Image Constraint Token. A copy deterrent feature of AACS that allows the content owner to set a flag on the disc that makes Blu-ray players downconvert digital video to 960×540 on analog outputs. Around the time Blu-ray was introduced, most Hollywood studios stated they had no plans to use ICT.

IDE Integrated drive electronics. An internal bus, or standard electronic interface between a computer and internal block storage devices. IDE was adopted as a standard by ANSI in November 1990. ANSI calls it Advanced Technology Attachment (ATA). Also see E-IDE and ATAPI.

IDTV 1) Improved-definition television. A television receiver that improves the apparent quality of the picture from a standard video signal by using techniques such as frame doubling, line doubling, and digital signal processing. 2) Integrated digital television (usually iDTV). A television incorporating features that would otherwise be in an external box, such as a DVR or digital cable tuner set-top box.

Glossary

IEC International Electrotechnical Commission.

IEEE Institute of Electrical and Electronics Engineers. An electronics standards body.

IEEE 1394 A standard for transmission of digital data between external peripherals, including consumer audio and video devices. Commonly used with DV camcorders. Also known as FireWire or i.Link.

IFE In-flight entertainment.

IFPI International Federation of the Phonographic Industry. An organization of sound recording producers and distributors. The IFPI focuses on fighting music piracy and protecting the economic well-being of the international recording industry

iHD The name originally given to the HD DVD advanced interactivity specification during development in the DVD forum. Not to be confused with iHD.org, Internet High Definition, a standards organization for Internet video. See Hdi.

I-MPEG Intraframe MPEG. An unofficial variation of MPEG video encoding that uses only intraframe DCT compression. I-MPEG is used by DV equipment.

import To convert media or data from one form to another.

in mux Elementary streams multiplexed into a single stream. Compare with out of mux.

INCITS International Committee for Information Technology Standards. Formerly known as NCITS (from 1996-2002) and Accredited Standards Committee X3, Information Technology (from 1961 to 1996). INCITS is accredited by the American National Standards Institute (ANSI) and is operated by the Information Technology Industry Council (ITI). Its mission is to produce market-driven, voluntary consensus standards in the area of information technology. (See SFF 8090).

integer transform An invertable, lossless, block transform used in VC-1 and for 4×4 blocks in AVC. Has advantages over DCT, which is lossy.

inter- A prefix meaning between or across. For example, interframe means spanning multiple frames.

interactive The capability of a format such as BD or DVD to respond to commands issued by a user and prompt a user to issue commands.

Interactive Audio Audio (LPCM only) in HDMV or BD-J that is mixed with Primary Audio and Secondary Audio.

Interactive Graphics One of the two graphic planes of BD. In HDMV mode, the Interactive Graphics data stream is used to generate menus. In BD-J mode, menus are generated by Java programs using the HAVi HGraphicsDevice model. Compare to Presentation Graphics.

interface jitter Deviation in timing of data transitions on an interface signal with respect to an ideal clock.

interframe Something that occurs between multiple frames of video. Interframe compression takes temporal redundancy into account. Contrast with intraframe.

interlace A video scanning system in which alternating lines are transmitted, so that half a picture is displayed each time the scanning beam moves down the screen. An interlaced frame is made of two fields.

interleave To arrange data in alternating chunks so that selected parts can be extracted while other parts are skipped over, or so that each chunk carries a piece of a different data stream. In DVD, used for seamless multi-angle and director's cut features, in which multiplexed streams are subsequently interleaved to allow seamless playback of alternate program material.

interpolate To increase the pixels, scan lines, or pictures when scaling an image or a video stream by averaging together adjacent pixels, lines, or frames to create additional inserted pixels or frames. This generally causes a softening of still images and a blurriness of motion images because no new information is created. Compare to filter.

intra- A prefix meaning inside or within. For example, intraframe means within a single frame.

Glossary

intraframe   Something that occurs within a single frame of video. Intraframe compression does not reduce temporal redundancy, but allows each frame to be independently manipulated or accessed. (See I picture.) Compare to interframe.

inverse telecine   The reverse of 2-3 pulldown, where the frames that were duplicated to create 60-field-per-second video from 24-frame-per-second film source are removed. MPEG-2 video encoders usually apply an inverse telecine process to remove duplicate fields from film-source video and add information enabling the decoder to recreate the 60-field-per-second display rate.

IRE   Institute of Radio Engineers (the predecessor to IEEE). IRE most commonly refers to the IRE scale of video signal amplitudes. One-volt peak-peak video is divided into 141 IRE units: 100 units of signal above blanking (0 volts) to peak white (0.714286 volts, 100 IRE), and 41 units of synchronization signal from blanking to -0.285714 volts (-40 IRE). Picture black is 7.5 IRE.

ISAN   International Standard Audiovisual Number. A 96-bit unique number for permanent identification of audiovisual works and related versions, developed by ISO (ISO 15706), similar to ISBN for books.

ISO   International Organization for Standardization (see www.iso.org).

ISO 10646   The ISO standard character set for the Universal Character Set (UCS) or Unicode, which can represent over 2 million characters. Annex D defines UTF-8 variable-length encoding for UCS.

ISO 2202   Information processing - ISO 7-bit and 8-bit coded character sets - Code extension techniques

ISO 3166   Codes for the representation of names of countries.

ISO 3901   Documentation - International Standard Recording Code (ISRC).

ISO 639   Codes for the representation of names of languages.

ISO 8859-1   Information processing - 8-bit single-byte coded graphic character sets. Standard encoding for western alphabet characters.

ISO 9660   The international standard for the file system used by CD-ROM. Allows file names of only 8 characters plus a 3-character extension.

ISO/IEC 11172   Information technology - coding of moving pictures and associated audio for digital storage media up to about 1.5 Mbps (MPEG-1).

ISO/IEC 13818   Information technology - generic coding of moving pictures and associated audio (MPEG-2).

ISO/IEC 14496   Information technology - Coding of audio-visual objects (MPEG-4).

ISRC   International Standard Recording Code. A unique identifier for a sound recording or a music video recording, which is encoded into a product to identify recordings for royalty payments. The ISRC system is administered by the IFPI. In DVD the ISRC is contained in the navigation pack (PCI) of an audio or video stream, or it can also be placed in a text data entry.

ITI   Information Technology Industry Council. Formerly Computer and Business Equipment Association (CBEMA). A trade association representing U.S. providers of information technology products and services. Operates NCITS.

ITU   International Telecommunication Union.

ITU-R BT.601   The international standard specifying the format of digital component video. Formerly known as CCIR 601.

ITU-T H.264   Advanced video coding for generic audiovisual services (AVC).

JAR   Java archive file. A collection of Java programs (classes) and metadata compressed using the Zip format into a single file. The included manifest file (MANIFEST.MF) is primarily used in BD-J for application signing. JAR files are stored in the BDMV/JAR directory on the disc. There is no limitation on the number of JAR files (aside from disc space).

Java   A highly portable, object-oriented programming language developed by Sun Microsystems. Not

Glossary

to be confused with **JavaScript** (below).
JavaScript   A programming language originally created by Netscape with specific features designed for use with the Internet and HTML, and syntax resembling that of Java and C++. Now standardized as ECMA-262, known as ECMAScript.
JCIC   Joint Committee on Intersociety Coordination.
JEC   Joint Engineering Committee of EIA and NCTA.
jewel box   The plastic clamshell case that holds a CD, DVD, or BD.
jitter   Temporal variation in a signal from an ideal reference clock. There are many kinds of jitter, including sample jitter, channel jitter, and interface jitter.
JPEG   Joint Photographic Experts Group. The international committee that created its namesake standard for compressing still images.
JScript   A proprietary Microsoft variant of JavaScript.
JTC   Joint Technical Committee. The group responsible for overseeing development of Blu-ray format specifications.
JVM   Java virtual machine. The software program running on a BD player processor chip (or computer processor) that executes BD-J applications.
JVT   Joint video team. The ISO/IEC MPEG Group and ITU-T Video Coding Experts Group joined together as JVT to design MPEG-4 AVC/H.264.

k byte   One thousand (10^3) bytes. Not to be confused with KB or kilobyte (2^{10} bytes). Note the small "k."
k or kilo   An SI prefix for denominations of one thousand (10^3). Also used, in capital form, for 1024 bytes of computer data (see **kilobyte**).
karaoke   Literally, "empty orchestra," from Japanese "kara" (empty) and "oke," the transliteration of "orchestra." The social sensation from Japan where sufficiently inebriated people embarrass themselves in public by singing along to a music track. Karaoke was largely responsible for the success of laserdisc in Japan, thus supporting it elsewhere. Karaoke is a feature of DVD-Video implemented by some DVD players, providing lyrics and guide vocals to assist the performer.
KB   Kilobyte.
kbps   Kilobits/second. Thousands (10^3) of bits per second.
key picture (or **key frame**)   A video picture containing the entire content of the image (intraframe encoding), rather than the difference between it and another image (interframe encoding). MPEG I pictures are key pictures. Contrast with delta picture.
kHz   Kilohertz. A unit of frequency measurement. One thousand cycles (repetitions) per second or 1000 hertz.
kilobyte   1024 (2^{10}) bytes. The progression of byte measurement is byte, kilobyte, megabyte, gigabyte, terabyte, petabyte, and so on.

land   The raised area of an optical disc.
laserdisc   A 12-inch (or 8-inch) optical disc that holds analog video (using an FM signal) and both analog and digital (PCM) audio. A precursor to DVD.
latency   The time delay from the moment something is begun until it begins to take effect or completes its action.
layer   The plane of a disc on which information is recorded in a pattern of microscopic pits. Each substrate of a disc can contain one or two layers.
layer 0   In a dual-layer disc, the layer closest to the optical pickup beam and the surface of the disc, and the first to be read when scanning from the beginning of the disc's data. For DVD, data on dual-layer discs is ten percent less dense than on single-layer discs to compensate for crosstalk between the layers.

Glossary

layer 1   In a dual-layer disc, the deeper of the two layers, and the second one to be read when scanning from the beginning of the disc's data.

layout   In the authoring process, layout occurs after multiplexing is complete and a fully qualified root structure of a disc is made. This layout can be played as a disc would play, but it is not yet an image intended for replication on an optical disc.

LBR   Low-bit-rate extension for DTS HD Audio. See DTS-HD Secondary Audio.

LCD   Liquid crystal display. Flat-panel, energy-efficient displays made of thousands of tiny crystal coils that control the amount of light passing through them. Used extensively on laptops. A popular technology for flat-panel, widescreen, high-definition television sets.

lead-in   The physical area 1.2 mm or wider preceding the data area on a disc. The lead-in contains sync sectors and control data including disc keys and other information.

lead-out   On a single-layer disc or PTP dual-layer disc, the physical area 1.0 mm or wider toward the outside of the disc following the data area. On an OTP dual-layer disc, the physical area 1.2 mm or wider at the inside of the disc following the recorded data area (which is read from the outside toward the inside on the second layer).

legacy   A term used to describe an older format disc that can be played in newer format player, such as a CD in DVD player or a DVD in a BD player.

letterbox   The process or form of video where black horizontal mattes are added to the top and bottom of the display area in order to create a frame in which to display video using an aspect ratio different than that of the display. The letterbox method preserves the entire video picture, as opposed to pan & scan. DVD-Video players can automatically letterbox an anamorphic widescreen picture for display on a standard 4:3 TV.

letterbox filter   Circuitry in a DVD player that reduces the vertical size of anamorphic widescreen video (combining every 4 lines into 3) and adds black mattes at the top and bottom. Also see filter.

level   In MPEG-2, levels specify parameters such as resolution, bit rate, and frame rate. Compare to profile.

line 21   The specific part of the NTSC video signal that carries the teletext information for closed captioning.

line doubler   A video processor that doubles the number of lines in the scanning system in order to create a display with scan lines that are less visible. Some line doublers convert from interlaced to progressive scan.

linear PCM   A coded representation of digital data that is not compressed. Linear PCM spreads values evenly across the range from highest to lowest, as opposed to nonlinear (companded) PCM that allocates more values to more important frequency ranges. PCM is most commonly used for digital encoding of audio signals.

lines of horizontal resolution   Sometimes abbreviated as TVL (TV lines) or LoHR. A common but subjective measurement of the visually resolvable horizontal detail of an analog video system, measured in half-cycles per picture height. Each cycle is a pair of vertical lines, one black and one white. The measurement is usually made by viewing a test pattern to determine where the black and white lines blur into gray. The resolution of VHS video is commonly gauged at 240 lines of horizontal resolution, broadcast video at 330, laserdisc at 425, and DVD at 500 to 540. Because the measurement is relative to picture height, the aspect ratio must be taken into account when determining the number of vertical units (roughly equivalent to pixels) that can be displayed across the width of the display. For example, an aspect ratio of 1.33 multiplied by 540 gives 720 pixels.

L_0/R_0   Left only/right only. Stereo signal (no matrixed surround information). Optional downmixing output in Dolby Digital decoders. Does not change phase, simply folds surround channels forward into Lf and Rf.

Glossary

Local Storage Memory in the BD player that can persistently hold data. Can take many forms, such as internal flash memory, external flash memory, internal hard drive, or external hard drive. Divided into the Application Data Area (Persistent Storage) that is mandatory for all BD players, and the optional Binding Unit Data Area in Profile 2 players.

locale See **region code**.

localization The process of translating information into another language. In terms of optical disc creation, menus and other textual displays are localized to be viewed in multiple countries.

logical An artificial structure or organization of information created for convenience of access or reference, usually different from the physical structure or organization. For example, the application specifications of DVD (the way information is organized and stored) are logical formats.

logical unit A physical or virtual peripheral device, such as a DVD-ROM drive.

lossless compression Compression techniques that allow the original data to be recreated without loss. Contrast with lossy compression.

lossy compression Compression techniques that achieve very high compression ratios by permanently removing data while preserving as much significant information as possible. Lossy compression includes perceptual coding techniques that attempt to limit the data loss to that which is least likely to be noticed by human perception.

lower third In television, this refers to the lower third of the screen that contains text. For example, when a person appears on the screen, their name and title will appear in the lower third. CNN runs news text in the lower third.

LP Long-playing record. An audio recording on a plastic platter turning at 33-1/3 rpm and read by a stylus.

LPCM See linear PCM.

L_t/R_t Left total/right total. Four surround channels matrixed into two channels. The mandatory down-mixing output in Dolby Digital decoders.

luma (Y′) The brightness component of a color video image (also called the grayscale, monochrome, or black-and-white component). Nonlinear luminance. The standard luma signal is computed from nonlinear RGB as Y' = 0.299 R' + 0.587 G' + 0.114 B'. Compare to chroma.

luma key Luminance keying, a technique used for PIP overlay where pixels darker than a certain level become transparent so that the underlying video remains visible.

luminance (Y) Loosely, the sum of RGB tristimulus values corresponding to brightness. May refer to a linear video signal or (incorrectly) a nonlinear signal. Compare to luma and chrominance.

M byte One million (10^6) bytes. Not to be confused with MB or megabyte (2^{20} bytes).

M or mega- An SI prefix for denominations of one million (10^6). Note the capital "M" (as lowercase "m" is used for micro.)

m or milli- An SI prefix for denominations of 1 thousandth (10^{-3}). Note the lowercase "m" (as uppercase "M" is used for mega-).

m2ts File extension for Clip AV Stream files containing MPEG-2 transport streams (in the BDMV/STREAM directory on the disc).

Mac OS The operating system used by Apple Macintosh computers.

macroblock In MPEG, AVC, and VC-1, a 16×16 area of a video frame formed of smaller blocks, usually one 8×8 block of luma information and two 8x8 blocks of chroma information.

macroblocking An artifact in block-based video encoding. See blocking.

Macrovision An anti-taping analog protection system (APS) that modifies a signal so that it appears unchanged on most televisions but is distorted and unwatchable when played back from a videotape recording. Macrovision takes advantage of characteristics of AGC circuits and burst decoder circuits in VCRs to interfere with the recording process.

Glossary

magneto-optical   Recordable disc technology using a laser to heat spots that are altered by a magnetic field. Other formats include dye-sublimation and phase-change.

main level (ML)   A range of proscribed picture parameters defined by the MPEG-2 video standard, with maximum resolution equivalent to ITU-R BT.601 (720×576×30). (Also see level.)

main profile (MP)   A subset of the syntax of the MPEG-2 video standard designed to be supported over a large range of mainstream applications such as digital cable TV, DVD, and digital satellite transmission. (Also see **profile**.)

managed copy   A feature of AACS where purchased discs can be legally copied to other devices or storage media (such as a hard drive in a PC or CE device, a portable media player, a mobile phone, or a recordable disc) using approved content protection technologies (such Windows Media DRM, CPRM, or encryption tied to a particular device) after authorization by the content owner. Managed copying is mandatory in the sense that all content providers must allow at least one copy of most of their titles, but the feature is optional for player manufacturers.

manifest   In BD, the Binding Unit Manifest defines how the files from disc and from Local Storage are bound together using the Virtual File System (VFS).

mark   The non-reflective area of a writable optical disc. Equivalent to a pit.

master   The metal disc used to stamp replicas of optical discs or to create additional stampers. For tape, the tape used to make additional recordings.

mastering   The process of creating the physical components needed to replicate optical discs. Usually includes glass master preparation, laser beam recording, glass master development and metallization, and electroforming of fathers, mothers, and stampers. Often used inaccurately to refer to premastering.

matrix encoding   The technique of combining additional surround-sound channels into a conventional stereo signal. Also see Dolby Surround.

matte   An area of a video display or motion picture that is covered (usually in black) or omitted in order to create a differently shaped area within the picture frame.

MB   Megabyte.

Mbps   Megabits per second. Millions (10^6) of bits per second.

MCOT   Managed copy output technology. A technology for protecting content in the AACS managed copy process.

MCS   Managed copy server. In AACS, an MCS is a Prepared Video Authorization Server (PVAS).

media   In the optical disc world, media means a physical disc, either prerecorded (replicated) or recordable (write-once or rewritable).

media key block (MKB)   A cryptographic element used in AACS, CPPM, CPRM, and other content protection systems for authenticating and revoking players based on their ability to decrypt portions of the MKB using a device key or set of device keys.

Media Transform   A scrambling process in BD+, applied to media streams on the disc so that they can't be played until the BD+ Content Code runs and untransforms them.

megabyte   1,048,576 (2^{20}) bytes.

megapixel   An image or display format with a resolution of approximately one million pixels.

memory   Data storage used by computers or other digital electronics systems. Read-only memory (ROM) permanently stores data or software program instructions. New data cannot be written to ROM. Random-access memory (RAM) temporarily stores data-including digital audio and video-while it is being manipulated, and holds software application programs while they are being executed. Data can be read from and written to RAM. Other long-term memory includes hard disks, floppy disks, digital CD formats (CD-ROM, CD-R, and CD-RW), and DVD formats (DVD-ROM, DVD-R, and DVD-RAM).

Glossary

menu state The condition of a player when a menu is presented.

Meridian lossless packing (MLP) A lossless compression technique (used by Blu-ray and DVD-Audio) that removes redundancy from PCM audio signals to achieve a compression ratio of about 2:1 while allowing the signal to be perfectly recreated by the MLP decoder. MLP was incorporated into Dolby TrueHD.

metadata As a general term, metadata means "information about data." For BD, metadata specifically refers to information about discs, titles, and libraries of discs, using Dublin Core elements such as name, subject, creator, publisher, year, actor, and so on. Metadata is used for the search feature that can be built into BD players.

MHP Multimedia home platform. Also DVB-MHP. A software ("middleware") layer developed by the DVB Project for set-top boxes and televisions to receive and execute interactive, Java-based applications. BD-J is based on the GEM subset of MHP (see GEM). Version 1.0.3 is standardized as ETSI TS 101 812 V1.3.1 (2003-06).

MHz One million (10⁶) Hz.

μ or micro- An SI prefix for denominations of 1 millionth (10-6).

Microsoft Windows The leading operating system for Intel CPU-based computers. Developed by Microsoft.

middle area Unused physical area that marks the transition from layer 0 to layer 1. Middle Area only exists in dual layer discs where the tracks of each layer are in opposite directions.

miniBD Usually refers to an 8-cm (3-inch) Blu-ray disc.

miniDVD Usually refers to an 8-cm (3-inch) DVD. Sometimes refers to a cDVD.

mixed mode A type of CD containing both Red Book audio and Yellow Book computer data tracks.

MKB See **Media Key Block**.

MLP See **Meridian Lossless Packing**.

MMC 1) Multimedia commands. A command set used by computers ("hosts") to read and write data on CD, DVD, and BD drives. Standardized by ANSI INCITS. See SFF 8090. 2) Mandatory Managed Copy. See managed copy.

MO Magneto-optical rewritable discs.

modulation Replacing patterns of bits with different (usually larger) patterns designed to control the characteristics of the data signal. BD uses 17PP modulation, where each set of 2 bits is replaced by a 3-bit code before being written onto the disc. DVD uses 8/16 modulation, where each set of 8 bits is replaced by a 16-bit code.

mosquitoes A term referring to the fuzzy dots that can appear around sharp edges (high spatial frequencies) after video compression. Also known as the Gibbs Effect.

mother The metal disc produced from mirror images of the father disc in the replication process. Mothers are used to make stampers, often called sons.

motion compensation In video decoding, the application of motion vectors to already-decoded blocks to construct a new picture.

motion estimation In video encoding, the process of analyzing previous or future frames to identify blocks that have not changed or have only changed location. Motion vectors are then stored in place of the blocks. This is very computation-intensive and can cause visual artifacts when subject to errors.

motion vector A two-dimensional spatial displacement vector used for MPEG motion compensation to provide an offset from the encoded position of a block in a reference (I or P) picture to the predicted position (in a P or B picture).

Movie Object An HDMV application and associated playback and user operation information corresponding to an HDMV Title.

Glossary

MP@ML   Main profile at main level. The common MPEG-2 format used by DVD (along with SP@SL).

MP3   MPEG-1 Layer III audio. A perceptual audio coding algorithm. Not supported in DVD-Video or DVD-Audio formats.

MPEG   Moving Pictures Expert Group. An international committee that developed the MPEG family of audio and video compression systems.

MPEG audio   Audio compressed according to the MPEG perceptual encoding system. MPEG-1 audio provides two channels, which can be in Dolby Surround format. MPEG-2 audio adds data to provide discrete multichannel audio.

MPEG video   Video compressed according to the MPEG encoding system. MPEG-1 is typically used for low data rate video such as on a Video CD. MPEG-2 is used for higher-quality video, especially interlaced video, such as on DVD, and high-definition video, such as on BD or HDTV.

MPEG-1 video   Video encoded in accordance with the ISO/IEC 11172 specification.

MPEG-2 video   Video encoded in accordance with the ISO/IEC 13818 specification.

MPEG-4 Part 10   See AVC.

mpls   File extension for BD PlayList files (in the BDMV/PLAYLIST directory on the disc).

MSK   Minimum-shift keying. A wobble modulation method where a digital signal is converted into a half-sinusoid waveform. BD combines MSK and STW.

Mt. Fuji   An industry group made up of optical disc drive manufacturers, operating system vendors, independent software developers, and other optical disc affiliated companies, formed to develop a command set for DVD drives. See SFF 8090.

MTBF   Mean time between failure. A measure of reliability for electronic equipment, usually determined in benchmark testing. The higher the MTBF, the more reliable the hardware.

multiangle   A BD or DVD-Video program containing multiple angles allowing different views of a scene to be selected during playback.

multichannel   Multiple channels of audio, usually containing different signals for different speakers in order to create a surround-sound effect.

multilanguage   A BD or DVD-Video program containing sound tracks and subtitle tracks for more than one language.

multilevel recording   A variation of optical disc storage technology that achieves volumetric recording by varying the depth of the pits, thus recording multiple values in a single spot as compared to the traditional binary value. Can be applied to CD, DVD, BD, and any other optical disc format, but has never achieved commercial success. Primarily championed by Calimetrics. Also called grayscale recording.

multimedia   Information in more than one form, such as text, still images, sound, animation, and video. Usually implies that the information is presented by a computer.

multiplexing   Combining multiple signals or data streams into a single signal or stream. Usually achieved by interleaving at a low level.

multisession   A technique in write-once recording technology that allows additional data to be appended after data written in an earlier session.

multistory   A configuration in which one title contains more than one story. That is, different versions of the program can be selected for viewing.

mux   Short for multiplex.

mux_rate   In MPEG, the combined rate of all packetized elementary streams (PES) of one program. The mux_rate of DVD is 10.08 Mbps. The mux_rate for BD is 48 Mbps.

n or **nano-**   An SI prefix for denominations of 1 billionth (10-9).

NAB   National Association of Broadcasters.

Glossary

native resolution   1) The resolution at which video was captured. 2) The number (horizontal and vertical) of actual pixels in a video display.

navigation   The process of operating a BD or DVD, choosing particular content and functions of the disc. Navigation is defined and programmed during the authoring process.

navigation command   Instruction used to enable the interactive capabilities of a BD or DVD.

NCITS   See **INCITS**.

NCTA   National Cable Television Association.

nighttime mode   Name for Dolby Digital dynamic range compression feature to allow low-volume nighttime listening without losing legibility of dialog.

noise   Irrelevant, meaningless, or erroneous information added to a signal by the recording or transmission medium or by an encoding/decoding process. An advantage of digital formats over analog formats is that noise can be completely eliminated (although new noise may be introduced by compression).

noise floor   The level of background noise in a signal or the level of noise introduced by equipment or storage media below which the signal can't be isolated from the noise.

non-drop-frame timecode   The method of time code computation where there are 30 numerical frames per second of video. "There are 30 frames of video per second," you say. Wrong. There are only 29.97 frames of video per second. In a mathematical hour, there would be 108,000 frames (30 frames per second × 60 seconds × 60 minutes). So, a mathematical hour is 108 frames longer than an hour of reality. See also drop-frame time code and time code.

NRZI   Non-return to zero, inverted. A method of coding binary data as waveform pulses. Each transition represents a one, while lack of a transition represents a run of zeros.

NTSC   National Television Systems Committee. A committee organized by the Electronic Industries Association (EIA) that developed commercial television broadcast standards for the United States. The group first established black-and-white TV standards in 1941, using a scanning system of 525 lines at 60 fields per second. The second committee standardized color enhancements using 525 lines at 59.94 fields per second. NTSC refers to the composite color-encoding system. The 525/59.94 scanning system (with a 3.58-MHz color subcarrier) is identified by the letter M, and is often incorrectly referred to as NTSC. The NTSC standard is also used in Canada, Japan, and other parts of the world. NTSC is facetiously referred to as meaning "Never The Same Color" because of the system's difficulty in maintaining color consistency. The other common TV systems are PAL and SECAM.

NTSC-4.43   A variation of NTSC where a 525/59.94 signal is encoded using the PAL subcarrier frequency and chroma modulation. Also called 60-Hz PAL.

numerical aperture (NA)   A unitless measure of the ability of a lens to gather and focus light. NA = n sin q, where q is the angle of the light as it narrows to the focal point. A numerical aperture of 1 implies no change in parallel light beams. The higher the number, the greater the focusing power and the smaller the spot.

OEM   Original equipment manufacturer. Usually refers to a computer maker.

open GOP   See GOP.

operating system   The primary software in a computer, containing general instructions for managing applications, communications, input/output, memory and other low-level tasks. DOS, Windows, Mac OS, and UNIX are examples of operating systems.

opposite track path (OTP)   Dual-layer disc where layer 0 and layer 1 have opposite track directions. Layer 0 reads from the inside to the outside of the disc, whereas layer 1 reads from the outside to the inside. The disc always spins clockwise, regardless of track structure or layers. OTP facilitates movie playback by allowing seamless (or near-seamless) transition from one layer to another. Also called RSDL (reverse spiral dual layer.) Contrast with parallel track path (PTP) on DVD.

Blu-ray Disc Demystified

Glossary

Orange Book   The document begun in 1990 that specifies the format of recordable CD. Three parts define magneto-optical erasable (MO) and write-once (WO), dye-sublimation write-once (CD-R), and phase-change rewritable (CD-RW) discs. Orange Book added multisession capabilities to the CD-ROM XA format.

OS   Operating system.

OSTA   Optical Storage Technology Association (see www.osta.org).

otf   File extension for text subtitle font data files (in the BDMV/AUXDATA directory on the disc). There can be up to 255 font files.

OTP   See **Opposite Track Path**.

out of band   In a place not normally accessible, such as data stored in sector headers.

out of mux   The technique whereby BD players can read an additional audio stream from internal buffers or from Local Storage and mix it into playback without interruption.

outer diameter   The width of a disc. 12 cm for standard-size CDs, DVDs, and BDs. 8 cm for small discs.

overscan   The area at the edges of a television display that is covered to hide possible video distortion. Overscan typically covers about four or five percent of the picture.

P or peta-   An SI prefix for denominations of 1015. Note the uppercase "P."

P picture (or **P frame**)   In MPEG video, a "predicted" picture based on difference from previous pictures. P pictures (along with I pictures) provide a reference for subsequent P pictures or B pictures.

pack   A group of MPEG packets in a program stream. In DVD, a pack is the size of one sector (2048 bytes).

packet   A low-level unit of MPEG data storage containing contiguous bytes of data belonging to a single elementary stream such as video, audio, control, and so forth. Packets are grouped into packs.

packetized elementary stream (PES)   The low-level stream of MPEG packets containing an elementary stream, such as audio or video.

PAL   Phase alternating line. An analog television standard used in Europe, Africa, Australia, and South America for composite color encoding. Various version of PAL use different scanning systems and color subcarrier frequencies (identified with letters B, D, G, H, I, M, and N), the most common being 625 lines at 50 fields per second, with a color subcarrier of 4.43 MHz. PAL is also said to mean "picture always lousy" or "perfect at last," depending on which side of the ocean the speaker comes from. The other common TV systems are NTSC and SECAM.

palette   A table of colors that identifies a subset from a larger range of colors. The small number of colors in the palette allows fewer bits to be used to represent each pixel. Also called a color look-up table (CLUT).

pan & scan   The technique of reframing a picture to conform to a different aspect ratio by cropping parts of the picture. DVD players can automatically create a 4:3 pan & scan version from widescreen video by using a horizontal offset encoded with the video, which allows the focus of attention to always be visible.

panning   1) Horizontally moving a camera or video picture. 2) Spreading a mono audio signal across a stereo or multichannel sound field. BD has a feature to pan a mono Secondary Audio stream by using gain values supplied for each channel. 3) Washing dirt in a pan in hopes of finding gold.

parallel track path (PTP)   Parallel track path. A variation of DVD dual-layer disc layout where readout begins at the center of the disc for both layers. Designed for separate programs (such as a widescreen and a pan & scan version on the same disc side) or programs with a variation on the second layer. Also most efficient for DVD-ROM random-access application. Contrast with opposite track path (OTP). BD does not have the PTP configuration.

Glossary

parental control Controlling playback by comparing the upper limit of permissible parental level playback preset by the user in the player with the parental level setting on a disc. In BD this is done by commands on the disc. In DVD this is usually done by the player (but can also be done by commands on the disc).

parental level A permissible level of content depending on the age of viewers and the nature of the content of a title.

patchiness A term referring to the occasional patchy appearance of compressed video.

PC Personal computer. A microcomputer designed for personal use. Originally referred to an IBM PC, often refers to any machine that runs the Microsoft Windows operating system.

PCM An uncompressed, digitally coded representation of an analog signal. The waveform is sampled at regular intervals and a series of pulses in coded form (usually quantized) are generated to represent the amplitude. PCM usually refers to linear PCM (LPCM).

PC-TV The merger of television and computers. A personal computer capable of displaying video as a television.

pel See **pixel**.

perceived resolution The apparent resolution of a display from the observer's point of view, based on viewing distance, viewing conditions, and physical resolution of the display.

perceptual coding Lossy compression techniques based on the study of human perception. Perceptual coding systems identify and remove information that is least likely to be missed by the average human observer.

Persistent Storage The small part of BD player Local Storage used for keeping settings such as high scores, bookmarks, and Easter egg status. Used only by BD-J. Called Application Data Area in technical documents.

PES See packetized elementary stream.

petabyte 1,125,899,906,842,624 (250) bytes. Sometimes 1,000,000,000,000,000 (1015) bytes. The progression of byte measurement is byte, kilobyte, megabyte, gigabyte, terabyte, petabyte, and so on.

PG textST Presentation Graphics Text Subtitles. Subtitles in text form to be rendered into the Presentation Graphics plane.

phase-change A technology for rewritable optical discs using a physical effect in which a laser beam heats a recording material to reversibly change an area from an amorphous state to a crystalline state, or vice versa. Continuous heat just above the melting point creates the crystalline state (an erasure), while high heat followed by rapid cooling creates the amorphous state (a mark). (Other recording technologies include dye-sublimation and magneto-optical.)

Photo CD Kodak's variation of compact disc format for storing 24-bit 4:2:0 YCC images hierarchically at resolutions of up to 3072×2048 pixels. Thumbnail image representation is also part of the Photo CD specification. Built upon CD-ROM XA.

physical format The low-level characteristics of the BD and DVD standards, including pits on the disc, location of data, and organization of data according to physical position. Usually called "part 1" in format specifications.

physical sector number Serial number assigned to physical sectors on a DVD disc. Serial incremented numbers are assigned to sectors from the head sector in the Data Area as 30000h from the start of the Lead In Area to the end of the Lead Out Area.

PICT A standard graphic format for Macintosh computers.

picture In video terms, a single still image or a sequence of moving images. Picture generally refers to a frame, but for interlaced frames may refer instead to a field of the frame. In a more general sense, picture refers to the entire image shown on a video display.

Glossary

picture in picture (PIP)   Inserting a second video image into a larger video image. For BD this specifically refers to presenting the Secondary Video within the Primary Video.

PiP PG textST   Presentation Graphics Text Subtitles to be rendered into the PIP (secondary video).

pit   A microscopic depression in the recording layer of a disc. Pits are usually 1/4 of the laser wavelength so as to cause cancellation of the beam by diffraction.

pit art   A pattern of pits to be stamped onto a disc to provide visual art rather than data. A cheaper alternative to a printed label.

pit length   Arc length of a pit along the direction of the track.

pixel   The smallest picture element of an image (one sample of each color component). A single dot of the array of dots that makes up a picture. Sometimes abbreviated to pel. The resolution of a digital display is typically specified in terms of pixels (width by height) and color depth (the number of bits required to represent each pixel).

pixel aspect ratio   The ratio of width to height of a single pixel. Often means sample pitch aspect ratio (when referring to sampled digital video). Pixel aspect ratio for a given raster can be calculated as y/x multiplied by w/h (where x and y are the raster horizontal pixel count and vertical pixel count, and w and h are the display aspect ratio width and height). Pixel aspect ratios are also confusingly calculated as x/y multiplied by w/h, giving a height-to-width ratio.

pixel depth   See **color depth**.

platform   A hardware or software environment in which applications can operate. Examples of a platform are Microsoft Windows XP, Mac OS X, the Sony PlayStation 3, Blu-ray players, Java, etcetera.

Playback Control Engine (PCE)   A core part of BD player software that provides PlayList playback and register access. Used by both HDMV and BD-J modules.

Player Status Register (PSR)   A defined value (32-bit unsigned integer) in a BD player that represent a player state or user setting, such as current playing title, parental level, or preferred audio language. (Similar to SPRMs in DVD-Video). BD players have 128 PSRs: 38 are defined, 8 are for saving state when jumping to a menu or other title, 16 are used for content protection, and 66 are undefined.

PlayItem   An entry in a PlayList that points to a single Clip or, for multi-angle playback, to multiple Clips.

PlayList   The basic playback unit for HDMV and BD-J. A PlayList is a sequential list of PlayItems, which are references to Clips (MPEG-2 transport stream segments) that will play back seamlessly. A Blu-ray disc may contain up to 2000 PlayLists. Each PlayList has a main path that defines the primary playback sequence and optional SubPaths that define alternative playback sequences.

PLUGE   Picture line-up generation equipment. A test pattern containing below-black (0 IRE), black (7.5 IRE) and near-black signals to assist in setting black level on a video monitor.

PMMA   Polymethylmethacrylate. A clear acrylic compound used in laserdiscs and as an intermediary in the surface transfer process (STP) for manufacturing dual-layer BDs and DVDs. PMMA is also sometimes used for disc substrates.

PMSN   Prerecorded media serial number. A unique identifier, usually recorded in the BDA, used by AACS for managed copy.

PNG   Portable network graphic. A compressed image file used for Blu-ray and other environments such as the World Wide Web. This format was not used in DVD.

POP   Picture outside picture. A feature of some widescreen displays that uses the unused area around a 4:3 picture to show additional pictures.

Popup Menu   A menu in HDMV that can be activated by the user appear on top of the video program as it plays.

premastering   The process of preparing data in the final format to create a disc image for mastering. Includes creating control and navigation data, multiplexing data streams together, generating error-

Glossary

correction codes, and performing channel modulation. Sometimes used in a more general sense to include the encoding and authoring process.

prepared video   Content very similar to what would be pre-recorded onto a manufactured disc, but designed for electronic distribution and subsequent recording onto recordable media. Used by AACS for applications such as electronic sell-through (EST) and manufacturing on demand (MOD).

Presentation Graphics   One of the two BD graphic place. The Presentation Graphics data stream ise used in both HDMV and BD-J more for subtitles and animated graphics. The primary video plane and the Presentation Graphics plane are combined by BD-J into the HAVi HVideoDevice. Compare to Interactive Graphics.

Primary Audio   The audio stream associated with the BD Primary Video stream. Comes from the disc or from Local Storage.

Primary Video   The video stream for the main program in BD playback, displayed full-screen.

PRML   Partial response maximum likelihood. A method for converting the analog signal from the pickup head of a BD drive into a digital signal. BD uses a Viterbi decoder for PRML.

professional mode   Mode used when an AC-3 encoded bitstream is output to a single channel of an AES/EBU two channel system.

profile   1) In MPEG-2, profiles specify syntax and processes such as picture types, scalability, and extensions. Compare to level. 2) In BD, profiles define minimum sets of player features. 3) In GEM, profiles define sets of features (and targets refer to deployment scenarios).

Profile 1   The original standard BD player class. Also known as Initial Standard Profile. Only allowed for players manufactured before November 1, 2007.

Profile 1.1   The standard BD player class. Also known as Final Standard Profile or Bonus View. Adds features such as PIP to Profile 1.

Profile 2   The enhanced BD player class. Also known as BD Live. Adds features such as a network connection and streaming video (Progressive PlayList) to Profile 1.1.

Profile 3   The audio-only BD player class. Although it comes after Profile 2, Profile 3 is a subset of Profile 1, designed for BD audio playback in environments without a video screen, such as a car or small, portable player.

Profile 4   Real-time recording and editing format (RREF) for BD-RE v3.

Progressive PlayList   A simple Internet streaming mechanism that uses a PlayList of small clips that are downloaded to the Binding Unit Data Area during playback and accessed through the Virtual File System.

progressive scan   A video scanning system that displays all lines of a frame in one pass. Contrast with interlaced scan.

prohibit RMTR   Prohibit repeated minimum transition run length. A data modulation technique that limits the number of consecutive minimum run lengths (2T) to six, which avoids low signal levels and improves readout performance.

psychoacoustic   See perceptual encoding.

PTP   See **Parallel Track Path**.

PTS   MPEG presentation time stamp. A time reference that tells the decoder when to present a video frame for display (or an audio block). PTS and DTS values reside in the packet header.

PUH   Pickup head. The assembly of optics and electronics that reads data from a disc.

PVAS   AACS Prepared Video Authorization Server.

QCIF   Quarter common intermediate format. Video resolution of 176×144.

quantization levels   The predetermined levels at which an analog signal can be sampled as determined by the resolution of the analog-to-digital converter (in bits per sample); or the number of bits stored for the sampled signal.

Glossary

quantize To convert a value or range of values into a smaller value or smaller range by integer division. Quantized values are converted back (by multiplying) to a value that is close to the original but may not be exactly the same. Quantization is a primary technique of lossless encoding.

QuickTime A digital video software standard developed by Apple Computer for Macintosh (Mac OS) and Windows operating systems. QuickTime is used to support audio and video from a DVD.

QXGA A video graphics resolution of 2048×1536.

RAM (random-access memory) Generally refers to solid-state chips that store data only when power is present. In the case of DVD-RAM, the term was borrowed to indicate ability to read and write at any point on the disc.

random access The ability to jump to a point on a storage medium.

random presentation A function of a player enabling programs in a title to play in any order determined by chance.

raster The pattern of parallel horizontal scan lines that makes up a video picture.

raw Usually refers to uncompressed or unprocessed data.

Red Book The document first published in 1982 that specifies the original compact disc digital audio format developed by Philips and Sony.

Reed-Solomon An error-correction encoding system that cycles data multiple times through a mathematical transformation in order to increase the effectiveness of the error correction, especially for burst errors (errors concentrated closely together, as from a scratch or physical defect). DVD and BD use rows and columns of Reed-Solomon encoding in a two-dimensional lattice, called Reed-Solomon product code (RS-PC).

reference picture or **reference frame** An encoded frame that is used as a reference point from which to build dependent frames. In MPEG-2, I pictures and P pictures are used as references.

reference player A player that defines the ideal behavior as defined by the format specification.

region code A code identifying one of the world regions for restricting BD or DVD-Video playback.

region playback control (RPC) A mandatory feature of BD and DVD-Video to restrict the playback of a disc to a specific geographical region. BD players, DVD players, and DVD-ROM drives are each set to a single region code, and each disc (or disc side for dual-sided DVDs) can specify in which regions it is allowed to be played. Region coding is optional - a disc without region codes will play in all players in all regions (although for DVD the player would need to convert between 525/60 and 625/50).

render farm A dedicated set of processing units (computers) and data storage (hard drives) with a software manager dedicated to processing video. A render farm may be used for encoding video. The processors and storage can be thought of as the farm, the software manager is the farmer, and the video encode is the crop.

replication 1) The reproduction of media such as optical discs by stamping (contrast this with duplication); 2) A process used to increase the size of an image by repeating pixels (to increase the horizontal size) and/or lines (to increase the vertical size) or to increase the display rate of a video stream by repeating frames. For example, a 360×240 pixel image can be displayed at 720×480 size by duplicating each pixel on each line and then duplicating each line. In this case, the resulting image contains blocks of four identical pixels. Obviously, image replication can cause blockiness. A 24-fps video signal can be displayed at 72 fps by repeating each frame three times. Frame replication can cause jerkiness of motion. Contrast this with decimation. See interpolate.

resampling The process of converting between different spatial resolutions or different temporal resolutions. This can be based on a sample of the source information at a higher or lower resolution or it can include interpolation to correct for the differences in pixel aspect ratios or to adjust for differences in display rates.

resolution 1) A measurement of the relative detail of a digital display, typically given in pixels of width

Glossary

and height; 2) The capability of an imaging system to make the details of an image clearly distinguishable or resolvable. This includes spatial resolution (the clarity of a single image), temporal resolution (the clarity of a moving image or moving object), and perceived resolution (the apparent resolution of a display from the observer's point of view). Analog video is often measured as a number of lines of horizontal resolution over the number of scan lines. Digital video is typically measured as a number of horizontal pixels by vertical pixels. Film is typically measured as a number of line pairs per millimeter; 3) The relative detail of any signal, such as an audio or video signal. See lines of horizontal resolution.

resume   A function used when a player returns from suspend or menu state.

RGB   Video information in the form of red, green, and blue tristimulus values. The combination of three values representing the intensity of each of the three colors can represent the entire range of visible light.

RLL   Run-length-limited encoding, which modulates data to prevent long sequences of bits without a transition (from 0-to-1 or 1-to-0). For optical drives and hard discs it is expressed as RLL(x, y), where x is the run length (the minimum bit length before a transition) and y is the run limit (the maximum bit length before a transition). BD uses RLL(1,7) parity preserve modulation.

ROM   Read-only memory.

root menu   The lowest level menu contained in a VTS, from which other menus may be accessed.

rpm   Revolutions per minute. A measure of rotational speed.

RREF   Blu-ray real-time recording and editing format. See **Profile 4**.

RS   Reed-Solomon. An error-correction encoding system that cycles data multiple times through a mathematical transformation in order to increase the effectiveness of the error correction. DVD uses rows and columns of Reed-Solomon encoding in a two-dimensional lattice, called Reed-Solomon product code (RS-PC).

RS-CIRC   See **CIRC**.

RSDL   Reverse-spiral dual-layer. See **OTP**.

RS-PC   Reed-Solomon product code. An error-correction encoding system used by DVD employing rows and columns of Reed-Solomon encoding to increase error-correction effectiveness.

RTFM   When all else fails, read the "fine" manual. Yeah, right.

run-length coding   Method of lossless compression that codes by analyzing adjacent samples with the same values.

R-Y, B-Y   The general term for color-difference video signals carrying red and blue color information, where the brightness (Y') has been subtracted from the red and blue RGB signals to create R'-Y' and B'-Y' color-difference signals. Refer to Chapter 3, Technology Primer.

S/N   See signal-to-noise ratio. Also called SNR.

S/P DIF   Sony/Philips digital interface. A consumer version of the AES/EBU digital audio transmission standard. Most DVD players include S/P DIF coaxial digital audio connectors providing PCM and encoded digital audio output.

sample   A single digital measurement of analog information or a snapshot in time of a continuous analog waveform. See **sampling**.

sample rate   The number of times a digital sample is taken, measured in samples per second, or Hertz. The more often samples are taken, the better a digital signal can represent the original analog signal. The sampling theory states that the sampling frequency must be more than twice the signal frequency in order to reproduce the signal without aliasing. DVD PCM audio enables sampling rates of 48 and 96 kHz.

sample size   The number of bits used to store a sample. Also called resolution. In general, the more bits allocated per sample, the better the reproduction of the original analog information. The audio sample size determines the dynamic range. DVD PCM audio uses sample sizes of 16, 20, or 24 bits.

Glossary

sampling   Converting analog information into a digital representation by measuring the value of the analog signal at regular intervals, called samples, and encoding these numerical values in digital form. Sampling is often based on specified quantization levels. Sampling can also be used to adjust for differences between different digital systems. See resampling and subsampling.

sampling frequency   The frequency used to convert an analog signal into digital data.

saturation   The intensity or vividness of a color.

scaling   Altering the spatial resolution of a single image to increase or reduce the size, or altering the temporal resolution of an image sequence to increase or decrease the rate of display. Techniques include decimation, interpolation, motion compensation, replication, resampling, and subsampling. Most scaling methods introduce artifacts.

scan line   A single horizontal line traced out by the scanning system of a video display unit. 525/60 (NTSC) video has 525 scan lines, about 480 of which contain the actual picture. 625/50 (PAL/SECAM) video has 625 scan lines, about 576 of which contain the actual picture.

scanning velocity   The speed at which the laser pickup head travels along the spiral track of a disc.

SCMS   The serial copy management system used by DAT, MiniDisc, and other digital recording systems to control copying and limit the number of copies that can be made from copies.

SCSI   Small Computer Systems Interface. An electronic interface and command set for attaching and controlling internal or external peripherals, such as a DVD-ROM drive, to a computer. The command set of SCSI was extended for DVD-ROM devices by the SFF 8090 specification.

SD   1) Standard definition. 2) Secure Digital. An SD card is a small, flash memory, data storage device, about the size of a US quarter, used in many consumer electronic products including BD players, digital cameras, and cell phones.

SDDI   Serial Digital Data Interface. A digital video interconnect designed for serial digital information to be carried over a standard SDI connection.

SDDS   Sony Dynamic Digital Sound. A perceptual audio-coding system developed by Sony for multichannel audio in theaters. A competitor to Dolby Digital and DTS, and an optional audio track format for DVD.

SDI   See **Serial Digital Interface**. Also Strategic Defense Initiative, aka Star Wars, which was finally released on DVD so fans can replace their bootleg copies.

SDMI   Secure Digital Music Initiative. A failed attempt at a specification to protect digital music.

SDTV   Standard-definition television. A term applied to traditional 4:3 television (in digital or analog form) with a resolution of about 700×480 (about 1/3 megapixel). Contrast this with HDTV.

seamless angle change   The ability of DVD-Video to change between multiple points of view within a scene without interrupting normal playback.

seamless playback   A feature of DVD-Video where a program can jump from place to place on the disc without any interruption of the video. This enables different versions of a program to be put on a single disc by sharing common parts.

SECAM   Séquentiel couleur à mémoire (sequential color with memory). An analog television standard used in France and in parts of Eastern Europe, Africa, and Asia. Various version of SECAM use different scanning systems and color subcarrier frequencies (identified with letters B, D, G, H, K, and L), the most common being 625 lines at 50 fields per second, with color subcarriers 4.2 or 4.4 MHz. It's very common for video to be carried in 625/50 (PAL) format and only converted to a SECAM signal before broadcast or output (as with DVD players with SECAM output). The other common TV systems are NTSC and PAL.

SECAMx   Sequential couleur avec mémoire/sequential color with memory. A composite color standard similar to PAL but currently used only as a transmission standard in France and a few other countries. Video is produced using the 625/50 PAL standard and is then transcoded to SECAM by the player or transmitter.

Glossary

Secondary Audio The audio stream associated with the BD Secondary Video stream. Comes from the disc or from Local Storage.

Secondary Video A supplemental video stream for picture-in-picture (PIP) display in the Primary Video. Comes from the disc or from Local Storage.

sector A logical or physical group of bytes recorded on the disc, the smallest addressable unit. A DVD sector contains 38,688 bits of channel data and 2,048 bytes of user data.

seek time The time it takes for the head in a drive to move to a data track.

SEI Supplemental enhancement information. Messages in an AVC bitstream that can be used for any variety of reasons to communicate information to a playback device

sequential presentation Playback of all Programs in a Title in a specified order determined in the authoring process.

Serial Digital Interface (SDI) The professional digital video connection format using a 270 Mbps transfer rate. A 10-bit, scrambled, polarity-independent interface, with common scrambling for both component ITU-R 601 and composite digital video and four groups each of four channels of embedded digital audio. SDI uses standard 75-ohm BNC connectors and coax cable.

setup The black level of a video signal.

SFF 8090 The specification number 8090 of the Small Form Factor Committee, an ad hoc group formed to promptly address disk industry needs and to develop recommendations to be passed on to standards organizations. SFF 8090 (also known as the Mt. Fuji specification) defines a command set for CD-ROM- and DVD-ROM-type devices, including implementation notes for ATAPI and SCSI.

shuffle presentation A function of a DVD-Video player enabling Programs in a Title to play in an order determined at random.

SI Système International (d'Unités). International System (of Units). A complete system of standardized units and prefixes for fundamental quantities of length, time, volume, mass, etc.

signal-to-noise ratio The ratio of pure signal to extraneous noise, such as tape hiss or video interference. Signal-to-noise ratio is measured in decibels (dB). Analog recordings almost always have noise. Digital recordings, when properly prefiltered and not compressed, have no noise.

simple profile (SP) A subset of the syntax of the MPEG-2 video standard designed for simple and inexpensive applications such as software. SP does not enable B pictures. See profile.

simulate To test the function of a DVD disc in the authoring system without actually formatting an image.

single-sided disc A type of DVD disc on which data is recorded on one side only.

SMPTE The Society of Motion Picture and Television Engineers. An international research and standards organization. This group developed the SMPTE time code, used for marking the position of audio or video in time.

SMPTE VC-1 See **VC-1**.

son The metal discs produced from mother discs in the replication process. Fathers or sons are used in molds to stamp discs.

SP@ML Simple profile at main level. The simplest MPEG-2 format used by DVD. Most discs use MP@ML. SP does not allow B pictures.

space The reflective area of a writable optical disc. Equivalent to a land.

spatial Relating to space, usually two-dimensional. Video can be defined by its spatial characteristics (information from the horizontal plane and vertical plane) and its temporal characteristics (information at different instances in time).

spatial resolution The clarity of a single image or the measure of detail in an image. See resolution.

SPDC Self-protecting digital content. An architecture that applies content protection mechanisms (code) on the content (disc) rather than on the playback device. Developed by Cryptography Research (CRI) and adapted for BD as BD+.

Glossary

squeezed picture or **squeezed video**   Refers to reducing 16:9 picture data horizontally to conform to a 4:3 image size. See **anamorphic**.

SRP   Sustained read performance. The highest read/write speed at which an optical disc can be operated. Often read as a multiple: 1X, 2X, 4X, etc.

stamping   The process of replicating optical discs by injecting liquid plastic into a mold containing a stamper (father or son). Also (inaccurately) called mastering.

still   In the authoring process, it is a single image not defined as video. In the player, it is a state in which video and graphics are frozen and audio is muted.

stop state   The condition in which the player is not executing a PlayList (or PGC for DVD).

storage media   Materials used to store data.

STP   Surface transfer process. A method of producing dual-layer discs that sputters the reflective (aluminum) layer onto a temporary substrate of PMMA and then transfers the metalized layer to the already-molded layer 0.

stream   A continuous flow of data, usually digitally encoded, designed to be processed sequentially. Also called a bitstream.

STW   Sawtooth wobble. A modulation method where the wobble is formed by combining the basic cosine wave and a sine wave of the doubled frequency (second harmonic with a quarter amplitude). BD combines STW and MSK.

SubPath   An alternative playback sequence for a BD PlayList. There are different types of SubPaths for audio in browsable slideshow, audio in a menu, text subtitles, and various in-mux or out-of-mux presentations.

subpicture   Graphic bitmap overlays used in DVD-Video to create subtitles, captions, karaoke lyrics, menu highlighting effects, etc.

subsampling   The process of reducing spatial resolution by taking samples that cover areas larger than the original samples, or the process of reducing temporal resolutions by taking samples that cover more time than the original samples. This is also called downsampling. See chroma subsampling.

substrate   A thin layer of an optical disc onto which data layers are stamped or deposited. May be clear polycarbonate, or in the case of BD, which doesn't define dual-sided discs, may be an opaque material such as paper.

subtitle   A textual representation of the spoken audio in a video program. Subtitles are often used with foreign languages and do not serve the same purpose as captions for the hearing impaired. See subpicture.

surround sound   A multichannel audio system with speakers in front of and behind the listener to create a surrounding envelope of sound and to simulate directional audio sources.

SVCD   Super Video Compact Disc. MPEG-2 video on CD. Used primarily in Asia.

SVGA   A video graphics resolution of 800×600 pixels.

S-VHS   Super VHS (Video Home System). An enhancement of the VHS videotape standard using better recording techniques and Y/C signals. The term S-VHS is often used incorrectly to refer to s-video signals and connectors.

s-video   A video interface standard that carries separate luma and chroma signals, usually on a four-pin mini-DIN connector. Also called Y/C. The quality of s-video is significantly better than composite video because it does not require a comb filter to separate the signals, but it's not quite as good as component video. Most high-end televisions have s-video inputs. S-video is often erroneously called S-VHS.

SXGA   A video graphics resolution of 1280×1024 pixels.

sync   A video signal (or component of a video signal) containing information necessary to synchronize the picture horizontally and vertically. Also, sync is specially formatted data on a disc that helps the readout system identify location and specific data structures.

syntax The rules governing the construction or formation of an orderly system of information. For example, the syntax of the MPEG video encoding specification defines how data and associated instructions are used by a decoder to create video pictures.

system menu The main menu of a DVD-Video disc, from which titles are selected. Also called the title selection menu or disc menu.

system parameters A set of conditions defined in the authoring process used to control basic DVD-Video player functions.

T or tera- An SI prefix for denominations of one trillion (10¹²). Note the uppercase "T."

TEG2 Technical Experts Group Two. One of five technical groups in the Joint Technical Committee (JTC) in the Blu-ray Disc Association (BDA). TEG2 is responsible for audio/video applications.

telecine The process (and the equipment) used to transfer film to video. The telecine machine performs 2-3 pulldown by scanning film frames in the proper sequence.

telecine artist The operator of a telecine machine. Also called a colorist.

temporal Relating to time. The temporal component of motion video is broken into individual still pictures. Because motion video can contain images (such as backgrounds) that do not change much over time, typical video has large amounts of temporal redundancy.

temporal resolution The clarity of a moving image or moving object, or the measurement of the rate of information change in motion video. See resolution.

terabyte 1,099,511,627,776 (2⁴⁰) bytes. Sometimes 1,000,000,000,000 (10¹²) bytes. The progression of byte measurement is byte, kilobyte, megabyte, gigabyte, terabyte, petabyte, and so on.

tilt A mechanical measurement of the warp of a disc. This is usually expressed in radial and tangential components, with radial indicating dishing and tangential indicating ripples in the perpendicular direction.

timecode Information recorded with audio or video to indicate a position in time. This usually consists of values for hours, minutes, seconds, and frames. It is also called SMPTE time-code. Some DVD-Video material includes information to enable the player to search to a specific timecode position. There are two types of timecode - non-drop frame and drop frame. Non-drop frame timecode is based on 30 frames of video per second. Drop frame timecode is based on 29.97 frames of video per second. Truth be told, there are only 29.97 frames of video per second. For short amounts of time, this discrepancy is inconsequential. For longer periods of time, however, it is important. One hour of non-drop frame timecode will be 108 frames longer than one hour of real time. See **non-drop frame timecode** and **drop frame timecode**.

title The largest unit of a BD-Video or DVD-Video disc (other than the entire volume or side). A title is usually a movie, TV program, music album, or so on. Titles are usually selected from a menu. Entire disc volumes are also commonly called titles.

track 1) A distinct element of audiovisual information, such as the picture, a sound track for a specific language, or the like. DVD-Video enables one track of video (with multiple angles), up to eight tracks of audio, and up to 32 tracks of subpicture; 2) One revolution of the continuous spiral channel of information recorded on a disc.

track buffer The circuitry (including memory) in a BD or DVD player that reads data from the disc and provides a variable stream of data to the system buffers and decoders.

track pitch The distance (in the radial direction) between the centers of two adjacent tracks on a disc. The DVD-ROM standard track pitch is 0.74 mm.

transfer rate The speed at which a certain volume of data is transferred from a device such as a BD-ROM drive to a host such as a personal computer. This is usually measured in bits per second or bytes per second. It is sometimes confusingly used to refer to the data rate, which is independent of the actual transfer system.

Glossary

transform The process or result of replacing a set of values with another set of values. It can also be a mapping of one information space onto another.

trick play Modes in digital video playback other than normal play, such as fast forward, rewind, slow motion, and so on.

trim See **crop**.

tristimulus A three-valued signal that can match nearly all the colors of visible light in human vision. This is possible because of the three types of photoreceptors in the eye. RGB, Y'CbCr, and similar signals are tristimulus and can be interchanged by using mathematical transformations (subject to a possible loss of information).

TVL Television line. See lines of horizontal resolution.

TWG Technical Working Group. A general term for an industry working group. Specifically, the predecessor to the CPTWG. It is usually an ad hoc group of representatives working together for a period of time to make recommendations or define standards.

UDF Universal Disc Format. A standard developed by the Optical Storage Technology Association designed to create a practical and usable subset of the ISO/IEC 13346 recordable, random-access file system and volume structure format.

UDF Bridge A combination of UDF and ISO 9660 file system formats that provides backward-compatibility with ISO 9660 readers while allowing the full use of the UDF standard.

USB Universal serial bus. A serial bus interface standard to connect devices such as keyboards, mice, and storage to a host such as a computer or BD player.

USB drive A small portable data storage device using flash memory or a hard disk that connects to a computer, BD player, or other device through a USB connection.

user The person operating a device such as a BD player or computer. Sometimes referred to as wetware, as in, the operation experienced a wetware breakdown.

user data The data recorded on a disc independent of modulation and error-correction overhead. Each disc sector contains 2,048 bytes of user data.

user operation (**UO** or **UOP**) A user operation (UO in BD, UOP in DVD) is a function that a user can perform, generally with a remote control. If a user operation is not locked or "masked," the user can employ it. Examples of user operations include stop, pause, fast forward, menu, and angle change. A title may have all user operations locked during copyright warnings, as studios desire that users fully view them. User operations may also be masked during previews of upcoming attractions that have often come and gone by the time the user views them. This is a function of insensitive, greedy studios that hope it will drive customers to buy more discs rather than drive them berserk.

UXGA A video graphics resolution of 1600×1200.

VBI Vertical blanking interval. The scan lines in a television signal that do not contain picture information. These lines are present to enable the electron scanning beam to return to the top, and they are used to contain auxiliary information such as closed captions.

VBR Variable bitrate. Data compressed into a stream with a fixed data rate. The amount of compression (such as quantization) is varied to match the allocated data rate, but as a result quality may suffer during high compression periods. In other words, data rate is held constant while quality is allowed to vary. Compare to VBR.

VBRx Variable bitrate. Data that can be read and processed at a volume that varies over time. A data compression technique that produces a data stream between a fixed minimum and maximum rate. A compression range is generally maintained, with the required bandwidth increasing or decreasing depending on the complexity (the amount of spatial and temporal energy) of the data being encoded. In other words, the data rate fluctuates while quality is maintained. Compare this to CBR.

VBV Video buffering verifier. A hypothetical decoder that is conceptually connected to the output

Glossary

of an MPEG video encoder. It provides a constraint on the variability of the data rate that an encoder can produce.

VC-1   VC-1 is a video compression standard used in BD. It was developed by Microsoft and standardized by SMPTE.

VCD   Video Compact Disc. Near-VHS-quality MPEG-1 video on CD. Used primarily in Asia.

verification   The process of ensuring that something built is in compliance with a specification. The BDA and the DVD Forum define verification processes and requirements carried out by verification labs. There are also software tools for specification verification of discs.

VFS   Virtual file system. A mechanism of BD that combines downloaded content in local storage with content from the optical disc, allowing files on the disc to be supplemented with or superseded by files in Local Storage.

VfW   See Video for Windows. Not to be confused with Veterans of Foreign Wars, a fine organization composed of persons who occasionally experience wetware problems when operating a VCR, DVD player, or computer.

VGA   (Video Graphics Array) A standard analog monitor interface for computers. It is also a video graphics resolution of 640×480 pixels.

VHS   Video Home System. The most popular system of videotape for home use. Developed by JVC.

Video CD   A CD extension based on MPEG-1 video and audio that enables the playback of near-VHS-quality video on a Video CD player, CD-i player, or computer with MPEG decoding capability.

Video Coding Experts Group (VCEG)   An ITU-T group that has been very instrumental in developing modern video coding mechanisms.

Video for Windows   The system software additions used for motion video playback in Microsoft Windows. Replaced in newer versions of Windows by DirectShow (formerly called ActiveMovie).

videophile   Someone with an avid interest in watching videos or in making video recordings. Videophiles are often very particular about audio quality, picture quality, and aspect ratio to the point of snobbishness. Videophiles never admit to a wetware failure.

VLC   Variable-length coding. See Huffman coding.

volume   A logical unit representing all the data on one side of a disc.

VSDA   Video Software Dealers Association.

WAEA   World Airline Entertainment Association. Discs produced for use in airplanes contain extra information in a WAEA directory. The in-flight entertainment working group of the WAEA petitioned the DVD Forum to assign region 8 to discs intended for in-flight use.

watermark   Information hidden as invisible noise or inaudible noise in a video or audio signal.

wetware   see **user**.

White Book   The document from Sony, Philips, and JVC begun in 1993 that extended the Red Book CD format to include digital video in MPEG-1 format. It is commonly called Video CD.

widescreen   A video image wider than the standard 1.33 (4:3) aspect ratio. When referring to BD, DVD, or HDTV, widescreen usually indicates a 1.78 (16:9) aspect ratio.

window   A usually rectangular section within an entire screen or picture.

Windows   See Microsoft Windows.

XA   See CD-ROM XA.

XBR   A DTS-HD extension stream for high-resolution audio at constant bit rates.

XDS   Extended data service information packets in NTSC line 21, as specified in EIA/CEA-608-B.

XGA   A video graphics resolution of 1024×768 pixels.

XLL   A DTS-HD extension stream for lossless encoding.

Glossary

XML Extensible Markup Language, a general-purpose specification for creating custom markup languages for organizing data by annotating text.

XVCD A non-standard variation of VCD.

XXCH A DTS-HD extension that goes beyond 6.1 channels.

Y The luma or luminance component of video, which is the brightness independent of color.

Y/C A video signal in which the brightness (luma, Y) and color (chroma, C) signals are separated. This is also called s-video.

$Y'C_bC_r$ A component digital video signal containing one luma and two chroma components. The chroma components are usually adjusted for digital transmission according to ITU-R BT.601. DVD-Video's MPEG-2 encoding is based on 4:2:0 $Y'C_bC_r$ signals. $Y'C_bC_r$ applies only to digital video, but it is often incorrectly used in reference to the $Y'P_bP_r$ analog component outputs of DVD players.

Yellow Book The document produced in 1985 by Sony and Philips that extended the Red Book CD format to include digital data for use by a computer. It is commonly called CDROM.

$Y'P_bP_r$ A component analog video signal containing one luma and two chroma components. It is often referred to loosely as YUV or Y', B'-Y', R'-Y'.

YUV In the general sense, any form of color-difference video signal containing one luma and two chroma components. Technically, YUV is applicable only to the process of encoding component video into composite video. See **$Y'C_bC_r$** and **$Y'P_bP_r$**.

Z The brilliant feature film directed by Constantinos Gavras, better known as Costa-Gavras, released in 1969 starring Jean-Louis Trintignant, Yves Montand, Irene Papas, and Jacques Perrin. At the time this book was written the film was available only on DVD.

We did not identify any terms that start with the last letter of the alphabet, but the Glossary looked unfinished without the character. Perhaps there will be more Z terms in the next edition. *-ed.*

Index

A

A/V file formats, 12–47
AACS. See Advanced Access Content System
AACS Licensing Agency, 4–11, 9–4
AACS Managed Copy, 4–8
AACS Protection for Prepared Video, 4–10
AC-3. See Dolby Digital audio coding
Access restrictions, 3–9
Acronyms, 6–3
Action-safe areas, 12–35
ADA. See Application Data Area
Adaptive Differential Pulse Code Modulation, 2–32
Address in pregroove, 5–6
Address units, 5–8
ADIP. See Address in pregroove
ADPCM. See Adaptive Differential Pulse Code Modulation
Adult content, 8–7 — 8–8
Advanced Access Content System, 1–16, 1–19 — 1–23, 3–12 — 3–15, 4–4 — 4–8, 4–25 — 4–26, 8–4, 9–2 — 9–4
Advanced Encryption Standard, 4–5
Advanced Optical Disc, 1–15
Advanced Television Systems Committee, 1–4 — 1–5, 2–10
Advanced video applications, 6–21 — 6–25
Advanced video codecs, 2–23 — 2–26
AES. See Advanced Encryption Standard
Aliasing, 8–11
Alternating current, 2–7
Anaglyph, 13–6
Analog stereo audio, 7–8
Analog sunset, 4–15
Analog surround, 7–9
Analog technology, 2–1 — 2–2, 8–6 — 8–7
Analog transmission, 4–15 — 4–16
Anamorphic process, 2–43, 2–45
Animation, 6–42 — 6–43, 8–12, 12–30
Animators, 12–6
AnyDVD, 1–23
AOD. See Advanced Optical Disc

Aperture plates, 2–45
APIs. See Application programming interfaces
Application cache, 6–36 — 6–37
Application compatibility, 7–24
Application Data Area, 6–45 — 6–46, 12–11 — 12–12
Application programming interfaces, 6–9, 6–12, 6–37 — 6–40
Application signing, 6–48 — 6–49
apt-X, 2–32
Artifacts, 8–11 — 8–12, 12–37 — 12–38
Aspect ratios, 2–39 — 2–45, 9–9 — 9–10
Asymmetric, 2–23
ATSC. See Advanced Television Systems Committee
AU. See Address units
Audio applications
Blu-ray Disc potential, I–1 — I–2
Audio compression, 2–26 — 2–27
Audio compressionist, 12–6
Audio connections, 7–13 — 7–16
Audio data reduction systems, 2–27 — 2–28
Audio formats, 6–26 — 6–29
Audio level, 8–13
Audio mixing, 6–44 — 6–45
Audio preparation, 12–30 — 12–32
Authentication, 4–6 — 4–7
Authoring, 12–38 — 12–39
Authorized copying, 3–11 — 3–13
Authorized recording, 3–13 — 3–15
Authors, 12–7
Autostereoscopic technology, 13–6 — 13–7
AV playback mode, 12–2 — 12–3
AVC. See Advanced video codecs

B

B frames, 2–20 — 2–21
Babbage, Charles, 1–1
Backup HDDVD, 1–22
Backward compatibility, 7–25
Baird, John Logie, 1–2
Banding, 8–11

Index

Barcode standards, 9–10
Bass effects, 7–15 — 7–16
BCA. See Burst cutting area
BD. See Blu-ray Discs™
BD+, 4–9 — 4–10, 8–4, 9–3
BD-5, 12–5
BD-9, 12–5
BD-25, 12–5
BD-50, 12–5
BD-J
  advanced features, 6–43 — 6–56
  application programming interface, 6–37 — 6–40
  application signing, 6–48 — 6–49
  compatibility, 8–4, 9–5 — 9–7
  credential process, 6–50
  downloading content, 6–46 — 6–47
  features, 6–9 — 6–10
  graphics drawing, 6–41 — 6–43
  local storage, 6–45 — 6–46
  memory overview, 6–37
  menus, 6–40 — 6–41
  module, 6–36 — 6–37
  multi-disc sets, 6–44
  navigation data, 6–12 — 6–16
  organizational structure, 6–11 — 6–12
  presentation planes, 6–10 — 6–11
  programmable audio mixing, 6–44 — 6–45
  programming language benefits, 6–35
  Progressive PlayList, 6–47
  resources, 6–36 — 6–37
  system overview, 6–36 — 6–40
  text, 6–43
  uploading content, 6–46 — 6–47
  user interaction, 6–16 — 6–18
  Virtual File System, 6–46
BD-J Mode programming, 12–3 — 12–5
BD-J objects, 6–12
BD+ Licensing LLC, 4–26
BD-Live, 3–9, 6–46, 8–2 — 8–4, 12–12— 12–14
D-R composition, 5–6 — 5–7
BD-R/RE
  defect management, 5–9
  recording, 5–8 — 5–9

BD-RE
  characteristics, 5–7
  composition, 5–6
  development of, 1–13
BD-ROM
  application types, 6–9 — 6–10
  composition, 5–5 — 5–6
  mastering, 5–5
  navigation data, 6–12 — 6–16
  organizational structure, 6–11 — 6–12
  presentation planes, 6–10 — 6–11
  production, 5–5—5–6, 12–45—12–47
  user interaction, 6–16—6–18
  virtual package, 6–56
BD-ROM Mark, 4–8 — 4–9, 4–26
BDA. See Blu-ray Disc Association
BDAV, 6–3 — 6–5, 12–1
BDCMF type P image, 12–46
BDMV, 6–5 — 6–8, 6–56
BDMV directory, 12–40
Bidirectional pictures, 2–20 — 2–21
Binding Nonce, 4–11
Binding Unit Data Area, 6–45, 12–11 — 12–12
Birthday scenario, 9–3
BIS. See Burst indicator subcode
Bit budgeting, 12–23—12–26, 12–46—12–47
Bit depth, 2–12
Bit shaving, 2–34
Bit starvation, 2–28
BITC. See Burned-in timecode
Blocks, 8–11
Blu-ray Disc Association, 1–16 — 1–19, 4–3, 4–20 — 4–21, 6–1, 8–3, 8–7 — 8–8
Blu-ray Disc Audio Visual. See BDAV
Blu-ray Disc Founders group, 1–14, 1–15
Blu-ray Disc Java. See BD-J
Blu-ray Disc Movie. See BDMV
Blu-ray Disc Plus. See BD+
Blu-ray disc recorders, 1–15 — 1–16, 7–7
Blu-ray Discs
  adapting DVD features, 3–2 — 3–9
  advantages of, 11–1 — 11–9
  capacities, 5–4
  cleaning, 5–13—5–14

Index

costs of, 8–7, 11–1 — 11–2
data modulation, 5–8
development of, I–1, 1–1 — 1–3, 1–14 —1–27
disadvantages of, 9–2—9–13
error correction, 5–7 — 5–8
forecasts for, 13–3 — 13–5, 13–8 — 13–9
handling, 5–12 — 5–13
high-definition video specifications, 2–11
media storage, 5–11 — 5–12
myths, 8–1—8–14
performance, 9–7
physical characteristics, 5–1 — 5–3
repairing, 5–14
sales rates, 13–2 — 13–3
stability of, 3–9, 5–11 — 5–12, 8–12 — 8–13
storage capacity, 5–14
storage of, 5–12—5–13
structure of, 5–4
Blu-ray read-only discs. See BD-ROM
Blu-ray rewritable discs. See BD-RE
Blurriness, 8–12
BNC connectors, 7–10
Board of Directors, 6–1 — 6–2
BoD. See Board of Directors
Bootstrap functionality, 6–47
Branch commands, 6–12 — 6–13
Browsable slideshows, 6–35
BUDA. See Binding Unit Data Area
Burned-in timecode, 12–34
Burst cutting area, 4–8, 5–10 — 5–11
Burst indicator subcode, 5–7 — 5–8
Bus encryption, 4–6
Bus Key, 4–6
Business applications
Blu-ray Disc potential, 11–1 — 11–9

C   CABAC. See Context-Adaptive Binary Arithmetic Coding
Camera angles, 3–5 — 3–6, 6–22 — 6–23
Capacitance electronic discs, 1–2
Cathode ray tubes, 2–45, 2–54 — 2–55
CBR. See Constant bitrate
CD. See Compact disc
CD-ROM
  development of, 1–5 — 1–6

CD-RW. See Compact Disc Rewritable
CDB. See Consumer download and burn
CEA. See Consumer Electronics Association
CED. See Capacitance electronic discs
CEMA. See Consumer Electronics Manufacturers Association
Chairiglione, Leonardo, 1–5
Channel Extension, 2–33
Channels, 9–12 — 9–13
Character strips, 6–43
Chroma, 2–17
Chroma crawl, 12–37
Chroma subsampling, 2–12
Chrominance, 2–12, 2–16
Cine-look, 2–8 — 2–9
Classroom education
  Blu-ray Disc potential, 11–9
Clips, 6–4 — 6–5, 6–12
Clusters, 5–7
CMF. See Cutting master format
CMI. See Content Management Information
Codec technologies, 2–23 — 2–26
Coherent Acoustics coding, 2–27, 2–32
Color crosstalk, 12–37
Color difference, 2–16
Color shift, 13–6
Colorists, 12–6
Colors, 12–35
Communications
  Blu-ray Disc potential, I–4, 11–6 — 11–7
Compact disc
  development of, 1–3
Compact Disc Digital Audio
  development of, 1–5
Compact Disc Interactive
  development of, 1–5—1–6
Compact Disc Rewritable, 1–6
Compare commands, 6–12—6–13
Compatibility issues, 7–22 — 7–26, 9–5 — 9–7, 9–10 — 9–11
Compliance Committee, 6–2
Component analog video, 7–9
Component video connections, 7–17

Index

Composite baseband video, 7–8
Composite video, 7–18, 12–30
Composition segment, 6–29
Compression
  digital video, 2–14 — 2–17
  moving pictures, 2–20 — 2–23
  single pictures, 2–17 — 2–19
Compression filters, 2–15
Computer compatibility, 9–10 — 9–11
Computer data
  Blu-ray Disc potential, 11–5 — 11–9
Computer software
  Blu-ray Disc potential, I–3, 11–5 — 11–9
Conditional access, 4–1—4–2
Conditional replenishment, 2–14
Cones, 2–15
Connection compatibility, 9–5
Connection types, 7–9 — 7–13
Constant bitrate, 2–30, 12–9, 12–28
Consumer download and burn, 3–15
Consumer electronics, 7–6
Consumer Electronics Association, 2–57
Consumer Electronics Manufacturers Association, 1–9
Content code, 4–9
Content Management Information, 4–12
Content participant agreement, 4–21
Content protection, 4–2 — 4–4
Content Protection for Prerecorded Media, 4–4
Content Protection Group, 4–3
Content protection obligations, 4–21
Content Scramble System, 1–10, 1–12 — 1–13, 4–4
Content service provider, 3–15
Context-Adaptive Binary Arithmetic Coding, 6–21
Conventions, I–9 — I–10
Conversions, I–7, 3–12
Copy Control Association, 1–10
Copy once flags, 3–14
Copy protection, 4–1, 9–2 — 9–4
Copy Protection Technical Working Group, 1–10
Copyable with restrictions content, 3–14

Copyright Treaty, 4–17, 9–2
Core+96k, 2–33
Cover layer, 5–1
CPA. See Content participant agreement
CPG. See Content Protection Group
CPO. See Content protection obligations
CPPM. See Content Protection for Prerecorded Media
CPTWG. See Copy Protection Technical Working Group
Crawlies, 8–12
Credential process, 6–50
Cropping, 2–47 — 2–48
CRTs. See Cathode ray tubes
CSP. See Content service provider
CSS. See Content Scramble System
CSS2 copy protection system, 1–13
Cuban, Mark, 13–8
Customization, 3–8
Cutting master format, 12–3
CyberLink PowerDVD software player, 1–23

D

D-Theater, 1–13—1–14
DAC. See Digital-to-analog converter
DAE. See Digital audio extraction
Data applications
  Blu-ray Disc potential, 11–5 — 11–9
Data-link jitter, 2–37
Data modulation, 5–8
Data streams, 2–12 — 2–13
DB-25 connectors, 7–11
DBS. See Direct broadcast digital satellite
DCT. See Discrete cosine transform
DD+. See Dolby Digital Plus
DDWG. See Digital Display Working Group
Decoders, 2–21 — 2–23
Decryption, 4–6 — 4–7
DeCSS, 1–12 — 1–13
Defect management, 5–9
Desktop HD video publishing, 11–3
Deterministic process, 2–21
Device Keys, 4–5
Dialog normalization, 7–13

Index

DigiRise DRA, 1–27
Digital artifacts, 8–11 — 8–12
Digital audio, 7–8, 7–13 — 7–14
Digital audio coding, 1–27
Digital audio extraction, 2–36n
Digital connections, 7–17
Digital copy, 3–12
Digital Display Working Group, 1–11, 4–13
Digital files, 3–12
Digital light processing, 255
Digital micromirror devices, 2–55
Digital Millennium Copyright Act, 1–9, 1–22, 4–17, 9–2
Digital Only Token, 4–15
Digital Recording Act of 1996, 1–9
Digital rights management, 3–12, 3–13
Digital signal processors, 1–21
Digital technology, 2–1—2–2
Digital Theater Systems, 2–27 — 2–29, 2–32
Digital-to-analog converter, 7–13
Digital Transmission Content Protection, –13
Digital Video Broadcast Project, 1–4
Digital video compression, 2–14 — 2–17
Digital Video Essentials, 7–21
Digital video noise reduction, 12–28
Digital Visual Interface, 1–11, 4–13 — 4–14, 7–9, 7–12
Direct broadcast digital satellite, 13–7
Direct draw, 6–41
Disc caddies, 2–6
Disc durability, 3–9, 5–11 — 5–12, 8–12
Disc ID, 6–49
Disc Keys, 4–4
Disc labeling, 12–43 — 12–44
Discovision Associates, 4–26
Discrete audio, 3–3
Discrete cosine transform, 2–18
Display technology, 2–54 — 2–56
DisplayPort, 7–4, 7–9, 7–12
Distribution process, 12–43 — 12–45
DLP. See Digital light processing
DMCA. See Digital Millennium Copyright Act
DMD. See Digital micromirror devices

Dolby, 4–25 — 4–26
Dolby Digital audio coding, 2–27 — 2–30, 6–26, 8–6, 8–13, 12–10
Dolby Digital Plus, 2–27, 2–30 — 2–31, 6–26, 6–29
Dolby TrueHD, 2–27, 2–31— 2–32, 6–26— 6–27, 6–54
DOT. See Digital Only Token
Downloading content, 6–46 — 6–47
Drive Revocation List, 4–5
DRL. See Drive Revocation List
DRM. See Digital rights management
DSPs. See Digital signal processors
DTCP. See Digital Transmission Content Protection
DTS, 4–25
DTS-96/24, 2–33
DTS audio coding, 2–27 — 2–29, 2–32, 6–26, 8–6, 8–13, 12–10
DTS-ES, 2–33
DTS-HD audio coding, 2–32, 6–26, 6–55
DTS-HD High Resolution, 2–27, 6–26 — 6–27
DTS-HD LBR, 6–29
DTS-HD Master Audio, 2–27, 2–33 — 2–34, 6–26 — 6–27
Dual-format players, 1–22
Dual-layer discs, 1–21
Duplication, 3–12, 12–43 — 12–45
DVB. See Digital Video Broadcast Project
DVD
  development of, 1–6 — 1–14
  features, 3–1 — 3–2
  sales rates, 13–1 — 13–4
DVD-Audio, 3–1 — 3–2, 3–4 — 3–5
DVD CCA, 1–10
DVD Consortium, 1–8, 1–10
DVD Forum, 1–10, 1–14, 1–16, 9–1
DVD Forum Steering Committee, 1–15
DVD-ROM, 3–1
DVD-Video, 3–1 — 3–2
DVI. See Digital Visual Interface
DVI connectors, 7–12
DVNR. See Digital video noise reduction
DVR Blue, 1–13
Dynamic range, 2–30

Blu-ray Disc Demystified

Index

 E-copy, 3–12 — 3–13
EAS. See Electronic article surveillance
EB. See Emergency brake
EBR. See Electron beam recorder
EDC. See Error-detection code
Edge enhancement, 8–11
EDTV. See Enhanced-definition television
Education applications
  Blu-ray Disc potential, I–3, 11–1 — 1–9
EICTA, 2–57
8-mm tape, 1–3
Electron beam recorder, 5–5
Electronic article surveillance, 12–43
Electronic program guides, 6–3
Electronic sell-through, 3–15, 4–10 — 4–11
Emergency brake, 5–9
Encoding formats, 9–11 — 9–12
Encoding process, 2–21 — 2–23
Enhanced AC-3. See Dolby Digital Plus
Enhanced-definition television, 1–4
Enhanced versatile disc, 1–15
EPG. See Electronic program guides
Erasable CD, 1–6
Error correction, 2–3 — 2–4, 2–6, 5–7—5–8
Error-detection code, 5–8
EST. See Electronic sell-through
Ethernet, 7–9
EVD. See Enhanced versatile disc
Expand mode, 2–46
Extension substreams, 2–32
External control standards, 9–10

 FAA. See Frame-accurate animation
Factory manufacturing on demand, 3–14
Family Entertainment and Copyright Act of 2005, 9–7
Family Movie Act, 9–7
FCC. See Federal Communications Commission
Federal Communications Commission, 1–4, 1–5, 2–7 — 2–8
File system compatibility, 7–24
FireWire, 4–13 — 4–14, 7–9, 7–11

Fix Up Table, 4–9
Flicker, 12–37
FLLA. See Format and logo license agreement
FLLC. See Format and Logo License Corporation
Font licensing, 6–30
Forensic Marking, 4–10, 4–12
Format and logo license agreement, 4–21
Format and Logo License Corporation, 4–26
Formatting, 12–39 — 12–40
Forward versatile disc, 1–15
Frame-accurate animation, 6–42
Frame rates, 2–8 — 2–10
Frame sizes, 12–10
Frames, 2–2
Frequency-domain error confinement, 2–28
Frequency masking, 2–26 — 2–27
Full high-definition resolution, 6–18
FUT. See Fix Up Table
FVD. See Forward versatile disc

 Game consoles, 7–6
Gates, Bill, 1–18, 1–19, 1–20, 13–8
Gaussian blur, 12–37
GEM. See Globally Executable MHP
General Purpose Registers, 6–16
Generational copies, 3–14
Gibbs effect, 8–11
Globally Executable MHP, 6–38
GPRs. See General Purpose Registers
Graphic artist, 12–6
Graphical user interface, 12–38
Graphics animations, 6–42 — 6–43
Graphics controller, 6–29
Graphics drawing, 6–41 — 6–43
Graphics object segment, 6–29
Graphics preparation, 12–35 — 12–37
GUI. See Graphical user interface

 Halos, 8–11
HD DVD
  development of, 1–15, 1–21
  discontinuation of, 1–24 — 1–27

Index

HD DVD9, 1–14
HD-MAC, 1–4
HD Movie. See HDMV
HD SDI. See HD Serial Digital Interface
HD Serial Digital Interface, 2–13
HDAV Mode programming, 12–3 — 12–5
HDCP. See High-bandwidth Digital Content Protection
HDMI. See High-definition multimedia interface
HDMI connectors, 7–12
HDMV
  browsable slideshow, 6–35
  features, 6–9 — 6–10
  graphics limitations, 6–34
  incompatibility, 8–4, 9–5 — 9–7
  interactive audio, 6–34
  multipage menus, 6–33 — 6–34
  navigation data, 6–12—6–16
  organizational structure, 6–11 — 6–12
  popup menus, 6–32 — 6–33
  presentation planes, 6–10 — 6–11
  user interaction, 6–16 — 6–18
HDTV. See High-definition television
Head-mounted displays, 13–6
High-bandwidth Digital Content Protection, 4–13 — 4–14, 9–3
High Bitrate Extension, 2–33
High-definition data streams, 2–12 — 2–13
High-definition multimedia interface, 7–3 — 7–4, 7–9, 7–12, 9–5
High-definition technology, 2–6
High-definition television
  aspect ratios, 2–39 — 2–45
  broadcast resolution, 2–56 — 2–57
  development of, 1–4 — 1–5
  formats, 2–10 — 2–11
High-density technology, 2–6
Hitachi, 4–20
HiVision, 1–4
Hollerith, Herman, 1–1
Hollywood Advisory Committee, 1–7
Hollywood studio revenue, 10–2
Holograms, 12–43
Holographic storage, 13–5 — 13–6

Host Revocation List, 4–5
HRL. See Host Revocation List
Hue, 2–16
Hybrid discs, 5–11, 12–47 — 12–48

I frames, 2–20 — 2–21
Icons, 12–44
ICT. See Image Constraint Token
IDCT. See Inverse discrete cosine transform
Identifiers, 12–44
IDTV. See Improved-definition television
IEEE 1394, 4–13 — 4–14, 7–9, 7–11
iHD, 1–19
Image Constraint Token, 4–15 — 4–16, 8–5, 9–3
Image dimensions, 12–35
Image frame-accurate animation, 6–42
Image resolution
  high definition, 2–10 — 2–11
Imaging Science Foundation, 7–21
Implementation compatibility, 7–25
Improved-definition television, 1–4
In-mux synchronous PIP, 6–25
Index table, 6–12
Industrial applications
  Blu-ray Disc potential, 11–8 — 11–9
InfiniFilm feature, 10–2
Information publishing
  Blu-ray Disc potential, I–3 — I–4
Instant access, 3–8
Integrated circuits interference, 2–35
Inter-hybrid discs, 5–11
Interactive audio, 6–34
Interactive Graphics, 6–8, 6–32, 6–34
Interactive user interface, 6–4
Interactive User Operations, 6–16 — 6–17
Interactivity, 3–8, 9–13, 10–1 — 10–9, 11–2 — 11–3
Interface jitter, 2–37
Interlaced display mode, 6–20
Interlaced scanning, 2–52 — 2–54
International Telecommunications Union, 2–24n
Internet connected players, 10–6 — 10–8
Intra-hybrid discs, 5–11

Index

Intra pictures, 2–20 — 2–21
Inverse discrete cosine transform, 2–21n
Inverse telecine, 2–53
ISF. See Imaging Science Foundation
ITU. See International Telecommunications Union
ITU H.264 Part 10, 2–24

J

Jacquard, Joseph-Marie, 1–1
Java Archives, 6–12
Java Media Framework, 6–38
Java programmers, 12–7
Java Virtual Machine, 6–9, 6–36, 9–6
JavaScript, 8–14
Jitter, 2–34 — 2–39
JMF. See Java Media Framework
Joint Photographic Experts Group, 2–17
Joint Technical Committee, 6–1
Joint Video Team, 2–24
JPEG compression, 2–17 — 2–18
JTC. See Joint Technical Committee
JVM. See Java Virtual Machine
JVT. See Joint Video Team

K

Karaoke, 3–5
Key events, 6–17 — 6–18
Key-selection Vector, 4–14
Key Variant Unit, 4–6
KSV. See Key-selection Vector

L

Labeling process, 12–43 — 12–44
Lands, 2–2 — 2–3
Language issues, 12–10 — 12–11
Laser beam recorder, 5–5
Laserdisc
  development of, 1–2 — 1–3
Layers, 2–4 — 2–5
LBR. See Laser beam recorder
LCD. See Liquid-crystal display
LCoS. See Liquid crystal on silicon
LDC. See Long-distance code
Letterboxing, 2–41, 2–47, 2–51
LFE. See Low-frequency effects
LG, 4–20
License preparer, 4–11

Licensing, 4–20 — 4–26
Licensing Agency, 4–11
Light bleed, 2–55
Linear pulse code modulation, 2–29, 6–26
Liquid-crystal display, 2–55 — 2–56
Liquid crystal on silicon, 2–56
Liquid immersion, 5–5
LLA. See Logo License Agreement
Local storage, 3–11, 6–45 — 6–46, 8–4, 10–8 — 10–9
Logo License Agreement, 4–21
Long-distance code, 5–7 — 5–8
Lossless compression, 2–15
Lossless Extension, 2–33
Lossy compression, 2–15
Low-frequency effects, 3–3n, 7–15 — 7–16, 8–14, 12–32
Luma, 2–17
Luma keys, 6–25
Luminance, 2–12, 2–16 — 2–17

M

Macroblocks, 2–20
Magnetic tape
  development of, 1–1 — 1–2
Magneto-optical technology, 1–3, 1–6
Managed Copy, 4–8
Managed copy, 3–12, 4–11, 8–10
Managed Copy Machine, 4–8
Managed copy output technologies, 3–13
Managed Copy Server, 4–8
Mandatory managed copy, 1–19, 3–12, 8–9
Manifest file, 6–46
Manufacturing on demand, 3–14, 4–10 — 4–11
Marketing
  Blu-ray Disc potential, I–4, 11–6
Marks, 2–2
Mastering, 5–5
Mastering Code, 5–11
Masters of Reverse Engineering, 1–12
Mattes, 2–41
MCM. See Managed Copy Machine
MCOT. See Managed copy output technologies
MCS. See Managed Copy Server

Index

Media convergence, 13–10 — 13–11
Media jitter, 2–37
Media Key Block, 4–5 — 4–6
Media Keys, 4–5
Media storage, 5–11 — 5–12
Media Transform, 4–9
Menu creation, 12–17 — 12–19
Menu design, 12–16 — 12–17
Menus, 3–7 — 3–8
Meridian Lossless Packing, 2–31, 3–4
Metadata, 6–31 — 6–32
MHP. See Multimedia Home Platform
Microsoft's Windows Media, 2–23
Midnight mode, 7–13
Minimum-shift-keying, 5–6
Miniphono connectors, 7–10
Mixed media, 11–3
MKB. See Media Key Block
MLP. See Meridian Lossless Packing
MMCD. See Multimedia CD
MO technology. See Magneto-optical technology
MOD. See Manufacturing on demand
Mold Code, 5–11
MoRE. See Masters of Reverse Engineering
Mosquitoes, 8–11
Motion-compensated prediction, 2–20
Motion estimation, 2–20
Motion judder artifacts, 2–53
Motion Picture Association of America, 1–9, 1–13, 6–31
Movie objects, 6–12
Movies
  Blu-ray Disc potential, I–1
Moving Picture Experts Group, 1–5
Moving pictures
  compression of, 2–20 — 2–23
Moving Pictures Expert Group, 2–20
MPAA. See Motion Picture Association of America
MPEG, 1–5, 1–6, 2–20 — 2–23
MPEG-2, 6–3, 6–18, 6–21, 6–51
MPEG-4 Advanced Video Codec, 2–24, 6–18, 6–21

MPEG audio coding, 2–27
MPEG LA, 4–25
MSK. See Minimum-shift-keying
Multi-disc sets, 6–44
Multichannel analog audio, 7–14 — 7–15
Multichannel configuration, 6–27 — 6–28
Multichannel digital audio, 7–14
Multilanguage issues, 12–10 — 12–11
Multimedia CD, 1–7 — 1–8
Multimedia Home Platform, 6–9, 6–38 — 6–39
Multipage menus, 6–33 — 6–34
Multiple surround audio tracks, 3–3 — 3–4
Multiplexing, 6–7 — 6–8
Multiplexors, 12–7
Multiregion issues, 12–10 — 12–11
Multistory seamless branching, 3–6
Music
  Blu-ray Disc potential, I–1 — I–2
Musical slideshows, 3–5

 NA. See Numeric aperture
National Television Systems Committee, 2–6 — 2–7
Navigation commands, 6–12 — 6–14
Navigation design, 12–20 — 12–23
Network access, 8–4
Network-connected features, 12–12 — 12–13
Network connection, 3–9 — 3–11
Network download, 3–15
New Line Home Video, 10–2
NexGuard, 4–13
NLE. See Nonlinear editing system
Noise, 8–11
Noise shaping, 2–28
Non-return to zero inverted, 5–8
Nonlinear editing system, 12–28
Notations, I–6 — I–8
NRZI. See Non-return to zero inverted
NTSC standard, 2–7, 9–4 — 9–5
Numeric aperture, 2–6

 Object buffer, 6–29
Offset printing, 12–43 — 12–44
OPC. See Optimum power control tasks

Index

Opposite track path, 2–4
Optical disc layers, 2–4 — 2–5
Optical Storage Technology Association, 1–8
Optimum power control tasks, 5–9
Organization ID, 6–49
Oscillator jitter, 2–36
OSTA. See Optical Storage Technology Association
OTP. See Opposite track path
Out-of-mux asynchronous PIP, 6–25
Out-of-mux synchronous PIP, 6–25
Outputting, 12–39 — 12–40

P P frames, 2–20 — 2–21
Package design, 12–44
PAL standard, 2–7, 9–4 — 9–5
Palette segment, 6–29
PALplus, 1–4
Pan and scan process, 2–41 — 2–42, 2–48, 2–51 — 2–52
Panasonic, 4–20
Parallel track path, 2–4
Parental lock, 3–6 — 3–7
Parental management, 6–31, 9–7 — 9–8
Parity preserve, 5–8
Patent licensing, 4–21 — 4–25
Paths, 6–8
PBP. See Personal Basis Profile
PC Playback, 3–9
PCM. See Pulse code modulation
PDP. See Plasma display device panel
Perceptual coding, 2–26, 2–27 — 2–29
Performances and Phonograms Treaty, 4–17, 9–2
Permanent information and control data zone, 5–8
Permission Request File, 6–48
Persistent storage, 3–11, 6–45, 12–11 — 12–12
Personal Basis Profile, 6–38 — 6–39
Phase Alternating Line, 2–7
Phase-change recording, 5–9 — 5–10
Phase-locked loop circuits, 2–38n

Phase noise, 2–35
Phase-transition metal, 5–5
Philips, 4–20, 4–25
Phono connectors, 7–10
Physical cluster, 5–8
Physical compatibility, 7–24
PIC. See Permanent information and control data zone
Picture compression, 2–17 — 2–19
Picture in picture, 3–3, 6–8, 6–23 — 6–25
Pillarbox, 2–46
Pioneer, 4–20
PIP. See Picture in picture
Pit art, 12–43 — 12–44
Pit jitter, 2–37
Pits, 2–2—2–3
Pixels, 2–2
Plane models, 6–10 — 6–11
Plasma display device panel, 2–55
Playback incompatibility, 9–5 — 9–7
Playback Control User Operations, 6–16 — 6–17
Player Status Registers, 6–14 — 6–16, 6–31
Players
  audio connections, 7–13 — 7–16
  Blu-ray disc recorders, 7–7
  compatibility, 7–22 — 7–26
  connection types, 7–9 — 7–13
  connections, 7–7 — 7–9
  DisplayPort interface standard, 7–4
  features, 7–1 — 7–3
  high-definition multimedia interface, 7–3 — 7–4
  Internet connected, 10–6 — 10–8
  profiles, 7–1 — 7–3
  remote control, 7–19 — 7–20
  types of, 7–6
  upscaling DVD, 7–5
  user tools, 7–19 — 7–22
  video connections, 7–16 — 7–18
  viewing distance, 7–21 — 7–22
Playlists, 6–4, 6–12, 6–47
PlayStation 3, 1–17, 1–20, 1–21, 1–26, 7–6
PlayStation© Portable, 1–17

Index

PLL. See Phase-locked loop circuits
PMSN. See Prerecorded Media Serial Number
PNG. See Portable Network Graphics
Polarization, 13–6
Popup menus, 6–32 — 6–33
Portability, 11–3
Portable Network Graphics, 6–41, 12–48
Posterization, 8–11
Predicted pictures, 2–20 — 2–21
Prefixes, I–6
Prepared Video Authorization Server, 4–11
Prepared Video Token, 4–11 — 4–12
Prerecorded Media Serial Number, 4–8
Presentation data, 6–7
Presentation Graphics, 6–29 — 6–30
PRF. See Permission Request File
Proactive software renewal, 4–6
Production processes
  asset management, 12–14
  asset preparation, 12–26 — 12–37
  audio asset preparation, 12–30 — 12–32
  authoring, 12–38 — 12–39
  BD-J mode, 12–3 — 12–5
  BD-ROM, 5–5 — 5–6, 12–45 — 12–47
  bit budgeting, 12–23 — 12–26
  distribution, 12–43 — 12–45
  duplication, 12–43 — 12–45
  formatting, 12–39 — 12–40
  graphics preparation, 12–35 — 12–37
  HDAV mode, 12–3 — 12–5
  hybrid discs, 12–47 — 12–48
  job positions, 12–5 — 12–7
  menu creation, 12–17 — 12–19
  menu design, 12–16 — 12–17
  navigation design, 12–20 — 12–23
  network-connected features, 12–12 — 12–13
  output, 12–39 — 12–40
  persistent storage, 12–11 — 12–12
  production decisions, 12–8 — 12–11
  programming titles, 12–1
  project design, 12–14 — 12–16
  projects, 12–2
  quality control, 12–40 — 12–43
  replication, 12–1, 12–43 — 12–45
  scheduling, 12–14
  simple AV playback mode, 12–2 — 12–3
  steps of, 12–7 — 12–8
  subtitles preparation, 12–33 — 12–34
  supplemental material, 12–13 — 12–14
  synchronization of audio and video, 12–33
  testing, 12–40 — 12–43
  video artifacts, 12–37 — 12–38
  video asset preparation, 12–27 — 12–30
Professional Disc format, 1–15
Profile 1.1 (BonusView) player, 1–23
Profile 2 players, 8–3 — 8–4
Profile 3 players, 8–3
Profile 4 players, 8–3
Profile compatibility, 7–24 — 7–25
Program Streams technology, 6–3
Programmable audio mixing, 6–44 — 6–45
Programmable read-only memory, 4–14
Programming titles, 12–1
Progressive display mode, 6–20
Progressive PlayList, 6–47
Progressive scanning, 2–52 — 2–54
Progressive segmented frame, 2–10
Prohibit repeated minimum transition run length, 5–8
Project design, 12–14 — 12–16
Project designers, 12–6
Project manager, 12–6
PROM. see Programmable read-only memory
Promotions Committee, 6–2
Protected distribution, 4–1 — 4–2
Protected storage, 4–1 — 4–2
Protected transmission, 4–1 — 4–2
Protection for Prepared Video, 4–10
PsF. See Progressive segmented frame
PSP. See PlayStation© Portable
PSRs. See Player Status Registers
Psychoacoustic coding, 2–26
Psychovisual encoding systems, 2–15
PTM. See Phase-transition metal
PTP. See Parallel track path
Public Key, 4–11

Index

Pulse code modulation, 2–29, 3–4, 12–10
PVAS. See Prepared Video Authorization Server
PVT. See Prepared Video Token

 Quality control, 12–7, 12–40 — 12–43
Quantization, 2–18 — 2–19, 2–28
Quantization noise, 2–28

 Radio-frequency audio/video, 7–9, 7–18
Radiofrequency interference, 2–35
RBG values, 2–16 — 2–17
RCA phono connectors, 7–9
Read Data Key, 4–6
Readout jitter, 2–37
Real playlists, 6–4
Recordable CDs, 1–6
Recordable discs
content protection, 4–10 — 4–12
Recorders, 1–15 — 1–16, 4–11, 7–7
Recording frame, 5–8
Recording unit block, 5–8
Reed-Solomon, 5–7
Region playback control, 4–17 — 4–20
Regional codes, 8–10
Regional management, 9–3 — 9–4
Remote control, 7–19 — 7–20, 12–22
Render farms, 6–21
Repeated minimum transition run length, 5–8
Replication, 12–1, 12–43 — 12–45
Retail managed recording, 3–14
Retail manufacturing on demand, 3–14
Reverse play, 9–9
Reverse spiral dual layer, 2–4
Revocation, 4–5
RF. See Radio-frequency audio/video
Ringing, 8–11
RJ-45 connectors, 7–12
RLE. See Run-length encoding
RMR. See Retail managed recording
RMTR. See Repeated minimum transition run length
ROM Mark, 1–18

Root certificate, 6–48
RPC. See Region playback control
RS. See Reed-Solomon
RSDL. See Reverse spiral dual layer
RUB. See Recording unit block
Run-length encoding, 2–15

 S-VHS, 1–3
S-video, 7–8
S-video connectors, 7–10, 7–18
Safe areas, 12–35 — 12–37
Sales
  Blu-ray Disc potential, 11–6
Sampling Frequency Extension, 2–33
Sampling jitter, 2–36 — 2–37
Samsung, 4–20
Saturation, 2–16
Saw-tooth wobble, 5–6
SCART connectors, 7–11
SD. See Super Disc Alliance
SDTV. See Standard definition television
Seamless branching, 3–6
Seamless multistory, 6–23
SECAM, 2–7
Second session, 3–12
Secondary audio, 6–28 — 6–29
Secure moves, 3–14
Secure Socket Layer, 6–46
Security code, 4–9
Security Virtual Machine, 4–9
SelectaVision, 1–2
Self-Protecting Digital Content, 4–9
Sequence Key Blocks, 4–6
Sequential Color with Memory, 2–7
Service bureaus, 12–8 — 12–10
Set commands, 6–12 — 6–14
SFAA. See Synchronized frame-accurate animation
SHA-1 hash, 6–48
Sharp, 4–20
Shutter, 13–6
SID. See Source identification codes
Silkscreen printing, 12–43 — 12–44
16:9 aspect ratio, 2–48 — 2–51
SKB. See Sequence Key Blocks

Index

SKU. See Stock-keeping units
Slideshows, 3–5, 6–35
SlySoft, 1–23
SMPTE. See Society of Motion Picture Television Engineers
SMPTE VC-1, 2–23 — 2–24, 6–18, 6–52
Snow, 8–11
Soft mattes, 2–41 — 2–42, 2–48
softDVDcrack, 1–11
Software
  Blu-ray Disc potential, I–3, 11–5 — 11–9
Software players, 7–6
Software renewal, 4–6
Sony, 4–20
Sony PlayStation 3, 1–17, 1–20, 1–21, 1–26, 7–6
Source identification codes, 5–11
Source mode, 6–41
Source-over mode, 6–42
Source tagging, 12–43
Spaces, 2–2
Spatial redundancy, 2–14 — 2–15
SPDC. See Self-Protecting Digital Content
Speakers, 2–34
Specialized user input, 10–9
Specification books, 6–2
SPS. See Start position shift
SRC. See Source mode
SRC_OVER. See Source-over mode
SSL. See Secure Socket Layer
Standard definition television, 2–6
Start position shift, 5–8
Stereo analog audio, 7–15
Stereovision, 13–6 — 13–7
Stock-keeping units, 12–11
STW. See Saw-tooth wobble
Subpaths, 6–8
Subpictures, 3–4 — 3–5
Subtitles, 3–4 — 3–5, 6–29 — 6–30, 12–33 — 12–34
Subwoofers, 12–32
Sun Microsystems, 4–26, 9–6
Super Disc Alliance, 1–7 — 1–8
Super VHS system, 1–3
Supplemental material, 12–13 — 12–14

Surround analog audio, 7–15
Surround sound, 3–3 — 3–4
SVM. See Security Virtual Machine
Synchronization, 12–33
Synchronized frame-accurate animation, 6–42

 TCKs. See Technology compatibility kits
Technical Expert Groups, 6–1 — 6–2
Technology compatibility kits, 4–26
TEGs. See Technical Expert Groups
Telecine machines, 2–53
Television
  Blu-ray Disc potential, I–1
  digital developments, 1–4 — 1–5
  digital format, 1–4 — 1–5
  systems, 2–6 — 2–8
Temporal-domain error confinement, 2–28
Testers, 12–7
Testing process, 12–40 — 12–43
Text-based subtitles, 6–30
TextST, 6–29
THD, 1–22
Theatrical mattes, 2–45
Thomson, 4–20
3D video, 13–6 — 13–7
Time jitter, 2–35
Title Control User Operations, 6–16 — 6–17
Title Keys, 4–4, 4–5
Title-safe areas, 12–35
Toslink fiberoptic connectors, 7–10
Total HD, 1–22
Training applications
  Blu-ray Disc potential, 11–7 — 11–8
Transacted recording, 3–14
Transmitted recording, 3–14
Transport jitter, 2–37
Transport Security Layer, 6–46
Transport Streams technology, 6–3
Trick play, 3–9
TSL. See Transport Security Layer
Twitter effect, 12–37
2-3 pulldown, 2–53
2-2 pulldown, 2–53

Index

Type F connectors, 7–11

U UDF. See Universal Disk Format
UDO. See Ultra Density Optical
Ultra Density Optical, 1–13
UMD. See Universal media disc
Units and notation, I–6 — I–8
Universal Disk Format, 1–8, 12–45 — 12–46
Universal media disc, 1–17, 1–20 — 1–21
Uploading content, 6–46 — 6–47
Upscaling DVD players, 7–5
User input, 10–9
User interaction, 6–16 — 6–18
User operations, 6–16 — 6–17
User tools, 7–19 — 7–22

V Variable bit rate encoding, 2–21, 2–31, 12–9, 12–28
VBR. See Variable bit rate encoding
VC-1 audio coding, 2–23 — 2–24, 6–18, 6–52
VESA. See Video Electronics Standards Association
VFS. See Virtual File System
VHD. See Video high density
VHRA. See Video Home Recording Act
VHS VCRs
  development of, 1–3
Via Licensing, 4–24, 4–26
Video applications
  Blu-ray Disc potential, 11–4—11–5
Video artifacts, 12–37—12–38
Video CD, 1–6
Video codecs, 2–23—2–26
Video compression, 2–14 — 2–17, 3–3, 8–12
Video compressionist, 12–6
Video connections, 7–16 — 7–18
Video editors, 12–6
Video Electronics Standards Association, 7–4
Video formats, 6–18 — 6–21, 8–8 — 8–9
Video games
  Blu-ray Disc potential, I–3
Video high density, 1–2 — 1–3
Video home disc, 1–2 — 1–3
Video Home Recording Act, 1–9

Video inputs, 9–12
Video preparation, 12–27 — 12–30
Video publishing, 11–3
Video training
  Blu-ray Disc potential, I–2 — I–3
Video transfers, 2–51 — 2–52
Videodisc
  development of, 1–2
Videotape recording
  development of, 1–2
Viewing distance, 7–21 — 7–22
Virtual asset, 6–47
Virtual File System, 6–25, 6–46, 12–12
Virtual Keys, 6–17 — 6–18
Virtual playlists, 6–4

 WAMO. See Warner Advanced Media Operations
Warner Advanced Media Operations, 1–11
Warner Bros. Company, 9–1
Watermarking, 4–2, 4–12 — 4–13, 9–3
Web standards, 9–11
Wide mode, 2–46
Widescreen displays, 2–45 — 2–48
Widescreen movies, 3–3
WiFi. See Wireless networking
Windows Media Digital Rights Management, 4–15
Windows Media Video HD, 2–23 — 2–24
WIPO. See World Intellectual Property Organization Copyright Treaty
WIPO Performances and Phonograms Treaty, 4–17, 9–2
Wireless networking, 7–13
Wireless video, 7–13
Wireless video area network, 7–13
WMDRM. See Windows Media Digital Rights Management
WMV-HD. See Windows Media Video HD
WO. See Write-once technology
World Intellectual Property Organization Copyright Treaty, 4–17, 9–2
Worms, 8–12
Write-once technology, 1–6

Index

 X96, 2–33
X-rated content, 8–7—8–8
Xbox 360, 1–18, 1–20, 1–23
XBR, 2–33
XCH, 2–33
Xlets, 6–36
XLL, 2–33—2–34
XXCH, 2–33

 Yasuda, Hiroshi, 1–5

SOFTWARE AND INFORMATION LICENSE

The content of this Blu-ray Disc™ (collectively referred to as the "Product") is the property of Television Production Services, Inc. ("TPS"), and is licensed to The McGraw-Hill Companies, Inc. ("McGraw-Hill"). The Product is protected by both United States copyright law and international copyright treaty provision.

TPS reserves the right to alter or modify the contents of the Product at any time.

This agreement is effective until terminated. The Agreement will terminate automatically without notice if you fail to comply with any provisions of this Agreement. In the event of termination by reason of your breach, you will destroy or erase all copies of the Product installed on any computer system or made for backup purposes and shall expunge the Product from your data storage facilities.

LIMITED WARRANTY

McGraw-Hill warrants the physical Blu-ray Disc enclosed herein to be free of defects in materials and workmanship for a period of sixty days from the purchase date. If McGraw-Hill receives written notification within the warranty period of defects in materials or workmanship, and such notification is determined by McGraw-Hill to be correct, McGraw-Hill will replace the defective Blu-ray Disc. Send request to:

Customer Service
McGraw-Hill
Gahanna Industrial Park
860 Taylor Station Road
Blacklick, OH 43004-9615

The entire and exclusive liability and remedy for breach of this Limited Warranty shall be limited to replacement of defective Blu-ray Disc and shall not include or extend to any claim for or right to cover any other damages, including but not limited to, loss of profit, data, or use of the software, or special, incidental, or consequential damages or other similar claims, even if McGraw-Hill has been specifically advised as to the possibility of such damages. In no event will McGraw-Hill's liability for any damages to you or any other person ever exceed the lower of suggested list price or actual price paid for the license to use the Product, regardless of any form of the claim.

THE McGRAW-HILL COMPANIES, INC. SPECIFICALLY DISCLAIMS ALL OTHER WARRANTIES, EXPRESS OR IMPLIED, INCLUDING BUT NOT LIMITED TO, ANY IMPLIED WARRANTY OF MERCHANTABILITY OR FITNESS FOR A PARTICULAR PURPOSE.

Specifically, McGraw-Hill makes no representation or warranty that the Product is fit for any particularpurpose and any implied warranty of merchantability is limited to the sixty day duration of the LimitedWarranty covering the physical Blu-ray Disc only (and not the contents) and is otherwise expressly and specifically disclaimed. This Limited Warranty gives you specific legal rights; you may have others which may vary from state to state. Some states do not allow the exclusion of incidental or consequential damages, or the limitation on how long an implied warranty lasts, so some of the above may not apply to you.

This Agreement constitutes the entire agreement between the parties relating to use of the Product. The terms of any purchase order shall have no effect on the terms of this Agreement. Failure of McGraw-Hill to insist at any time on strict compliance with this Agreement shall not constitute a waiver of any rights under this Agreement. This Agreement shall be construed and governed in accordance with the laws of New York. If any provision of this Agreement is held to be contrary to law, that provision will be enforced to the maximum extent permissible and the remaining provisions will remain in force and effect.